Y0-AAY-439

Natural Terpenoids as Messengers

The cover represents a pentagram, depicting the relationship between bacteria, fungi, plants, insects and vertebrates (including humans). This pentagram is used throughout the book to illustrate messenger functions of terpenoids in the relationships between biological kingdoms.

Natural Terpenoids as Messengers

A multidisciplinary study of their production,
biological functions and practical applications

by

Paul Harrewijn
*Plant Research International,
Wageningen, The Netherlands*

Adriaan M. van Oosten
*TNO Industrial Technology,
Eindhoven, The Netherlands*

and

Paul G.M. Piron
*Plant Research International,
Wageningen, The Netherlands*

KLUWER ACADEMIC PUBLISHERS
DORDRECHT / BOSTON / LONDON

Library of Congress Cataloging-in-Publication Data

Harrewijn, P. (Paul), 1937-
   Natural terpenoids as messengers : a multidisciplinary study of their production,
biological functions, and practical applications / by Paul Harrewijn, Adriaan M. van
Oosten, Paul G.M. Piron.
     p. cm.
   Includes bibliographical references and index.
   ISBN 0-7923-6891-6 (hb : alk. paper)
   1. Terpenes--Physiological effect. 2. Second messengers (Biochemistry) I. Oosten,
Adriaan M. van. II. Piron, Paul G. M. III. Title.

QP752.T47 H37 2001
572'.5--dc21

2001023448

ISBN 0-7923-6891-6

Published by Kluwer Academic Publishers,
P.O. Box 17, 3300 AA Dordrecht, The Netherlands.

Sold and distributed in North, Central and South America
by Kluwer Academic Publishers,
101 Philip Drive, Norwell, MA 02061, U.S.A.

In all other countries, sold and distributed
by Kluwer Academic Publishers,
P.O. Box 322, 3300 AH Dordrecht, The Netherlands.

Printed on acid-free paper

All Rights Reserved
© 2001 Kluwer Academic Publishers
No part of the material protected by this copyright notice may be reproduced or
utilized in any form or by any means, electronic or mechanical,
including photocopying, recording or by any information storage and
retrieval system, without written permission from the copyright owner.

Printed in the Netherlands.

Contents

	page
Preface	ix
1 Introduction	1
2 Production of terpenes and terpenoids	11
2.1 Biosynthesis of isoprenoids	11
2.1.1 Bacteria and algae	12
2.1.2 Fungi	17
2.1.3 Green plants	18
2.1.4 Insects	24
2.1.5 Vertebrates	25
2.1.6 Evolution of isoprenoid biosynthesis	26
2.2 Production of terpenoids by different organisms	30
2.2.1 Monoterpenes	30
2.2.2 Sesquiterpenes and diterpenes	39
2.2.3 Triterpenes, sterols, steroid hormones and carotenoids	49
2.2.4 Complex terpenoids and isoprenoid structures	54
3 The origin and evolution of terpenoid messengers	59
3.1 Selection of key enzymes in isoprenoid biosynthesis	59
3.1.1 Key enzymes in MAI and MAD routes	60
3.1.2 The development of regulatory mechanisms	69
3.2 Feedback mechanisms	74
3.2.1 Key enzyme regulation in non-sterologenic cells	74

	3.2.2	Key enzyme regulation in sterologenic cells	77
3.3	**Other types of feedback regulation**		81
	3.3.1	Internal factors affecting feedback mechanisms	81
	3.3.2	External factors affecting feedback mechanisms	83
3.4	**Site of action of the regulatory mechanisms**		86
	3.4.1	Transcriptional and post-transcriptional regulation	86
	3.4.2	Oxysterols	91
	3.4.3	Non-sterol terpenoids	94
3.5	**Interference with DNA synthesis: prenylated proteins**		96
	3.5.1	The mechanism of prenylation	97
	3.5.2	Prenylation of Ras proteins	102
	3.5.3	Inhibition of protein prenylation	105

4	**Specific properties of terpenoids**		109
4.1	**Messengers on the cellular level**		109
	4.1.1	Messengers within a cell	110
	4.1.2	Messengers within organisms	113
	4.1.3	Homeostasis of cell numbers	129
	4.1.4	Effects of terpenoids on Ras proteins	135
	4.1.5	Cancer research	141
	4.1.6	"To kill or cure,….."	153
	4.1.7	Terpenoids with different effects on cancer cells	160
	4.1.8	Impairment of mevalonic acid synthesis in tumours	166
	4.1.9	Terpenoids with unexpected effects	168
4.2	**Direct toxic effects**		171
	4.2.1	Anti-microbial effects	171
	4.2.2	Toxic effects on insects and other animals	176
	4.2.3	How to overcome autotoxicity	178

5	**Functions of natural terpenoids in the interrelationships between organisms**		181
5.1	**Semiochemicals**		182
	5.1.1	Sex pheromones and other semiochemicals with impact on behaviour	183
	5.1.2	Alarm pheromones and defensive secretions	191
	5.1.3	Repellents and deterrents: defensive metabolites by plants	197
	5.1.4	Kairomones	216
	5.1.5	Synomones	222

	5.2	**Relationship between structure and function**	228
		5.2.1 Similarities between receptors	228
		5.2.2 The role of terpenoids in the evolution of insect behaviour	233
		5.2.3 Signal transduction and neurological adaptation	243
6	**Terpenoids in practice**		253
	6.1	**Natural terpenoids and their sources**	253
		6.1.1 Terpenoids in land plants and simple isolation procedures	253
		6.1.2 More complex isolation techniques	264
	6.2	**Utilization of terpenoids in biology and agriculture**	269
		6.2.1 Beneficial insects	270
		6.2.2 Insect control	272
		6.2.3 Resistance breeding	286
		6.2.4 Other applications	290
7	**Natural terpenoids to the benefit of human health**		297
	7.1	**Applications in medicine**	297
		7.1.1 Anti-microbials	298
		7.1.2 Analgesics	299
		7.1.3 Cholesterolemia, vascular problems, PVD	300
		7.1.4 Tracheal and bronchial disorders (COPD)	304
		7.1.5 Arthritis, rheumatism, inflammatory disorders	306
		7.1.6 Intestinal disorders	307
		7.1.7 Stress-related problems, sedatives	308
		7.1.8 Cancer therapies	309
	7.2	**Cosmetics**	329
		7.2.1 Sex attractants	329
		7.2.2 Dermatological preparations	331
	7.3	**Terpenoid analogues and derivatives**	333
		7.3.1 Applications in agriculture	333
		7.3.2 Applications in medicine	335
8	**Prospectus and suggestions for further research**		339
	8.1	**Search for new bioactive terpenoids**	339
		8.1.1 Incorporation of terpenoids into new methodologies	340
		8.1.2 Future of terpenoids in registration requirements	345
	8.2	**A multidisciplinary approach**	348

8.2.1	Bacteria, fungi, plants, insects and humans	348
8.2.2	Terpenoids and model systems	351

Epilogue 357

References 361

General reading 417

Glossary 421

Index 425

Preface

No other group of metabolites shows such diversity, with so many functions and produced by so many organisms than terpenoids. One cannot avoid being in contact with these carbon-based compounds. Step into your car, and you drive on terpenoids: tires are made of rubber and rubber consists of isoprene units, collected from trees and chemically altered with sulphur by the industry. Excellent literature on rubber production is available, and the same holds for many terpenoids which have drawn the attention of mankind because of their usefulness. Textbooks exist on several classes of terpenoids, their presence and structure. Detailed studies and monographs are available on specific terpenoids, e.g. on azadirachtin, a triterpene produced by the neem tree, *Azadirachta indica*, which among other applications is used as an insecticide.

Over the years a huge amount of data on terpenoids has been collected and most of it is easily accessible, to mention the websites of the internet as an example. So why should we write a book on terpenoids? Those involved in research on terpenoids inevitably had to specialize and became involved in a small group of terpenoids in a narrow area of biology or chemistry. Because natural terpenoids have a role in internal or external communication in all thinkable organisms, a particular terpenoid has been studied by different groups, unaware of the fact that the same compound has profound effects in undreamed forms of life. It happened to ourselves: in the 1980s we studied semiochemicals of aphids such as farnesenes, to find that mevalonate metabolism of these insects produces an array of terpenoids with unknown functions in Homoptera. We came into contact with colleagues working in different disciplines and found that it came as a surprise to many scientists that the terpenoids they were investigating in a particular organism were also present in unrelated organisms, where they appear to have unthought functions. The same terpenoid can have different targets, depending on the receptor (and the organisms).

To discuss all known natural terpenoids in all disciplines of science and industry is an impossible task. During our studies of insect-plant relations we eventually realized that plants, invertebrates and vertebrates produce the same isoprenoid structures and that insects could ideally be utilized in a model system to investigate regulatory functions of terpenoids. Not that the idea of using the insect as a medium for the study of physiology was new: No one less than Sir Vincent B. Wigglesworth has stressed that "rich veins of gold await the real specialist who cares to utilize the insect as a medium for the advancement of physiology" in chapter 10 of his collected essays: Insects and the life of man (1976). Not that the use of terpenoids is new either: terpenoids have been utilized since ancient times, and old cultures like in Asiatic countries use essential oils for many purposes, often because there simply is not a cheap alternative. And alternatives do not always offer a definite solution. We are not surprised to find that we are coming back to ways of thinking, which were general in the first half of the 20th century, but had been abandoned with the development of potent synthetic insecticides, antibiotics and (at least) the hope for a rapid improvement of cures for fatal diseases such as cancer.

As one of us was specialised in mevalonate metabolites, one in insect-plant relations, and one with a medical background became involved in aphids, we were struck by the role of terpenoids in the interrelationships between micro-organisms, plants, insects, and vertebrates, in particular human beings. We began to look at the collected material with other eyes: when terpenoids are involved, plant defense is often based on a number of compounds (allomones) aimed at several targets and directed toward different phases of the attack. We became convinced that this principle is of great value to combat complex pathological disorders in mammals (and man), such as developing cancers. Combination and multi-stage therapies (with terpenoids already known to have a function in e.g. plant-insect relationships) show a great deal of analogy with the strategies utilized by plants to defend themselves to attack by micro-organisms and herbivores.

When focussing on the relationship between plants, insects and man, as did the great Wigglesworth, we might be able to explain why so many functions of these compounds were selected early in evolution and why they are omnipresent in the living world. The present comparative study of terpenoids is not devoted to mere generalities but outlines the fascinating ways along which Nature has utilized isoprene units, and how they could serve to our own benefit.

We realize that much of the available information cannot be presented within this compass, but we hope that the book offers a good deal of

Preface   *xi*

scientific and practical surprises. When you step out of your car and have a look if the tyres should be replaced, you may not realize that old tyres can produce useful terpenes, even those, which can be applied in the treatment of cancer. Kirk Manfredi of the University of Northern Iowa heats tyres up to 382°C under exclusion of oxygen and distils limonene, a monoterpene that is found throughout this volume.

We hope that the reader after finishing the book has become aware of the enormous potential of terpenoid structures, waiting to be discovered for novel applications. What we hope most is that the book will serve to stimulate multidisciplinary studies on terpenoids, resulting in an optimal utilization of these "green" chemicals.

It is quite impossible to thank all those who have contributed to the completion of this study. We have been in contact with many specialists in different disciplines to update and check our understanding of the complex relationship between terpenoid messengers and their targets, and for scientific assistance in editorial matters. We wish to make an exception for a few colleagues who were so unfortunate to be at close range, and to express our gratitude for their readiness to incline their critical mind to conflicting opinions with regard to several subjects of the material treated in this volume. In alphabetical order: Drs G.J. Bollen, A.K. Minks and J.H. Visser.

P. Harrewijn, Wageningen, The Netherlands
A. M. van Oosten, Eindhoven, The Netherlands
P.G.M. Piron, Wageningen, The Netherlands

Chapter 1

Introduction

Throughout the Plant and Animal kingdom, terpenoids are known to have a wide range of functions. They can act as defensive substances in plants (allomones) and animals, they can be used by plants to deter herbivores or to inform conspecifics, or attract natural enemies of herbivores (synomones). Plant hormones are often derivatives of terpenoids, such as cytokinins, gibberellins and abscisic acid. The steroid hormones of mammals are terpenoids with an advanced but not very complex structure.

Terpenes have a unique structure: they consist of an integral number of five-carbon (5C or isoprene) units. Two such units can form a monoterpene (C-10), but sesquiterpenes (C-15), diterpenes (C-20), triterpenes (C-30), tetraterpenes (C-40) and polyterpenoids (>40C) are also possible. A message may consist of a blend of terpenoid types, enabling many permutations with respect to a "terpenoid code". Many terpenoids are produced via the mevalonic acid pathway that probably has its origin early during development of life on this planet. Other terpenoids are biosynthesized via a recently discovered pathway, a mevalonate independent route.

The words terpenes, isoprenoids and terpenoids are often used for the same structural formulas. To avoid confusion, we will keep to the following definitions of terms frequently used in this volume:

DEFINITIONS

Terpenes are regarded to be primary derivatives of isoprenic units with a biological function and terpenoids are secondary or condensed mevalonate metabolites with additional groups, such as more complicated derivatives of sesqui-, di- and triterpenes. Isoprenoid is a common name for a compound obeying the isoprene rules: examples are carotenoids, steroids and dolichols.

An explanation of most terms can be found in the Glossary. Each discipline is characterized by its own terminology and the same word may have a different meaning in another discipline. Throughout this book the terms *transcription* and *translation* are used. As it is difficult to explain their meaning in a few words the Glossary refers to them on this page.

Transcription is a process in the living cell to transmit specific information from the genetic code of a gene to an intermediate in the protein biosynthesis. In this process a specific part of the DNA (strand) forms a template for the synthesis of the intermediate: a complementary messenger RNA unit. In short, it is the information transfer from DNA to protein structure:

DNA transcription → RNA translation → proteins.

Translation is a successive process in the living cell connected to the process of transcription. In this following part of protein synthesis, the code of the base of the nucleic acids has to be transformed into the amino acid code of the proteins. In this way a "scale of four" code is changed into a "scale of twenty" code.

Isomery – In isomers molecules can exist with a different orientation of their atoms in space. This phenomenon is called stereoisomerism. Chemical compounds can have two types of stereoisomers: geometrical and optical isomers. Geometrical isomers differ in how a carbon-carbon double bond is positioned in the carbon-hydrogen chain of the molecule. This difference in position is caused by a hindered rotation about carbon-carbon double bonds. Most of the substances treated in this book have carbon-hydrogen chains. The two possible isomers are called: *cis-trans* isomers or in a modern way, according to the IUPAC rules respectively Z and E isomers (from German: Zusammen and Entgegen).

In the case of a terminal carbon atom with a double bond there is no geometrical isomerism, because hydrogen atoms take the final position. A double bond of this type has possibilities to change its position in space. A compound with this type of double bond consists of conformational isomers, which can be transferred into each other by rotation about single bonds.

The simplest form of optical isomerism of a compound is *stereoisomerism*. In that case, two isomeric forms of the compound exist, which are mirror images of each other and cannot be superimposed. It is like right- and left-handed gloves. These isomers are defined *enantiomers*. Also isomers exist named *diastereomers*; they are not mirror images.

The enantiomers have differences in rotation of the plane of polarized light. There can be right-handed rotation, called a D(+) isomer (dextrorotatory) or left-handed rotation, called an L(-) isomer (levorotatory). Asymmetry is involved in the phenomenon of optical symmetry.

Chapter 1. Introduction

There are several conditions of asymmetry for a compound to be optically active. There can be one or more asymmetric carbon atoms in an optically active compound, with a corresponding complex terminology. An exception of this rule is formed by compounds like allenes, spiranes, biphenyls and cycloalkenes.

With the help of a correct structure formula, a specific isomer can be reproduced in drawing by using fat rods and dotted lines, respectively representing atomic bonds reaching out of the plane of reproduction and atomic bonds positioned backwards (Fig. 1.1). Of course this type of reproduction does not give information on the angles between the various atoms in space, although most of these angles are known. The following procedure has been proposed to describe a particular configuration of a defined isomer upon which the structure can be called up before ones' mind: Choose a specific asymmetric carbon atom (the central globe of the figures in Fig. 1.1) and the four groups of the molecular structure (or atoms) attached to it. In case of an asymmetric carbon atom, those four atoms are different. Look at the atomic number of these atoms and place the atom with the lowest priority backwards. The three remaining atoms are in the front and determine the character of the configuration. On creating a circle running from the atom with the highest priority to the priority next in rank and then to the lowest one, the circle is moving clockwise or counter-clockwise (Fig. 1.1.). This gives the isomer the nomination of respectively R (from Latin: Rectus, right) or S (from Latin: Sinister, left).

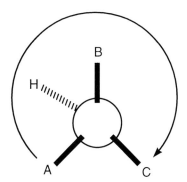 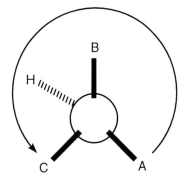

Fig. 1.1. Scheme to define the position of groups or atoms bound to an asymmetric carbon atom with S or R configuration in specific isomers. Central globe symbolizes the carbon atom. In this example, H = hydrogen and A, B and C are other groups or atoms (not necessarily carbon).

In nature isomers and enantiomers of many terpenes are produced, depending on the enzyme systems involved. They can be stored in the same compartments. For industrial purposes some terpenoids are chemically synthesized and again the endproduct is often accompanied by a certain proportion of isomers, some of which are undesirable, unless precautions are taken with selected precursors or special methods (see chapter 7.3). In specific environments and under certain conditions terpenes can be converted into more stabile isomers.

Unfortunately, all kind of notations are found in publications, e.g. in the order of sequence of R and S, (+) and (-), (Z) and (E). There is a differential use of brackets and dashes. To make it worse the symbols R and S (*R* and *S*) versus D and L (*D* and *L*, sometimes d and l) can be expected in one book volume, even in one paper. The same holds for (Z) and (E) and *cis* and *trans* (also cis and trans). Special attention has to be given to the symbols R and S as a more modern replacement for D and L. R does not automatically stand for D. As an example: (R)-(+)-limonene = D(+)-limonene; however, (S)-(+)-carvone = D(+)-carvone!

Chemists utilize both (S)-(-)-X and S(-)-X (X being the name of a compound). The authors of this book prefer the first notation. However, in changing all notations quoted into the preferred one, we are confronted with the fact that our text is not in agreement with the original title of a paper as the reader finds it in the References. To avoid confusion, in referring to a certain study we have often followed the notation of the authors in question; in more extended discussions we have utilized our preferred notation.

Terpenoids can have a simple aliphatic structure, or a cyclic one. Monocyclic, bicyclic, tricyclic and polycyclic structures exist and many of them can be polymerized. Strong acids, such as nitric acid, UV-light, temperature, oxygen and co-polymers can artificially induce polymerization that can result in complicated structures to be found in chemistry journals.

From ancient times, humans have utilized the messenger functions of natural terpenoids for several purposes without knowledge of their structure. This had to wait until the second half of the 19th century. Table 1.1 gives an overview of the development of our knowledge on terpenoids. Since its identification in 1956, mevalonic acid has been recognized as a key substance in the biosynthesis of a wide range of isoprenoids, including terpenoids. Mevalonic acid is produced in many organisms from acetate via a generally occurring enzyme, acetyl co-enzyme A. Terpenoids form a group of isoprenoids that are vital for various biological functions. Endproducts of the mevalonic pathway include sterols, such as cholesterol, involved in membrane structure; haem A and ubiquinone, active in electron transport; dolichol, required for glycoprotein synthesis; carotenoids with many

Chapter 1. Introduction

functions; steroid hormones in animals; hormones in insects and isopentenyl adenine and isoprenoid proteins, both involved in DNA synthesis.

A MULTIDISCIPLINARY STUDY

By the transit to the 21st century we have discovered that many basic processes occur in all living creatures and that eukaryotes share a number of energy producing and energy consuming processes in which regulatory systems operate with basic molecular structures.

Table 1.1. Landmarks in the development of our knowledge of terpenoids

2000+	Genetical modification of terpene synthesis in plants. Usage as repellents to insects and in pharmacy. Production of perillyl alcohol, limonene etc.
2000	Increased understanding of synergistic effects of combinations of terpenoids on biological systems.
1990s	Terpenoids applied in cancer therapy.
1980s	Treatment of cholesterolemia with competitive inhibitors of a key enzyme.
1970s	Terpenes recognized as allomones and kairomones.
1965	Barnes synthesizes an analogue of insect juvenile hormone (sesquiterpene).
1963	Karlson *et al.* identify the steroid structure of ecdysone, the major moulting hormone of insects.
1956	Wright *et al.* elucidate the structure of mevalonic acid. Williams isolates insect juvenile hormone.
1930s	Discovery of steroid hormones and their function.
1866	Kékulé introduces the term "terpenes" (empirically).
1839	Goodyear invents the vulcanization procedure of latex (natural polyisoprene).
1803	Hoffman isolates camphor.
1796	Gren discovers that essential oils consist of volatile compounds.
1750	Separation and recombination of odour components for perfumes etc.
1680	Van Leeuwenhoek utilizes clove oil against storage insects.
1500	Paracelsus prescribes extracts of *Melissa* spp. as a sedative and of *Artemisia* spp. to treat malaria.
1488	Brunschwig isolates chamazulene from chamomile.
1200	Essential oils utilized to treat venereal diseases (Al-Raud Al-Ahir).
≈200	Aromatic plants utilized in pharmacy (e.g. pulegone as an abortivum).
50 BC	Varro protects stored cereals with *Coriandrum* and *Artemisia* spp.
500 BC	Hippocrates uses extracts of chamomile for several purposes.
3000 BC	Aromatic oils utilized in dynastic Egypt.

Complex systems of feedback regulation are involved in biosynthesis of organic substances and in developmental processes. The same regulatory compounds are found in completely different organisms, yet they may have different targets, depending on the receptor (and the organisms). Nature has selected either the most efficient or the most secure way to produce these carbon-based compounds early in evolution and this may explain their omnipresence in the living world.

What makes terpenoids most interesting is the fact that they have a role in the regulation of isoprenoid metabolism and signal transduction and as such can exert a profound effect on cell growth, differentiation, apoptosis and multiplication. No other biochemical group of secondary metabolites has such a potential to interfere with processes ranging from cell level to ecological interactions. Moreover, the lower terpenes are rather volatile, an essential physical property for air-borne long distance effects. Nature uses terpenes in a "chemical language" between plants, insects, vertebrates and even humans (Penuelas *et al.*, 1995). Terpenes and other isoprenoids have also important functions as messengers on a shorter range, different structures being active within organisms, within organs and within the cell body, in particular between the cell surface and the cell nucleus.

A study of a basic function of a messenger molecule in a particular organism increases our understanding of regulatory systems in distant taxa. The more these systems are involved in basic processes (e.g. gene expression), the more scientific disciplines will benefit from such a study. Specific targets of messenger molecules in fully developed organisms are usually studied by specialists, unaware of the same compound having profound effects in organisms belonging to other groups, although the knowledge is somewhere amidst a wealth of accessible information. Terpenoids are such compounds.

The present volume is the result of a multidisciplinary study. This may appear a hard job for only three authors in a scientific world with its continuously increasing degree of specialization. We could have chosen to struggle through the contributions of 20, 30 authors with inevitable overlap of some sections and a strongly reduced number of cross-references. Each of them should have been fully aware of the view of all others, and possess a broad vision, in fact, should be a generalist and a specialist in one person. Unfortunately, this is the period of the decline of the generalist (Seitz, 2000). After some preliminary skirmish we decided to write the volume ourselves. Our backgrounds are: biochemistry, biology, medicine and agronomy with specializations in insect physiology, insect-host plant relationships and the effects of secondary plant metabolites (especially terpenoids) on herbivores, including human beings.

Chapter 1. Introduction 7

A wealth of monoterpenes is found in essential oils of angiosperms; gymnosperms, especially conifers, produce both monoterpenes and diterpenes in the form of resins (acids and related alcohols and aldehydes). Sesquiterpenes and diterpenes also occur in (marine) plants, insects and animals.

Fig. 1.2 gives the major defensive and attractive functions of terpenoid messengers in the relations between bacteria, fungi, plants, insects and vertebrates and illustrates the object of this book, as outlined in the Preface. During evolution, terpenoids have been selected to bring either good or bad tidings. They contribute to the message: "flowers have opened their petals!" or: "stay away from this plant!" They have acquired functions as semiochemicals, e.g. to invite sexual partners or to keep enemies on a distance. All these functions were there when *Homo sapiens* developed ways to utilize natural products that are only available because of a delicate equilibrium between the plant and animal kingdom, in which isoprenoids have a role we are only beginning to understand.

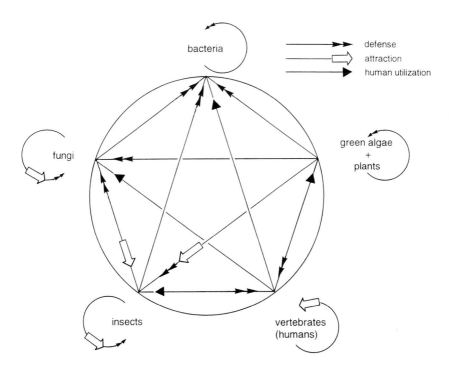

Fig. 1.2. Major functions of terpenoids in the relationships between five biological kingdoms.

We need insects to pollinate our crops, but they also destroy them and defoliate our forests, not to speak of many epidemic diseases both for livestock and humans. Terpenoids have become of widespread importance in perfumery, detergents, foods and beverages, chemical manufacturing industries, pharmacy and biotechnology, yet knowledge of their potential utilization is in its infancy.

HOW TO READ THIS BOOK

This is not a textbook on all thinkable messengers with an isoprenoid structure, providing endless tables and lists with effects of certain terpenoids on a number of organisms. As an example of what can be found elsewhere: weevils (Curculionidae), agriculturally important Coleoptera, constitute the largest family of beetles with at least 40,000 species. Many of their pheromones have terpenoid structures. An excellent overview of these structures has been made by R.J. Bartelt as chapter 5 in: Pheromones of non-lepidopteran insects associated with agricultural plants (eds J. Hardie and A.K. Minks, 1999). Our book focusses on the evolutionary, comparative and physiological backgrounds of messenger functions of terpenoids. Why does carvone inhibit germination of seeds of wild plants and can it be utilized to inhibit germination of stored potato tubers? Why is this monoterpene a repellent to certain insects and does it attract other species? Has a particular function been selected at the level of the primary target (the receptor) or at a higher level? Understanding of the relationship between terpenoid structure and receptors may open doors to completely different disciplines e.g. cancer research.

Following this trend of mind, we will not bring in a condensed form what others have published on specific topics in an extended way. In the figures, only those structures are shown that represent a chemical configuration essential in the relation between structure and function. The same procedure drawing chemical structures has been followed throughout the book. Final methyl groups that are most common are represented by a simple rod or "zebra" line: | or ≡. In contrast, the single final H atoms that occur less frequent have been put into the figure. The reader seeking to find the exact sense in which a term is used is invited to consult the Glossary as a first step. Those who follow the logical sequence of the sections of this book will automatically be referred to additional, more detailed information on a subject. Those who do not, can use the Index to trace (most of) the requested subjects.

This book contains many views and opinions not commonly found in a volume that on first sight resembles a textbook. We apologize, but hope the reader upon absorbing the facts provided will follow our deductions, conclusions and hypotheses to a level permitting a novel view on the

Chapter 1. Introduction 9

material presented. To facilitate this process the reader is addressed with "we, our" and "us". This, of course, is not in the sense of *pluralis majestatis*, nor *pluralis modestiae*, but, without the intention to escape from critical readers, in the sense of *pluralis concordiae,* to arrive at a tuning of the mental constitution of those who read the book in those who have written it.

This book contains boxes, but their number is limited. Boxes provide background information on which there is a consensus of opinion. They are easy to find because of their greyish shade. They facilitate reading, but may not be essential to specialists in a certain discipline. In contrast, most figures reflect our view on a particular topic, resulting from a critical review of the presented research and may contain details on which there are conflicting opinions.

A study of the fascinating world of terpenoids is best made starting with the structure of terpenoids and their synthesis via the major pathways, as treated in chapter 2.

Chapter 2

Production of terpenes and terpenoids

2.1 Biosynthesis of isoprenoids

The best-studied routes of isoprenoid biosynthesis are those of vertebrates, where terpenes and terpenoids are generally produced via the mevalonate metabolism. Throughout this book we will refer to it as the *mevalonate-dependent* route (MAD route). The biosynthesis of mevalonic acid, which is their precursor, is governed by at least four different enzymes: acetyl-CoA thiolase, HMG-CoA synthase, HMG-CoA cyclase and HMG-CoA reductase. The latter enzyme that reduces β-hydroxy-β-methylglutaryl CoA to mevalonic acid (MVA) (Box 2.1) is a key enzyme in the production of MVA and isoprenoid groups.

Chapter 3 reveals how finely tuned mechanisms regulate the biosynthesis of mevalonate. In many organisms, endproducts in which isoprenoids are incorporated can reduce the activity of HMG-CoA reductase(s) with a feedback or degradation system, in this way achieving e.g. cholesterol homeostasis. Among the compounds that interfere with mevalonate synthesis are terpenes and terpenoids themselves. MVA, however, is not the only possible precursor for the isoprene structure. Especially in lower organisms alternative routes have found to be involved, suggesting that what originally was thought to be a common biosynthetic process, may have developed later in evolution. Recently it was found that MAI routes may be of more than marginal importance in higher plants (see 2.1.1 and 2.1.3), making isoprenoid biosynthesis an intriguing subject of evolutionary biochemistry.

How did the MAD route of isoprenoid biosynthesis evolve? Did different pathways precede it and do these pathways still exist? Chapter 3 will discuss the various biosystems in which isoprenoids may have been produced, such as Archaebacteria and archeukaryotes. In this more general chapter on biosynthesis we selected those phyla in an evolutionary sequence, in which the present knowledge is substantially enough to contribute to a better understanding of isoprenoid biosynthesis in plants, insects and vertebrates.

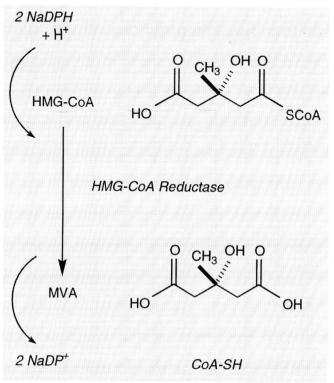

Box 2.1. Conversion of HMG-CoA into mevalonic acid

2.1.1 Bacteria and algae

Until the end of the 20th century, major isoprenoid biosynthetic routes in bacteria did seem to operate excluding HMG-CoA and mevalonate as precursors. Exogenous acetate was never directly incorporated into isoprenoids in *Zymomonas mobilis* that belongs to the Eubacteria (Rohmer and Bisseret, 1994). Formation of the C-5 skeleton of isoprene units probably results from condensation of a C-3 sub-unit derived from pyruvate decarboxylation on the C-2 carbonyl group of a triose phosphate derivative, and a transposition step (Fig. 2.1). Comparable to the biosynthesis of L-valine, these reactions are followed by phosphorylation, H$_2$O elimination and

Chapter 2. Production of terpenes and terpenoids

reductions. This pathway, which we call *mevalonate-independent* (MAI), appeared to be present in both Gram-negative and Gram-positive bacteria, including *Escherichia coli*. This type of synthetic route is known as the deoxy-xylulose phosphate (DOXP) pathway, which was partly elucidated by Rohmer and of which details were reviewed by Bach (1995). The DOXP pathway, which seems to be widely distributed in photosynthetic organisms and is not restricted to algae (Schwender *et al.*, 1997) biosynthesizes the two building blocks of terpenoids, isopentenyl-PP and dimethylallyl-PP. In labelling experiments with 1-[2H(1)]deoxy-D-xylulose efficient incorporation was observed in red and green algae and a higher plant. Now, at the onset of the 21st century, we know that the deoxy-xylulose pathway is almost certainly involved in common MAI routes in green land plants (see 2.1.3).

Schwender *et al.* (1996) have elucidated much of the MAI route of isoprenoid biosynthesis in the green alga *Scenedesmus*. In contrast to the MAD route, the primary substrate is not acetic acid, but the combination pyruvate and glyceraldehyde-3-phosphate (GA-3-P). The transketolase reaction, in which a C-2-unit is transferred to GA-3-P, is thiamine-dependent (Fig. 2.1 (DOXP pathway)). The resulting 1-deoxy-D-xylulose-5-P needs further intramolecular rearrangements of carbon atoms to arrive at isopentenyl diphosphate (IPP), the basic active C-5 unit for isoprenoid formation. Not every single step of the DOXP pathway is known, as only part of the concerned enzymes have been elucidated. The enzyme 1-deoxy-xylulose-5-P synthase could be a key enzyme in the MAI route of isoprenoid biosynthesis. The enzyme 1-deoxy-D-xylulose-5-phosphate reductoisomerase (DXR) converts this first intermediate, DOXP into 2-C-methyl-D-erythritol-4-phosphate (MEP). The DXR enzyme of plants appeared to have a high degree of homology with the DXR enzyme of *E. coli* (Schwender *et al.*, 1999).

A subsequent intramolecular rearrangement is involved in the formation of the branched isoprene skeleton. This pathway appears to be widespread in bacteria, including those related to pathogens and it has an important role in plants (Rohmer, 1998), were it is involved in the formation of essential chloroplast isoprenoids such as carotenoids and phytol and probably in the formation of other plastid-related isoprenoids as well, such as mono- and diterpenoids. So, the first example of isoprenoid synthesis in bacteria leads us at the same time to higher organisms, suggesting that MAI synthetic routes can operate independently from MAD routes in certain compartments.

Not long ago it was found that HMG-CoA reductase is active in Archaebacteria, which makes it plausible that they can synthesize isoprenoids by reactions analogous to those of eukaryotes. Genome enzyme sequencing of true bacteria revealed two unsuspected classes of this enzyme

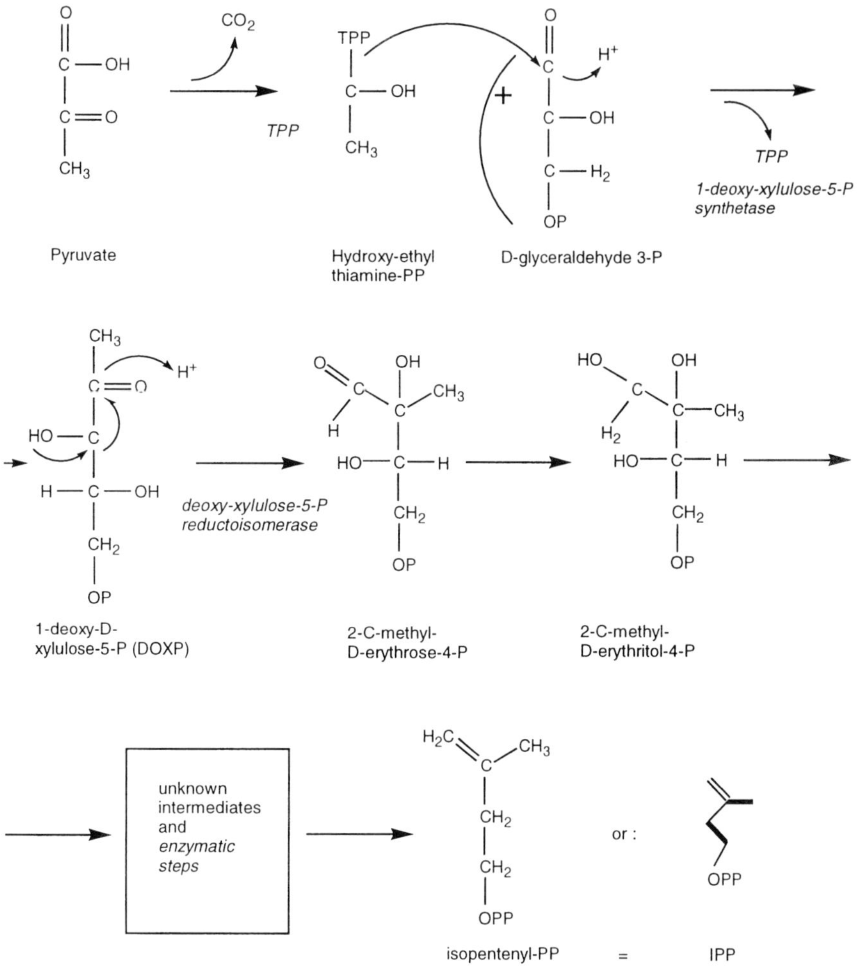

Fig. 2.1. The deoxy-xylulose phosphate (DOXP) pathway according to our present knowledge. TPP is thiamine-PP.

(Bochar et al., 1999). Takahashi et al. (1999) purified and characterized an HMG-CoA reductase from a strain of the eubacterium *Streptomyces* sp. They cloned a gene responsible for HMG-CoA reductase. The amino acid sequence of the enzyme revealed several limited motifs, which were highly conserved and common to the eukaryotic and archaebacterial enzymes. Dairi et al. (2000) report cloning of the gene encoding HMG-CoA reductase from *Streptomyces* strains. In this genus there exist both MAI and MAD routes. These strains produce terpenoid antibiotics. From eubacterial, archaebacterial and eukaryotic sources of HMG-CoA reductases the authors

Chapter 2. Production of terpenes and terpenoids

conclude that eubacterial HMG-CoA reductases form one cluster, distinct from eukaryotic and archaebacterial HMG-CoA reductases.

It remains to be solved if bacterial HMG-CoA reductase is rate-limiting as in humans (see 3.2.2), but the fact that both MAD and MAI routes are operative in these bacteria demonstrates the biological feasibility of isoprenoid biosynthesis from acetate or from pyruvate early in evolution, leaving the question whether they developed independently or in cooperation (exchange of IPP pools). To our present knowledge, the DOXP pathway is missing in fungi, yeast and in vertebrates including humans, placing these organisms in a special group. In contrast, unicellular organisms exist with a metabolic dichotomy as in higher plants: mevalonate and DOXP pathways occur in one organism (Disch *et al.* 1998a).

Where did the enzymes needed for isoprenoid formation arise? The ancient hyper-thermophilic Archaea (3.5 billion years or more), living in total darkness and anaerobic conditions may have been dependent on inorganic redox reactions for positive entropy and can be characterized as chemolithotrophs. Early development of photosynthetic systems is indicated by the fundamental enzyme ribulose bisphosphate carboxylase oxygenase (RuBisCO), found in purple sulphur bacteria and cyanobacteria. This is a phylogenetically interesting, old enzyme, catalyzing photosynthetic fixation of CO_2 by combining it with a phosphorylated sugar (Watson and Tabita, 1997). This enzyme may have been a key factor in the transition from anaerobic redox exchange to the production of oxygen by cyanobacteria in the Precambrian.

Members of the Methanoarchaea, a phylogenetically diverse group of the Archaea can produce acetate out of CO_2. Other members can synthesize acetyl-CoA out of simple carbon elements. Some of these live in deep sea under anaerobic conditions (Abbanat and Ferry, 1990; Ferry, 1999). It would seem that acetyl-CoA synthase is an early-developed enzyme, which together with CO dehydrogenase is a basis for the production of methane from acetate by methanogenic bacteria.

The neutral lipid compositions of a number of methanogenic and thermoacidophilic Archaebacteria contain a range of acyclic isoprenoids with chains from C-14 to C-30. Many members of this group exist in environmental conditions as described for evolutionary stages of archaean ecology and resemble the isoprenoid distribution isolated from ancient sediments and petroleum (Tornabene *et al.*, 1999).

Both MAI and MAD routes need their own enzymatic machineries. When the DOXP pathway would have been widespread in early evolution, basic enzymatic processes should be able to operate under anaerobic conditions. Most enzyme reactions have not yet been elucidated, but for the formation of thiamine diphosphate (TPP)-acetaldehyde, which is catalyzed by pyruvate decarboxylase. This is a typical enzyme of anaerobic metabolism, present in yeast.

The sterol pathway is thought to originate with bacterial heterotrophs that developed enzymes capable of epoxidizing squalene to squalene oxide. This ancestral epoxidase probably produced a mixture of stereoisomeric squalene oxides. These isomers were confronted with the ancestral cyclase, resulting in anaerobic cyclization into tetracyclic and pentacyclic products (Nes and Venkatramesh, 1994).

Nitrogen is a component of DNA and of enzymes. Biosynthesis of isoprenoids needs activity of enzymes, therefore the availability of N, although there is no N in the isoprenoid skeleton. When amino acids are linked into polymers, a molecule of water is removed at each link in the chain with a N atom in the peptide bond. When isoprene units are linked, the bond is C to C. Polymer formation can be based on different types of bond; some are autocatalytic as there mere presence favours the formation of identical molecules. Such is the case with the double-strand helix of DNA. The first protoenzymes may have been polymer chains with autocatalytic functions. Enzymes of subsequent generations were not merely catalysts but add in addition a directing or coupling function, ensuring that the chemical free energy released by one reaction is utilized by another reaction rather then being dissipated as heat. The processes of chain elongation of isoprenoids are discussed in chapter 2.2. Basically a new C to C bond is realized by attaching a highly reactive electron-deficient carbo-cation to an electron-rich carbon-carbon double bond. A simple chemical process, however, dependent on the generation of carbo-cations, with energy-rich pyrophosphates as a driving force.

The invention of photosynthesis enabled living organisms to become the primary producers of energy-rich molecules instead of consuming those provided by non-biological processes. Where green and purple sulphur bacteria use H_2S as a hydrogen source to convert CO_2 into glucose, cyanobacteria and green algae obtain hydrogen from water releasing oxygen. Miller *et al.* (1999) have performed experiments to prove the existence of the DOXP pathway in cyanobacteria. Gene amplification and DNA sequencing resulted in identification of a DOXP synthase gene in *Synechococcus leopoliensis.*

Living charophycean algae possess several biosynthetic attributes also present among land plants, such as the production of sporopollenin, cutin, phenolic compounds and activity of the glycolate oxidase pathway. This makes studies of green algae particularly valuable, as they can contribute to our understanding of the transition of aquatic life forms to those adapted to terrestrial environments. From a metabolic point of view, such life forms had

Chapter 2. Production of terpenes and terpenoids 17

to cope with different osmotic and oxygen conditions as well as availability of a few essential elements for organic molecules like carbon and nitrogen.

2.1.2 Fungi

In fungi, terpenoids are produced from MVA. Birch *et al.* (1958) demonstrated that 2-^{14}C-labelled MVA was incorporated into the 10-carbon terpenoid side chain of mycelianamide, produced by *Penicillium griseofulvum*. In the same year Tamura and Folkers reported hiochic acid, a growth factor of bacteria, to be identical with MVA. This study proved that MVA is not specific for fungi, as just two years before the group of Skeggs had shown an acetate-replacing growth factor in a strain of *Lactobacillus acidophilus* (and became designated MVA) to be distinct from lipoic acid and of all other compounds thought to contribute to microbial growth (named "vitamin B13" in the 1940s). Another study of MVA metabolism of *Phytophthora cactorum* is by Richards and Hemming (1972).

In the same decade it was demonstrated that fungi can incorporate 2-^{14}C-MVA into β-carotene and that the ^{14}C isotope from 2-^{14}C-MVA becomes incorporated into alkaloids, such as agroclavine and elymoclavine, produced by the fungus *Claviceps purpurea*, a parasitic fungus of rye.

Fungi show an extraordinary richness of life forms and in many life stages resemble more to animals than to plants. From this point of view, it is not so amazing that they have developed a MAD route of the synthesis of isoprenoids. However, fungi are known that use both MAI and MAD metabolism. *Streptomyces* spp. synthesize IPP via both routes; some terpenoids are MAI, others are MAD. *S. griseolosporeus* produces the important antibiotic terpentecin via MAD. A mutant lacking the gene encoding HMG-CoA reductase does not produce this terpenoid (Dairi *et al.*, 2000).

The production of terpenoids by fungi is not an exception. *Ceratocystis* spp. were found to synthesize a series of monoterpenes, such as citronellol and geraniol (Fig. 2.5) (Collins and Halim, 1970). *Penicillium decumbens* produces an aromatic oil when sporulating, containing thujopsene and nerolidol, and *Phellinus* spp. apart from benzoates and phenolic compounds are rich in linalool (Collins, 1976). This author found already in 1970 that the specific odour of many fungi is based upon the dominating terpenes in their essential oil.

Terpenoids are carbon-based. The early appearance of fungi capable of decomposing lignin (belonging to Ascomycetes and Basidiomycetes) may have been an important contribution to the recycling of organic carbon, which together with atmospheric carbon could have stimulated the development of functional secondary metabolites along the isoprenoid route.

Every year about 5% of the 700 billion tons of carbon in the atmosphere are converted into plant material. Fungi possess the most important machinery to recycle carbon of dead plants and give it back to the atmosphere in the form of CO_2. In addition, the major energy flow into animal components of early terrestrial ecosystems was probably through decomposition of organic debris rather than direct herbivory (Di Michele *et al.*, 1992, discussed in Kenrick and Crane, 1997).

Fungi, like vascular plants, not only found a way to overcome the problems involved with surface tension and mechanical stability in above-ground systems, in addition some of them (forming lichens) have their share in utilizing air by way of symbiosis with single-celled algae that biosynthesize carbohydrates from CO_2 and water, and with bacteria which bind nitrogen from the air. This is not to be neglected, as about 8% of the terrestrial surface is covered with lichens.

Fungi that may be regarded as the third kingdom, apart from Plants and Animals have a profound impact on plant and animal life forms. Therefore, their MAD routes of terpenoid biosynthesis and their regulation of MVA metabolism (see chapter 3.2) are of great interest in a study of the origin and evolution of terpenoid messengers.

2.1.3 Green plants

Two years after the "discovery" of MVA by Wright and his cooperators, Stanley (1958) showed that seedlings of *Pinus attenuata* incorporate 2-^{14}C-MVA into a monoterpene fraction, mainly consisting of α-pinene (Fig. 2.7). This suggests that monoterpenes in conifers are produced via the MAD route. With the present knowledge, this seems to be more of an exception than a rule, or it may be an example of exchange of precursors. A wealth of terpenoids is produced by vascular plants, but the classic textbook scheme as shown in Box 2.1 is not the only route along which higher plants produce terpenoids. Bach (1995) discussed two alternatives for the production of IPP by vascular plants: the already mentioned Rohmer pathway and the so-called Hano pathway along which hemiterpene moieties could be produced from sucrose either by the pentose phosphate pathway or by glycolysis. The commonly accepted MAI route, which is now described as the deoxy-xylulose phosphate (DOXP) pathway, originates from this work.

The studies by Schwender *et al.* (1996, 1997) and Lichtenthaler (1998) have greatly contributed to our knowledge of MAI isoprenoid biosynthesis in vascular plants. An extended review of the importance of the DOXP pathway is made by Lichtenthaler (1999). The classic acetate/MVA pathway is related to the cytosol and produces sterols and possibly sesquiterpenes, triterpenes and polyterpenes. The DOXP pathway is located in the plastids

Chapter 2. Production of terpenes and terpenoids

and delivers mono- and diterpenoids, carotenoids, phytol and plastoquinone. There may be some exchange of IPP from plastids to the cytosol, and in this way sterol production would not be entirely dependent on mevalonate. The production of ubiquinone (Q-9 and Q-10) would seem to proceed in mitochondria from cytosolic IPP.

Land plants may have originated from charophycean green algae with their possible MAI routes (see 2.1.1), or from single-celled algae which evolved into mosses that became adapted to less humid conditions (Brown, 1999). A palaeobotanical study of mosses and liverworts adds to our view that early diversification of land plants pre-dates the late Silurian (Kenrick and Crane, 1997). At the end of the Silurian regression of big seas might have forced sea life to probe on land. The transgression of the Silurian to Devonian was marked by the siege of the land by plants with novel adaptations. They were usually small, but bigger examples are known, such as Rhynophyta (obsolete: Psilophyta), with the likely ancestral forms *Cooksonia* and *Rhynia*. Fig. 2.2 depicts a setting of the early Devonian with its possible terpenoid chemistry. The early land plants may have produced isoprenoids along both MAD and MAI routes. Carbon-based metabolism could easily be realized, availability of CO_2 was no problem. It is assumed that CO_2 levels were more than 10 times higher than at present. Plants did not need big leaves and the stems harboured the stomata regulating gas exchange.

There is growing evidence that different machineries for the synthesis of a particular isoprenoid precursor, be it xylulose or MVA still exist (Bach, 1995). If it is more rule than exception that prokaryotic terpenoid pathways have been consolidated in land plants, the reasons why are a fine challenge to coming studies. The possible interactions between compartmentized enzymes and the symbiotic origins of organelles (plastids) could contribute to the consolidation of different strategies of isoprenoid production. The great diversity in structure and function of the terpenoid endproducts may be another reason. There is a world of difference between the substantial production and release of monoterpenes and the finely tuned feedback systems as involved in e.g. photosynthesis and cell cycle regulation, yet these processes are active within one organism (see 2.2).

Evidence for the compartmentation and complete enzymatic independence of the biosynthesis of long-chain all-*trans* polyprenols in mitochondria and in chloroplasts is provided by Disch *et al.* (1998b). Studies of the incorporation of ^{13}C-labelled glucose or pyruvate into the isoprenoids of tobacco BY-2-cells allowed these authors to determine the origin of IPP. Sterols synthesized in the cytoplasm and the prenyl chain of ubiquinone Q-10 located in mitochondria appeared to be derived from the same IPP pool, synthesized from acetyl-CoA through mevalonate, whereas the prenyl chain

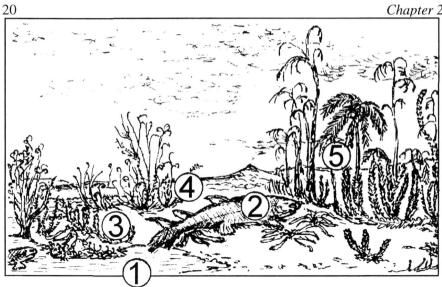

1	MAI	*Scenedesmus* (green algae)	phytol (diterpene)
2	MAD	*Ichtyostegas* (vertebrates)	cholesterol (sterol)
3	MAD	*Aspergillus* (fungi)	γ-cadinene (sesquiterpene)
4	MAI	Lycopodia (club mosses)	bornyl acetate (monoterpene derivative)
5	MAI MAD	Pteridophytae (ferns)	taraxastane (diterpene)

Chapter 2. Production of terpenes and terpenoids 21

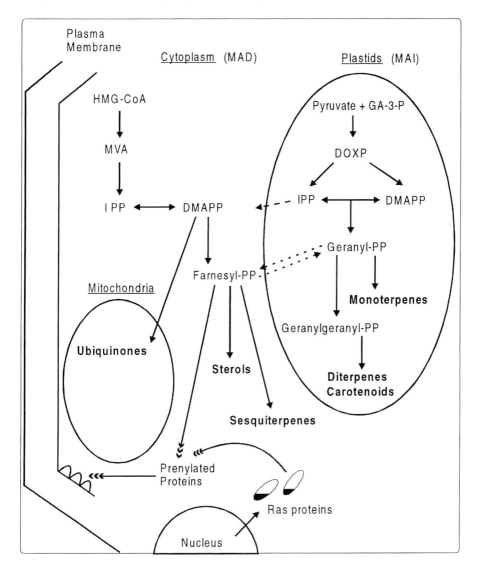

Fig.2.3. Biosynthesis of isoprenoids via the MAI and MAD routes in different compartments of a plant cell with some of the basic terpenoids produced with specific enzymatic machinery. Partly based upon Lichtenthaler (1999). Dotted arrows stand for possible exchange of precursors.

Fig. 2.2. A scene of the middle Devonian depicting wetlands inhabited by the first terrestrial vertebrates. Advanced terpenoid structures were produced via both MAI and MAD routes.

of plastoquinone was obtained from the DOXP pathway, like all chloroplast isoprenoids from higher plants. Fig. 2.3 shows the compartmentation of MAI and MAD routes in a plant cell.

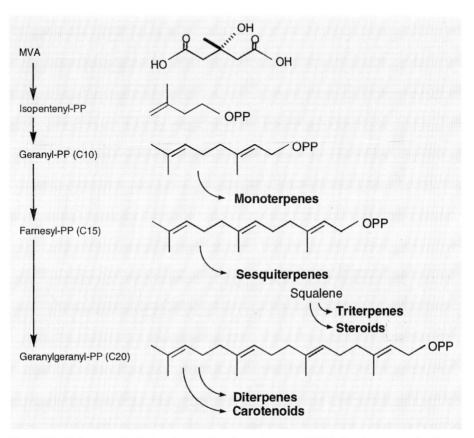

Box 2.2. Scheme of chain elongation based upon MVA in the cytosol of a plant cell. OPP stands for pyrophosphate.

According to Arigoni *et al.* (1999) dimethylallyl-PP is not the committed precursor of IPP in the MAI route. Isotope labelling patterns of phytol and lutein derived from 1-deoxy-xylulose revealed that there is a direct preferentially labelling of IPP and not of DMAPP (see Fig. 2.3). IPP is the physiologically relevant, terminal intermediate of the DOXP pathway (Lange and Croteau, 1999) with a possible exchange between MAI and MAD compartments.

The occurrence of both MAI and MAD routes of isoprenoid biosynthesis in vascular plants may have resulted from the development of chloroplasts from cyanobacteria, a view which is supported by Lichtenthaler (1999), followed by some cytosolic and plastidic specialization with respect to endproduct release.

Provided the proper enzyme systems are available, a multi-branched isoprenoid pathway can develop, enabling the production of several classes of terpenoids, as treated in section 2.2 and with functions as messengers as discussed in chapter 3.

Dimethylallyl-PP is a simple 5-carbon molecule, geraniol a 10-carbon molecule, farnesol (C-15) and geranylgeraniol (C-20) are constructed with increasing isoprene units, by addition of the allylic moiety to the double bond in isopentenyl diphosphate. In living cells, geranyl, farnesyl and geranylgeranyl pyrophosphate (coded as: GG-PP) can be bound to enzymes, the primary cyclases, which are involved in the production of carbon skeletons for the synthesis of an enormous variety of terpenes and terpenoids.

According to Sacchettini and Poulter (1997) many thousands of enzymes are involved in isoprenoid chain elongation and cyclization, of which at best a few hundreds have been studied. The various isoprenoid structures often derived from the same substrate may arise from differences in folding of the substrate in the active site, or may depend on how carbo-cationic intermediates are stabilized to induce or discourage rearrangements, or on how positive charge will be quenched once the product is formed.

In general, biosynthesis of terpenoids in higher plants is accompanied by the synthesis of more complex terpenoids, such as carotenoids, steroids and ubiquinones, which does not say that they are all derived from MVA, as different enzyme systems operate in plastids, mitochondria and the cytosol. A summary of mevalonate metabolism in higher plants is given by Spurgeon and Porter (1983). Whether derived via the MVA or DOXP pathway, the fundamental chain elongation of terpene biosynthesis leads to the formation of the main classes of endproducts. This chain elongation is achieved by condensation of isopentenyl-PP (coded as IPP) and carbon of an allylic pyrophosphate, catalyzed by prenyltransferase, in the 5-C homologue of the allylic pyrophosphate in question. Geranyl-PP (coded as G-PP) is such a condensation product. After chain elongation with another isopentenyl-PP (IPP), farnesyl-PP (coded as F-PP) is formed. It is an intermediate for many mevalonate metabolites in plants and animals (Box 2.2 and Fig. 2.3). It is situated on a very important branch point of the metabolism.

The sterols produced by plants, such as ergosterol, are close to cholesterol, which in its turn is the basic structure of steroid hormones, not only of plants but also of vertebrates (Box 2.2). The vitamins of the D group still incorporate the aliphatic side chain as in ergosterol, but their function is quite different. Vitamin D2 (calciferol) can be obtained with UV radiation of ergosterol. This study is not a textbook on steroid endproducts, but these examples illustrate that in all classes of terpenoids numerous variations exist on additions to the basic structure, variations that often have different biological functions. Evolutionary aspects of origin and function of steroids are discussed in chapter 3.

2.1.4 Insects

Hydrocarbons in general serve many functions in insects. Insects do not synthesize squalene or sterols, but are able to biosynthesize isoprenoids and produce a wealth of terpenoids. In concert with research on MVA metabolism in vertebrates and higher plants, most studies on terpenoid production in insects were based on the assumption that isoprene units are constructed from MVA with HMG-CoA reductase as a key enzyme. Brown *et al.* (1983) demonstrated that isolated embryonic *Drosophila* cells exhibit HMG-CoA reductase activity, which is membrane-bound. Havel *et al.* (1986) continued this study of IPP synthesis in embryonic *Drosophila* cells. Using [1-^{14}C]mevalonate they found that the insect cells "shunted" a significant fraction of post-IPP carbon at the level of C-5-C-15 prenyl 1-pyrophosphates for oxidative catabolism to acetyl-CoA. This pointed to a possible regulation of HMG-CoA reductase activity, suggesting a feedback regulation of HMG-CoA reductase as is found in mammals (see 3.2). This is not necessarily the same type of regulation as in mammals, as sterols can take no part in it and, as may be expected, Brown *et al.* (1983) found their *Drosophila* cells to be not responsive to feedback inhibition by sterols. They realized that those eukaryotic cells, which do not produce sterols from MVA, would eliminate a major demand for mevalonate carbon, offering excellent possibilities to study MVA metabolism with non-sterol endproducts. We will discuss the regulation of MVA metabolism of insects in section 3.2.1.

According to the present knowledge the MAI route does not operate in insects or perhaps in exceptional situations. Roughly speaking a bit more than half the world's insects are either parasitoids or predators or feed on dead organic material, leaving a substantial number of herbivorous species, especially among the orders of Hemiptera, Thysanoptera, Lepidoptera, Hymenoptera and Coleoptera.

As precursors of many terpenoid semiochemicals produced by herbivorous insects are present in their host plants, one could expect that these insects need only those biosynthetic steps which are essential for a specific endproduct, or for the release of a semiochemical from a specific site to which a precursor is transported prior to being upgraded to the final product. The Californian pine bark beetles *Ips paraconfusus* and *Ips pini* produce the acyclic monoterpenes ipsenol and ipsdienol (Fig. 5.4) as components with a function in aggregation behaviour. The monoterpene myrcene is a direct precursor to ipsdienol and, ultimately, to ipsenol. These beetles tunnel and feed in phloem tissues of pines, and myrcene is present in xylem and phloem oleoresin that adjoins and pervades these tissues. As the aggregation pheromone accumulates in the hindgut and a number of studies have failed to demonstrate a hindgut-associated pheromone gland, it would seem an adaptive phenomenon that the beetles use plant-derived myrcene for the final steps towards ipsdienol. However, studies with radiolabelled acetate and mevalonate derivatives have provided direct evidence for *de novo* biosynthesis of ipsenol, ipsdienol and amitol by bark beetles (Seybold *et al.*, 1995; Seybold, 1998) and the key regulatory enzyme is HMG-CoA reductase, indicating that the essential isoprenoids are biosynthesized via the MAD route. Only a minor proportion of the semiochemicals are derived from myrcene, that contributes to less than 3% of ipsdienol, accumulated in the hindgut. As we know from recent work on plant-produced monoterpenes, this fraction is not derived from MVA at all or at best in minor quantities.

In contrast to plants, MAD routes seem to be of major importance in isoprenoid biosynthesis in insects, although it is not necessarily the only possible way along which isoprenoids are produced within one individual.

Herbivorous insects utilize fungi and bacteria to overcome the biochemical barrier to herbivory. Symbiotic micro-organisms, depending on their position in the intestinal tracts or as endosymbionts, are able to detoxify plant allelochemicals, degrade cellulose and biosynthesize specific nutrients. These fungi and bacteria produce their own isoprenoids via the MAI route. In this way, both MAD and MAI routes can operate within a herbivorous insect. This puts herbivorous insects into a unique position, deserving a more extended discussion in the coming chapters. One reason is that this fact can rise novel types of insect control, by affecting the MAI metabolism of the endosymbionts.

2.1.5 Vertebrates

Vertebrates produce terpenes and terpenoids via the mevalonate pathway. The biosynthetic MAD route has already been mentioned at the beginning of this chapter, as it is by far the best-studied isoprenoid synthesis of all, because of the pharmacological interest in the regulation of cholesterol production and steroid hormones. In lower vertebrates these compounds are produced via the so-called squalene route, but in vertebrates with higher levels of development there is a complex relationship between the production and release of lipoproteins and sterols. Complex feedback regulation and synthesis from more than one precursor have caused many research groups in this world to resume their experiments and adjust their conclusions (see 3.2 and 3.3). These studies have greatly contributed to our knowledge of MVA metabolism in vertebrates. It would seem that the biosynthesis of the isoprenoid skeleton in vertebrates greatly resembles that of insects.

Most studies on MVA biosynthesis and pathways to isoprenoid endproducts were done with mammalian liver, which is chiefly responsible for maintaining blood levels of cholesterol. A number of isoprenoid intermediates and endproducts arise from MVA in vertebrates: squalenes, dolichols, sesquiterpenes, lipids, sterols and ultimately steroid hormones. They will be discussed in chapter 2.2, but all have a common precursor: F-PP (see Box 2.2). The conversion of MVA to F-PP is not restricted to the liver, however, the enzyme systems in those cells or tissues are often limited. Blood platelets need MVA as a substrate to produce lanosterol, which they share with the skin, testis, bone marrow and blood vessels that are able to synthesize cholesterol.

The same system of chain elongation as shown in Box 2.2 applies to animal cells, although the endproducts may be different: mammals, for instance, do not produce monoterpenes. HMG-CoA is the rate-limiting enzyme in MVA biosynthesis (Box 2.1). Probably the evolution of brown plants, fungi and animals has a common radiation point, which makes them more related to each other than to green plants (Brown, 1999). This puts isoprenoid synthesis of this biochemical clade via a MAD route in a specific position.

Not that the endproducts of the MVA metabolism of vertebrates are always very different from those of organisms developed earlier in evolution. As this book is dedicated to the interrelationships of plants, insects and humans it is intriguing to know that the moulting hormone of insects is derived from cholesterol, to give one example. For a better understanding of the production of mevalonate metabolites in vertebrates we refer to section 3.2.2, where regulatory processes of the synthesis of more complex isoprenoids are discussed.

2.1.6 Evolution of isoprenoid biosynthesis

Most evolutionary theories on the development of eukaryotes agree in their assumption that they are fundamentally chimeric (Katz, 1999) with the Archaebacteria (or Archaea) possessing the closest prokaryotic nucleocytoplastic lineage of eukaryotes. A fusion with an eubacterium that became the mitochondrion, offered the heterotrophic host novel possibilities for energy metabolism: instead of glycolysis in which one mol glucose is oxidized to pyruvate, yielding only two mol ATP, mitochondriate eukaryotes can further oxidize pyruvate through the pyruvate dehydrogenase complex. The Krebs cycle with uptake of O_2 and production of CO_2 and water yields an additional 34-36 mol ATP per mol glucose. There is no agreement, however, on how the ancestral, organelle-lacking eukaryote might have arisen. It may find its roots in the Archaezoa (many of which are probably extinct) and evolved a nucleus, a primitive cytoskeleton, and endocytosis, followed by endocytosis of an eubacterium (mitochondrion). Another possibility is an early fusion between an archaebacterium and an eubacterium, which means a primary endosymbiosis of the nucleus and a secondary symbiosis in connection with the acquisition of the mitochondrion. According to Roger (1999) there is substantial evidence that the mitochondrial endosymbiosis took place prior to the divergence of all extant eukaryotes. The evolutionary trees based on ribosomal RNA leave many questions unsolved. Doolittle (1998) refers to the puzzling fact that the vast majority of gene products from the Archaea most resemble counterparts among the Eubacteria and not eukaryotes. Another question he puts is: where are the precursors of tubulin and actin and other cytoskeletal proteins?

It should be noticed that a prokaryotic cell-division protein, related to actin, is not found in any of the archaeal genomes, indicating that the origin of the cytoskeleton cannot be the result of simple chimeric mergers of an eubacterium and an archaebacterium. W.F. Doolittle (1996, 1998) suggests that the suggested merger of an eubacterium with an archaebacterium took place within a third cell type, which already possessed a cytoskeleton. This hypothetical ancestor may have lost its identity, being overrun by the trimeric entity.

Whenever the marriage between an archaebacterium and an eubacterium happened, Martin and Müller (1998) provide striking evidence that eukaryotes developed from symbiosis of a cyanobacterium (Archaebacteria) with an eubacterium. The host (the archaebacterium) was autotrophic, dependent on H_2 for O_2 production. The eubacterium, transforming into the mitochondrion, produced H_2 and CO_2. The archaebacterium could have had a methanogenic ancestry.

So far, these hypotheses do not bring a very satisfying solution to the question of the origin of the nucleus. Karlin *et al.* (1999) provide data to support their hypothesis that the eubacterium was *Clostridium*-like, and the archaebacterium was *Sulfolobus*-like. The genome signatures of these microorganisms are compatible and each has significantly more similarity in genome signatures with animal Mt sequences than do all other available prokaryotes. Our question would be: does this also hold for fungi? Anyhow, Karlin *et al.* (1999) put forward that the molecular apparatus of endospore formation in *Clostridium* could serve as raw material for the development of the nucleus and cytoplasm of the eukaryotic cell. According to Roger (1999) it is very difficult to discern whether the nucleus evolved before or after the mitochondrial endobyosis. We agree with this author that our current picture

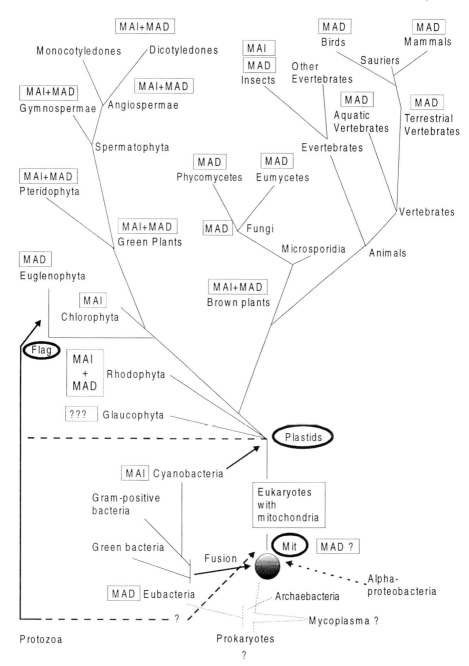

Fig. 2.4. The evolution of MAI and MAD routes in the Plant and Animal kingdom (partly tentative). Fat arrows: fusion between two organisms resulting in additional organelles and novel biosynthetic pathways. ☉ = nucleus; Mit = mitochondria; Flag = Flagelata.

of early eukaryotic evolution is in a state of flux; to the best of our knowledge we have positioned the nucleus and the mitochondria close together in Fig. 2.4, with a possible line to the α-proteobacteria.

We tried to construct an evolutionary scenario of isoprenoid synthesis, culminating in Fig. 2.4. Referring to a methanogenic metabolism, the earliest enzymes may have been polymeric metal complexes, able to catalyse reactions in protected compartments. Later self-replicating peptides developed, along with co-factors with bound metal ions. They still exist. In eukaryotic cells we find e.g. cytochrome p450, a redox enzyme with metal-associated co factors. Cytochrome p450 has a function in sterol biosynthesis, with its isoprenoid basic structure. It is involved in cyanogenic glycoside biosynthesis. With the Cyanobacteria in retrospect, biosynthesis of p450 could have existed before oxygen became established in the atmosphere.

Members of the Archaea can synthesize acetyl-CoA out of 1-C elements and CoA (Abbanat and Ferry, 1990; Ferrante *et al.*, 1990). We assume that the early eukaryotes harboured a MAI isoprenoid route. Isopentenyl diphosphate is an important C-5 precursor of isoprenoids. In *E. coli* the compound is synthesized from the DOXP pathway. This pathway in its turn is derived from glycolysis and the Krebs cyclus (Harker and Bramley, 1999). Several living Eubacteria including *E. coli* use this pathway (Fig. 2.1), which runs along 2-C-methyl-D-erythritol 4-phosphate via a single step with an enzyme designated 1-deoxy-D-xylulose-5-phosphate reductoisomerase by intramolecular rearrangement and reduction. The gene that expresses this enzyme is responsible for terpenoid biosynthesis (Takahashi *et al.*, 1998).

Cavalier-Smith (2000) opines that the introduction of mitochondria and chloroplasts were rare evolutionary phenomena with a great momentum on eukaryotic evolution. This author stresses that it is now widely agreed that chloroplasts originated only once and that the plastids of green plants depicted in Fig. 2.4 almost certainly originated from an ancestral cyanobacterium. Cyanobacteria (one exception) have three different membranes, like chloroplasts. Subsequent lateral transfers of plastids formed more complex photosynthetic chimeras. According to Ourisson and Nakatani (1994) primitive membranes could initially have been formed by simple terpenoids, and vesicles formed from these membranes may have evolved into progressively more complex units, more and more similar to protocells. In more evoluated systems they acquired functions in signal transduction by attachment to membranes in a protein-bound form.

Fig. 2.4 depicts the origin of terpenoid biosynthesis in eukaryotes and brings us to the onset of the phylogenetic tree with green plants, fungi and animals. Fungi and animals share the MAD route of isoprenoid biosynthesis. The radiation point may be the Brown plants. Isoprenoids produced via the MAD route and membrane-bound MVA biosynthesis could acquire specific tasks as messengers, because of the localization of enzyme systems, a prerequisite for fine tuning and endproduct regulation.

2.2 Production of terpenoids by different organisms

2.2.1 Monoterpenes

Monoterpenes can have an open chain or a cyclic or bicyclic structure. Their names are derived from variations in their chemistry: -ol means an alcoholic group (C-OH), -one points at a ketone group (C=O), -ene means absence of these groups. As an example: geraniol is an open chain monoterpene alcohol, pulegone a monoterpene ketone, myrcene does not end in an oxigenated group (Fig. 2.5). The isoprenoid pathway probably presents the greatest structural diversity in terms of molecular recognition based on stereochemistry and functionality. In addition to variations in basic structure, isomers (Fig. 2.6) and enantiomers (see the Introduction for terminology) of monoterpenes make an almost infinite number of combinations, not unlike the many permutations of a fine tumbler lock, with a few elements only: carbon, hydrogen and oxygen.

The process of cyclization of monoterpenes and sesquiterpenes is catalyzed with terpenoid cyclases. They cyclizise the universal acyclic precursors G-PP and F-PP to cyclic terpenoids. They consist of moderately lipophilic proteins of a molecular weight of 40,000-100,000. Their co-factors are usually a divalent metal, mostly Mg^{2+} and occasionally Mn^{2+}. Terpenoid cyclases have been found in the protozoon *Tertrahymena pyriformis* (Renoux and Rohmer, 1986). These cyclases operate in cell-free systems and enable the synthesis of complex cyclic isoprenoids. These compounds are analogous to those identified in numerous sediments. The role of metal ions and the structure of these enzymes are further discussed in section 2.2.2.

Fig. 2.3 and Box 2.2 show how G-PP is synthesized. G-PP is the basis for acyclic and cyclic monoterpenes. The precursor of monoterpenes, isopentenyl-PP (IPP) can enter a completely different branch by isomerization and subsequent reaction with adenosine monophosphate (ADP), followed by conversion to isopentenyl adenine. This molecule can be further converted to the three major forms of naturally occurring cytokinins, well known for their role in cell division of plants. Their link with terpenoids is the early presence of an isoprenoid group; the rest of the molecule that contains nitrogen does not have a terpenoid structure but rather that of a purine base derivative. Their dependence on a simple isoprenoid structure

Chapter 2. Production of terpenes and terpenoids

makes that cytokinins can be regulated (at least in part) by common factors active in isoprenoid production (see chapter 3.2).

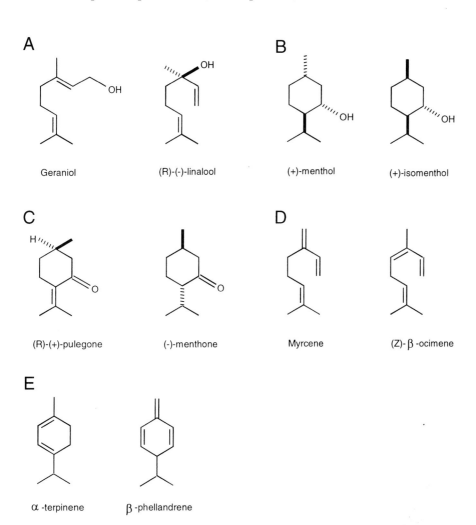

Fig. 2.5. Examples of monoterpenes. **A** = open chain (alcohols); **B** = cyclic with one OH group; **C** = monoterpene ketones; **D** = open chain without an oxigenated group; **E** = cyclic, without an oxigenated group.

Fungi have a skeleton of chitin, like insects, and biosynthesize monoterpenes via the MAD route, which potentially has an adequate feedback regulation, avoiding overproduction when there is no need for big quantities.

Monoterpenes alone form a cornucopia of secondary plant metabolites. Together with other volatiles, monoterpenes (and sesquiterpenes) are ideally suited for a "chemical language" between individuals of the same or different species, as low molecular weight essential for aerial dispersion is combined with a physical requirement for perception: high lipophilicity. To a better understanding of the messenger function of terpenoids we should know which compounds are produced by our objects of study with special attention to plants, insects and vertebrates, where terpenoids are stored, how they are transported within an organism, and how they are released.

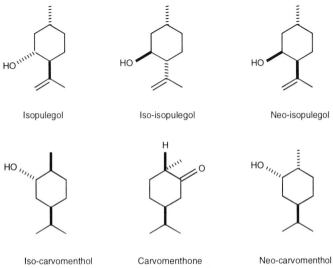

Fig. 2.6. Examples of monoterpene isomers.

PLANTS

Anyone who has made big walks in forests with predominating conifers recognizes the typical smell of needle-leaved trees. Monoterpenes such as pinenes, phellandrenes, terpinenes, cineole, limonene, myrcene, camphene and santalene (Figs 2.5 and 2.7) are main components of the odour. Apart from large variations in the relative amounts of monoterpenes and their isomers in conifers, the enantiomeric compositions both within parts of the trees and among species show a great deal of variations. In *Picea abies*, the Christmas tree, the (-)-enantiomers of α-pinene, limonene and camphene in the needles dominate over their corresponding (+)-enantiomers (Borg-Karlson *et al.*, 1996). High concentrations of monoterpenes are usually toxic: many micro-organisms use them as defensive compounds, at the same time risking autotoxicity. There are several solutions to this problem (see 4.2.3), that may be found in the anatomy and membrane structure of the organisms in question.

In liverworts monoterpenes are most probably synthesized via the DOXP pathway, whereas the isoprene building blocks of sesquiterpenes and sterols are dependent on MVA. Studies on *Ricciocarpos natans* and *Conocephalum conicum* (Adam *et al.*, 1998) have made it plausible that in liverworts both isopentenyl diphosphate biosynthetic pathways are involved in different cellular compartments.

In Gymnospermae, like the Christmas tree, monoterpenes are most likely produced via the MAI route and stored at the site of production in numerous small oil glands. In pine trees, G-PP is the source of an array of monoterpenes, such as α-pinene, already mentioned to be found by Stanley in 1958, and in addition β-pinene, 3-carene, β-phellandrene and limonene (Fig. 2.7) each produced with their own monoterpene synthases (Savage and Croteau, 1993; Savage *et al.*, 1994).

Monoterpene cyclases produce cyclic monoterpenes through a multistep mechanism involving a universal intermediate, a terpinyl cation that can be transformed into several compounds. In the liverwort *C. conicum*, mentioned above, (-)-sabinene is the major monoterpene produced by a European strain. The monoterpene cyclase that is responsible for the cyclization of G-PP to sabinene has been characterized as a soluble enzyme of M(r) 65,000 with a pH optimum at 7.5 (Adam and Croteau, 1998). This enzyme requires a divalent metal ion as the only co-factor. The preferred one is Mg^{2+}, which seems to occur in many plant cyclases (see 2.2.2). The general properties of the sabinene synthase from *C. conicum* resemble those of other monoterpene cyclases from gymnosperms and angiosperms. Other metal ions present in monoterpene cyclases are Zn^{2+} and Mn^{2+}, the latter in S-linalool synthase of *Clarkia breweri*, converting G-PP to S-linalool, which is abundant in stigmata of freshly opened flowers.

Plants which produce a number of monoterpenes may do this in a ratio that changes in time. This can be realized by storage of intermediates in different compartments. In caraway (*Carum carvi*), which is an important source of carvone, limonene is predominantly found in younger stages. Limonene and carvone are synthesized via a three-step pathway. According to Bouwmeester *et al.* (1998) the first step is cyclization of G-PP to (+)-limonene by a monoterpene synthase. This intermediate is stored in the essential oil ducts. Later, the enzyme limonene-6-hydroxylase converts limonene to (+)-*trans*-carveol. In the third step this monoterpene is oxidized by a dehydrogenase to (+)-carvone, which accumulates in the seeds. The oxygen atom of oxygenated monoterpenes is commonly derived from water. This process has a.o. been studied by Croteau *et al.* (1994) with 1,8-cineole synthase as a model. The enzyme 1,8-cineole cyclase catalyzes the

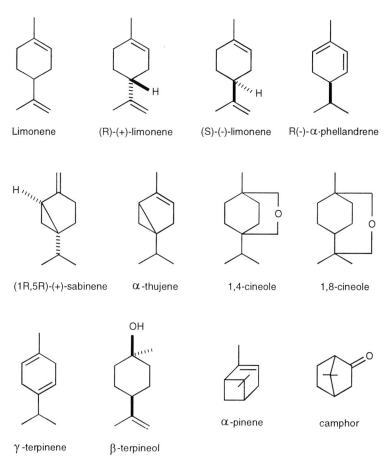

Fig. 2.7. Monoterpenes of gymnosperms.

conversion of G-PP to the monoterpene ether 1,8-cineole (Fig. 2.7) by initial isomerization of the substrate to linalyl-PP, and cyclization of this bound intermediate to the α-terpinyl carbo-cation that is subsequently captured by water and undergoes heterocyclization to the remaining double bond. The authors have isolated the enzyme from secretory cells of the glandular trichomes of *Salvia officinalis* (Lamiaceae).

The oxidative pathways involved in the synthesis of oxygen-containing monoterpenes may be of archaic origin. The presence of several oxidative pathways in Proteobacteria has been confirmed by several physiological studies. The formation of *p*-cymene from monoterpenes, as an example, could be detected in methanogenic enrichment cultures (Hylemon and Harder, 1998).

Chapter 2. Production of terpenes and terpenoids

Conifer oleoresins are typical mixtures of volatile monoterpene hydrocarbons and non-volatile diterpene resin acids. As a result, wound stimulation of oleoresin biosynthesis involves translational or transcriptional expression of multiple genes encoding terpenoid pathway enzymes. The grand fir (*Abies grandis*) has been developed as a model system for the study of oleoresin production, particularly in response to wounding and insect attack. The turpentine fraction consists of 87.5% monoterpenes and 12.5% sesquiterpenes and they alone comprise 38 different structures! Some of the synthases produce more than one terpene. Steele *et al.* (1998) found clustered regions of significant apparent homology between the enzymes of this gymnosperm and angiosperm species, particularly with respect to the sesquiterpene synthases, which are among the most complex terpenoid cyclases described (see also 2.2.2).

The production of monoterpenes by trees is not restricted to Gymnospermae. *Quercus ilex* (Angiospermae) is the most common oak species in Mediterranean forests. Its leaves emit many monoterpenes, among which α-terpinene, sabinene and myrcene are the most abundant. The carbon of all four monoterpenes was completely labelled with ^{13}C after 20-min exposition to 350 ppm $^{13}CO_2$ (Loreto *et al.*, 1996). The uniform labelling pattern suggests that the monoterpenes are formed from photosynthesis intermediates and share the same synthetic pathway with isoprene, that is emitted by other *Quercus* species: the DOXP pathway.

Monoterpenes are not found in the leaves alone. In xylem samples of *Picea abies* the (-)-enantiomers of α-pinene, β-pinene and β-phellandrene are prevalent. In *Pinus sylvestris*, however, (-)-limonene dominates over (-)-α-pinene, but in this common pine tree xylem samples mainly (+)-α-pinene is found and less (-)-α-pinene (Persson *et al.*, 1996). Obviously, terpenoids can be transported in vascular systems to be released, stored or upgraded in the roots. The group of Croteau (1988) has revealed that monoterpenes can be converted into water-soluble glucosides, that may be translocated to other plant parts e.g. the rhizome. In spearmint (*Mentha spicata*) leaves, part of menthone is reduced to *neo*-menthol and subsequently converted to the water-soluble glucoside (Fig. 2.8).

McConkey *et al.* (2000) determined the rate of biosynthesis of the principal monoterpene (-)-menthol in the glandular trichomes of peppermint (*Mentha piperita*) and obtained evidence for a developmental regulation of monoterpene biosynthesis at the level of gene expression. Transcriptional and immediate translational activity occurs during a relatively short period of leaf development that is followed by turnover of the corresponding enzymes accompanied by the cessation of monoterpene production.

The roots of vascular plants are a largely unexplored biological frontier with the ability to synthesize a diversity of secondary metabolites, some of

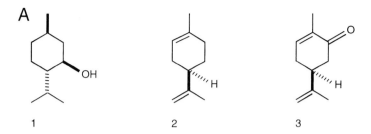

Fig. 2.8. **A**: basic structures of main components of spearmint oil. 1 = L(-)-menthol; 2 = (S)-(-)-limonene; 3 = (S)-(+)-carvone. **B**: conversion of menthone to a water-soluble glucoside. 4 = (-)-menthone; 5 = (+)-*neo*-menthol; 6 = *neo*-menthyl glucoside.

which are of biological significance in root-rhizosphere interactions (Flores et al., 1999). Croteau already elegantly demonstrated in 1988 that hydrolized cyclic monoterpenes upon arrival in the rhizome can be hydrolized and the aglycone oxidized back, followed by a lactonization reaction resulting in a ring opening that allows β-oxidative degradation, generating acetyl-CoA, fatty acids and sterols.

Aromatic herbs emit the monoterpenes mentioned above, together with other monoterpenes and non-terpenoid volatiles with an endless variation in ratio. Aromatic oils rich in monoterpenes produced by herbs such as found in regions with a Mediterranean climate are utilized for many purposes. This has a long history and covers many features of human life. Their antimicrobial effects have long ago been appreciated (albeit empirically) in pharmacy and conservation. Hysop (Hebr.: ezob), mentioned in the Bible, is probably not *Hyssopus officinalis*, which is an old medicinal plant, but the herb a.o. referred to in Kor. 4:33, Lev. 14:4,6 and Lev. 14:49-51 as a disinfectant herb and utilized in embalm procedures is most probably *Origanum maru*, common in Palestine. *Pulegium vulgare* (c.f. *Mentha pulegium*) was already used in the second century as a human abortivum

2. Production of terpenes and terpenoids

(Holland, 1994). Pulegone (Fig. 2.5) is the main compound of this herb and it may be an eye-opener that it inhibits oviposition in the aphid *Acyrthosiphon pisum*, in combination with the sesquiterpene (E)-β-farnesene (Harrewijn *et al.*, 1994), produced by both plants and aphids. Chapters 3, 4 and 5 will discuss the diversity of functions of a particular terpenoid, and in this book it is the first example of a monoterpene which can be simply toxic or exert a function as messenger, depending on the target and dose.

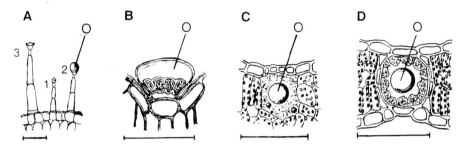

Fig. 2.9. **A**: glandular hairs of *Primula sinensis* showing three phases of excretion. 1 = young cell; 2 = cell filled with an essential oil; 3 = with ruptured cuticular vesicle. **B**: epidermal gland of *Thymus vulgaris* with cuticula pushed upward. **C**: lysigenic oil gland of *Dictammus albus* with disintegrated cell walls. **D**: schizogenic oil gland of *Hypericum perforatum* with diverted secretory cells. O = essential oil. Bar = 100 μm (redrawn after De Bary, Haberlandt and Rauter).

In aromatic plants monoterpenes with great structural diversity are predominantly found in fruits, seeds and special oil glands on leaves, sometimes as glandular hairs, as in the Lamiaceae (Fig. 2.9). In vegetative parts they can also be located internally in special oil glands or so-called idioblasts, as in Lauraceae. Cells producing monoterpenes often have to die because of the toxic effects. In cells filled with monoterpenes protoplasts can collapse and the cell walls can become reinforced with cork. Other structures to concentrate essential oils are found in lysigenic oil glands: secretory cells which fuse while their cell walls and protoplasts disintegrate to form a minicontainer of essential oil, as happens in citric fruits, in many Rutaceae and Myrtaceae. Plants have developed other systems, like the schizogenic secretory cells, which divert to make an intercellular space filled with essential oil. A few examples are shown in Fig. 2.9. Phellandrene, (the β isomer) for instance, is the main compound in the oil of *Oenanthe phellandrium*, an umbelliferous water plant. Essential oils can contain many compounds, of which a limited number of monoterpenes form the major part. To stay with the Lauraceae, an example is *Cinnamomum bejolghota* (obs. *C. obtusifolium*). This tree, a.o. found in Northeast India, Assam and the Himalayas can occur up to 2100 m above sea level. Choudhury *et al.* (1998) hydrodistilled the leaves from plants located in different regions to extract the essential oils and investigated their composition by GC and GC/MS, a common procedure. The oils were found to contain not less than 67 and 41 compounds, respectively, with significant differences in the ratio of the terpenoids. The major constituents of both oils appeared to be linalool (36% and 52%), 1,8-cineole (14% and 6%) and bicyclogermacrene (10% and 7%). Some compounds, such as camphor, nerol, geraniol and cubebene were present in plant material of one location and totally absent in the other one. We will meet this phenomenon throughout this book, especially in chapters 3 and 4 and it is not connected to monoterpenes alone. There is an almost endless variation in ratio and content of terpenoids in essential oils among the Angiospermae, many of which have remained unstudied at the early part of the 21st century.

The practice of hydrodistillation of plant parts of course does not result in detailed information on the compartmentization of terpenoids. Storage in oil glands or other specific organelles is one possibility, but a more diffuse presence in various plant parts is obtained by conversion into a glycoside, reducing toxicity, or by binding to proteins. Complex formation of proteins and terpenoids may have been an essential step in the evolution of the regulation of the cell cycle, and details will be discussed in chapter 3.

In generative plant parts monoterpenes have an important role in pollination, as they act as kairomones in many entomophilic angiosperms (see 4.2.2). We experience the smell of most gymnosperms and angiosperms as pleasant, but that is an anthropomorphic attitude. Chapter 1 has already revealed that volatile terpenoids, like non-volatile ones, are produced for different reasons of which attraction to insects is only one.

INSECTS

Insects synthesize monoterpenes via a MAD route, unless endosymbionts are involved (see 2.1.4). They can produce monoterpenes in different organs and store them in the hemolymph or, like plants, in glands specialized in attraction or defense. The approaches to address the site of production have evolved from the application of classical biochemical techniques such as radiolabelling studies with ^{14}C-acetate and ^{14}C-mevalonate to more sophisticated techniques, such as PCR amplification of relevant genes, followed by Northern blot analysis of messenger RNA to assay for levels of gene transcription. Seybold (1998) could locate the production of ipsdienol and ipsenol by bark beetles to be most probably in the thorax, by analysis of messenger RNA for HMG-CoA reductase from pheromone producing individuals.

Especially members of the Homoptera and Thysanoptera, Hymenoptera and Coleoptera produce monoterpenes that often have a function as semiochemicals (see 5.1). Quite often they come in combination with sesquiterpenes or with compounds of different origin. They can be released from the gut, by specific glands or by expelling a small amount of body fluid into the air, as aphids do. Some species have developed a highly specialized apparatus to do so (see 4.2).

VERTEBRATES

Although monoterpenes exert profound effects at the cellular level in vertebrates, they do not form a major group of mevalonate endproducts in these animals. Messengers within and between vertebrates are mostly more complex terpenoids, although simple isoprene compounds such as F-PP and G-PP are essential intermediates in cholesterol biosynthesis. However, these isoprenoids cannot be regarded as endproducts.

Proteins of vertebrates can be isoprenoid-modified, as is discussed in chapter 3.5, and a geranylgeranyl-modification of HeLa cell proteins was shown to occur before the biological significance was elucidated (Farnsworth *et al.*, 1990). Therefore, monoterpenes can be bound to proteins of vertebrates, as happens in other organisms, but this does not indicate any functions of monoterpenes in unbound form.

2.2.2 Sesquiterpenes and diterpenes

Further chain elongation with the aid of prenyltransferases leads to sesquiterpenes that are particularly well represented in angiosperms and insects. Sesquiterpenes, like monoterpenes, are volatile compounds that have been selected as semiochemicals among many species of plants, insects and even vertebrates. However, they are also produced by micro-organisms. Germacrene D is one of the volatile metabolites produced by Actinomycetes. The same compound is a sex pheromone mimetic for cockroaches (see 5.1.5), whereas germacrene A is an alarm pheromone for an aphid species (see 5.1.1 and 5.1.2). Diterpenes share a common carbon skeleton, based upon geranylgeraniol. Various biosynthetic pathways lead to different families of diterpenes, with functions ranging from anti-microbials to plant and insect growth regulators.

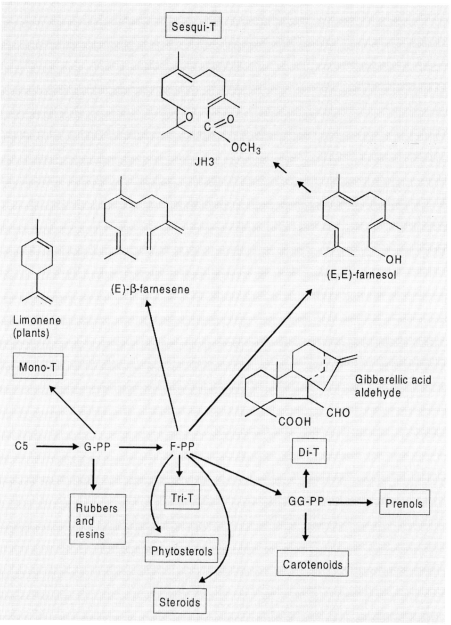

Box 2.3. Biosynthesis of sesquiterpenes in relation to more complex terpenoids.

PLANTS

Sesquiterpenes often present at the surface of leaves and fruits are farnesenes. α-Farnesene is found in the wax layer of apples (Murray, 1969).

Sesquiterpenes are also present in vegetative plant parts. Ginger, the preserved root of *Zingiber officinale*, contains the monocyclic sesquiterpene zingiberene. Sesquiterpenes with their C-15 structure result from a reaction of G-PP (Box 2.2) with IPP, which results in F-PP. Box 2.3 gives the basics of the biosynthesis of sesquiterpenes, including farnesenes and farnesol.

As an example, the acyclic sesquiterpene (E)-β-farnesene is one of the components of the essential oil of *Mentha piperita* (peppermint). It can be synthesized from F-PP by a cell-free extract of secretory gland cells of this labiate plant. Crock *et al.* (1997) have cloned a cDNA from peppermint encoding (E)-β-farnesene synthase by random sequencing of an oil gland library and succeeded in expression in *Escherichia coli*. The enzyme requires a divalent cation for catalysis, either Mg^{2+} or Mn^{2+}; the ratio of farnesene isomers produced depending on the type of cation.

To obtain a greater diversity of terpenoids the chain length should increase and cyclization reactions should be realized. Isoprenyl-PP synthases catalyze the consecutive condensation of IPP with allylic diphosphates to arrive at a variety of prenyl diphosphates with well-defined chain lengths. Wang and Ohnuma (1999) combined site-directed mutagenesis with X-ray crystallography and identified specific amino acid residues responsible for chain length determination. Combinations of these residues confer product specificities to all isoprenyl-PP synthases and represent structural features that reflect the evolutionary course of this enzyme family.

Terpene cyclases catalyze the synthesis of cyclic terpenes with 10-, 15- and 20-C acyclic diphosphates as substrates. Those that have F-PP as a substrate convert this compound into sesquiterpenes. When the enzyme binds to the farnesyl group loss of the diphosphate results in an allylic carbo-cation that electrophilically attacks a double bond further down the terpene chain to induce the first ring closure. Starks *et al.*, (1997) have elucidated the crystal structures of 5-epi-aristolochene synthase (TEAS), a sesquiterpene from tobacco. This study provides a basis for understanding the stereochemical selectivity displayed by other cyclases, together with more insight in the involvement of aromatic quadrupoles in carbo-cation stabilization. The crystallographic model of TEAS suggests that the specificity of the cyclases depends on the presence of a particular active site conformation determined by the surrounding layers. If the active site topology is in some way affected (e.g. by combination of interactions from wild-type enzymes), the ability to select a single conformation could be lost, resulting in multiproduct chimeric enzymes. In that case, they do no longer give rise to a single product. Mau and West (1994) used cDNA-cloning techniques to compare casbene synthase of bean seedlings with 5-epi-aristolochene synthase from tobacco and limonene synthase from spearmint.

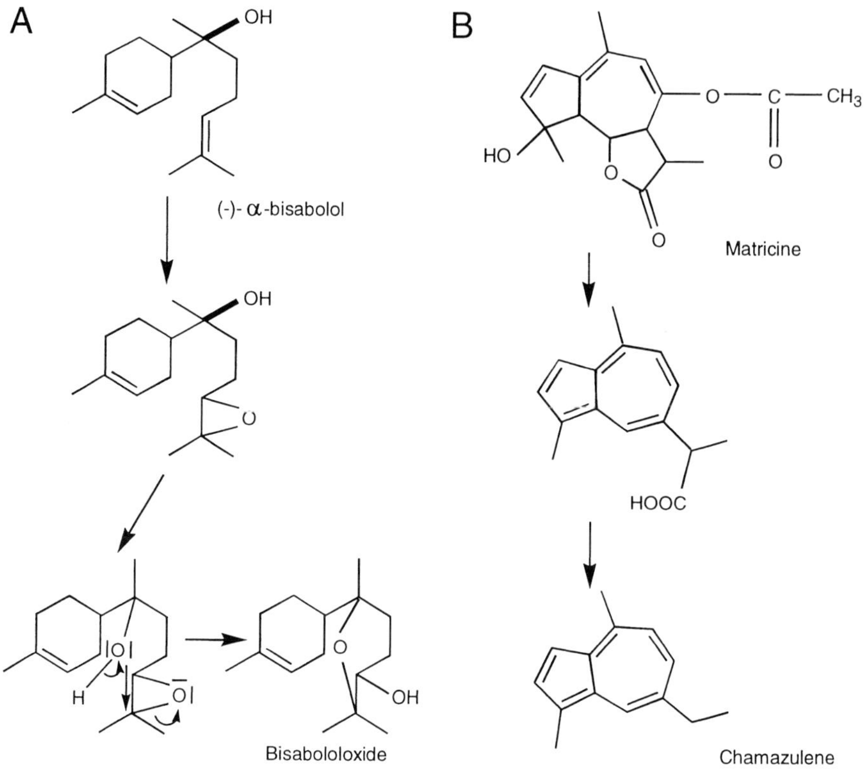

Fig. 2.10. **A**: possible biosynthesis of bisabololoxide from bisabolol. **B**: formation of chamazulene from matricine.

Their analyses of the alignment of the three proteins suggest that both primary and secondary structural elements have been conserved. These similarities suggest that cyclization enzymes of this type in angiosperms have undergone divergent evolution from an ancestral progenitor gene.

How did novel terpene cyclases arise? Back and Chapell (1996) combined the two functional domains of the tobacco TEAS gene and a *Hyoscyamus muticus* vetispiradiene synthase (HVS) gene and characterized the resulting chimeric enzymes expressed in bacteria. Combination of the two functional domains of the TEAS and HVS genes resulted in a novel enzyme capable of synthesizing reaction products reflective of both parent enzymes.

The studies mentioned above may serve to propose that the evolution of terpenoid cyclases is based on recombinations of functional domains of genes coding for cyclases. The development of multiproduct enzymes may not have been an exception. The essential oil of many angiosperms contain

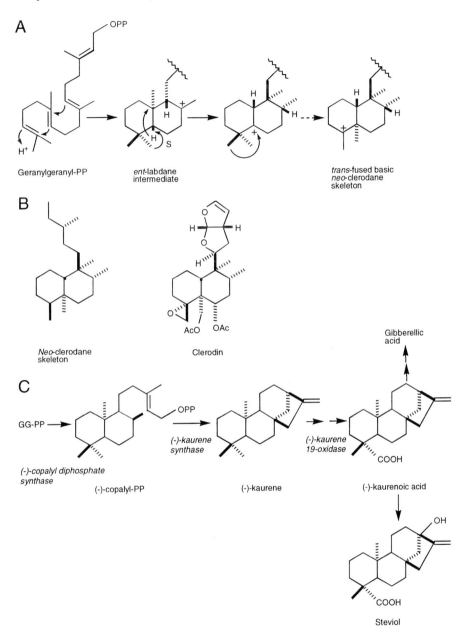

Fig. 2.11. **A**: Proposed biosynthetic pathway from geranylgeranyl-PP (GG-PP) towards the *trans*-fused *neo*-clerodane skeleton (after klein Gebbinck, 1999). **B**: The *neo*-clerodane skeleton and the diterpene clerodin. **C**: Proposed conversion of GG-PP into the diterpene steviol according to Richman *et al.*, 1999. Kaurenoic acid is the precursor of gibberellic acid (compare with Box 2.3).

an array of related terpenes that may not immediately have acquired a function and nevertheless remained produced.

It is beyond the compass of this book to discuss in detail all known enzymatic mechanisms of the isoprenoid metabolism such as electrophilic cyclizations, hydride transfers and Wagner-Meerwein rearrangements. The same issue of Science, in which the study of Starks *et al.* (1997) has been published, brings crystal structures and mechanistic insights of other isoprenoid cyclases. The picture arises that all plant cyclases share similar three-dimensional structures with an important role of Mg^{2+} ions. Alteration of active site residues may result in the formation of what we would call aberrant cyclization products. The enzymes, however, seem to serve as templates to channel conformation and stereochemistry during cyclization and elegantly protect and stabilize reactive carbo-cation intermediates.

In some *Matricaria* species the essential oil can consist for more than 50% of α-bisabolol, a sesquiterpene that easily forms two 5,6-epoxide isomers (Fig. 2.10A). In the early flower stage a sesquiterpene glycoside is synthesized via the MAI route. This glycoside is transported to the oil glands and glandular hairs. Metabolites are farnesol and farnesene (Box 2.3). Farnesene isomers are active constituents of fresh *Matricaria* flowers and are also present at lower levels in leaves and roots (Harrewijn *et al.*, 1993, 1994). After closing of the ring structure both α-bisabolol and matricine are produced. The second compound can be transferred into chamazulene (Fig. 2.10B), which was isolated already in 1863 by Pierse, who named it azulene. Both α-bisabolol and chamazulene have bactericidal properties (Schilcher, 1987). Essential oils, which are blue or blue-green, may contain chamazulene. Hot water extracts of chamomile flowers (pharm.: Matricariae flos) are used to treat inflammation of epidermis and mucosa, and gastro-intestinal disorders (see 7.1.6). Preparations of chamomile were already mentioned by Hippokrates 500 BC.

Biosynthesis of diterpenes is based upon an addition of IPP to F-PP. The resulting GG-PP has a C-20 structure and is the basis for many diterpenes, including polycyclic diterpenes such as gibberellic acid (Box 2.3). Another possibility is to arrive at a labdane structure via proton initiated cyclization (Fig. 2.11A).

Diterpenes have a noticeable position in the triangle of plants, insects and humans. They can have anti-microbial, insecticidal, analgesic and antioxidative properties (Alcaraz and Rios, 1991). They are less volatile than monoterpenes and therefore are less repellent to insects, but together with triterpenes they have more specific functions, among which as feeding deterrents (with toxic properties, see 5.1.3).

The essential oil of *Juniperus communis* and *J. oxycedrus* contains the sesquiterpene cadinene. The fruits of these endemic gymnosperms are collected and are used in the classic distilling process of gin and "genever" and have long been appreciated to ease digestive disorders. Although the oil is toxic at higher doses, there is no harm to be expected from the low terpene concentration in these beverages.

Diterpenes with a range of biological functions are the clerodane diterpenes. According to the IUPAC nomenclature on natural compounds (see Herz *et al.*, 1977) their names are derived from the clerodane skeleton as the parent structure. Depending on the configuration of the majority of chiral centres of the carbon skeleton, there is a prefix *neo* or *ent-neo*. (some use *ent* – and no prefix, but this is somewhat confusing). An overview of natural (and synthetic) clerodane diterpenes is found in klein Gebbinck, 1999. Fig. 2.11B shows the full *neo*-clerodane skeleton and the structure of clerodin, an antifeedant from the Indian bhat tree, *Clerodendron infortunatum* (Verbenaceae).

The plant hormone gibberellic acid is a polycyclic diterpene, derived from GG-PP. Plant hormones, like all hormones, need fine-tuning of their synthesis and activity to do their job properly. The regulation of gene expression in gibberellic acid synthesis is discussed in section 4.1.2. At this site it can be noted that the biosynthetic pathway of the diterpene steviol in *Stevia rebaudiana* (Asteraceae) has a close relationship to that of the intermediate steps in gibberellin biosynthesis (Richman *et al.*, 1999). Some of the enzymes have been elucidated by this group: GG-PP is converted into (-)-kaurene in two steps, with the enzyme (-)-copalyl diphosphate synthase and (+)-kaurene synthase and possibly (-)-kaurene 19 oxidase. The resulting (-)- kaurenoic acid is the basis of gibberellic acid 12 aldehyde or steviol (Fig. 2.11C). Richman *et al.* (1999) suggest that the synthesis of steviol glycosides by *S. rebaudiana* is a secondary metabolic development, with a production independent from a highly regulated primary route previously committed to hormone biosynthesis.

Many species of Angiospermae arrange isoprenoids into both monoterpenes and diterpenes, sometimes triterpenes. Members of the genus *Salvia* not only produce bicyclic and cyclic monoterpenes such as respectively camphor and 1,8-cineole but also diterpenes. *Salvia canariensis* (Lamiaceae), known from the Canary Islands, synthesizes a.o. the specific diterpene 16-acetoxycarnosol. This is an acetylated version of carnosol, which together with carnosic acid and 11-acetoxycarnosic acid are produced by young plantlets on a culture medium. In older plants there is a shift towards differently structured diterpenes (Luis *et al.*, 1992).

Diterpenes are not necessarily produced in photosynthesizing systems. Recent progress in growing roots in isolation has revealed that the synthesis of diterpenes by roots is no exception. As an example, the diterpene ginkgolides of *Ginkgo biloba* (the maidenhair tree, Ginkgoinae) are the major active principles of root extracts with pharmacological properties (Flores *et al.*, 1999).

Chapter 5 goes into more details, including effects on insects.

INSECTS

Insects and mammals produce and utilize a range of sesqui- and diterpenes. The sex pheromones of some Diptera contain a.o. farnesene isomers.

An isomer of α-farnesene, (E)-β-farnesene (Box 2.3), is the alarm pheromone of a number of aphid species. Aphids biosynthesize many terpenoid endproducts, including sesquiterpenes (Fig. 2.12). The same isomers are produced as in plants, but their ratio depends on internal and external factors. The site of production of these sesquiterpenes in aphids is not well known. Although the endosymbionts of aphids are supposed to be asterologenic bacteria of the *Buchnera* type, (see section 5.1.3), we found that they incorporate relatively large amounts of farnesenes (Harrewijn *et al.*, 1991; van Oosten *et al.*, 1990). The presence of farnesenes is independent of a host plant, as the symbionts incorporated the same amount of farnesenes in aphids reared on chemically defined diets. In several aphid species three farnesene isomers are transported in the hemolymph of both sexual and parthenogenetic forms (Gut and van Oosten, 1985) and are released from the siphunculi within a droplet of hemolymph as part of the alarm reaction. The farnesene isomers are present in a ratio that is form-specific.

The head capsules of soldiers of the subterranean termite *Reticulitermes lucifugus* contain more than 90% (E)-β-farnesene, that of workers less than 20% (Fig. 2.13). These high quantities in the heads of soldiers suggest their use as a chemical weapon against natural enemies. Sesquiterpene derivatives are common among soldiers of termites. The development of new soldiers seems to be suppressed by a pheromone from the soldiers' frontal gland and controlled by juvenile hormone (JH) or its analogues. At high rates of predation more soldiers may be recruited by increasing overall JH levels in a colony (Kaib, 1999). Mixtures of monoterpenes, sesquiterpenes and diterpenes are found in the frontal gland of the genus *Nasutitermes*, that may have a function in attraction of ants (Kaib, 1999). The soldiers of some termite species incorporate diterpenes in their defensive secretions. They can be present in specific glands in the form of propionate esters (Prestwich *et al.*, 1980). Diterpenes are also found in the mandibular glands of ants. It is not surprising that the glandular epithelium is composed of different types of secretory cells. As shown for plants in Fig. 2.9, a modified cuticle can form the reservoir for the terpenes in termites. Selection of terpenoids with messenger functions in insects is discussed in chapter 5.1.

Chapter 2. Production of terpenes and terpenoids

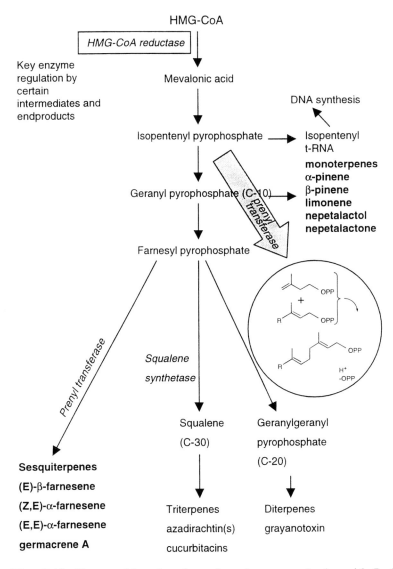

Fig. 2.12. Terpenoid endproducts based on mevalonic acid. In bold those known to be produced by aphids.

Sesquiterpenes with specific functions in insects and already mentioned in the former paragraph are the juvenile hormones (Fig. 2.14), that are produced in the *corpora allata* (CA). They are derived from farnesol via farnesal into farnesoic acid (Box 2.3). The endproduct JH3 is epoxy methyl

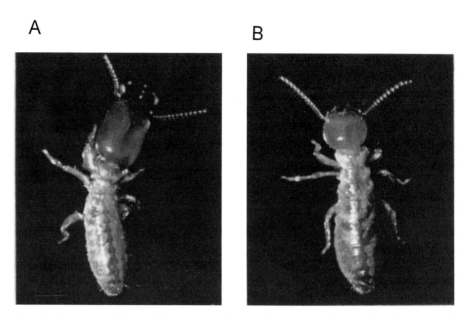

Fig. 2.13. **A**: a soldier of the termite *Reticulitermes lucifugus* showing a head capsule containing >90% (E)-β-farnesene; **B**: a worker of the same species with a head capsule containg <20% of this sesquiterpene.

farnesoate. The enzyme farnesal dehydrogenase requires NAD^+ as an electron acceptor in the CA. The JHs are relatively simple sesquiterpenes regulating development and reproduction and are involved in diapause, polymorphism and pheromone production (see section 4.1.2). JH1 and JH2 are mainly morphogenetic hormones, whereas JH3 has predominantly effects on the reproductive system. In the adult insect, it promotes the formation of

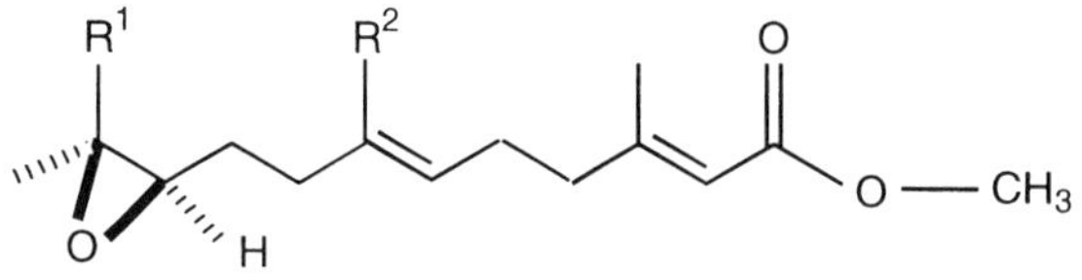

JH1: $R^1 = R^2 = C_2H_5$

JH2: $R^1 = C_2H_5, R^2 = CH_3$

JH3: $R^1 = R^2 = CH_3$

Fig. 2.14. The three main types of insect juvenile hormones.

protein yolk within the oocytes. The JH titre is controlled both by synthesis in the CA and by breakdown with esterases (de Kort and Granger, 1981). JH biosynthesis of small larval insects is often close to or below the limit of detection, necessitating more individuals for an assay. The claimed threefold higher titre of JH3 in the aphids *Aphis fabae* and *Megoura viciae* reared under long-day conditions compared with short-day JH3 titres (associated with the production of sexuals) by Hardie *et al.* in 1985, is still open to discussion, because of similar detection limitations. In the tomato moth *Lacanobia oleracea* (Lepidoptera) the JH synthesis is known to be dependent on methionine concentration (Audsley *et al.*, 1999).

In fact complex feedback systems operate in development, polymorphism and behaviour in which the CA have proven to be a two-lobed master gland, also involved in diapause. It is interesting to note that both the prothoracic glands and the CA are derived from embryonic ectoderm, even from adjacent body segments.

VERTEBRATES
The dorsal cutaneous secretion of the antelope *Antidorcas marsupialis*, commonly known as the springbok, contains a number of isoprenoid and terpenoid hydrocarbons and ketones, among which β-springene, a diterpene analogue of β-farnesene. Other diterpenes can be found in different antelope species. Although mammalian pheromones widely differ in chemical nature, terpenes are involved in mammalian conspecific chemical communication. The accessory sex glands of male mice are a source of farnesene isomers and could have a function in social behaviour (Harvey *et al.*, 1989).

2.2.3 Triterpenes, sterols, steroid hormones and carotenoids

Triterpenes (hopanoids) are found in several bacteria, where they originate via the DOXP pathway. Simonin *et al.* (1996) report C35 triterpenoids in the prochlorophyte *Prochlorothrix hollandica*, a prokaryotic oxigenic phototroph of the Cyanobacteria. The hopanoids are present in the cell wall and thylakoid membrane. With respect to intracellular localization, triterpenic membrane stabilizers of the hopane type were described for the first time in a cyanobacterium by Jurgens *et al.* (1992). They were detected in the cell wall and thylakoid membrane. In purple bacteria, such as *Rhodopseudomonas* spp. more complex hopanoids have been detected, with side chains based upon a D-ribose derivative (Neunlist *et al.*, 1988).

An early function of polyterpenoids may have been membrane reinforcement (Rohmer and Bisseret, 1994) and possibly interactions with nucleic acids (see 3.1) Ourisson *et al.* (1987) envisaged a phylogenetic tree, comprising membrane-bound lipids, bacterial carotenoids, cycloartenol, and finally cholesterol. At that time, however, it was not yet known that carotenoids in higher plants are biosynthesized via the DOXP pathway of the MAI route.

Carotenoids are plastidic isoprenoids, so it is not surprising that they are biosynthesized via the MAI route. Sterols and steroid hormones are biosynthesized via MVA and squalene. Insects lack the enzyme squalene synthase (Fig. 2.12) and cannot produce sterols. They also lack the enzyme to synthesize carotenoids out of GG-PP by dimerizing the geranylgeranyl group into phytoene.

PLANTS

Plants harbour two classes of natural carotenoids: the proper carotenoids, with a C-40 chain (8 isoprene units), and those possessing less than 40 carbon atoms. Well-known carotenoids are α-, β- and γ-carotene, related to lycopene, the red pigment of tomato. Many green, yellow and even red pigments are proper carotenes; an example of a carotenoid with less than 40 carbon atoms is crocetine (with two COOH groups), the yellow pigment from anthers of *Crocus* spp., used as saffron in kitchen and pharmacy.

The extensive conjugated system of the carotenoids offers many possibilities for *cis-trans* isomerism. Because of the greater stability of the all-*trans* configuration, it is not surprising that this is the most common structure in carotenoids. Carotenoids have required considerable interest as dietary supplements in the prevention of cardiovascular disease and cancer (see 4.1.5). Because of the reflection of wavelengths visible to insects, plant carotenoids have acquired a function in host plant recognition, apart from their major function in photosynthesis, like chlorophyll. As this is beyond the scope of this book, we refer to textbooks on plant physiology for light absorption and detailed structure of carotenoids. The biosynthesis of phytosterols in higher plants is mainly associated with the microsomes. Triterpenes, polyterpenes and sterols are produced via a MAD route (Fig. 2.3). Phytoecdysteroids are a major group of the steroids found in plants. Their basic structure is that of a polyhydroxysteroid containing a 5β-H-7-ene-6-one system, mostly hydroxylated at sites 2β, 3β, 14α, and usually 20R, 22R. Additional functional groups can be hydroxyls, esters and lactones. There is substantial literature on steroid hormones: we restrict ourselves to main structures, which are essential for later discussions, e.g. in chapter 4. One glance at Fig. 2.15 shows the striking similarity with ecdysone, well known from insects and with human steroid hormones. The same occurred to the group of Nakanishi, who isolated three polyhydroxysteroids, including 20-hydroxyecdysone from the Chinese medicinal plant *Podocarpus nakaii* (Podocarpaceae) in the 1960s.

2. Production of terpenes and terpenoids

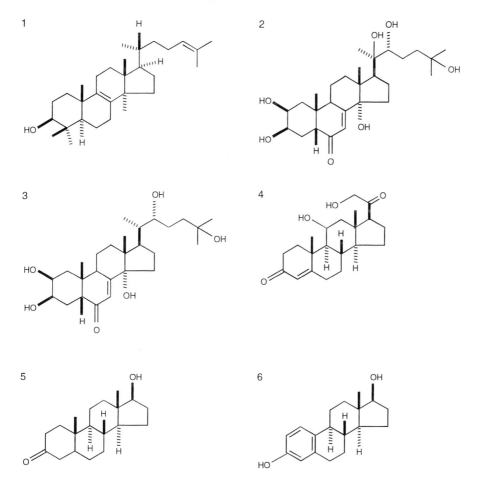

Fig. 2.15. Examples of sterols and steroids produced by plants, insects and vertebrates. 1 = lanosterol (plants); 2 = 20-hydroxyecdysone (plants and insects); 3 = ecdysone (insects); 4 = corticosterone (vertebrates); 5 = dihydrotestosterone (vertebrates); 6 = estradiol (vertebrates). Numbers 4, 5 and 6 are human steroid hormones with a high degree of similarity with insect and plant hormones.

Many phytoecdysteroids cause moulting cycle failure in herbivorous insects by interfering with their own moulting hormone, ecdysone. These effects can easily be obtained with crude methanolic leaf extracts: a standard plant being *Ajuga remota*. Isolation and identification of phytosterol, biologically active in insects is treated in section 6.1.1. More than 50 phytoecdysteroids have now been identified.

In spinach (*Spinacia oleracea*) biosynthesis of phytoecdysteroids occurs both in roots and shoots. Schmelz *et al.* (1999) demonstrated that wound- and herbivore-induced accumulation of 20-hydroxyecdysone in roots is an active process that requires increase in *de novo* biosynthesis, based on MVA.

Phytosterols can be transported in plants. Conversion to a glycoside makes terpenes water-soluble and in this form they can be easily translocated. The leaves of *Kalanchoë blossfeldiana* (Crassulaceae) for example, contain much larger amounts of cholesterol in bound, water soluble form than in its free state (Pryce, 1971). Transport processes for secondary metabolites allow alternative sites of accumulation, and this also holds for terpenoids. Biosynthesis of phytoecdysteroids can be exclusive in the roots, as occurs in *A. reptans*. An alternative for their production could be conversion of previously synthesized membrane sterols. According to Schmelz *et al.* (1999) this is an unlikely mechanism in spinach, since levels of possible precursors such as spinasterol and 22-dihydrospinasterol are extremely low.

INSECTS

Many pigments of insects are based on carotenoids. Their colours range from yellow to green and brown. Insects lack the enzyme systems to biosynthesize carotenoids and sterols. The carotenoids of insects belong to the C-40 polyenes with a high level of conjugation as is known from micro-organisms, so they could be produced by the symbionts. The visual pigments of the eyes of insects are carotenoids, e.g. retinal, which is the aldehyde of vitamin A, a chemical instrumentation related to that of vertebrates (see 4.1). The major carotenoid pigments of the aphid *Sitobion avenae* have been elucidated by Jenkins *et al.* (1999), who also discuss their possible role in protection against damaging solar radiation.

Veerman (1980) demonstrated that in spider mites, belonging to the arthropods, carotenoids are involved in photoperiodic induction of diapause.

Although insects need a dietary source of sterols if not delivered by their symbionts, they metabolize dietary phytosteroids to internal messengers (Svoboda *et al.*, 1994). There is a close connection between the steroid hormones of Crustacea and those of Arthropoda. Ecdysone was the first hormone of insects to be isolated in pure crystalline form. Fig. 2.15 shows that this structure resembles that of cholesterol. The elucidation of the structure of ecdysone from the silkworm *Bombyx mori* by Huber and Hoppe in 1965 came just one year before the group of Nakanishi discovered 20-hydroxyecdysone in a plant. Ecdysone that is produced in the prothoracic glands is generally known as the moulting hormone and has an essential role in the quinone tanning process of the cuticular proteins. A key enzyme for

N-acetyl-dopamine is dopa decarboxylase, and this enzyme is induced by ecdysone.

VERTEBRATES

Derivatives of carotenoids such as retinoids are important signal transducers in vertebrates. Their functions are discussed in chapters 3.1 and 4.1. Carotenes serve as provitamines for the production of vitamin A. Theoretically one molecule of β-carotene gives rise to two molecules of vitamin A, although the conversion in the human liver comes close to 1:1. Plants are an important source of carotenes, especially carrots, spinach, celery and apricots. In humans resorption is greatly improved by fat intake.

In vertebrates triterpenes and sterols are produced from squalene, via the MAD route. In mammals, where steroid hormones have been most intensively studied, lipoproteins, bile acids, vitamin D and steroid hormones all are synthesized from cholesterol, in itself an important product of the squalene route (Box 2.2, Figs 2.3 and 2.15). In higher animals cholesterol is not only synthesized within the cell, but can also be derived from plasma low-density lipoprotein (LDL), which enters the cell by receptor-mediated endocytosis. Each cell must balance these internal and external sources to sustain mevalonate synthesis while avoiding sterol overaccumulation. This balance is achieved through a complicated feedback regulation of the sequential enzymes in mevalonate synthesis (see 3.2 and 3.3). Studies on steroid synthesis have rapidly progressed. A few years after the discovery of the importance of MVA, experiments with ^{14}C-labelled acetate had elucidated the biosynthetic route of cholesterol via G-PP, F-PP and squalene (Box 2.2). Surprisingly for those days, the biosynthetic pathway from MVA to squalene in the liver of mammals appeared to be similar to that in yeast. In the early 60s Wagner and Folkers, pioneers in the field of MVA metabolism, wrote a review on its isolation, structure identification and configuration, its synthesis and biological importance. This review with 62 references stressed the biosynthetic significance of MVA in other terpene fields, such as the incorporation of MVA into pinenes, the major monoterpenes of *Pinus* species. The authors were unaware of the later discovered MAI route.

In vertebrates cholesterol biosynthesis from MVA can occur in cells lacking cytosolic components, e.g. in permeabilized cells. The enzymes that catalyze the individual steps between mevalonate kinase and F-PP synthase are localized in the peroxisomes (Biardi and Krisans, 1996). This holds for phosphomevalonate kinases, mevalonate diphosphate kinases, and isopentenyl-PP isomerase. Their possible role as key enzymes is discussed in section 3.1.1. It follows that at least parts of the MAD route are compartmentized and that biosynthesis of MVA in higher organisms may be regulated in a complex way.

Although steroid hormones are derived from sterols with their isoprenoid structure, this review does not primarily focus on their direct effects, on which there is a wealth of literature, but rather discusses the impact of natural terpenoids on synthesis and mode of action of sterols and hormones. Plants produce many of these terpenoids. As in animals, plant sterol biosynthesis is mainly associated with the microsomes, but studies on the regulatory mechanisms of sterols may overlook the mobility of these phytosterols.

A comparison between Figs 2.15,3 and 2.15,5 shows that ecdysone, apart from differences in hydroxyl groups has a longer side chain than the mammalian steroid hormone dihydrotestosterone. Their targets and functions are species-specific. The structure of estradiol (Fig. 2.15,6) and dehydro-testosterone are much more similar but again their target cells have different binding proteins. The steroid hormone testosterone can travel to all parts of the body and can be converted in a.o. the hypothalamus and the amygdala into both estradiol and dihydrotestosterone. Although estradiol may be seen as a "female" hormone, it is well known that it exerts strong male effects in gene expression of particular cells where transformation occurs. A comparison of targets and functions of these steroid hormones is found in section 4.1.2.

2.2.4 Complex terpenoids and isoprenoid structures

The isoprene skeleton is the basis for numerous complex terpenoids. The IPP unit can lead to polymerised isoprenes that are the basis of resins and natural rubber. There are many textbooks on rubber production and chemistry. As they have no explicit functions as messengers rubber is beyond the context of this book. Natural resins can have various defensive functions and have been utilized since ancient times, e.g. myrrh, which is a mixture of resins mainly of members of the Burseraceae such as *Commiphora myrrha*, plants from the Middle East and India.

Routes for synthesis of sesquiterpene dialdehydes have been elucidated by Jansen (1993). Muzigadial, polygodial and warburganal (Fig. 2.16A) are drimane sesquiterpenes containing a common α,β-unsaturated 1,4-dialdehyde moiety and some adducts, and are potent antifeedants. A more complex bioactive terpenoid based on the clerodin skeleton is azadirachtin, (Fig. 2.16B), present in the leaves and fruits of *Azadirachta indica*, the neem tree. This tetranortriterpene, the structure of which was elucidated in 1985, appeared to be a very active antifeedant against a broad range of insect pests, and its history and mode of action is discussed in chapters 4 and 5.

In recent years, many thousands of complicated terpenoids have been isolated and their structure completely or partly elucidated. These

Chapter 2. Production of terpenes and terpenoids 55

compounds exert all thinkable kinds of actions on cell systems and complete organisms. Many terpenoids affect the production of other terpenoids via feedback systems or by competitive inhibition. In the course of this study several examples are discussed within an evolutionary context, examples ranging from monoterpenes to complex isoprenoids. Special attention is given to the cascade of messengers within cells and relationship between structure of terpenoids and anatomy and physiology of their receptors.

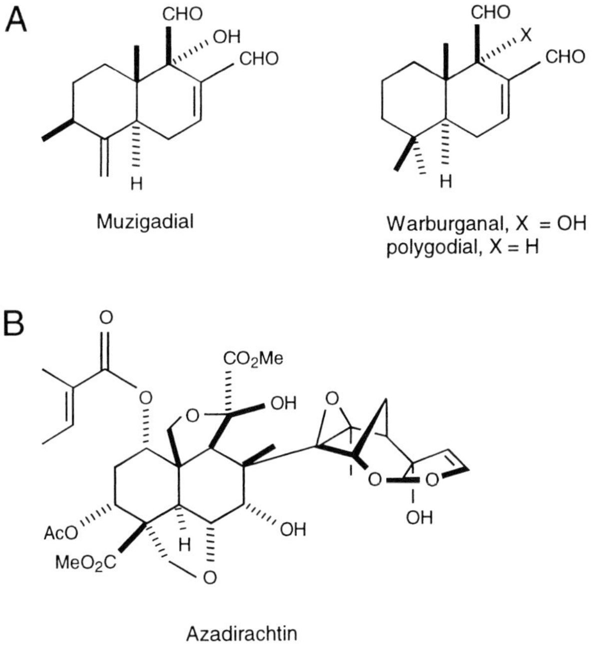

Fig. 2.16. Natural terpenoids acting as insect antifeedants. **A**: drimane sesquiterpenes with α,β-unsaturated 1,4-dialdehyde moiety. **B**: azadirachtin, a major allomone of the tropical neem tree.

To conclude this chapter we notice that a unique decision seems to have been made at the bifurcation of Eukaryota leading to green plants at one side, and fungi and animals at the other side (Fig. 2.4). Plants have retained MAI routes of isoprenoids biosynthesis, operating in their plastids providing specific terpenoids among which monoterpenes. The MAD routes, operating at different sites in the cell, provide sterols, phytohormones, ubiquinones and much more.

At first sight fungi and animals would seem to be devoid of MAI metabolism; they lack cell organelles like plastids that perform this task. On second consideration, however, we are not so sure that MAI metabolism is absent from fungi and animals. Many vertebrates harbour an active intestinal bacterial life with its MAI metabolism. The intestinal microflora may be there in the first place for their MAI route-derived isoprenoids, although the anti-microbial effects of monoterpenes could contribute to an adequate balance of intestinal symbionts. Insects, in particular herbivorous insects have developed a symbiotic relationship with specific micro-organisms that are often housed in mycetocytes. In this way insects have regained MAI metabolism, not by fusion, but by offering board and lodging to organisms, which in return act as metabolic brokers or have other useful functions (Douglas, 1992). For that reason both MAI and MAD metabolism (but in separate frames) have been depicted in Fig. 2.4.

What about the plastids? They are not completely lost either. Lichens, with their epiphytic way of life, thrive in symbiosis with algae. The fungi (mostly Ascomycetes) can dehydrate for months without apparent damage; the algae need water to have an active MAI metabolism. The relative importance of MAI- and MAD-derived products in lichens would be an intriguing object of study.

Astonishing enough, even plastids as isolated organelles have made their come back! As an example, sea slugs belonging to the Ascoglossa (Gastropoda) feed on algal cells and imbibe their cell contents. The plastids are not destroyed but are incorporated into digestive cells and utilized as photosynthetic factories (Rumpho *et al.*, 2000), just beneath the epidermis. The plastids capture enough light to fuel CO_2 fixation, a process that not only sustains the slugs in the absence of food, but also serves to produce allomones (terpenes?) for their own defense. Undoubtedly there are many more examples to be discovered of the return of the MAI metabolism in animals in a way as inventive as that of plastid phagocythosis of cells of the digestive tract of the sea slugs.

The coming chapters illustrate the enormous diversity of messenger functions of terpenoids. To be produced, they always need genes encoding enzymes essential for their biosynthesis. Enzymes, being proteins, contain nitrogen. Therefore terpenoids are dependent on nitrogen although this element does not participate in the chemical structure. Terpenoids are not self-reproducing like RNA molecules but perform essential functions as messengers within and between organisms.

The cornucopia of monoterpenes alone suggests a flexible encoding of synthases to arrive at the structural diversity of today. We have argued in sections 2.2.1 and 2.2.2 that small changes of amino acid residues of synthases can result in novel oxidized and cyclisized terpenes. From an evolutionary point of view the abundance of isoprenoid structures would seem to be almost inevitable. When synthases are not highly specific one can expect some variation in endproduct structure; this could explain the presence of minor quantities of related monoterpenes in essential oils. Many synthases, however, appear to be remarkably selective. As an example S-linalool synthase in *Clarkia* spp. has an almost 100% selectivity (Dudareva *et al.*, 1996). One of the authors of this paper, E. Pichersky, opines (pers. comm., 2000) that gene duplication may create novel genes that do not *exactly* encode for the same endproduct. Messenger functions of these terpenoid endproducts can only become consolidated if their biosynthesis is reproducible and accurate. Exact replicas of enzyme molecules can only be produced with exact copies of RNA. Therefore, for functions of terpenoids to become consolidated, there was a need for an accurate RNA replication. Self-replicating RNA may have been the very basis of life on this planet enabling the development of prokaryotes, assumed to be the oldest existing form of life. According to Poole *et al.* (1999) there was an absolute requirement for high-accuracy RNA replicases even before proteins evolved. These authors argue that the ribosome of eukaryotes is so large that a prior function before protein synthesis is most probable and that eukaryotes have retained more vestiges of the RNA world. The significance of RNA in early evolution fits well in our discussion of a reproducible production of isoprenoids.

The advantage of the utilization of hydrocarbons as messenger molecules between organisms is that nitrogen is spared during production. A disadvantage, however, is that the costs of production are relatively high due to the need for extensive chemical reduction and since terpenoid biosynthetic enzymes are not shared with other metabolic pathways (Gershenzon, 1994). This author (1994) concluded at a symposium on Chemical Ecology of Terpenoids (19th Annual Meeting of ISCE in 1993; see "General reading") that not all of the processes involved in terpenoid accumulation require large investments of resources. We should realize that only a few years ago the importance of MAI metabolism was not yet understood. The direct contribution of photosynthesis in the fusion of five-carbon units leading to the production of lower terpenes makes this picture more complicated.

In eukaryotes overproduction of terpenoids is limited by efficient feedback systems ranging from simple endproduct inhibition to complex multi-factor systems in vertebrates. These mechanisms are discussed in chapter 3.

Chapter 3

The origin and evolution of terpenoid messengers

3.1 Selection of key enzymes in isoprenoid biosynthesis

Within organisms, terpenoids are known for their physiological activities on cell level. They can influence cell stage and mitosis, resulting in changes in morphology and differentiation. Terpenoid endproducts and certain precursors of the mevalonate metabolism and other exogenous terpenoids can affect the synthesis of endproducts and intermediates of other branches or interfere with gene expression, or more directly, act as enzyme or key enzyme regulators. The preciseness with which the enzymes of the various steps of the biosynthesis of terpenoid endproducts operate and the effective feedback systems acting on enzymes and key enzymes are the result of a long evolutionary process. Between organisms, functions of specific terpenoids as messengers are likewise the outcome of a long process of selection of adequate production and reception. This chapter covers the basic regulatory processes involved in the biosynthesis of isoprene units in plants, insects and vertebrates. As in many biochemical processes, overproduction of endproducts is in some way prevented. Once produced in an efficient way, the same terpenoid can acquire a number of messenger functions in completely different organisms. Chapter 3 will discuss the development of regulatory mechanisms during evolution, whereas chapter 4 focusses on the relationship between chemical and physical characteristics of terpenoids and their effects on the receptor sites.

When did the early activation and inactivation of mevalonic acid, isopentenyl-PP and other steps of isoprenoid synthesis arise? Where did the essential genes come from and what is the degree of conservation of the enzyme systems? How did the sequential mechanisms of substrate binding develop? What is the position of eukaryotic cells compared with prokaryotic cells? We will deal with these and similar questions starting with the prokaryotes and follow, as far as our present knowledge permits, the evolutionary steps in the selection of the major enzymes and key enzymes with their feedback regulation.

3.1.1 Key enzymes in MAI and MAD routes

Whether via the MAI or MAD route, biosynthesis of metabolites build up with isoprene units is driven by enzymatic steps. Enzymes are proteins and the amino acid sequence of proteins is based upon a genetic code. Early prokaryotes may have had a precursor system of the genetic code. Predecessors of such a system could have been self-replicating proteins. The "hallmark" organelles of most eukaryotes: mitochondria and peroxisomes do not exist in facultative anaerobic protists belonging to old groups, such as Trichomonads. Instead of mitochondria, they contain hydrogenosomes that produce ATP from pyruvate (Bui *et al*., 1996). Hydrogenosomes lack DNA, the cytochromes and the Krebs cycle. Based upon biochemical analyses of hydrosomal heat shock proteins the authors conclude that mitochondria and hydrogenosomes have a common eubacterial ancestor.

Peroxisomes contain a number of enzymes involved in cholesterol biosynthesis that previously were considered cytosolic or located in the endoplasmic reticulum. Peroxisomes contain HMG-CoA reductase, although its structure has yet to be determined. In addition, peroxisomes contain mevalonate kinase, phosphomevalonate kinase, phosphomevalonate decarboxylase, IPP isomerase and farnesyl-PP synthase (Aboushadi *et al*., 1999). The question arises if peroxisomes still have an essential role in isoprenoid biosynthesis or if the enzyme systems of peroxisomes are "living chemical fossils".

Various studies indicate that peroxisomes, like mitochondria, are of endosymbiotic origin. In general, Archaezoa lack mitochondria and peroxisomes, and the same holds for Archamoebae, possibly the most primitive extant phylum of eukaryotes. Still, these ancient organisms possessed substrates like pyruvate, energy-rich phosphates and phosphate binding sites, elements for isoprenoid biosynthesis. However, there were constraints: lack of peroxisomes does not permit the production of farnesylated proteins (Bui *et al*., 1996; Kammerer *et al*., 1997) that have an important function in DNA synthesis (see chapter 3.5). According to Aboushadi *et al*. (1999) the reduced HMG-CoA reductase levels in peroxisomal-deficient mammalian cell mutants support the vision that peroxisomes play an essential role in isoprenoid biosynthesis.

Chapter 3. The origin and evolution of terpenoid messengers

MAI ROUTE

Pyruvate is the substrate for the DOXP pathway in a MAI route. The first reaction of the DOXP pathway consists of the condensation of (hydroxyethyl) thiamine, derived from pyruvate, with the C-1 aldehyde group of glyceraldehyde 3-P to arrive at 1-deox-yxylulose 5-P (Fig. 2.1). Lois *et al.* (1998) reported the molecular cloning and characterization of a gene from *E. coli* that encodes the essential enzyme in this step: 1-deoxy-xylulose-5-P synthase. The gene, designated dxs, was identified as part of an operon that also contains ispA, the gene that encodes farnesyl-PP synthase. DOXP is a precursor for the biosynthesis of thiamine and pyridoxol (vitamins of the B group).

DOXP synthase belongs to a group of transketolase-like proteins that are highly conserved in evolution. The enzyme has been cloned from *Mentha* spp. followed by heterologous expression in *E. coli* (Lange *et al.*, 1998). This research group speaks of a novel class of highly conserved transketolases that could be the fundament of plastid-derived isoprenoids. Thinking along these lines, in plastids, once incorporated into eukaryotes and originating in the Eubacteria (Fig. 2.4), genes encoding the first enzymatic steps of the DOXP pathway were already present.

Key enzymes are not the only enzymes that govern the rate of production of terpenoids (Fig. 3.1). Once geranyl-PP is formed in plants, several enzymatic steps can give rise to a number of terpenes. Low activity of these enzymes limits the production of some of these terpenes. As an example, the group of Bouwmeester in the Netherlands (see Bouwmeester *et al.*, 1998; Toxopeus and Bouwmeester, 1993) revealed that in caraway young plants produce only limonene (which needs only a monoterpene synthase). The biosynthesis of carvone, however, needs two additional steps: the production of (+)-*trans*-carveol with the enzyme limonene-6-hydroxylase, and the step (+)-*trans*-carveol to (+)-carvone with the enzyme *trans*-carveol dehydrogenase. Of these two enzymes, limonene-6-hydroxylase is not active in younger stages. In a later stage, this enzyme becomes activated resulting in about equal quantities of limonene and carvone in the seeds.

According to Sacchettini and Poulter (1997) diverse families of isoprenoid structures, often formed from the same substrate in an enzyme-specific manner, could have arisen from differences in folding of the active site of the substrate. Another factor is the degree of stabilization of carbo-cationic intermediates, leading to further rearrangements. The biosynthetic reactions of isoprenoid metabolism are electrophilic alkylations in which a new carbon to carbon single bond is formed by attaching a highly reactive electron-deficient carbo-cation to an electron-rich carbon to carbon double bond. As discussed above, the activity of additional enzymatic steps can further encourage or discourage chain elongation or cyclization.

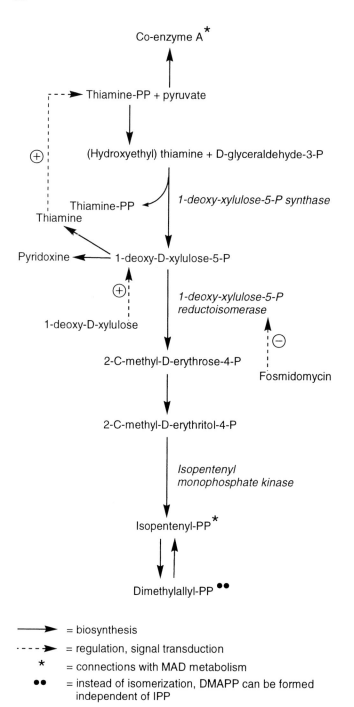

Fig. 3.1. MAI metabolism with possible regulations (with a thiamine loop in *Escherichia coli*).

Chapter 3. The origin and evolution of terpenoid messengers   63

Finally, the feasibility of new structures depends on how positive charge is quenched when the substance is formed. Tables 3.1 and 3.2 present those enzymes in MAI and MAD routes, which are known or proposed key enzymes.

The structural similarities of the isoprenoid synthases support an evolutionary relationship with respect to chain elongation and cyclase enzymes (see chapter 2.2). The enzymes that use diphosphate-containing substrates share highly conserved aspartate-rich motifs supposed to be Mg^{2+} binding sites (Sacchettini and Poulter, 1997). The active site of the enzyme is located in a cavity formed by the helix structures and the aspartate-rich motifs interact with the substrates via Mg^{2+}.

Table 3.1. Regulating enzymes of the MAI route in different organisms up to geranyl diphosphate

Enzyme	Eub.	Arch.	Yeast/ fungi	Plants	Insects ○	Vertebrates ●
DXS [1]	★	–	–/–	★	–	–
DXR [2]	★	–	–/–	★	–	–
IPK [3]	★	–	–/–	★	–	–

[1] 1-deoxy-D-xylulose-5-phosphate synthase (DOXP-S)
[2] 1-deoxy-D-xylulose-5-phosphate reductoisomerase (DOXP-RI)
[3] isopentenyl monophosphate kinase
★ = presence is known
– = presence is not known or not yet investigated
○ = presence in insect symbionts is possible
● = presence in vertebrate parasites and pathogenic bacteria is possible

Whether or not the isoprenoid enzymes can catalyze the formation of products with a certain chain length e.g. geranylgeranyl-PP (GG-PP) is not a matter of endproduct regulation, but is dependent on the position of certain amino acids. According to Wang and Ohnuma (1999), the fifth amino acid N-terminal to the first aspartate-rich motif is significant for the determination of the chain length in the endproduct. Substitution of aromatic amino acids by alanine results in greater chain lengths: from farnesyl-PP synthase to GG-PP synthase or from this enzyme to C-30 - C-50 synthase. An illustration of the evolution of the MAI route is the fact that both plant and bacterial GG-PP synthases have alanine residues.

The extensive family of terpenes and terpenoids including carotenoids and steroids are all based upon the condensation of isopentenyl-PP and dimethylallyl-PP, resulting in the substrate geranyl-PP. Therefore, DOXP synthase is an important enzyme, the activity of which governs the biosynthesis of all plastid-derived isoprenoids. Whether this enzyme acquired a status as key enzyme in early evolution depends on the possible regulation of terpene synthases, in particular the cyclases. During evolution of isoprenoid biosynthesis, fine-tuned mechanisms for product specificity have developed, based upon specific substrate interactions. The common 3D structure of the terpenoid synthase fold enables a number of variations, such as the amino acids present in the aspartate-rich motifs, giving rise to an array of terpene structures. The synthases and cyclases may be selective, but that does not make them key enzymes. Nevertheless, the activity of many of the "higher chain" enzymes must be regulated in some way. As an example, in a developing plant that reacts adequately to changes in the environment, the production of gibberellic acid and of carotenoids depends on different internal and external factors. Both can be biosynthesized by plastids. If only one enzyme, say DOXP synthase, would govern their presence and activity, rather inefficient ways of activity regulation present themselves: compartmentation and break down. In that case fine-tuning of the release of bioactive terpenoids should be based on a dynamic process of membrane permeability or on down-regulation of degrading enzymes (Fig. 3.1). With respect to the MAI route, these are almost uncharted waters.

When the enzymes of the isoprenoid metabolism evolved sequentially, as is the common opinion, every step along this way could have provided a solution to a particular problem. Some sesquiterpenes became involved in chemical defense against fungi and are classified as fytoalexins. Those that did may have acquired multiple adaptive roles, e.g. as semiochemicals. It would seem that multiple factors have contributed to the consolidation of enzymatic steps of terpenoid biosynthesis. In green algae and plants, a study of possible major enzymes of the MAI route should give equal attention to the MAD route as operating in the cytosol, as the two are partly interdependent (Fig. 2.3).

MAD ROUTE

In the MAD route, HMG-CoA reductase is a key enzyme in higher organisms. One should realize that this enzyme belongs to a family of proteins that acquired either biodegradative or biosynthetic functions. Of the few known bacterial HMG-CoA reductases, that of *Pseudomonas mevalonii* is strictly biodegradative. In contrast, the HmgA gene in the thermophilic archaeon *Sulfolobus solfataricus* encodes a true biosynthetic reductase. This gene was shown to exhibit a high degree of sequence identity to the HMG-CoA reductase of the halophilic archaeon *Haloferax volcanii* (Bochar *et al.*,

1997). Phylogenetic analyses of HMG-CoA protein sequence suggested that the two archaeal genes are distant homologues of eukaryotic genes.

Table 3.2. Regulating enzymes of the MAD route in different organisms up to squalene and the diterpenes

Enzyme	Eub.	Arch.	Yeast/fungi	Plants	Insects	Vertebrates
HMGS [1] ▲	–	–	★/★	★	★	★
HMGR [2] ▲	–	★	★/★	★	★	★
HMGL [3]	–	–	–/–	★	–	★
MK [4]	★	★	★/★	★	★	★
PMK [5]	★	–	★/–	★	–	★
MPPD [6]	–	–	★/–	★	–	★
IPPI [7]	–	★	★/★	★	–	★
GPPS [8]	–	–	–/–	★	●	–
GPPT [9] ▲	★	★	★/★	★	–	–
FPPS [10] ▲	★	★	★/★	★	★	★
FPPT [11] ▲	★	★	★/★	★	★	★
Pase [12] ▲	★	★	★/★	★	★	★
GGPPS [13]	★	★	★/★	★	–	★
GGPPT [14]	★	★	★/★	★	–	★

[1] HMG-CoA synthase
[2] HMG-CoA reductase
[3] HMG-CoA lyase
[4] mevalonate kinase
[5] phosphomevalonate kinase
[6] mevalonate diphosphate decarboxylase
[7] isopentenyl diphosphate isomerase
[8] geranyl diphosphate synthase
[9] geranyl diphosphate transferase
[10] farnesyl diphosphate synthase
[11] farnesyl diphosphate transferase
[12] phosphatase
[13] geranylgeranyl diphosphate synthase
[14] geranylgeranyl diphosphate transferase

▲ = enzymes consisting of two or more types. [10] belongs to the group of [9]; [13] belongs to the group of [11]
★ = presence is known
– = presence is not known or not yet investigated
● = insect symbionts can express GPPS

In *Trypanosoma (Schizotrypanum) cruzi*, about 80% of the activity of HMG-CoA reductase is associated with the glycosomes. These are microbody-like organelles unique to kinetoplastid protozoa. Glycosomes contain most of the enzymes of the glycolytic pathway. The remaining 20% of this key enzyme was found in the cytoplasmatic fraction and the glycosome-associated enzyme is not membrane-bound (Concepcion *et al.*, 1998). The authors demonstrated for the first time that a soluble eukaryotic HMG-CoA reductase is present in glycosomes. As has been discussed in section 2.2.3 other essential enzymes of the MAD route are present in peroxisomes (Biardi and Krisans, 1996). A good part of the enzymes of the MAD route that have been considered to be cytoplasmic, are in fact compartmentized. When this view is correct, there must be an intensive traffic of intermediates with increasing complexity between different organelles, like a freight train picking up additional carriages at every station.

Given the fact that different types of HMG-CoA reductase can exist in one organism, one might expect multiple genes encoding these proteins. This is a common phenomenon in evolution, although it is difficult to reconstruct when gene duplication took place. To say it in other words: did duplication of genes encoding HMG-CoA reductase occur after branching from a common ancestor or in the ancestors themselves? In yeast, maybe the first scenario is the most likely one. The gene encoding HMG-CoA reductase in the fission yeast *Schizosaccharomyces pombe* was isolated based on its ability to confer resistance to lovastatin, a competitive inhibitor of HMG-CoA reductase. Gene disruption analysis showed that the hmgl+ gene is essential. Lum *et al.*, (1996) provided evidence that unlike *Saccharomyces cerevisiae*, *S. pombe* contained only a single functional HMG-CoA reductase gene. The authors discovered a previously undescribed "feed-forward" regulation in which elevated levels of HMG-CoA synthase, that catalyzes the synthesis of HMG-CoA reductase substrate, induced elevated levels of hmgl+ protein in the cell and conferred partial resistance to lovastatin. Compared with human HMG-CoA reductase, the amino acid sequences of yeast were highly conserved in catalytic domains, although they were highly divergent in the membrane domains. Phylogenetic analyses of available HMG-CoA reductase sequences suggested that the lineage of *S. pombe* and *S. cerevisiae* diverged approximately 420 million years ago. However, the duplication event resulting in two HMG-CoA reductase genes occurred some 56 million years ago, leaving *S. pombe* with a single HMG-CoA reductase gene and a proven activity of HMG-CoA synthase (Table 3.2). Fig. 3.2 gives the position of key enzymes in MAD isoprenoid biosynthesis.

When more than one gene encodes HMG-CoA reductase, the transcription levels of the genes may be different and dependent on the compartment here the genes are expressed. Maldonado-Mendoza *et al.* (1997) isolated three HMG-CoA reductase gene family members and characterized the expression of all three genes in *Camptotheca acuminata* (Nyggaceae), a Chinese tree

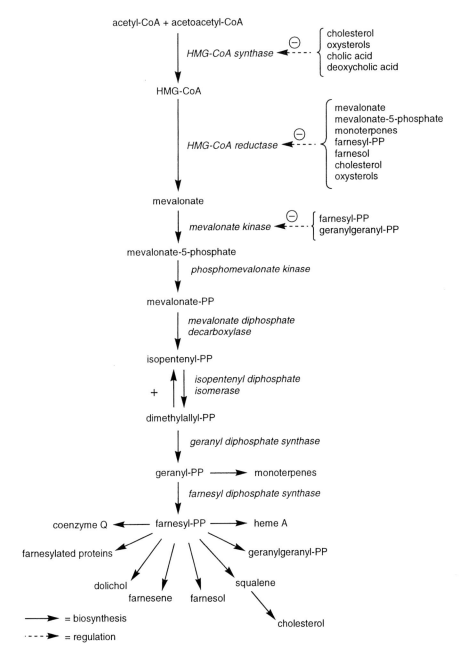

Fig. 3.2. Regulating enzymes and metabolites of MAD metabolism.

that produces the anti-tumour monoterpenoid alkaloid camptothecin. The three key enzymes appeared to be differentially expressed in various tissues under different physiological conditions. The level of expression of genes encoding key enzymes in isoprenoid biosynthesis may contribute to the regulation of the MAD route.

The two ascomycetous fungi *Gibberella fujikuroi* and *Sphaceloma manihoticola* produce gibberellin to induce elongation of infected host plants, one of them even plant tumours. Analyses of the amino acid sequences of HMG-CoA reductases of these fungi revealed a high degree of similarity between the deduced amino acid sequences in the C-terminal catalytic domains of all known HMG-CoA reductases, but the highest degree was found between the sequences of the two ascomycetes (Woitek *et al.*, 1997). These fungi possess only one copy of this gene, and this must be sufficient for the production of quite different terpenoids: sterols, possibly secondary defensive terpenes, the plant hormone gibberellin and quinones. The enzyme seems to operate straightforward as either glucose or ammonium excess does not influence gene expression.

There is a noticeable similarity between the GG-PP synthases from yeast and humans. All known members of this subgroup have glutamic acid in the FARM motif. Wang and Ohnuma (1999) stress that glutamic acid might affect substrate specificity, as these types of GG-PP synthase do not accept dimethylallyl-PP as a substrate. This is not so strange as we are dealing with the MAD route with its own enzyme systems and yeast, as we discussed in section 2.1.2 may be regarded to be closer to animals than to plants.

Apart from HMG-CoA reductase, in birds mevalonate kinases inducing phosphorylation of MVA are active. Mevalonate kinase and mevalonate 5-phosphate kinase are mevalonate-activating enzymes. They are dependent on nucleotides, mostly ATP and IPP. Mevalonate 5-PP is probably the first substrate to bind to decarboxylase.

In males of the bark beetle *Ips paraconfusus* JH3 induces terpenoid pheromone production. Tittiger *et al.* (1999) showed that topical application of JH3 increases HMG-CoA transcript levels in a dose- and time-dependent manner. This indicates that HMG-CoA reductase is a key enzyme in MAD metabolism of insects. It should be noted, however, that symbionts could have a role in bark beetle pheromone biosynthesis, which evokes the question if JH3 could affect the symbiotic MAI route. Nevertheless, the fact that *de novo* biosynthesis of monoterpenes in *Ips duplicatus* is inhibited by compactin, and that this inhibition is not an indirect one (by blocking the synthesis of JH3) speaks for a key enzyme function of HMG-CoA reductase. Moreover, the effect of compactin was counteracted by methoprene, a JH analogue (Schlyter and Birgersson, 1999). A comprehensive overview of biosynthesis and endocrine regulation of insect pheromones (including terpenoids) is given by Tillman *et al.* (1999). It would seem that insects have evolved some tissue-specific auxiliary or modified enzymes to transform the

Chapter 3. The origin and evolution of terpenoid messengers

products of basic isoprenoid metabolism to specific pheromone compounds with their own stereochemical and quantitative specificity.

In chickens, the major product of mevalonate metabolism is cholesterol, although other terpenoids are formed, such as dolichol, squalene and its oxides, and lanosterol. After hatching, the hepatic synthesis of different cholesterol precursors emerged sequentially in the young birds. In the liver, a great increase of HMG-CoA reductase was observed, resulting in the increase of both cholesterol precursors and lanosterol derivatives (Aguilera *et al.*, 1987). Cholesterol has proven to be an important factor in feedback regulation of HMG-CoA reductase, not only in birds, but also in mammals including humans (see 3.2.2). The regulation of this key enzyme in sterologenic cells, especially of humans, is one of the best-studied feedback mechanisms of MVA metabolism.

To our present knowledge in vertebrates, three compartments are involved in the biosynthesis of MVA: the cytosol, the peroxisomes and the mitochondria. Cytosolic and mitochondrial HMG-CoA synthases were recognized as different entities already in 1975, when they were purified and characterized by Lane's group. Hegardt (1999) has made a concise review of the developments over the last quarter of the 20th century that have led to our current understanding of mitochondrial HMG-CoA synthase and the role of this enzyme in ketogenesis in the liver and small intestine of mammals, as well as the expression of mitochondrial HMG-CoA synthase in cells deficient in cytosolic HMG-CoA synthase. Regulation of this enzyme is discussed in sections 3.2.2 and 3.3.1.

3.1.2 The development of regulatory mechanisms

Both farnesyl-PP and farnesol are present in the neutral lipid fraction of Archaebacteria. Farnesol inhibits incorporation of acetate, but not mevalonate, into the lipid fraction of *Haloferax volcanii*, suggesting that farnesol inhibits the pathway from acetyl-CoA to MVA (Tachibana *et al.*, 1996). Only 2 µM in rich medium and 50 nM in minimal medium resulted in a strong inhibition of growth. Other isoprenoid alcohols such as isopentenol, dimethylallyl alcohol, geraniol and geranylgeraniol in concentrations up to 500 µM did not affect growth. The authors conclude that farnesol regulates isoprenoid synthesis in *H. volcanii*. Their results indicate that in Archaebacteria feedback systems were operative in a MAD route.

Such feedback systems could have originated as a form of self-protection because of cytostatic or cytotoxic effects of farnesol. Farnesol that still has a role in feedback regulation of MVA biosynthesis in mammals (see 3.2.2 and

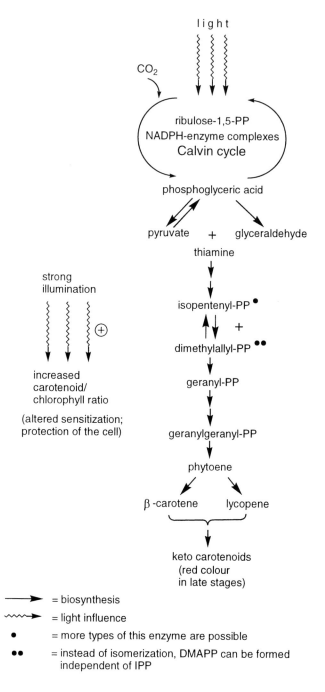

Fig. 3.3. Tentative early exogenous regulation of the MAI metabolism in an unicellular chlorophyte at the level of the Calvin cycle.

Chapter 3. The origin and evolution of terpenoid messengers 71

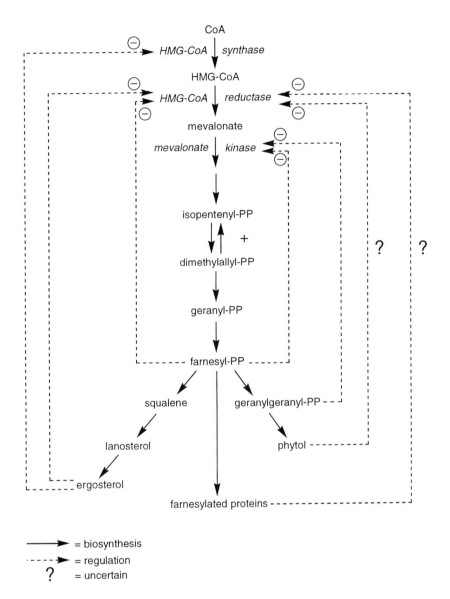

Fig. 3.4. Tentative early endogenous regulation of the MAD metabolism in an euglenophyte.

3.4.3) may have a long history in the control of the production of MVA metabolites.

Cell-free extracts from *H. volcanii* can phosphorylate farnesol with ATP to generate farnesyl-monophosphate and farnesyl-PP (Tachibana *et al.*,

1996). Therefore, it cannot be excluded that farnesyl-PP acts as the messenger molecule, as it does in yeast (Gardner and Hampton, 1999). Farnesyl-PP can be converted into farnesene in plants and insects, and it is involved in feedback control (see 3.3.2 and 3.4.3). However, farnesene cannot be converted into farnesyl-PP.

Gardner and Hampton (1999) demonstrated that a farnesyl-PP-derived messenger molecule serves as a positive signal for HMG-CoA reductase degradation in both yeast and mammalian cells. This strategy for regulation of HMG-CoA reductase could be a conserved regulatory mechanism in eukaryotes. In an earlier study, Lutz *et al.* (1992) showed that farnesyl-PP inhibits conversion of MVA into farnesyl-PP and that geranylgeranyl-PP inhibits its own synthesis in rabbit reticulocyte cytosol, indicating that these isoprenyl pyrophosphates can down-regulate their own synthesis at least *in vitro*.

There has been a long way from the development of novel enzyme systems with their corresponding genes to the thousands of isoprenoid molecules derived from MVA produced by the living world of today. At the onset of the 21st century, we know that the regulation of HMG-CoA reductase alone is complex and includes feedback control, cross-regulation by independent biochemical processes and contra-regulation of separate isoenzymes. Early regulatory systems, however, may have developed to control the biosynthesis of bioactive terpenoids close to IPP and DMAPP. This may have been the case in MAI metabolism (Fig. 3.3). Farnesyl-PP, geranyl-PP and simple monoterpenes and sesquiterpenes are candidates for early feedback control of their own activity (Fig. 3.4).

As discussed in chapter 2, substantial evidence exists that more complex terpenoids evolved rapidly, e.g. sterols in yeast. The mechanism regulating the sterol-synthesizing MVA pathway partly operates through endoplasmatic reticulum degradation of HMG-CoA reductase. In yeast, ubiquitination of the Hmg2p isozyme is involved in this type of regulation. It requires the UBC7 gene that is specific for Hmg2p isozyme (Hampton and Bhakta, 1997). The ubiquitination and degradation of Hmg2p is controlled by a feedback signal derived from farnesyl-PP. According to these authors, this implicates conservation of an HMG-CoA reductase degradation signal between yeast and mammals. This would mean that farnesyl-PP could master the production of more complex terpenoids than of its own. Another possibility is that we see only part of the regulatory cascade, and that the farnesyl-PP-derived signal is under command of another isoprenoid endproduct, e.g. a sterol or steroid. This could either be stimulation or inactivation, depending on the "stand by" status of the signal. Estrogens, as an example, known to influence cholesterol metabolism, could do so by

Chapter 3. The origin and evolution of terpenoid messengers 73

regulating the HMG-CoA reductase gene promoter in tissue-specific estrogen-responsive regions (Di Croce *et al.*, 1999).

Degradation of HMG-CoA reductase follows the biological rules by which regulated degradation of specific proteins is mostly mediated, which involves recognition of specific features of the enzyme. This part of the machinery is programmed in discrete regions of primary sequence known as degrons. However, degrading enzymes may recognize their target protein by reacting to more diffuse characteristics, resembling quality control of proteins, where degradation is heralded by a number of structural features common to damaged or misfolded proteins. Gardner and Hampton (1999) found this mode of targeting to be involved in degradation of the yeast HMG-CoA reductase isozyme Hmg2p. The recognition of Hmg2p is based on dispersed structural features rather than primary sequence motifs. So, key enzyme regulation can be based on at least two different machineries, of which the one utilizing dispersed structural features must have developed concurrently with the increasing complexity of proteins.

Gene promoters, isoprenoid intermediates and endproducts, secondary metabolites of different classes, they can all be part of the regulatory cascade resulting in the appropriate level of activity of degrading enzymes. In addition, different molecular classes of activators and inactivators vary in their effective intracellular concentrations. McAdams and Arkin (1999) have entitled their paper on genetic regulatory circuits "It's a noisy business! Genetic regulation at the nanomolar scale". This paper discusses how regulatory molecules achieve reliability in spite of the *noise,* the fluctuations in reaction rates in regulatory circuits operating with extremely low intracellular concentrations of the inducing molecules. To that end, cells use redundancy in genes as well as redundancy and extensive feedback in regulatory pathways. In analogy, key enzyme regulation of biosynthesis of terpenoids may have developed into the complex, multi-branched systems part of which is known today. McAdams and Arkin (1999) state that some regulatory mechanisms exploit noise to randomize outcomes where variability is advantageous. This strategy differs from the consolidation of a limited selectivity of terpenoid synthases and cyclases that could offer a choice of novel endproducts. As discussed at the end of chapter 2, it would seem that in the struggle for life priority lies more in the creation of a diversity of terpenoid synthases and cyclases than in a limited selectivity of these biosynthetic enzymes. In addition the diverse strategies of regulatory mechanisms as discussed in the coming sections of this chapter contribute to variability with a high degree of adaptability.

3.2 Feedback mechanisms

3.2.1 Key enzyme regulation in non-sterologenic cells

In non-sterologenic cells as present in insects, the regulatory mechanism is not necessarily the same as in the sterologenic cells of other organisms and mammals. Nevertheless, non-sterologenic cell systems are regulated by isoprenoid metabolites and endproducts. An example is the suppression of juvenile hormone 3 synthesis after exogenous application of JH3, which is an isoprenoid endproduct (Feyereisen, 1984). As JH is the endproduct of MVA in the corpora allata (CA), as steroids are in sterologenic cells, this is a form of feedback regulation. This is not the only regulative function of JH. There can be an effect on genes and specific nucleic receptors (see 4.1.2). Studies on pheromone production by bark beetles have demonstrated that JH3 raises HMG-CoA transcript levels, resulting in increased activity of the MVA pathway with corresponding pheromone production (Tittiger *et al.*, 1999).

In an earlier study, Venkataraman *et al.* (1994) had shown that JH can evoke its own degradation by induction of transcription of the gene responsible for JH esterase, the JH degrading enzyme. It may be questioned if both synthases and esterases are key enzymes in non-sterologenic cells. In insects proper timing of JH degradation is critical for development, therefore we categorize JH among terpenoids involved in feedback regulation. This vision is supported by the fact that juvenile hormone esterase levels in *Trichoplusia ni*, a well-studied cabbage looper, are primarily due to changes in the transcription rate of the esterase gene (Venkataraman *et al.*, 1994). Hormonal regulation of JH esterase is also present in *Choristoneura fumiferana*, the spruce budworm. In this insect, JH1 activates the expression of a gene responsible for rapid inactivation of JH1 in the transition from pupal stadia to the end of the first larval stadium (Feng *et al.*, 1999). Cloning of cDNA revealed similarity of the deduced amino acid sequences with those of *Heliothis virescens* JH esterases. In *Drosophila melanogaster*, found in many laboratories of genetics and insect physiology, the *in vitro* synthesis of total detectable JH3 was reversibly inhibited by JH3 bisepoxide and JH3, but not by methyl farnesoate (Richard and Gilbert, 1991).

Another important feedback loop is formed by interaction of JH with the ovary. In the viviparous *Diploptera punctata*, there is a well-defined correlation between the CA activity and the development of the oocyte. Stimulatory and inhibitory factors from the ovary, apparently born by the hemolymph, seem to be responsible for the fine correlation between gland activity and oocyte development. According to Khan (1988) 20-

Chapter 3. The origin and evolution of terpenoid messengers 75

hydroxyecdysone is such an inhibitory ovarian factor, not necessarily acting on HMG-CoA reductase.

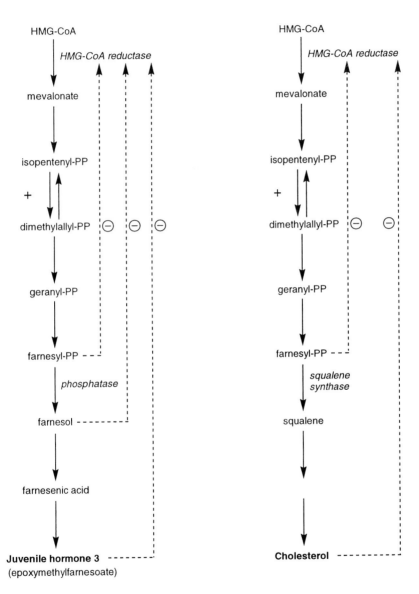

Fig. 3.5. Similarity between juvenile hormone 3 biosynthesis in insects and cholesterol biosynthesis in vertebrates.

The activity of the CA is regulated by various external and internal factors. Environmental factors such as the photoperiod and food quality exert their effects through the brain, which is directly connected to the CA through nerves. The brain can also command the CA in a humoral way via the hemolymph. The rate of JH biosynthesis could therefore be dependent on HMG-CoA reductase, giving it the status of a key enzyme.

A major effect of the brain on JH biosynthesis is inhibition. The compounds involved, allostatins, are neuropeptides, thought to be released from the axon terminals. In this way, the brain could exert a short-term control of CA activity, whereas long-term effects could be reserved for humoral factors. It is questionable if this dual mode of regulation is a general rule (Khan, 1988; Richard and Gilbert, 1991). The first quoted author suggests that JH3 could effect a feedback regulation in association with lipoproteins.

After synthesis JH is not stored in the gland to an appreciable extend but is secreted almost immediately. This makes it sensible that the major regulatory mechanism of this endocrine process occurs at the level of synthesis. Fig. 3.5 illustrates the biochemical chain of JH synthesis in insects, compared with that of cholesterol in vertebrates, both operating via a MAD route. Apart from HMG-CoA reductase, major enzymes such as methyl transferase and epoxidase could be rate limiting. However, experiments with addition of farnesoic acid *in vitro*, combined with radio-chemical assays indicate that these enzymes are not rate limiting (Feyereisen, 1984; Khan, 1988).

Couillaud and Rossignol (1991) measured the conversion of acetyl-CoA to HMG-CoA in locust CA and found that the HMG-CoA synthase was not in concert with the pattern of JH3 biosynthesis. In other words, HMG-CoA synthase activity did not reflect the asymmetry in the rates of JH biosynthesis observed in both halves of the CA. This suggests that the effect of the brain on JH activity is not expressed via HMG-CoA, leaving JH esterase as a possible regulator: break down, and not synthesis. This still does not exclude the possibility of a regulation of JH synthesis by inactivation of methyl transferase or epoxidase. Recent literature states that there exists a degrading cytochrome enzyme and an esterase system. Omega-hydroxylation of farnesol by a cytochrome p450 enzyme to a more polar product inhibits JH3 synthesis. See also 4.1.2.

Interestingly, compactin, a substance that inhibits HMG-CoA reductase in sterologenic cells (see 3.2.2) inhibits this enzyme in insects, pointing to similar phosphorylation reactions of known kinases. In contrast with sterologenic cells, HMG-CoA reductase in insects is not inhibited by 25-hydroxycholesterol.

3.2.2 Key enzyme regulation in sterologenic cells

Sterols are biosynthesized in the cells of plants, fungi and vertebrates, but not in insects, as discussed in chapter 2.2.3. Fungi produce sterols via MVA, which implies a MAD route (see Fig. 2.4). The fungus *Phycomyces blakesleeanus* does not grow in the presence of 1 µM lovastatin, an inhibitor of HMG-CoA reductase in mammals. In this fungus, lovastatin inhibits the synthesis of sterols and carotenoids (Bejarano and Cerdá-Olmedo, 1992). In contrast, the fungus *Gibberella fujikuroi* grows and produces carotenoids and all essential terpenoids including sterols. However, gibberellin production is strongly inhibited (Giordano *et al.*, 1999). This may be due to a compartmentation different from other fungi. The effects are much stronger *in vitro* than *in vivo*. This speaks for a subcellular separation of the terpenoid pathways in *G. fujikuroi*. Other explanations would be that lovastatin specifically inhibits a further step in gibberellin biosynthesis, or inhibits only one of two different HMG-CoA reductases. The latter hypothesis is not in concordance with our present knowledge of enzymes from numerous organisms.

In yeast, HMG-CoA reductase is a key enzyme of the MVA pathway and this enzyme is sensitive to feedback regulation. The ability of the yeast isozyme Hmg2p to be recognized for degradation depends on small regions of the Hmg2p transmembrane domain. Upon replacement of these regions by corresponding regions of a stable yeast HMG-CoA reductase isozyme Hmg1p, the enzyme is not recognized and not degraded. This depends on the first 26 amino acids (or maybe less) of Hmg2p (Gardner *et al.*, 1998). Replacement of amino acid residues 27-54 resulted in degradation. However, the degradation rate was poorly regulated. The authors stress that independent processes may be involved in Hmg2p degradation and its regulation. We have to envisage that the biosynthesis of terpenoids in sterologenic cells is regulated at the level of the synthases, at that of the reductases (independently on enzyme stability and mode of regulation), at the level of lyases and MVA kinases. To make things more complicated, at least HMG-CoA reductase could be regulated by a multi-branch system with partly unknown cascades.

In cotton, two full-length genes (hmg1 and hmg2) were characterized encoding HMG-CoA reductase. Although hmgr genes exhibit features typical of other plant genes, hmg2 seems to encode the largest of known HMG-CoA reductase genes. One of the characteristics is a sequence of 42 amino acids located in the region separating the N-terminal domain and C-terminal catalytic domain (Loguercio *et al.*, 1999). During the ontogeny of terpenoid-containing pigment glands in embryos hmg1 and hmg2 are differentially expressed in time and space. This is an example of time-dependent compartmentation of a MAD key enzyme.

Apparently, mesodermic cells of all mammals synthesise sterols *in vivo*, even though cholesterol may be present in extra-cellular fluids. In the absence of serum lipids, the rate of sterol synthesis is high, especially in proliferating cells such as present in the intestinal mucosas, in the developing brain, and in spontaneous, transplantable tumours.

The production of mevalonic acid and cholesterol is regulated with a multivalent feedback system, especially active on the key enzyme HMG-CoA reductase (Goldstein and Brown, 1990; Fig. 3.6). This is a glycoprotein associated with the endoplasmatic reticulum. The step from HMG-CoA to MVA is irreversible. Hepatic sterol synthesis is rapidly suppressed by dietary cholesterol, but in other tissues oxygenated derivatives of cholesterol, such as 25-hydroxycholesterol and 7-ketocholesterol, are up to 100 times more potent than cholesterol itself. 25-hydroxycholesterol is not a stable cholesterol derivative, but potent reductase inhibitors derived from fungi, such as compactin and lovastatin (mevinolin) completely inhibit the enzyme at concentrations of less than 10^{-8} M. This still leaves the possibility of cholesterol production from plasma (external low-density lipoprotein (LDL)), which enters the cell by receptor-mediated endocytosis. This source of cholesterol is also under cellular control. When cellular sterols increase, the LDL receptor gene is repressed. Other endproducts of the mevalonate pathway, such as ubiquinone and dolichol, probably act as non-sterol regulators of HMG-CoA reductase (Brown and Goldstein, 1980; Omkumar *et al.*, 1992). A disturbed regulation of cholesterol biosynthesis is found in tumour cell systems and in the case of cholesterolemia (Zakharová, 1988; Siperstein, 1984; Brown and Goldstein, 1980).

The enzyme farnesyl pyrophosphatase (F-PPase) acts at a key branch of the isoprenoid pathway for the biosynthesis of mevalonate products such as cholesterol, ubiquinone, heme-A and farnesol. Farnesol is produced from farnesyl pyrophosphate (F-PP) through one enzymatic reaction: the -PP group is removed to yield farnesol. Farnesol may be metabolized further but can be an endproduct secreted from cells, to have functions at other sites. A different enzyme (GG-PPase) is responsible for a comparable conversion of geranylgeranyl-PP to geranylgeraniol, which has many important functions, as we shall see. A nonhydrolyzable F-PP analogue, an isoprenoid phosphonate, inhibits microsomal F-PPase in cultured cells, and in this way, it inhibits the accelerated degradation of HMG-CoA reductase. Moreover, this inhibition is reversed by added farnesol. In other words: farnesol activates inhibition of HMG-CoA reductase, thus preventing overaccumulation of F-PP when isoprenoid pathway flow is high (Meigs and Simoni, 1997). From an evolutionary point of view, farnesol production

Chapter 3. The origin and evolution of terpenoid messengers

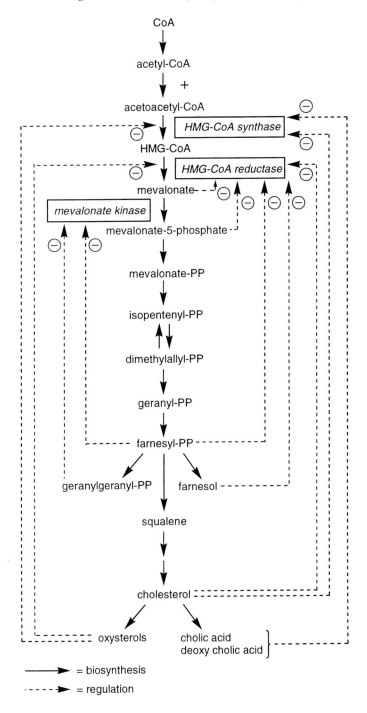

Fig. 3.6. More detailed multivalent regulation systems of steroid biosynthesis of the MAD metabolism in vertebrates.

may have had a primary purpose of reducing levels of F-PP, that are not harmless, as F-PP is an amphipathic molecule which could act as a detergent.

Other research groups looked at farnesol as *the* non-sterol regulator, but there is more to it than a common regulator. Farnesol could interfere with a phosphatidylinositol-type signalling which is involved in cell cycle progression as a cause of farnesol-induced growth inhibition as seen in yeast cells. The growth of *Saccharomyces cerevisiae* was inhibited in a medium containing 25 µM farnesol (Machida *et al.*, 1999).

Undoubtedly, the regulation of HMG-CoA reductase is a complex phenomenon and maximal inhibition may be reached by a combination of a steroid and isoprenoid factors (Hohl and Lewis, 1995; Panini *et al.*, 1989; Peffley and Gayen, 1997; Yokoyama *et al.*, 1999). Moreover, HMG-CoA synthase, essential in the production of HMG-CoA, has a control function in ketogenesis. It is regulated by two systems: succinylation and desuccinylation in the short term, and transcriptional regulation in the long term (Hegardt, 1999). Both control mechanisms are influenced by nutritional and hormonal factors, which will be discussed in 3.3.2.

One class of HMG-CoA reductase-effective steroids is formed by steroid hormones. Estrogens are known to exert an effect on cholesterol metabolism. Di Croce *et al.* (1999) studied the potential regulation of the key enzyme gene promoter by estrogens. The promoter contains an estrogen-responsive element-like sequence at position −93 (termed Red-ERE). Red-ERE can mediate hormonal regulation of the HMG-CoA reductase gene in tissues, which respond to estrogens and do so with enhanced cell proliferation, such as the rat breast cancer cell line MCF-7. However, the expression of the promoter was not induced by estrogen treatment in hepatic cell lines. In other words: the promoter of the rat HMG-CoA reductase gene contains a tissue-specific estrogen-responsive region, enabling different tissues to respond adequately to a steroid hormone. Tissue-specific regions of genes expressing key enzymes are another form of compartmentation of a MAD key enzyme.

Inhibitors of the synthesis of intermediates acting on specific synthases can induce compensatory increases in HMG-CoA reductase activity. An example is the inhibition of squalene synthase by zaragozic acid A. An inhibition of 75% of the activity of squalene synthase resulted in a 40-fold increase of HMG-CoA reductase activity (Petras *et al.*, 1999). Inhibitors of the production of isoprenoids upstream of MVA by post-transcriptional mechanisms (see section 3.4.1) often stimulate activity of HMG-CoA reductase, a positive feedback on this key enzyme.

As in key enzyme regulation, inhibition of cell growth and proliferation is probably regulated by multiple factors (Havel *et al.*, 1986; Watson *et al.*,

1985), which can be of the terpenoid type. It is therefore not surprising that already in an early stage a search was made for plant compounds, which exert cytostatic, cytotoxic and anti-tumour properties with respect to mammalian cell systems and that active terpenoids were found, often with complicated structures (Bucknall *et al.*, 1973; Fuchs and Johnson, 1978; Kupchan, 1970; Manfredi *et al.*, 1982). This search continues (e.g. Takasaki *et al.*, 1995; see also 4.1.8).

3.3 Other types of feedback regulation

The previous sections have shown that although HMG-CoA reductase is an important key enzyme, it is not the only enzyme regulating mevalonic endproducts. Dependent on the cell systems regulation is not necessarily realized by the endproducts alone. HepG2 cells for instance, are regulated by coordination of HMG-CoA synthase, HMG-CoA reductase and farnesyl-PP synthase (prenyltransferase (Rosser *et al.*, 1989)). This is a multivalent regulation system and multivalent regulation systems can be affected both by specific endproducts and additional factors. Such factors are not necessarily derived from the cell in which the key enzyme is operating: we distinguish between internal and external factors that can affect feedback mechanisms.

3.3.1 Internal factors affecting feedback mechanisms

To keep this section concise we give a few examples. In the pear cultivar Bartlett there appeared to be a remarkable regulation system with cyclic production of α-farnesene and volatile esters via acetyl-CoA (Jennings and Tressl, 1974). This is an internal regulation system, again with a role for α-farnesene. Comparable, and not less remarkable feedback systems in which farnesenes are involved, are operating in aphids. We found endogenous feedback systems with regard to the presence of farnesene isomers in different forms of aphids. In these forms, completely different ratios of isomers exist. Winged males and gynoparae of *Aphis fabae* and *Myzus persicae* contain relatively high amounts of (E,E)-α-farnesene. At the same time, amounts of (E)-β-farnesene were remarkably low (van Oosten *et al.*, 1990). This suggests a shift in synthesis of farnesene isomers by the aphids, probably caused by the endocrine activity in the aphids. It should be kept in mind that JH3, being a sesquiterpene, has (E,E)-farnesol as its precursor, like farnesenes branched from farnesyl-PP, where the amount of farnesenes is some magnitudes higher then farnesol itself. We found relatively high amounts of farnesenes in the hemolymph and the endosymbionts. If the

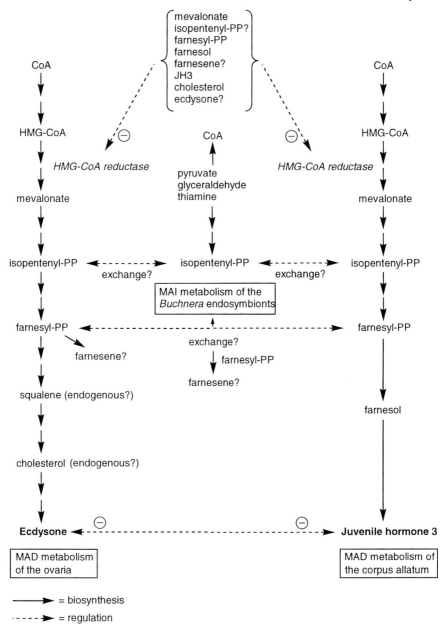

Fig. 3.7. Isoprenoid metabolism in aphids with possible interactions.

endosymbionts are involved in biosynthesis of farnesenes, in aphids MAI metabolism could play a role in association with MAD metabolism.

One of the mevalonate products, squalene, also reduced the presence of (E)-β-farnesene and α-farnesenes in the hemolymph. Exogenous squalene prevented moulting of nymphs, which may be explained by inhibition of sterol (and ecdysone) production. Fig. 3.7 gives a (partly) tentative internal feedback network on biosynthesis of isoprenoid endproducts in aphids. Some questions remain unsolved. We still neither know where the steps from farnesyl-PP to squalene, leading to steroids are compartmentized in aphids, nor if steroids are produced at all in the aphids, with or without the aid of micro-organisms other than bacteria.

The enzymes mevalonate kinase, phosphomevalonate kinase and mevalonate diphosphate decarboxylase (Table 3.2), active in peroxisomes, could be inhibited by several phosphorylated and non-phosphorylated isoprenes. Hinson *et al.* (1997) found only MVA kinase to be inhibited through competitive interaction at the ATP-binding site. The inhibitory capacities by the intermediate isoprenes were: geranylgeranyl-PP > farnesyl-PP > geranyl-PP > isopentenyl-PP. MVA-PP, geraniol and dolichol were not effective. This is additional evidence that MVA kinase has a central regulatory role in the feedback control of cholesterol and non-sterol isoprene biosynthesis.

3.3.2 External factors affecting feedback mechanisms

We have to restrict ourselves to a few examples. Carotenoids and diterpenes are synthesized via GG-PP, not via farnesyl-PP (Box 2.2) and their synthesis can be regulated by external factors. The biosynthesis of carotenoids in the yeast *Rhodotorula minuta*, is regulated by light. In the presence of mevalonate, the light dose given to the cells determined the amount of carotenoids produced during the dark incubation after illumination (Tada *et al.*, 1990). These results suggest that at least one additional enzyme is involved in carotenoid formation beyond mevalonate.

An intriguing feedback system associated with the regulation of the isoprenoid synthesis in plant cells is operating in ripening apples and demonstrates the omnipresence of mevalonate metabolites. (E,E)-α-farnesene (Fig. 3.8) is an isoprenoid endproduct, present in the wax layer of fruits, that seems to protect against rot. Cultivars with a low α-farnesene content are particularly sensitive to rot, especially when temperatures are lower than 5°C, probably because biosynthesis of α-farnesene is strongly reduced at low temperatures. This is called "low-temperature breakdown". The picture is complex, because apples of which the wax layer is rich in α-farnesene can show a different process of rot, "superficial scald". This is probably caused by oxidation products of α-farnesene. Injections of geraniol into the fruits effectively reduce the low-temperature breakdown in Jonathan

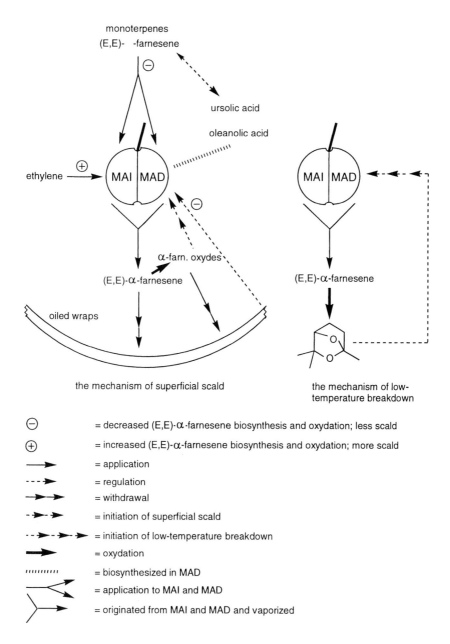

Fig. 3.8. Regulating systems in apples, influencing superficial scald and low-temperature breakdown.

apples. After a short period, geraniol can no longer be traced. It was assumed to be transferred into geranyl-PP, in this way inhibiting isoprenoid biosynthesis via a feedback regulation (Huelin and Coggiola, 1968, 1970; Wills et al., 1977; Wills and Scriven, 1979). With the present knowledge we have to consider the involvement of MAI metabolism, as α-farnesene can only be synthesized via a MAD route. However, the first steps of farnesene synthesis can occur via a MAI route in the plastids (Lichtenthaler, 1999). Geraniol applied to the fruits can reach the MAI and MAD metabolism and can be transformed into geranyl-PP and farnesyl-PP. Farnesyl-PP can inhibit HMG-CoA reductase but also deliver farnesene via farnesene synthase. The question is: what is the overall effect on the biosynthesis of α-farnesene? Obviously, the feedback inhibition is less than the formation of farnesene, because the low-temperature breakdown is diminished on application of geraniol.

Superficial scald can be prevented with good ventilation that removes the volatile oxidation products of α-farnesene. It can also be done with oiled paper, into which the products dissolve. An interesting feedback system was found by applying exogenous α-farnesene. As a result, the amount of endogenous α-farnesene decreased. The function of α-farnesene in the ripening of apples is a natural process. A cyclic oxidation compound released by decomposition of α-farnesene triggers the decay of the fruit, resulting in a disposition of the seeds (Stanley et al., 1986). The applied α-farnesene could inhibit a key enzyme, in this way modulating the expression of genes or influencing receptors. Ju and Curray (2000) showed that lovastatin can inhibit α-farnesene biosynthesis, indicating that this key enzyme is HMG-CoA reductase. These facts illustrate that increasing knowledge of these regulating systems can have consequences not only for storage and preservation of fruits, but also for our understanding of MAI and MAD routes and their interrelationships.

There is a negative feedback regulation of exogenous JH3 and JH bisepoxide (JHB3) on total detectable JH (JHB3 + JH3 + methyl farnesoate) by the CA portion of the isolated ring gland of *Drosophila melanogaster in vitro*. Methyl farnesoate does not have such an effect. Both JH3 and JHB3 inhibited the in vitro production of ecdysteroids by ring gland and brain-ring gland complexes of this insect (Richard and Gilbert, 1991). These data indicate that JH3 and some closely related terpenoids act on two different, though interrelated endocrine systems.

The dorid nudibranch *Cadlina luteomarginata*, living in the northeastern Pacific Ocean makes a drimane sesquiterpene and two degraded sesterterpenes (peculiar C-25 terpenes) that may have defensive functions a.o. as fish antifeedants. Kubanek et al. (2000) obtained evidence that *de novo* biosynthesis of these putative defensive terpenes is down-regulated when there is an abundant source of dietary compounds, in particularly allomones which could reduce the need for allomone production by *Cadlina*.

The authors suggest that biosynthesis of these terpenes by a nudibranch could be modulated by habitat-specific external factors, which warrants further investigation of the regulatory mechanism.

Potent inhibitors of squalene synthase are the zaragozic acids. In zaragozic acid A-treated mice farnesol formation can no longer be detected, indicating that farnesyl-PP is directly converted to farnesoic acid and dicarboxylic acids in the liver (Vaidya *et al.*, 1998). The effects of zaragozic acids are more complex, as they can stimulate HMG-CoA reductase and still inhibit squalene synthase (Petras *et al.*, 1999).

3.4 Site of action of the regulatory mechanisms

3.4.1 Transcriptional and post-transcriptional regulation

The facts presented in chapters 3.2 and 3.3 leave no doubt that isoprenoid metabolites both affect biosynthesis of mevalonate metabolites and the cell reproductive apparatus in the nucleus. Both HMG-CoA synthase and HMG-CoA reductase are regulated by sterols. In fact, the genes responsible for these enzymes are repressed by steroids via receptors and an initiated cascade system of successive cellular events. This is transcriptional regulation, which operates on the *production rate* of the enzymes (see Goldstein and Brown, 1990). A different way of regulation is achieved with non-sterol isoprenoids and terpenes, sometimes with relatively small molecules. This is post-transcriptional regulation, in which *activity* of the enzymes is affected. HMG-CoA reductase is a protein with a complex structure, with several hundreds of amino acids, bound to the outer membranes of the nucleus and to the smooth endoplasmatic reticulum. Its activity can probably be reduced by cytosolic proteins that preferentially bind oxysterols, potent metabolic regulators.

TRANSCRIPTIONAL REGULATION
Fatty acids can act in the transcription factors on the nuclear-receptor-responsive element of the mitochondrial HMG-CoA synthase promoter and on the DNA-binding properties of the peroxisome-proliferator-activated receptor (PPAR). Fasting and fatty acids increase the transcriptional rate of mitochondrial and cytosolic HMG-CoA synthases, although they are encoded by different genes. Feeding and insulin repress transcription of this enzyme (Hegardt, 1999).

Blattella germanica (the German cockroach) has two HMG-CoA synthase genes that are differently expressed during development, with a higher expression in ovary and fat body in adult tissues (Buesa *et al.*, 1994). In case environmental factors can influence gene expression, multiple synthase genes enable fine-tuning of transcriptional regulation. It has been mentioned before (see 3.1.1) that topical application of JH3 increases HMG-CoA reductases levels in male *Ips paraconfusus* (Tittiger *et al.*, 1999). This is an example of feedforward transcriptional regulation of a key enzyme involved in pheromone production.

Feedforward effects on the activity of synthases are no exception. In conifers a number of wound-inducible monoterpene and sesquiterpene synthases are known, the production of which is dependent on external factors acting on the level of transcription. Bohlmann *et al.* (1999) found the deduced amino acid sequences of the monoterpene synthases of *Abies grandis* (the grand fir) to be 618-637 residues in length and to be translated as preprotein with an amino-terminal plastid targeting sequence of 50-60 residues. These monoterpene synthases, the translation of which can be induced, resemble sesquiterpene and diterpene synthases of gymnosperms more closely than mechanistically related monoterpene synthases from angiosperms.

In a review on the role of terpenoid synthases in terpenoid defenses in conifers Bohlmann and Croteau (1999) stress that several events of gene duplication and functional specialization of novel synthases occurred during the evolution of terpenoid biosynthesis in the grand fir. The monoterpene synthases utilize geranyl-PP and other prenyl diphosphates as a substrate to generate an enormous diversity of ancient and successful defensive terpenoids against pathogens and herbivores. The role of defensive terpenes in gymnosperms is discussed in chapters 4 and 5.

Potent inhibitors of HMG-CoA reductases may not exert a direct effect on the transcription of this enzyme, but could in fact unmask transcriptional regulation of a terpenoid endproduct such as a sterol. Lovastatin, as an example, does not decrease the rate of transcription of hepatic HMG-CoA reductase in rats if there is not a dietary supplementation of cholesterol (Lopez *et al.*, 1997). In response to dietary cholesterol, the levels of reduction in mRNA were decreased to about 10% of the controls. Farnesol, known as a non-sterol regulator of reductases, was much less effective. Other studies have revealed that cholesterol exerts feedback regulation of hepatic HMG-CoA reductases gene expression primarily at the level of transcription. Examples of possible regulation of HMG-CoA reductase by lovastatin and other statins with subsequent effect on the cell cycle are presented in Fig. 3.9.

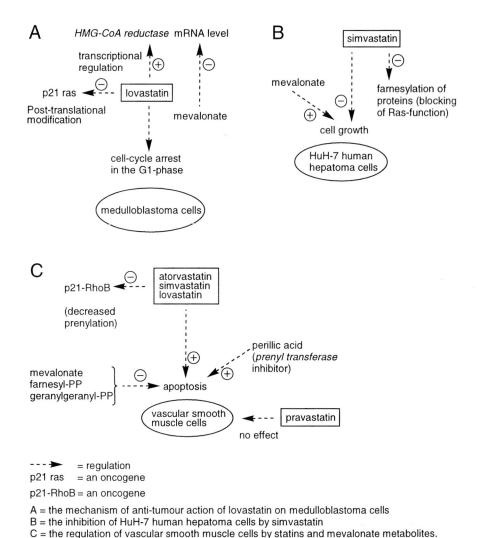

Fig. 3.9. Regulation of the key enzyme HMG-CoA reductase and signal transduction systems by lovastatin and other statins.

A group of nuclear receptors known as orphan receptors possess the structural features of hormone receptors, but lack known ligands. They respond to farnesol and its metabolites. These intermediary metabolites have a comparable transcriptional regulatory function in bacteria and yeast (Forman et al., 1995). These facts support our view of an increasing complexity of key enzyme regulation evolving from prokaryotes to eukaryotes.

Chapter 3. The origin and evolution of terpenoid messengers

In bacteria several transcription factors, belonging to different regulating pathways may operate in the same promoter. Nevertheless, bacteria seem to have been particularly inventive in adapting gene expression regulation to survive in a diversity of environments. According to Vicente *et al.* (1999) they have done so by exploiting the interplay of effectors and the properties of the recruitment of the effector-RNA polymerases to the promoter, developing complexities that may match those found in eukaryotic cells.

Substances that influence membranes can have additional effects on parts of these processes, such as membrane affinity. Small terpene molecules have such a property, used in medicine. Combination of an antibiotic compound (tetracycline) with a terpene causes more rapid absorption by the cells of a target organ (Piccinini, 1972; Villani *et al.*, 1974). This is reminiscent of the specific action of some oxidized monoterpenes against insect acetylcholinesterase as first demonstrated by Ryan and Byrne (1988). See also section 4.3.2 and Fig. 4.14. As both transcriptional and post-transcriptional regulation is involved in the activity of HMG-CoA reductase it is not astonishing that maximal inhibition can be reached with a combination of steroid and other isoprenoid factors (Hohl and Lewis, 1995; Panini *et al.*, 1989; Peffley and Gayen, 1997). Consequently, the rate of DNA synthesis also declines markedly. Apparently, oxygenated sterols block cell division in the G1 phase of the cell cycle (Boxes 3.1 and 3.2, see 3.4.2).

Clofibrate accelerated epidermal barrier development as did the FXR activator farnesol (Hanley *et al.*, 1997). This suggests a direct effect of mevalonate metabolites on gene transcription. Besides, there are retinoid receptors (RXR) and vitamin D3 receptors (VDR) that are involved with transcriptional regulation. Generally, repressors of the transcription machinery, whether or not they are isoprenoids, function globally to maintain genes in a repressed state. In eukaryotes this depends on the relative *inactivity* of promoters, because genes must be activated for the process of transcription (Struhl, 1999).

POST-TRANSCRIPTIONAL REGULATION

The biosynthesis of many classes of plant secondary metabolites including isoprenoids involves cytochrome p450 enzymes. One of these is geraniol/nerol 10-hydroxylase that catalyzes a key step in the biosynthesis of the iridoid class of plant terpenes. According to Hallahan and West (1995) cytochrome p450 (normal abbreviation cyp450) from *Catharanthus roseus* and *Nepeta racemosa* are capable of hydroxylating geraniol, nerol and citronellol. In other plant species, the iridoid monoterpenes are synthesized from only one of these precursors.

In post-transcriptional regulation, acting on cyp450 hydroxylation, the biosynthesis of either one or a group of iridoid monoterpenes could be affected by a particular external factor. This is of importance for cyp450-linked detoxification systems in animals, e.g. herbivorous insects (see 4.2.2).

In some genotypes of *Perilla frutescens* more than one copy of spearmint limonene synthase exists. In these plants, the corresponding mRNA accumulated in the aerial parts of the plants, particularly in the leaves. In other genotypes the mRNA was detected in much smaller amounts (Yuba *et al.*, 1996). Post-transcriptional regulating factors should reach those plant parts that are the main site of synthesis of the terpenoid.

The regulated degradation of HMG-CoA reductase associated with an excess of sterols can be seen as an early-studied example of post-transcriptional regulation. Degradation of this key enzyme in membranes of the endoplasmatic reticulum (ER) pre-treated with sterols is accelerated. The proteases responsible for reductase degradation are stably associated with the ER membrane. Of the many studies on this subject we mention those of McGee *et al.* (1996), who showed that the ubiquitin system for protein degradation, essential for degradation of other ER membrane proteins, is not required for the degradation of HMG-CoA reductase. The authors postulate that MVA and sterols evoke increased susceptibility of the reductases to ER proteases rather than inducing a new specific proteolytic activity. It remains to be seen of this also holds for other isoprenoids, such as oxysterols and non-sterol isoprenoids derived from MVA in the same organism.

The complex and tightly regulation of HMG-CoA reductase (see 3.2.2) is in part through post-transcriptional mechanisms that are mediated by non-sterol products of MVA metabolism. Hinson *et al.* (1997) analyzed several phosphorylated and non-phosphorylated isoprenoid intermediates as inhibitors of the enzymes MVA kinases, phosphoMVA kinases and MVA diphosphate decarboxylases. Only MVA kinases were inhibited at the ATP-binding site. The C-20 intermediates GG-PP and the C-15 intermediate farnesyl-PP were most effective. There seems to be an important role of MVA kinases in the control of terpenoid biosynthesis suggesting that *the* key enzyme HMG-CoA reductase is only part of the multi-branched regulatory network (Fig. 3.2).

Another form of post-transcriptional regulation is the compensatory increase in HMG-CoA reductase activity of squalene synthase and HMG-CoA reductase inhibitors as discussed in section 3.2.2. A variety of structurally distinct reversible and competitive squalene synthase inhibitors strongly reduced cholesterol synthesis with no appreciable increases in HMG-CoA reductase activity (Petras *et al.*, 1999). This study suggests that non-sterol-mediated post-transcriptional regulation of HMG-CoA reductase

Chapter 3. The origin and evolution of terpenoid messengers

activity is independent of squalene synthase and does not depend on squalene and its metabolites as possible mediators.

Tumour cells can have a strongly reduced sensitivity to primary regulation of HMG-CoA reductase, but may retain high sensitivity to isoprenoid-mediated secondary regulation. One such regulation is post-transcriptional proteolytic degradation of this key enzyme. A combination of terpenes with additive or even synergistic effects can strongly reduce HMG-CoA reductase in tumours (Elson et al., 1999). This in its turn reduces the pool of isoprenoid intermediates that provide lipophilic anchors essential for membrane attachment and biological activity of growth hormone receptors, nuclear lamins A and B and oncogenic Ras, all dangerous with respect to tumour growth. The effects of terpenes on key enzymes important in malignant cell growth are treated in chapter 4, but serve in this section to illustrate the multi-component aspect of post-transcriptional regulation.

The monoterpene perillyl alcohol is known to inhibit ubiquinone synthesis from MVA and to block the conversion of lathosterol to cholesterol in mammary carcinomas (Ren and Gould, 1994). Limonene and perillyl alcohol exert other effects upstream of MVA, such as post-transcriptional modification of cellular proteins in cell proliferation (see 3.5).

There exists a nuclear receptor-signalling pathway for oxysterols. This receptor, named LXRα may be important as a sensor of cholesterol metabolites (Janowski et al., 1996). It should be kept in mind that there is also a farnesoid type of receptor in cell nuclei (FXR). The nuclear hormone receptors PPARα and FXR are members of the superfamily of nuclear hormone receptors that are obligate heterodimeric partners of the retinoid X receptor. Activators of PPARα such as oleic acid, linoleic acid and clofibrate induce an increase in the transcription of the mitochondrial HMG-CoA synthase gene. This is an example of combinatorial control of gene-specific activators and general negative regulators. It is beyond the scope of this book to discuss the association of repressor molecules with their target genes, in particular the relationship between nucleosomes and repressors. A comprehensive review on gene repression entitled "Repression: targeting the heart of the matter" has been written by Maldonado et al., in Cell (1999). Two major groups interfering with the production of isoprenoid metabolites at the post-transcriptional level are discussed separately: oxysterols and non-sterol terpenoids.

3.4.2 Oxysterols

Oxysterols (oxidized derivatives of cholesterol) are known to affect a variety of biological functions. Among others, they are potent regulators of cell sterol levels, causing down-regulation of both the activity of HMG-CoA reductase and enhancing cellular cholesterol esterification by activation of acyl-CoA: cholesterol acyltransferase. They also reduce the number of low-density lipoprotein receptors in cells. The down-regulation of HMG-CoA reductase by oxysterols is not based on direct binding to the enzyme, but could be due to rapid turnover rates, comparable with a reduction of the integrity of death-sparing activity of e.g. Bcl-2 protein (Nishio and Watanabe, 1996).

Table 3.3. Mode of action of oxysterols in the regulation of different cell types

Type of oxysterol	Mode of action
7-ketocholesterol (non-enzymatically formed in human atherosclerotic plaque by free radical oxydation)	Apoptosis in vascular smooth muscle cells (dietary 7-ketocholesterol is rapidly metabolized *in vivo*)
15-ketosterols	Regulation in HepG2 cells. Modulation of the mRNA level of HMG-CoA reductase and the low-density lipoprotein receptor (independent regulation).
25-hydroxycholesterol (enzymatically derived from cholesterol)	Transcriptional and post-transcriptional regulation of HMG-CoA reductase. Apoptosis in the human leukemic cell line CEM and the murine T-cell lymphoma line BW-5147. Inhibition of the growth of lymphocytes and macrophages. Apoptosis in vascular smooth muscle cells.

Oxysterols decrease the level of HMG-CoA reductases and reduction of oxysterol levels up-regulates this reductase in liver cells (Tamasawa *et al.*, 1997). Cholesterol itself is a mild inhibitor of HMG-CoA synthase activity: cholic acid and deoxycholic acid are more active (Honda *et al.*, 1998). 25-Hydroxycholesterol is more potent in decreasing the level of HMG-CoA reductase than 15-ketosubstituted sterols (Table 3.3).

Oxysterols are important activators of the nuclear LXR receptors, mentioned in 3.4.1 (Box 3.1). The identification of a specific class of oxidized derivatives of cholesterol as ligands for the LXRs greatly improved

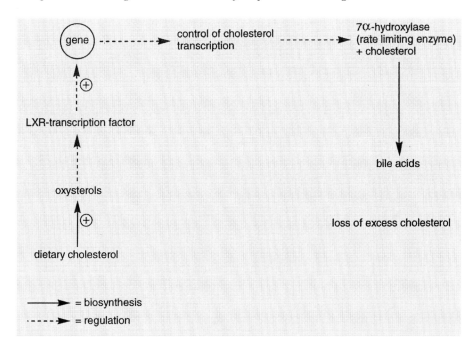

Box 3.1. Regulation of the cholesterol metabolism via oxysterols. The LXR-transcription factor regulates the catabolic degradation of cholesterol.

our understanding of the function of these receptors and opened a door that gives on to the field of regulation of lipid metabolism. The LXRs have a critical role in the regulation of cholesterol homeostasis. We now realize that the multivalent feedback system operating at the key enzymes of MVA metabolism is one mechanism of regulation, another system being an isoprenoid response pathway via the nuclear LXR receptors. Janowski et al. (1996) discovered that oxysterols act as positive transcription signalling molecules through the LXRα receptor and that transactivation by oxysterols occurs at concentrations that are realistic *in vivo*. They found that the most potent activators were in fact intermediary substrates in the rate-limiting steps of three important metabolic pathways: conversion of lanosterol to cholesterol, steroid hormone synthesis and bile acid synthesis.

Two orphan members of the nuclear receptor superfamily, LXRα and LXRβ are activated by 24(S),25-epoxycholesterol and 24(S)-hydroxy-cholesterol. In addition, there exists an LXR response element in the promoter region of the 7α-hydroxylase gene (Lehmann et al., 1997). The enzyme 7α-hydroxylase is responsible for the removal of cholesterol by increased turnover to bile acids. Oxysterols can reverse the inhibition of LXRα. In contrast, geranylgeraniol inhibits activity of this receptor, suggesting that LXRα represents a central component of a signalling pathway that is both positively and negatively regulated by mevalonate metabolites (Forman *et al.*, 1997). LXRα in its turn plays a role in the modulatory peroxisome-activated receptor α (PPARα) and the 9-*cis*-retinoic acid receptor RXRα.

In addition, cytotoxic and antiproliferative effects of oxysterols are known and both 25-hydroxycholesterol and 7-ketocholesterol can induce apoptosis in vascular smooth muscle cells. The compounds are present in human atherosclerotic lesions and their relation with exogenous cholesterol has been studied by Nishio and Watanabe (1996). The authors found that cholesterol, which itself does not exert an apoptotic activity, inhibits 25-hydroxycholesterol-induced apoptosis but not 7-ketocholesterol-induced apoptosis. This suggests that oxysterol-induced apoptosis in the vascular muscle cells is mediated through different mechanisms. It should be considered, that usually high concentrations of cholesterol exist in atherosclerotic lesions and that 7-ketocholesterol can be produced by free radical oxidation in the lesions, and inactivate proteins that function as an inhibitor of apoptosis or stimulate proteases involved in apoptosis. Box 3.1 and Fig. 3.6 provide schemes of the role of oxysterols in feedback regulation of MAD metabolism. The evolution of the oxysterol pathway is discussed by Parish (1994).

3.4.3 Non-sterol terpenoids

In mammals the degradation of HMG-CoA reductase is dependent on both sterol and non-sterol derivatives of MVA. Correll and Edwards (1994) demonstrated for the first time that accelerated degradation of this key enzyme can be initiated *in vitro*, and that farnesol is specifically required for degradation of the reductases in digitonin-permeabilized cells. Later studies implicated that the squalene synthase products presqualene-PP and squalene or their metabolites are also candidates. However, Petras *et al.* (1999) found that non-sterol regulator production is not dependent of squalene synthase action, which makes a function of terpenoids, based on squalene synthase less probable.

According to Peffley and Gayen (1997) MVA-derived non-sterols synthesized between squalene and lanosterol decrease HMG-CoA reductase synthesis at a translational level. Inhibition of squalene synthase overruled these effects, but inhibition of squalene cyclase did not. This would mean that non-sterol terpenoids are active in MVA-mediated suppression of the reductase synthesis at a post-transcriptional level.

Just like oxysterols, terpenes probably do not directly bind to the HMG-CoA enzymes themselves. The oxy- and hydroxy-substituted monoterpenes cineole and menthol (Figs 2.5 and 2.7) seem to mimic the effect of oxy-substituted cholesterol derivatives. Intake of 3.0 mM/kg body weight resulted in a 70% reduction of HMG-CoA reductase activity within 20 h. Immunotitration of HMG-CoA reductase pointed to a diminished amount of the enzyme. The synthesis is inhibited or the degradation is increased. It was mentioned that a "crippled enzyme" was resulting (Clegg et al., 1982).

The loss of HMG-CoA reductase activity was not associated with increased phosphorylation of the enzyme. Nakanishi et al. (1988) suggest that sterol derivatives suppress the enzyme (but not completely) by repressing gene transcription, and that non-sterol products of the mevalonate pathway (such as terpenes) reduce the enzyme by inhibiting translation of the mRNA (that is: post-transcriptional). The simple sesquiterpene ester farnesyl acetate can inhibit HMG-CoA reductase (Bradfute and Simoni, 1994). Together, sterols and non-sterol products accelerate degradation of reductase protein. The combination enables a regulation in enzyme activity over a several 100-fold range.

In plants, where the DOXP pathway produces isoprenoids in plastids, the situation is far from clear and may be complicated because of possible exchange of IPP, DMAPP and geranyl-PP from plastids with other cell organelles (Fig. 2.3). In this way non-sterol terpenoids could exert feedback effects both on enzymes of the DOXP pathway as well as on key enzymes of the MAD route outside the plastids.

The effects of plant-derived monoterpenes on feedback regulation of key enzymes in mammals can be unexpected. An example is in the fourfold increase in incorporation of ^{14}C-acetate into MVA in colon adenocarcinoma cell line SW480 in cells treated with 1 mM perillyl alcohol. The step from MVA to cholesterol, however, was not stimulated. Whereas lovastatin (mevinolin) inhibited cholesterol synthesis from acetate, concurrent addition of lovastatin and perillyl alcohol did not significantly alter the stimulatory effects of perillyl alcohol (Cerda et al., 1999). These results suggest that perillyl alcohol up-regulates HMG-CoA reductase, but that appeared not to be the case. We have discussed the compensatory increase of HMG-CoA reductases upon treatment with lovastatin before, so we should not be surprised if lovastatin stimulated this key enzyme in the present study. Indeed a more than 12-fold increase of HMG-CoA reductase activity was measured; concurrent treatment with 1 mM perillyl alcohol attenuated this to about fivefold increase. The authors then evaluated the effects of perillyl alcohol on low-density lipoprotein (LDL) receptor activity. The internalization of LDL decreased to 50% and lovastatin reduced this effect. The conclusion is that perillyl alcohol affects cholesterol metabolism not because more cholesterol is derived from MVA, but via some mechanism of action that is insensitive to the HMG-CoA inhibitor lovastatin. If only the first part of the experiments had been done, the conclusion might have been that perillyl alcohol stimulates sterol synthesis from acetate. This example serves to illustrate that many questions with respect to non-sterol regulation of terpenoid production remain as yet unsolved. Whenever lovastatin (mevinolin) is combined with terpenoids, the overall effect on isoprenoid metabolism *in vitro* and *in vivo* has to be determined. Figs 3.2, 3.6 and Table 3.2 deal with feedback regulation of non-sterol terpenoids in MAD metabolism of mammalian cells.

The possible effects of perillyl alcohol on insects are open just as well. What would be the effect on ecdysone biosynthesis or other metabolites and endproducts, let alone effects on MAI metabolism of symbionts?

Monoterpenes such as limonene and perillyl alcohol have chemotherapeutic properties because they affect isoprenylation of proteins which forms an important link in signal transduction between cells. The effects on protein isoprenylation will be discussed in chapter 3.5.

3.5 Interference with DNA synthesis: prenylated proteins

In the 1970s peptide mating factors secreted by fungi were shown to contain a farnesyl group attached in thioether linkage to the C-terminal cysteine. Later, the group of Glomset discovered that cultured mammalian cells incorporate mevalonate into covalent linkage with proteins with a mass range of 20K-30K and 50K-70K. The interest in isoprenoid metabolites was revitalized because it was found that farnesyl or geranylgeranyl-groups can be build into proteins which play a role in the inhibition of DNA synthesis (Farnsworth *et al.*, 1990; Rilling *et al.*, 1990; Schafer *et al.*, 1989). Attachment of chemical groups to proteins during their translation from mRNA is a common procedure in eukaryotes. These groups have various functions and can serve as a chemical switch or to locate a protein to a membrane.

Prenylation of proteins is a particular form of protein decoration, usually followed by other modifications, such as the removal of the last three amino acids from the terminal COOH-group. Prenylated proteins are signal transduction proteins that function on cell membranes and have an important function in cell division and a.o. in the cascade of messengers involved in blood platelet aggregation in protection against injury of vascular endothelial cells.

3.5.1 The mechanism of prenylation

In prenylation of proteins a farnesyl group, a geranylgeranyl group or a larger isoprenoid is attached to the terminal CAAX-group of a protein. C stands for cysteine, A is an aliphatic amino acid, and X is the C-terminal residue, preferentially one out of the following five: cysteine, serine, methionine, glutamine or alanine. The enzyme farnesyltransferase (in this book further referred to as FTase) links the farnesyl group to the proteins. In mammals, this enzyme uses farnesyl-PP as the prenyl donor, first purified from liver cells. Usually the amino acid cysteine is decorated with a 15-C or 20-C isoprenoid. More than one isoprenoid can be attached to a protein, often denominated "double prenylation", in which e.g. two geranylgeranyl groups become attached to specific proteins. As in the biosynthesis of isoprenoids with increasing size from MVA, the enzymes involved in prenylation have binding sites adapted to the prenyl group they carry, with the proper distance of the electrophilic carbon of the prenyl group to the sulfhydryl group of the prenyl acceptor of the enzyme. In this way, a degree of selectivity is attained depending on the length of the isoprene group. These valuable studies on prenylating enzymes, a.o. done by Gelb and Yokoyama (reviewed by Gelb, 1997) help to develop selective and potent inhibitors of protein prenylation (see 3.5.3).

The prenylated proteins undergo further modifications: proteolysis of the C-terminal –AAX tripeptide, methylation of the carboxyl group of prenylcysteine, which is now at the C-terminal position, and palmitoylation. This is shown for Ras proteins in Box 3.3. A recently purified prenyl protein protease from yeast and a human homologue of the proteases were produced in insect Sf9 cells by infection with a recombinant baculovirus, upon which membrane preparations derived from the infected Sf9 cells exhibited a high level of prenyl protease activity (Otto *et al.*, 1999). What is more: the recombinant enzyme recognized both farnesylated and geranylgeranylated proteins as substrates and even prenylated Ras proteins. This study indicates that yeast and mammals have prenyl-protein proteases with a high degree of similarity, suggesting that they may have obtained an early function in MAD metabolism, selected in the processing of CAAX-type prenylated proteins. For proper understanding: these proteases do not disable the prenylated proteins, but serve in a process of proteolytic maturation.

Freije *et al.* (1999) identified two human genes encoding enzymes potentially involved in this process of maturation. These enzymes, reacting with farnesylated proteins are integral membrane proteins, belonging to distinct families of metalloproteinases. Both genes have been demonstrated to be expressed in a variety of human tissues and one of them is located at

11q13, a region known to be amplified in human carcinomas and lymphomas.

Baculovirus-infected insect Sf9 cells, discussed above, were used to study prenylation of heterotrimeric G-protein subunits that have an important role in signal transduction in cells. G-proteins will frequently return in the coming sections of this book. Their major functions are in sensing and communication systems, ranging from mating in fungi to immune responses and vision and olfaction in mammals and they have an important role in pharmacology. The G-proteins are associated with a receptor just inside the cell membrane. The secret of G-proteins is their composition of three subunits, smaller proteins classified α, β and γ subunits (Box 3.2). Upon ligand binding, subunits (mostly β and γ) are released into the cell interior and can initiate transcription factors in the nucleus. According to Roush (1996), who reviewed the studies done on regulation of G-protein signalling (especially the RGS proteins), which were in its beginning during the first half of the 1990s, the G-proteins are part of an ancient communication system.

Myung *et al.* (1999) studied the relation between the nature of the prenyl group on the γ subunit of the G-protein and the activity of the β,γ-effector complex. They overexpressed six recombinant β,γ-dimers in baculovirus-infected Sf9 cells and examined for their ability to stimulate phospholipase C-β isozymes and type II adenylyl cyclase. The γ subunit modified with geranylgeranyl activated type II adenylyl cyclase about 12-fold, whereas subunits modified with farnesyl or with double farnesyl remained inactive.

Three covalent attachments anchor heterotrimeric G-proteins to cellular membranes. According to Morales *et al.* (1998), the α subunits are myristoylated and/or palmitoylated, whereas the γ chain is prenylated. Palmitoylation that confers on α,z-specific localization at the plasma membrane is suggested to be facilitated by myristate that promotes stable association with membranes not only by providing hydrophobicity, but also by stabilizing attachment of palmitate.

Palmitoylation is a modification of prenylated proteins that could serve to an optimal positioning of the protein to the plasma membrane. Prenylated proteins have hydrophobic prenyl groups and this may help to locate cellular membranes. Palmitoylation could be required to trap the molecule in the plasma membrane. Understanding of the separate roles of farnesyl or geranylgeranyl groups compared with palmitoylated prenylated proteins is in its beginning, although some studies support the idea of a two-step localization. Because palmitate is not attached without prior prenylation, it is difficult to discriminate between the contributions of either group. Booden *et al.* (1999) constructed novel C-terminal mutants of Ha-ras that retained their sites for palmitoylation, but replaced the C-terminal residue of the CAAX

Chapter 3. The origin and evolution of terpenoid messengers

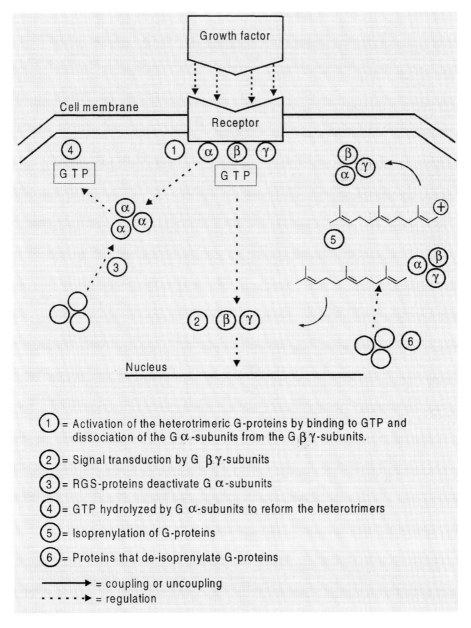

Box 3.2. Scheme of the heterotrimeric G-protein signal transduction pathway.

signal for prenylation with six lysines. The palmitoylated Ha-ras mutants bound about 40% as well to membranes as native Ha-ras and appeared to be potent transformators of NIH 3T3 cells and to initiate differentiation of PC12 cells. Thus, the roles of palmitate can be independent of and distinct from those of farnesyl. Moreover, it would seem that a farnesyl group is not needed for palmitate attachment, but that may depend on the cell or membrane type.

With respect to higher plants, we have to consider that both MAI and MAD metabolism is present, with their possible exchange of isoprenoids. In green plants protein prenylation is certainly present, although our knowledge is limited and a review on this subject is bound to be "a greasy tale" (Nambara and McCourt, 1999). Our understanding of the role of prenylated proteins in developmental and physiological processes is in its infancy. In spinach cotyledons, prenylation of proteins can be divided into two categories. One is the conventional prenylation involving farnesyl and geranylgeranyl groups, bound as we may expect, to cysteine residues via thioether linkages, and depending on nuclear gene expression. The other category represents a type of prenylation confined to plastids. This does not involve a thioether bond. Upon GC-MS analysis, the modifying isoprenoid was identified as phytol (Parmryd *et al.*, 1999). These authors could derive phytol from farnesol radiolabelled as [^{3}H] and taken up by the cotyledons.

The question is if phytol is the only isoprenoid involved in protein prenylation in plastids. Thai *et al.* (1999) demonstrated that microsomal fractions of *Nicotiana tabacum* cells catalyzed the enzymatic conversion of geranylgeraniol to geranylgeranyl monophosphate and GG-PP in reactions dependent on particular prenylated proteins (CTPs). This type of CTP was obtained upon incubation of microsomes with [^{3}H]CDP and either farnesyl-PP or GG-PP, but not farnesyl monophosphate. This study documents the presence of at least two CTP-mediated kinases in tobacco cells that are involved in protein prenylation. These facts add to understand the basics of the mechanism of prenylation in MAD metabolism of plants, but leave enough to speculate on possible prenylation reactions in MAI metabolism as present in plastids.

ORIGIN OF PRENYLATION

The human farnesylated protein-converting enzymes belong to families of metalloproteinases (Freije *et al.*, 1999). Their resemblance to yeast proteases suggests an ancient role in the functions of prenylated proteins. In *Giardia lamblia*, protein prenylation by farnesyl and geranylgeranyl isoprenoids was first published by Lujan *et al.* (1995), who demonstrated that this post-translational protein modification is an essential step in the regulation of growth of this eukaryote. Inhibitors of HMG-CoA reductase blocked cell growth (and protein prenylation), and cell growth was irreversibly blocked by inhibitors of subsequent steps in the modification of prenylated proteins. Most prenylated proteins are members of signal transduction cascades and

Chapter 3. The origin and evolution of terpenoid messengers

would seem to be involved in a highly conserved mechanism of co-localization of components to membranes, increasing their interactions by at least a factor 10^6, in this way dictating specificity among signal transduction pathways (Gelb, 1997). The constructive elements of RNA and DNA, nucleotides and amino acids all contain nitrogen, whereas prenylation occurs with isoprenoids without nitrogen. As has been discussed in chapter 2, metabolic routes for the synthesis of isoprenoids probably co-developed with those of metalloproteins and nucleoproteins.

OVEREXPRESSION OF BINDING ENZYMES – WHEN THINGS GO WRONG

Overexpression of FTase that can be based on aberrant gene expression may not only increase the expression of the α and β subunit proteins of FTase, but may also result in enhanced farnesylation of Ras proteins. Nagase *et al.* (1999) found that overexpression of FTase in NIH3T3 cells induced by transfection with cDNA constructs of human FTase resulted in enhanced DNA synthesis and anchorage-dependent growth, in a dose-dependent manner. These FTase-transfectants developed progressive tumours in nude mice. The tumours were distinct from v-ras-induced tumours and were characteristic of fibrosarcoma. The authors conclude that overexpression of FTase in NIH3T3 cells amplifies growth factor-mediated cell growth and malignant transformation. These effects are opposite to those of FTase inhibitors that suppress the growth of Ras-transformed cells *in vitro* and *in vivo* (see 3.5.3).

Lovastatin, an inhibitor of cholesterol production, can also inhibit prenylation of several key regulating proteins including Ras and this is why it is among the possible Ras protein inhibitors tested in cancer research, as will be discussed in chapter 4.

We should realize that regulation of gene expression in prokaryotes is different from that in eukaryotes. According to Struhl (1999), the ground state in prokaryotes is non-restrictive, whereas it is restrictive in eukaryotes. In prokaryotes, RNA polymerases recognize promoters via specific sequences immediately upstream of the initiation site. When the quality of the promoter sequence is sufficient, transcription is initiated, a process limited by the rate of RNA polymerase elongation, upon which the promoter is set free, permitting the next cycle of initiation/transcription. It follows that in prokaryotes there is no inherent restriction on the ability of RNA polymerase in its access to the DNA template and initiation of transcription: the ground state for transcription is *non-restrictive*. Repression in prokaryotes can act via DNA loops blocking access of RNA polymerase to the DNA template, or by occlusion of RNA polymerase binding to the promoter.

In contrast, RNA polymerase in eukaryotes has limited DNA sequence specificity and a strong core promoter must interact with RNA polymerase, general transcription factors and other associated proteins. The core promoter is basically *inactive*. Therefore, in eukaryotes genes do not have to be *repressed* for low activity, but to be *activated* for efficient transcription. The complex gene regulation in eukaryotes includes combined action of promoters, which is a protection against overexpression.

According to Struhl (1999), one function of activators is recruiting chromatin-modifying activities to promoters. These activities can occur in several steps that can be temporally separated and independently regulated. A big difference with prokaryotes is that the generator of an active chromatin structure can be distinct from the process of transcriptional activation *per se*. Still, this remarkably diverse and precise mechanism of gene regulation can show aberrant gene expression that must be silenced. To that end, there is a repression mechanism in which a DNA-binding protein occupies selected sites of the genome, thereby creating a domain of modified chromatin structure, unable to be transcribed. This is quite different from inhibition of the transcription machinery itself. Prenylation of the DNA-binding protein may be an essential factor in target-location of the repressor. Failure of the repressor to silence gene expression can result in unlimited cell division and other pathological events in the cell cycle.

3.5.2 Prenylation of Ras proteins

Ras proteins are critical components of signalling pathways leading from cell-surface receptors to the control of cellular differentiation and proliferation, or apoptosis (programmed cell death). Ras proteins must be activated to exert their effects, and this occurs by initial activation of a cell-surface receptor, receptor-associated (tyrosine) kinases, or so-called G-protein-coupled receptors. Upon activation, Ras stimulates a cascade of serine/threonine kinases to initiate transcriptional activation of genes (see 3.4.1). The interaction of Ras proteins with protein acceptors in the cytoplasmatic membrane is thought to be dependent on prenylation (farnesylation or geranylgeranylation). Fig. 3.10 is a scheme of the role of terpenoids in protein prenylation and Ras gene activation/deactivation.

Within the family of Ras proteins there are differences in further modifications. Various types of post-translational processing have comprehensively been reviewed by Rebollo and Martínez-A. (1999). For those studying differentiation and development it is of interest to know that non-farnesylated H-ras still can be palmitoylated, which may be sufficient for H-ras membrane targeting and to initiate differentiation.

Chapter 3. The origin and evolution of terpenoid messengers

Once arrived at their target, Ras proteins need to be bound to special G-proteins. This is not the end of the cascade, as this step is followed by activity of the Raf kinases that phosphorylate cytoplasmatic targets and other kinases. Substantial research has been done on the serine/threonine kinase Raf. This in its turn activates a.o. ERK kinases that phosphorylate cytoplasmatic targets and translocate to the nucleus, where a variety of transcription factors is stimulated.

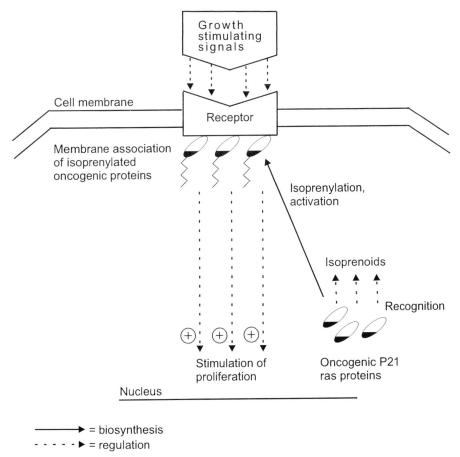

Fig. 3.10. A simplified scheme of the prenylation and activation of oncogenic proteins in a vertebrate cell.

An important messenger molecule such as Ras must become unbound upon upstream initiation of the messenger cascade. To that end the active GTP-bound can be broken with GTPases or by exchange activities, the latter resulting in inactivation by binding to guanine nucleotide (GDP-bound state). Ras proteins possess intrinsic activity of these enzymes, but according to Rebollo and Martínez-A. (1999) they are not powerful enough to do the job, so they are assisted by GTPase activating proteins, known as GAPs, belonging to the RGS subfamily. We will meet Gaps and RGS in sections 4.1.3 and 4.1.5 because they have an important role in mitosis and apoptosis.

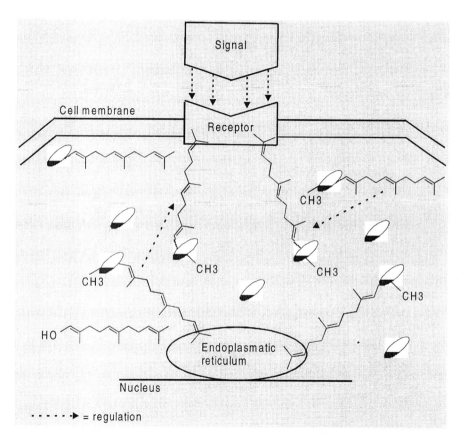

Box 3.3. Methylation and farnesylation, a form of prenylation of Ras proteins in a cell. Regulation: the methylated and farnesylated Ras protein is transferred to the receptor. The incoming signal is not bound to a normal G-protein, but to a modified Ras protein. Compare with Box 3.2.

The non-sterol isoprenoid allows mitotic cells to enter the S phase, that is the DNA synthetic phase. The smaller mevalonate-labelled proteins belong to a family of GTP-binding and hydrolizing proteins that regulate cell growth when bound to the inner surface of the plasma membrane. They are known to be among others p21 ras proteins, production of which can lead to uncontrolled cell division in mammals, that is if at the same time the tumour suppressor gene p53 is mutated, or if other oncogenes, growth factors and/or cytokines have altered expressions.

The step to cancer research is small. Mutated Ras genes are oncogenes and can be responsible for tumour formation in certain cases. Ras proteins known as small G-proteins occur in several organisms and play a role in signal transduction (Boxes 3.2 and 3.3). They are modified by isoprenoid lipids (Clarke, 1992; Marshall, 1993). GTPases can have essential roles in cell cycle progression through G1 (Olson *et al.*, 1995). FTases deserve special attention and are discussed in section 3.5.3.

3.5.3 Inhibition of protein prenylation

Post-translational farnesylation is required for Ras activation and this holds for mutated Ras genes responsible for malignant cell development as well (Fig. 3.11). During this prenylation process farnesol and geranylgeraniol are involved as in healthy cells. Novel developed isoprenoid analogues can selectively inhibit protein prenylation and can do so in transformed cells. Gibbs *et al.* (1999) showed that 3-allyl farnesol inhibits protein farnesylation of Ras-transformed cells. A geranylgeraniol analogue has a different mode of action, discussed in section 4.1.5.

In an earlier study, Schulz *et al.* (1994) demonstrated that D-limonene, perillic acid and perillyl alcohol (0.5-5 mM) selectively inhibit prenylation of proteins in the proliferation of human lymphocytes. The isoprenoids exerted a dose-dependent inhibition of DNA synthesis. The mode of action resembled that of the HMG-CoA reductase inhibitor compactin and likewise arrested cells in G1 and prevented them to enter the S phase. However, the effects of perillic acid could not be overruled with additional MVA. Obviously, MVA metabolites required for lymphocyte proliferation include one (or more) prenylated protein(s) essential for cell cycle progression. Many more studies in this rapidly developing field increased our understanding of the site of action of isoprenoids interfering with prenylation of proteins. Schulz *et al.* (1994) had a choice of possible modes of action that all could compete with normal prenylation:
- A competitive inhibitor of farnesyl-PPase
- An inhibitor of FTase
- Feedback inhibition of a key enzyme
- Prenylation of a DNA-binding protein resulting in a more efficient repression to silence gene expression
- Accelerating Ras protein degradation by upgrading recognition by proteolytic enzymes.

We could think of more possibilities but refer to the study of Hohl (1996) who found that limonene and perillyl alcohol decrease farnesylated Ras protein levels in human myeloid and lymphoid leukemia cell lines, not by depleting cells of farnesyl-PP or inhibiting FTase (the first two options). The monoterpenes were thought to alter Ras protein synthesis or degradation.

It should be noted that this study revealed that lovastatin reduced the amount of farnesyl-PP (whether or not by inhibiting HMG-CoA reductase) and that levels of farnesylated Ras decreased. At the same time, levels of non-farnesylated Ras increased, suggesting (to our opinion) that lovastatin did not accelerate Ras protein degradation.

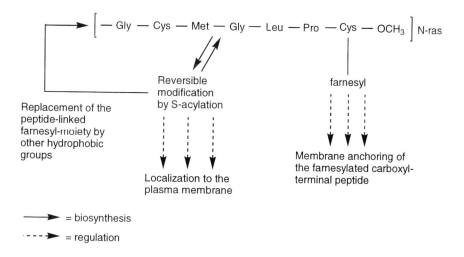

Fig. 3.11. The principle of farnesylation of an oncogenic protein (N-ras of mammalian fibroblasts) with possible regulation of its activity.

There is already a search for inhibitors of FTase, the protein prenylating enzyme, with respect to cancer therapy (Cox *et al.*, 1995; James *et al.*, 1993; Kohl *et al.*, 1993, 1994; Tamanoi, 1993). Many more inhibitors have been found, among which terpenoids (Gibbs *et al.*, 1999; Jayasuriya *et al.*, 1998; Ren *et al.*, 1997). See also sections 4.1.4 and 4.1.5. The action of monoterpenes against Ras-farnesylation is different from the HMG-CoA reductase inhibitors such as lovastatin. The overall effect is a decrease of farnesylated Ras protein levels, but that is not achieved in the case of the monoterpenes by inhibiting FTase. The effect is specific and has to be elucidated (Hohl and Lewis, 1995; Hohl *et al.*, 1998). Non-inhibitory regulations are also known: the diterpenoid alcohol geranylfarnesol can cause redifferentiation in mouse tumour cells (Ishikura *et al.*, 1984).

Chapter 3. The origin and evolution of terpenoid messengers 107

Geranylgeranyl acetone is also known to cause differentiation in myeloid leukemia cell lines (Sakai et al., 1993). See also chapter 4.1 and section 7.1.8.

Referring to the paper of Schulz et al. (1994), Hardcastle et al. (1999) in another study come to the conclusion that R- and S-limonene *are* inhibitors of FTase, and that their metabolites, perillic acid and perillyl alcohol are more effective, but all of them have a greater affinity to geranylgeranylTase than to FTase. This does not mean that selective FTase inhibitors do not exist; however, they mainly belong to a new generation of tricyclic compounds with a different mode of action (compare with 4.1.4). Another example is clavaric acid from the fungus *Clavariadelphus truncatus*, which specifically inhibits FTase. This nitrogen-free compound has no effect on HMG-CoA reductase (Lingham et al., 1998).

The search for isoprenoid FTase inhibitors continues intensively. Cohen *et al.* (1999) have tested several farnesyl-PP analogues for their specificity in inhibiting FTase, geranylgeranylTase-1 and squalene synthase. The research was done with human smooth muscle cells in culture. A strong inhibitor of FTase (IC50 value: 340 nM) encoded TR006 not only inhibited farnesylation of Ras in a Ha-ras transfected cell line, but slowed down cell growth. Proliferation of the smooth muscle cells was inhibited with 25 µM TR006. Growth factor-induced DNA synthesis was decreased with 100 µM, indicating that the inhibitor exerts its effect in an early phase of the cell cycle, probably by preventing protein prenylation.

However, other farnesyl-PP analogues decreased growth-factor-induced DNA synthesis without affecting cell growth. The question arises if the most potent FTase inhibitor selected by Cohen et al. (1999) has other effects on transformed cells distinct from prenylation. What would be the effect of TR006 on major nuclear farnesylated proteins, such as the lamins? These *in vitro* studies may overlook possible cytochrome c release from mitochondria and induction of caspases (see 4.1.3). Such effects could be comparable to those of manumycin A, that exerts an effect on protein prenylation in *Trypanosoma brucei*, although it does not affect DNA synthesis and cell cycle progression, but caused significant mitochondrial damage (Bassam *et al.*, 1999).

These examples illustrate that *in vitro* studies with inhibitors of prenylation should be regarded as a first screening and should be followed by *in vivo* experiments with attention to possible additional effects on factors inducing apoptosis. This also holds for isoprenoids and will be treated in sections dealing with apoptosis (sections 4.1.3 to 4.1.6).

Chapter 4

Specific properties of terpenoids

Chapter 3 has treated the functions of terpenoids in key enzyme regulation, in multi-branched enzyme regulating systems and in prenylation of proteins. In all these cases, terpenoids or their isoprenoid intermediates have specific binding sites to membranes and to terminal groups of a protein, in which the three-dimensional structure of the functional group is in close relationship with the dimensions of the isoprenoid structure of the messenger molecule. The many functions of terpenoids rise the question what the receptors of the thousands of terpenoid messengers look like. What is the difference between prenylation of a Ras protein and a receptor of a volatile terpene on an insect antenna? In this chapter we compare the modes of action of messengers on the cellular level with those between different organisms (either of the same or another species) with special attention to the receptor sites. From now on, we will discuss practical aspects of the material presented.

4.1   Messengers on the cellular level

Terpenoids can act as internal messengers, as semiochemicals, as defensive substances or as volatiles directing herbivores or natural enemies to their host. Often their effects are additive or even synergistic with other mevalonate metabolites, or they are inhibitors of parts of the mevalonate pathway. Therefore, their mode of action should be viewed with respect to the role of other mevalonate metabolites in growth, development and behaviour of the organism studied.

4.1.1 Messengers within a cell

Isoprenoid intermediates and terpenoids are an important part of the signal transduction system in cells. Prenylation of proteins is an essential part of the process (see 3.5). The prenylated proteins are anchored in the cell membrane and transport the signal to the next cascade step. The processes appear to be comparable in yeast, fungi and mammals, including humans (Caldwell *et al.*, 1995). Both internal and externally administered terpenoids and their metabolites influence these systems in various ways. Especially farnesol and geranylgeraniol are involved in the prenylation of proteins. The responsible enzymes are farnesyltransferase (FTase) and geranylgeranyl-transferase (GG-Tase) respectively. Studies with rat C6 glial cells that were incubated with [^{3}H]geranylgeraniol suggested that C6 glial cells are capable of converting geranylgeraniol to geranylgeranyl-PP, or possibly an "activated" form of the alkyl isoprenol, that can be utilized for protein isoprenylation reactions (Crick *et al.*, 1994). Recently, the processes discussed above were extensively studied in *Saccharomyces cerevisiae* (Greenberg and Lopes, 1996). In fact, most functions of isoprenoids as discussed in chapter 3.2 – 3.5 can be classified as messengers within a cell and will be treated from a different point of view in this chapter, especially in sections 4.1.3, 4.1.4 and 4.1.5.

There may be more to prenylation than membrane anchoring. Some prenylated proteins are in the plasma membrane and others are bound to internal cell membranes. Prenylated proteins are subject to COOH-terminal proteolysis and methylation and one may ask if these modifications are necessary to functioning of the protein. According to Gelb (1997) these proteases could be essential for the localization of e.g. Ras-2p, as lack of the enzymes results in misallocation of Ras-2p to the interior of the cell. Gelb argues that the prenyl protein-specific protease and methyltransferase, like FTase, may be targets for anti-oncogenetic therapeutics (see 4.1.5), taking in mind that yeast lacking prenyl protein-specific protease activity is still viable. The mechanism of prenylation has been discussed in section 3.5.1, although it leaves enough questions as to positioning and anchoring of a protein decorated with an isoprenoid on the membrane of a cell organelle.

The isoprenoid chain of messenger molecules within a cell is not necessarily a short one. Longer chains are present in the quinones, already mentioned in chapter 2.2. The quinones are involved in many more processes than mere electron transport. In animal cells they are not only present in the inner mitochondrial membrane but also in the plasma membrane, the Golgi apparatus, lysosomes and peroxisomes.

FUNCTIONS OF QUINONES

Quinones can have a number of isoprenoid units in the side chain, denominated prenyl quinones. One group is that of coenzyme Q. Their members have a number of isoprenoid units in the side chain, 10 in the case of coenzyme Q10 (Fig. 4.1). Animals can biosynthesize this compound in the mitochondria (Fig. 2.3). Vitamin K2 (menaquinone) belongs to the second group, which animals obtain from their diet.

A review of prokaryotic and eukaryotic ubiquinones focussing on their possible role in gene regulation, respiration and oxidative stress is by Søballe and Poole (1999). In ubiquinone (UQ) biosynthesis the nucleus of the molecule is derived from 4-hydroxybenzoate and the isoprenoid side chain starts with condensation of IPP with polyprenyl-PP (enzyme: polyprenyl-PP synthase). Specific polyprenyl synthases determine the length of the side chain. In yeast and other eukaryotes UQ synthesis is a MAD process, in prokaryotes IPP (via DOXP) and polyprenyl-PP likewise condense, but "poly" may stand for farnesyl-, geranyl- and solanenyl-PP, as the polyprenyl-PP synthase is relatively non-specific.

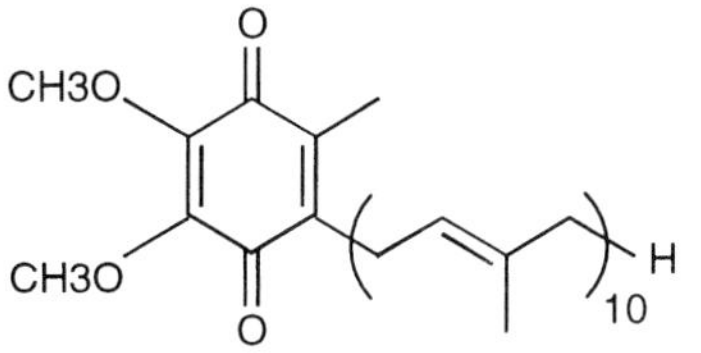

coenzyme Q10 (ubiquinone-10)

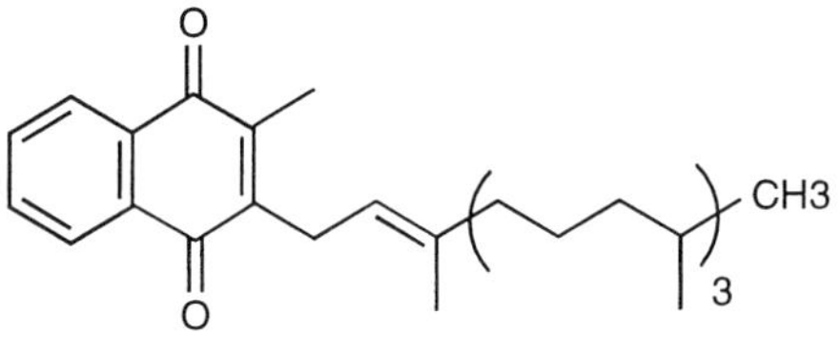

vitamin K1 (phylloquinone)

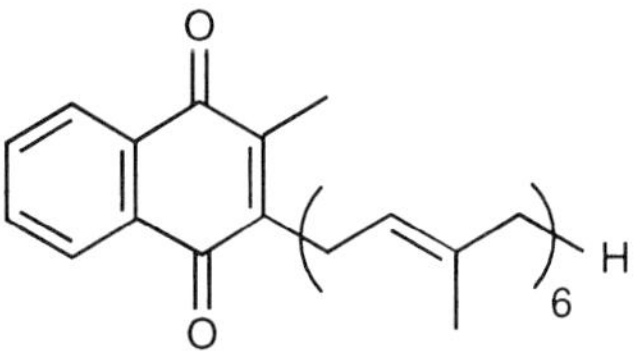

vitamin K2 (menaquinone)

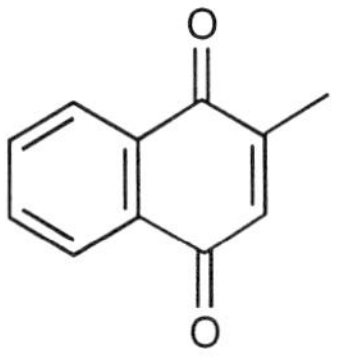

vitamin K3 (menadione)

Fig. 4.1. Structures of quinones with isoprenyl side chains.

A number of regulatory genes, required for UQ biosynthesis has been identified in *Escherichia coli*. In mutants of *E. coli* absence of UQ was suggested to cause derepression of a nitrate reductase, converting chlorate into chlorite, which is highly toxic. However, more research is needed before a role for UQ in gene regulation of microbes can be established.

There is general agreement on a function of UQ as a redox mediator, such as in cyt. b oxidation in aerobic conditions, while vitamin K2 acts mostly under anaerobic conditions. UQ and vitamin K2 can accept electrons from e.g. NADH, glycerol-3-phosphate, hydrogen and formate (they are more selective for other electron acceptors). In fact, the quinones act as receptors that are dependent on the structural constraints of the quinone-binding sites of dehydrogenases and oxidoreductases such as succinyl dehydrogenase. The isoprenoid side chain is an essential element in the positioning of the quinone molecule in the functional group of the enzyme.

The actual process of reduction/oxidation occurs at the quinone nucleus, by addition or release of two single H^+: =O becomes -OH (Fig. 4.1). We may state that UQ functions as it does because of the combination of the isoprenoid side chain with the quinone nucleus, resulting in hydrophobic properties that allow free movement between its partner reductors/oxidators in membranes of cell organelles.

Because UQ (coenzyme Q) receives a good deal of attention due to its supposed pharmacological properties, its role in oxidative stress should be discussed. Active oxygen anions such as ($O_2^{\bullet-}$) are regarded to be able to cause oxidative damage in living cells and UQ is a candidate to act as an antioxidant. According to Søballe and Poole (1999) it can only do this in the reduced state, as UQH_2. It can react with the superoxide anion and becomes $UQ^{\bullet-}$ under production of H_2O_2. However, *in vitro* this step is followed by reaction with normal oxygen, giving UQ and $O_2^{\bullet-}$. When this happens at the same ratio as the reaction of UQH_2 with the superoxide anion, no great protection of UQH_2 is to be expected. Additional functions of UQH_2 to prevent autoxidation of unsaturated fatty acids and to scavenge nitric oxide radicals ($NO^{\bullet}$) have been suggested, but as yet little is known of the biological significance.

As UQ/UQH_2 is the only lipid-soluble antioxidant known to be biosynthesized by mammalian cells, extra supply of UQ-10 has obtained attention. One useful effect is that it can regenerate vitamin E from its oxidized form (the isoprenoids vitamin E and β-carotene must be derived from the diet). UQ-10 is an efficient protector against lipid peroxidation and it decreases the ratio LDL/UQ, thereby enhancing HDL/LDL. This may explain why it can be prescribed to treat ischaemic heart disease, and possibly developing angina pectoris (see 7.1.3). However, to our opinion the present knowledge does not justify abundant advertising as food supplements. Its role should be seen in combination with other antioxidants e.g. ascorbic acid. Moreover, there are situations in which additional UQ inhibits oxidation of substances, which become active upon oxidation only.

Chapter 4. Specific properties of terpenoids 113

Inhibition of oxidation of Ω-3-fatty acids, for example, may strongly reduce their therapeutic effects on metastatic colon cancer and other specific tumour cell types (see 7.1.8). There exist other antioxidant (and oxidant) defense systems in organisms, belonging to the Phase I and Phase II enzymes. These enzymes like glutathion peroxidase, superoxide dismutase and DT-diaphorase are also able to scavenge free radicals and other reactive oxygen species. Their role in multi-drug resistance (MDR) of certain tumour cell systems is very important (details in 7.1.8).

4.1.2 Messengers within organisms

Within organisms, terpenoid endproducts ranging from small terpenes to more complicated terpenoids have messenger functions in a.o. development, metabolism and reproduction. In addition, they are involved in the perception of external stimuli such as light and odours. Probably the best-studied isoprenoids are the steroid hormones in mammals, on which there is a wealth of literature regarding biosynthesis, target specificity and pharmacology. A list of recommended textbooks is positioned right after the References in "General reading".

This chapter focusses on the relation of the molecular structure of the terpenoid messenger and its receptor, comparing organisms from quite different taxa.

ISOPRENOID HORMONES OF ANIMALS

Already in 1960 Schneiderman and Gilbert mentioned the occurrence of high concentrations of substances with JH activity in the eyestalk of lobsters, which is an endocrine centre in Crustacea. They referred to their demonstration of JH activity in the adrenal cortex of vertebrates (Gilbert and Schneiderman, 1958), micro-organisms and plants (Schneiderman *et al.*, 1960) at a time when the structure of the JHs had not yet been elucidated.

Insect JHs do not have effects on insects alone. Early studies of which we mention Zielinska and Laskowska-Bozek (1980) revealed that JH can influence growth and proliferation on mammalian cells. JHs, of which the biosynthesis and general functions have been discussed in section 2.2.2, can either activate or suppress gene transcription. As an example, in Coleoptera JH3 acts at the transcriptional level by increasing the amount of mRNA for HMG-CoA reductase, a key enzyme in pheromone biosynthesis (Tillman *et al.*, 1999). In other insect orders such as Lepidoptera this occurs with a neuropeptide (PBAN). JH can transcriptionally suppress proteins of the hemocyanin superfamily, involved in larval-pupal metamorphosis, e.g. in *Trichoplusia ni* (Jones *et al.*, 1993). High JH titres decreased the stability of the mRNA for these proteins.

The effect of JH (and this holds for the group of JH-active compounds: JH1-3, JH bisepoxide, methoprene, methyl farnesoate) is linked with that of the ecdysteroids, here treated as ecdysone. Our understanding of the effects of these ligands on their receptors is in its beginning. Following the line of thoughts as in 3.5.2, JH and ecdysone could enter a signalling pathway that resembles the G-protein pathways of mammalian cells. The G-protein-signalling pathway is regarded to be a major signalling cascade to transduce extra-cellular signals including sensorous stimuli to eukaryotic cells (Box 3.2). The recently discovered RGS family with their GTPase-activating proteins (GAPs) has also been mentioned above. JH and ecdysone seem to bind to a receptor that is a heterodimer of two proteins: ultraspiracle (USP) and ecdysone receptor (EcR). DNA is able to bind to these receptors, but DNA binding is localized in a different domain.

Lezzi *et al.* (1999) have reviewed recent studies on the signalling cascade in which USP and EcR are involved, entitled: "The ecdysone receptor puzzle". Let us try to put some pieces of this puzzle on their places. Because JH often counteracts ecdysone, the process of transactivation of ecdysone may serve as a primary model. Upon ligand binding of this terpenoid, its receptor (E) domain establishes a new interface to which the coactivator complex will bind. As we may expect, this consists of many proteins, carrying a so-called signature motif that mediates their binding to nuclear receptors. Some coactivator proteins belong to the bHLH-Pas family, members of which act as a receptor of dioxin or of methoprene (a JH analogue). Members of the coactivator complex recognize specific DNA sequences of their targets and are active as histone acetyl transferases, which, in an as yet unexplained way, change the structure of nucleosomes and chromatin, resulting in enhanced transcription.

It is beyond the scope of this chapter to comment on the effects of photoperiod and possibly of nutritional status in accessibility of the nucleosomal DNA to the coactivator (discussed in Lezzi *et al.*, 1999). To our purpose we follow these authors in their suggestion that JH has its effect in the EcR/USP heterodimer. Whether or not JH binds directly to USP awaits further research, but the heterodimeric nature seems to be typical for non-steroid receptors and may differ from the G-proteins involved in steroid reception. Nature may have selected for the best possible control of signalling pathways rather than for the best binding. This is of interest with respect to interference of terpenoids with the messenger cascade: other ligands than JH or ecdysone could operate in ligand-controlled transactivation and gene expression.

Homologues of a particular EcR isoform have been discovered in almost all insect species investigated. It is still intriguing that JHs occur in Protozoa, Fungi, Crustacea and mammals (Schneiderman *et al.*, 1960). It should be noted that crabs respond to retinoids during regeneration. The heterodimer partner of EcR in Crustacea has a domain closely related to the insect USP, but also an E domain with a sequence homology close to a vertebrate RXR receptor.

The interaction of JH with ecdysone, or more specific of JH1 to 20-hydroxyecdysone in the regulation of transcription factors (Zhou *et al.*, 1998) suggest that JH1 increases sensitivity to 20-hydroxyecdysone in *Manduca sexta*. In these cases it could suppress transcription of ecdysone-specific GTPase-like proteins, shifting the balance towards activity of the ecdysone-induced transcription factor E75A, expressed in *M. sexta* and also known from *Drosophila melanogaster*.

Nuclear steroid hormone receptors have a tripartite action not unlike the nuclear receptors acting upon contact with terpenoid insect hormones: these are the ligands, the receptor itself, and its coregulator proteins. We take the steroid hormone estradiol (Fig. 2.15) as an example. In mammals the sex hormones, the hormones of the adrenal cortex, not to forget the bile acids all are produced by modifying cholesterol in cells of mesodermic origin (see 2.2.3) and exert their effects in a variety of target organs. Estrogen receptors (ERs) are discussed within this context not so much because of the relevance of estrogens to the structure of the mucosa of the uterus or the induction of progesterone production in the corpus luteum, but because terpenoids are involved in suppression of ER-mediated gene expression and cell proliferation, including malignancies such as mammary cancer.

Comparable to the EcR receptors discussed above, the Ers possess a C-terminal helix 12 region of the ligand-binding domain. In their search for coregulators, Montano *et al.* (1999) traced a 37-kDa protein, which is a repressor of ER activity (not to be confused with Endoplasmatic Reticulum), denoted REA that represses ER transcriptional activity. It not only inhibits estrogen activity by suppressing estrogen-occupied Ers, in this way desensitizing cells to estrogens, but also directly interacts with liganded ER resulting in enhanced potency of anti-estrogens. This makes it an important mediator of anti-estrogens such as tamoxifen (see 4.1.4 and 4.1.9).

From an evolutionary perspective it may be noted that Montano *et al.* (1999) point to the high degree of identity of REA with soluble murine B cell receptor-associated protein BAP-37, present in other cells and tissues examined so far. The authors state that its role in inhibiting activity of the ER is an example of the increasingly common theme in functional genomics of novel functions exerted by previously known proteins.

In insects, JH synthesis can be suppressed by a cyp450 terpenoid hydroxylase. This enzyme metabolizes (2E,6E)-farnesol into (10E)-12-hydroxyfarnesol that converts JH3 to 12-*trans*-hydroxy JH3 and does the same with other JH-like sesquiterpenes. This is inactivation of an isoprenoid hormone before it can reach its target and act as a ligand.

This book has a limited number of pages, a reason to restrict ourselves to a few examples of isoprenoid hormone receptors. However, they may serve to illustrate that terpenoids have more that one mode of action with respect to these receptors: they can both act as ligands and as suppressors of steroid hormone-mediated gene expression, and even interfere with the messenger before it reaches its target.

ISOPRENOID HORMONES OF PLANTS

According to Zibareva (2000), there is no other genus of any plant family comprising such a large number of known ecdysteroid-containing species as the genus *Silene* (Caryophyllaceae). This author demonstrated that ecdysteroid levels undergo fluctuations during plant development. Higher levels are found in aerial parts during periods of intensive growth, in some species at the budding stage. The role of phytoecdysteroids in plant defense is discussed in section 5.1.3. Maximal expression of the genes encoding key enzymes in ecdysteroid biosynthesis in growing parts of a plant is in accordance to their defensive properties.

Stem elongation and development of reproductive organs are known to be associated with activity of plant hormones. The enzyme gibberellin 3β-hydroxylase catalyses the final step in the biosynthetic pathway to gibberellin (GA). Itoh *et al.* (1999) isolated a cDNA clone of a 3β-hydroxylase from tobacco that encoded an active GA 3β-hydroxylase to investigate the expression pattern within developing plants. GA may be translocated to growing regions, but upon histological analyses of GA 3β-hydroxylase the authors propose that the enzyme be controlled in time and space by expression of the gene encoding this hydroxylase. The expression was restricted to specific regions, where cells actively divide and elongate. If not translocation of GA but spatial and temporal regulation of GA 3β-hydroxylase expression underlies GA activity, there should be a messenger that controls expression of the encoding gene.

ISOPRENOID RECEPTORS IN ONTOGENY

An intensively studied group of low molecular weight, lipophilic derivatives of vitamin A, the *retinoids,* has a profound effect on the formation of embryonic structures, in invertebrates as well as in vertebrates. A general picture arises from various studies of hydroids, ascidians, developing limb buds in chickens, or regenerating amphibian limbs and tails (axolotls are famous for their cut and regenerated extremities). There is a striking uniformity in the effect of retinoids on pattern formation. Is this because there is uniformity in the retinoid receptors?

Two families of retinoid receptors exist, the retinoic acid receptors (RARs) and the retinoid X receptors (RXRs) that form transcriptionally active RAR/RXR (heterodimeric) or RXR/RXR (homodimeric) complexes on DNA. The RARs are subdivided into α, β, etc. receptors, in which the level of expression induced by the same ligand, e.g. retinoic acid, depends on the tissue. As an example, Ghosh *et al.* (1996) found that in urodeles RARα was expressed at a significant level in wound epidermis, but not in blastemal cells. These authors report on different responses to retinoids of regenerating facial structure in embryos, larvae and adults of newts.

Similar effects have been reported by Dolle *et al.* (1990) in RAR transcription during mouse organogenesis. This rises the question in which way fine tuning of gene expression during the development of diverse embryonic structures is organized. A certain gradient of retinol and retinoic acid (RA) may be created by cytoplasmic binding proteins that can act as a storage/release system, as is known from mammalian embryos as a protection against vitamin A deficiency, or to sequester the retinoids when low levels are required (Dolle *et al.*, 1990). A retinol binding protein is synthesized in the yolk sac placenta of rodent embryos and in the syncytiotrophoblast of the human placenta, which is essential for access of retinoids to the embryo (Moriss-Kay and Sokolova, 1996).

It would seem that the mode of action of retinoids is not quite the same as that of JHs and steroid hormones discussed above. Retinoids operate in gradients, and their concentration depends on the region. Scadding and Maden (1994) measured a four times higher concentration of retinoic acid in posterior quarters of limb regeneration blastema in axolotls compared to anterior quarters. Anterioposterior gradients have been reported in many more tissues.

One obtains the impression that in embryogenesis the retinoids, their receptors, their gradients, cover only part of the fine tuned process of differentiation. When applied at the wrong moment, retinoic acid is a powerful teratogen that can induce severe malformations, one of the reasons why one has to be careful with nutritional additives of vitamin A. Obviously sequestering systems do not sufficiently cope with overproduction of retinoic acid, and in these pathological events RARα receptors seem to be involved (Ghosh *et al.*, 1996).

A common research line is to work with receptor null mutants. Those missing RARα and RXRα, show regional pattern-specific resistance to heterogenic levels of retinoic acid (Morris-Kay and Sokolova, 1996), again indicating that supranormal doses of vitamin A and retinoids exert their effects via RAR and RXR receptors, in down-regulation of certain homeobox-containing genes and possibly prenylation of metalloprotease genes, depending on the target cell type (Ballock *et al.*, 1994).

Synthesis of retinoic acid can be inhibited with citral (Fig. 4.2). Treatment of developing wing buds of chickens that were grafted to the stump of host embryos formed truncated cartilage elements. These defects could be restored by simultaneous treatment with retinoic acid (Tanaka *et al.*, 1996). From an evolutionary point of view, it is interesting that a comparative inhibition of endogenous retinoid biosynthesis in the axolotl *Ambystoma mexicanum* results upon exogenous treatment with citral (Scadding, 1999). Limb regeneration was inhibited; the animals showed hypomorphic limbs and pattern irregularities. It should be noted that citral is in fact a mixture of geranial and neral (Fig. 4.2). Because the authors discuss the effects as caused by a single compound we do not know if both contributors have the same mode of action, or even if one of them competitively interferes with the retinoid receptor.

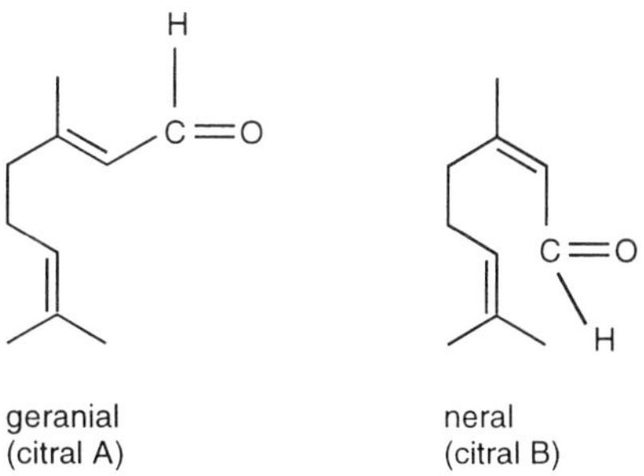

geranial
(citral A)

neral
(citral B)

Fig. 4.2. The structure of "citral".

Summarizing, the various messenger functions of terpenoids within cells or organisms (excluding their role as antioxidants as that is not a messenger function *per se*) isoprenoids and terpenoids are involved in:
- Redox mediators
- Compartmentation of plant growth hormones
- Prenylation of proteins: initiation of gene expression
- Interference with protein prenylation: inhibition of gene expression
- Hormones in lower and higher organisms
- Inhibition of ligand production
- Inhibition of ligand binding.

Terpenoids not only have functions within organisms, they are also mediators of external stimuli such as light and odours. As in photosynthesis, special light-sensitive substances, chromophores, convert photons into chemical energy.

TERPENOIDS IN VISION

The chromophore is situated at the protein rhodopsin that forms a proteinaceous membrane. The chromophore is 11-*cis*-retinal that is a derivative of vitamin A, via oxidation and isomerization reactions. In the tissues of the retina retinoids are active in rhodopsin metabolism. Photons induce isomerization of 11-*cis*-retinal from the *cis* to the *trans* position (Fig. 4.3A and B). *Trans*-retinal has a specially folded, energy-rich configuration which enables it to maintain its active position within rhodopsin, originally developed around 11-*cis*-retinal. In the following cascade G-proteins are involved.

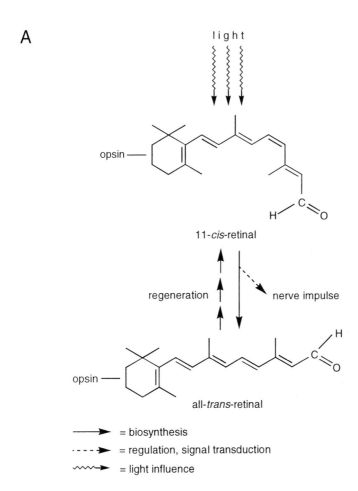

Fig. 4.3**A**. The isomerization of an important terpenoid in the process of vision.

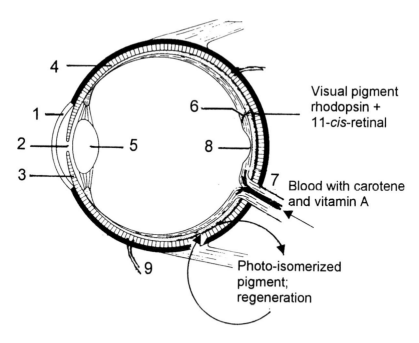

Fig. 4.3B. Diagram of a vertebrate eye with the principles of light perception. 1 = cornea; 2 = pupil; 3 = iris; 4 = choroid + epichoroid; 5 = lens; 6 = retina with optic nerve fibres, third and secondary neurones and layer of rods and cones, covered by the pigment epithelium; 7 = optic nerve with central artery; 8 = fovea; 9 = vortex vein.

The light-sensitive system of the retina seems to follow the familiar way of signal transduction with prenylation of proteins as has been discussed in section 3.5.1. Attenuation of receptor activation requires farnesylation of the negative regulator rhodopsin kinase (Inglese *et al.*, 1992+b). In the retina of vertebrates farnesoic acid is present, a precursor of insect JH3 (epoxymethyl farnesoate). It is intriguing that our vision is based upon an insect hormone!

The whole process of light perception, from the triggering of retinal through a series of intermediates to the final ion shifts in the nerve cells, including regeneration via RGS inactivators runs with high speed. The human brain can resolve 18 separate pictures per second, and flickering of light becomes inperceivable between 50 and 100 Hz, so the cycles of the signalling cascade in the retina should have at least this speed. Isomerization of 11-*cis*-retinal is extremely rapid: within 200 femtosec, but it is astonishing that the complete signalling cascade (of which rhodopsin kinase is only one

Chapter 4. Specific properties of terpenoids

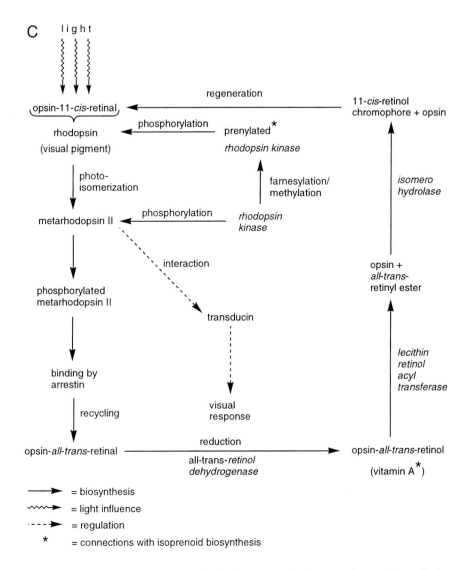

Fig. 4.3C. Role of terpenoids in light perception in vertebrates. Regulation stands for activation of transducin and initiation of the visual response.

enzyme activity) takes less than 10 msec. From an evolutionary perspective this is not so strange as vision in a rapidly moving animal should meet these demands. Figs 4.3C and D give an impression of the role of terpenes in light perception and the involvement of the G-protein transducin.

A rapid cycle of activation/deactivation/regeneration is also required in odour perception. An insect in search for a host plant must be able to

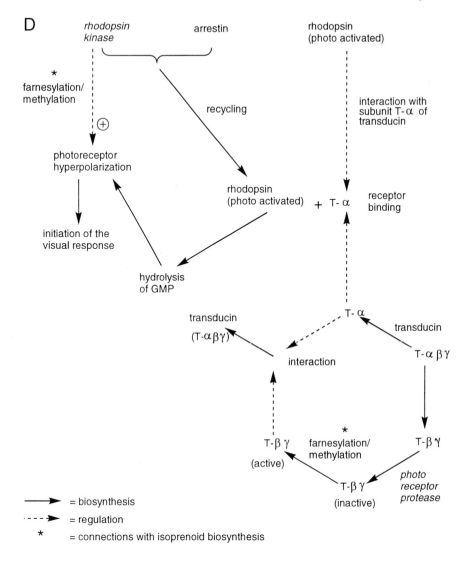

Fig. 4.3D. The role of the rhodopsin-transducin (with its G-protein subunits) interaction in the process of vision in vertebrates.

discriminate a particular odour from the chemical noise that due to air turbulence is continuously changing the ratio of its constituents.

TERPENOIDS IN OLFACTION AND GUSTATION

This section does not discuss the effects of terpenoids on behaviour, but their role in the process of reception. Whereas in the anatomy of the eye there exist two basic structures, the compound eye as in insects and eyes

Chapter 4. Specific properties of terpenoids

with a single lens, there exist a variety of "noses" positioned at all thinkable sites of animals. However, the process of olfaction is based on a signalling cascade with a great deal of similarity, though not quite the same, as in the receptors involved in internal messengers. Kurahashi *et al.* (1994) stressed that odour perception is the only sensorial contact with the environment where mevalonate products are not involved in the signal transduction, except for their role as initiators. The olfactory signalling system is deviating from the general enzyme cascade system. Recent literature states, however, that these deviating data can be explained by the properties of the G-proteins involved (Ebrahimi and Chess, 1998; Roush, 1996).

Within the scope of this book the role of terpenoid volatiles in the relation between plants and insects stands out, although vertebrates will not be overlooked. Odour perception in insects has a high degree of perfection and an incredible sensitivity, especially to semiochemicals. Insects have a relatively simple brain but must have solved the problem of information processing to recognize essential odours from the background noise. There is a wealth of literature on odour perception, especially on pheromone detection, e.g. in Lepidoptera (textbooks edited by Cardé and Minks, 1996 and by Ridgway, Silverstein and Inscoe, 1990), that lack terpenoid pheromones, or Coleoptera (textbook by Hardie and Minks, 1999). Herbivorous insects, of great interest within the context of this book, must discriminate between pheromones and host plant odours. Are there special receptors for terpenes? Let us try to compare the anatomy and physiology of odour perception of vertebrates and insects in a nutshell before shifting our attention to terpenes.

In vertebrates the olfactory epithelium, locked in a specialized mucosa, is pseudostratified columnar, shows no goblet cells and lacks a distinct basal lamina. Compared to other epithelia the cells are very tall (Fig. 4.4A). The receptor cells are evenly distributed between supporting sustentacular cells. They are bipolar nerve cells with a dendrite extending to the surface and terminating in small, bulb-like olfactory vesicles, from which olfactory hairs are protruding into a fluid film secreted by the glands of Bowman. The olfactory hairs are probably conserved structures, modified (but non-mobile) hairs as in Ciliata, and function as the actual receptor elements. Many receptor cells are grouped in the epithelium; their axons are collected in small bundles that run to the olfactory bulb of the brain.

In contrast to this, in insects the sensilla are positioned on or in the antennae and have a different grouping. The sensilla often have two or more receptor cells (Fig. 4.4B) of which a fluid film in the sensillum covers the dendrites. This fluid film is protected by the protruding cuticle in which there are pores through which odour molecules, in our case terpenes, can attain the dendrite membrane of the sensory cell. Variations on this theme are possible, e.g. in circumfila as present on the antennae of male and female *Dasineura tetensi* (Diptera). This type of looped sense organs are common to the Cecidomyiidae, to which this species and a.o. the Hessian fly, *Mayetiola destructor*, belongs. A circumfilum has only a single neurone (receptor cell), but this branches already in the hairshaft into numerous dendrites, that run parallel up the length of the stalk. These branches lead to or may interact with those of an adjacent circumfilum stalk (Crook and Mordue, 1999).

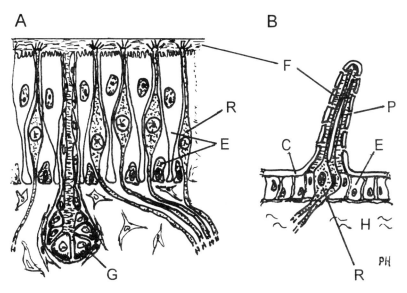

Fig. 4.4. **A**: diagram of an olfactory epithelium of a vertebrate. E = epidermal sustentacular and basal cells; R = receptor cell; G = gland of Bowman; F = fluid film. **B**: diagram of an insect olfactory hair. E, R and F similar as in A; C = cuticula; H = hemolymph; P = pore.

There is more variation in sensilla among insects. In aphids (and not in aphids alone) two types of sensilla are situated on the antennae: the primary rhinaria, positioned on the distal segments of the flagellum of the antennae, and the secondary rhinaria, mostly placoid sensilla, starting on the third antennal segment. Many of them can be seen with a handlens, sometimes their form is elongate or there are ring structures around the antenna. At times that there was only an inkling of odour perception in aphids, the secondary rhinaria have been taken for photoreceptors. We now know that the primary rhinaria are receptors of plant odours and alarm pheromones, whilst the secondary rhinaria of males are the main receptors of sex

Chapter 4. Specific properties of terpenoids 125

pheromones (Anderson and Bromley, 1987). Both alarm pheromones and sex pheromones are terpenes (see 5.1.1 and 5.1.2).

Maybe the reader begins to become confused at this point in finding so many variations in odour perception of insects. In trying to understand in which way herbivorous insects succeed in finding a host plant or their partners by selecting the adequate odours (in our case terpenes) some basics are needed, in order to follow the poor insect that moves in a chemical world, most of which is background noise. If we had such an olfactory system! Maybe we should be glad that we can find our way in a supermarket, only distracted by the joyless sounds usually descending from the ceiling. Insects have to cope with the following problems:

- Background noise (such as "green volatiles")
- Discrimination of allomones from kairomones, that is defensive odours from attractive ones
- Discrimination of plant odours from sex pheromones
- Discrimination of sex pheromones from alarm pheromones
- Continuously changing odour plumes with accordingly changing ratio of odour components
- Discrimination of enantiomers. For terpenes this is important, as different enantiomers can have completely different effects.

Study of these mechanisms and of related mechanisms in insects can contribute to novel insecticides and cytostatics. The fact that terpenoids are involved in basic functions may be the reason that cell systems and organisms cannot easily overcome the effect of terpenoids. Simple molecules like monoterpenoids are still toxic to certain insects after a long period of evolution, although there is a remarkable difference between toxicological effects on several insect genera and even species (Tsao *et al.*, 1995, section 5.1.3). The variety of structures and forms in terpenoids is so great, that this contributes to prevent the building up of resistance in insects. To overcome the impact of terpenoids the organisms would have to alter their own fundamental cell physiology.

ODOUR TRANSDUCTION

Signal transduction of odours across phyla resembles that of G-protein-coupled second-messenger pathways as discussed a.o. in chapter 3.5 (Box 3.2) and earlier in this section. These G-proteins contain hydrophobic domains that may act as transmembrane helices. In insects there exist odorant binding proteins, small (about 15 kDA), soluble proteins that transport odorant molecules to the receptor. This is not quite the same process as prenylation of proteins with their specific transferases that are specialized on isoprenoid molecules. Different classes of odorant molecules can rapidly associate with the binding proteins. In the case of terpenes protein binding followed by a G-protein cascade reminds us to the function of terpenes in the cell cycle, in particular the G1 phase.

The odour molecules carried by the binding proteins are supposed to activate olfactory receptor proteins in the dendritic membrane, followed by a G-protein cascade. In vertebrates as well as insects the messengers activate cation channels, resulting in an electrical response. In vertebrates cAMP is a second messenger, in insects this can be inositol triphosphate (IP3), sometimes cAMP as in vertebrates, but this may hold more fore general odours than for pheromones (Krieger *et al.*, 1999).

A rise of IP3 concentration sufficient to induce "firing" of the sensillum takes less than 50 msec. Electroantennogram techniques serve to measure the frequency and amplitude of the sensillum potential and are common practice (see a.o. Frazier and Hanson in a textbook edited by Miller and Miller, 1986). According to Laurent *et al.* (1999) spike tuning and synchronization are important parameters for odour sensation and perception.

What happens when a mixture of a terpenoid pheromone and another terpenoid plant odour arrive at the antenna? We know that there are specific pheromone binding proteins (PBPs) and more general odorant binding proteins (GOBPs) that may operate in aphids, discussed above. In these insects pheromones and plant odours will bind to different sites of the antenna. One can imagine that two or more attractive odours evoke firing of the same type of sensilla with additive effect, providing spike tuning and synchronization can match the decoding by downstream neural networks (Laurent *et al.*, 1999). These authors demonstrated with an intoxication technique that desynchronization impairs the discrimination of molecularly similar odours, but not that of dissimilar ones. This may be a reason why we found synergistic effects of mixtures of terpenes on aphid behaviour (see 5.1.3).

What happens when an attractive and a repellent terpene meet the antenna at the same moment? This depends on the mode of action of a repellent and the organization of the olfactory pathways in the insect. As Laurent *et al.* (1999) state, deciphering of olfactory codes would contribute to our understanding of one of the most remarkable pattern recognition devices ever studied. Already in the antennal lobe some neurones inhibit other ones and this inhibition can be caused by a repellent.

Although greatly simplified, the scheme as presented in a paper by Visser and de Jong (1988), based on studies of Boeck, Matsumoto, Hildebrand and Ernst (referred in that paper) is one of the best to support this discussion (Box 4.1). This Box shows that the first relay station is in the glomeruli of the antennal lobe. The network of interneurones is presented in its utmost simplicity but illustrates how two attractants, a pheromone and a host odour

Chapter 4. Specific properties of terpenoids

(kairomone) can cooperate via interrelation of two glomeruli (A and B with C) in the antennal lobe. More general receptors can add to the olfactory code. When we replace a host odour by a repellent that acts on different receptors, glomerulus C can inhibit A or/and B.

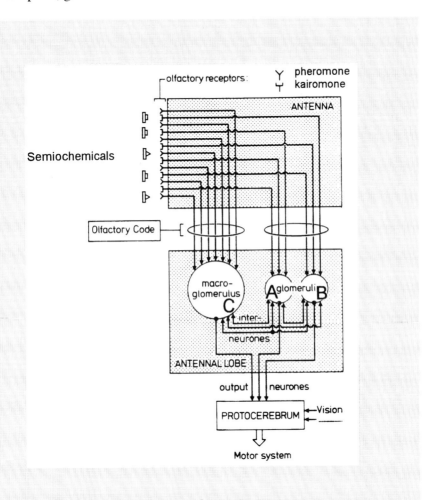

Box 4.1. Scheme of the olfactory pathway in insects. For more details see text (with courtesy of J.H. Visser).

This is not the only possible mode of action of a repellent. It could interfere with the binding protein of receptors specialized in attractive odours, or it may inhibit the G-protein cascade. However, when we realize that in odour perception very low concentrations of odorant molecules are sufficient, repellents that have the latter mode of action need much higher concentrations. This would be the case in competitive inhibition of kairomones and indeed many repellents must be present in higher concentrations than attractive odours, but many does not mean all!

Box 4.1 also shows that there is an input convergence in the antennal lobe, resulting in an amplification of the original signal, as well as in a reduction of background noise. At the level of the antennal lobe the activities of individual neurones are modified to stimulus levels 2 or 3 orders of magnitude higher than that of the peripheral receptors. Depending on the size of the glomeruli, the output signals for pheromones, kairomones and allomones can greatly differ from the original signals from the receptor cells. Similar processes operate in mammals (Laska *et al.*, 1999b; Mori *et al.*, 1999).

Last but not least, there is the protocerebrum. Signals arriving here are interpreted and combined with those of other stimuli, such as vision. The brain decides how to react on the combination of attractant and repellent odours.

TASTE PERCEPTION

Basically, the processes in gustation show some similarities to those of olfaction. Chemoreceptors are specialized in the perception of non-volatile molecules. However, there is quite a difference between specialization of taste perceptors in mammals and e.g. insects. Gustation in humans is a combination of a simple gustatory system discriminating between bitter and sweet substances, salt and acids plus olfaction. We need odours to complete our taste. In insects receptor cells for bitter substances can be amidst those for specific compounds. There is also a great variation in the position of gustatory organs in insects, some of which taste with their feet (the tarsi), although in many insect species taste receptors are present on mouthparts. The chemical world of insects comprises not only olfaction but also gustation. There are chemosensilla for sugars, salts, bitter substances, amino acids, and complex plant metabolites, even for water.

Taste chemosensilla of humans are located in taste buds in the tongue and possess, as in olfaction, supporting or sustentacular cells and neuroepithelial taste cells, but these are not sensitive to terpenes. In insects the chemosensilla resemble those for olfaction, albeit that the cuticle runs uninterrupted to a pore at the tip of the sensillum. Two or three receptor cells, mostly unipolar neurones, are present in one sensillum. The axons run to the protocerebrum. For techniques of recording chemosensory potentials we refer to a.o. Frazier and Hanson in a textbook edited by Miller and Miller (1986).

The antifeedant effect of clerodane terpenes, discussed in section 2.2.2 and their derivatives have been tested on larvae of *Pieris brassicae*, (the

Chapter 4. Specific properties of terpenoids 129

cabbage white butterfly), on larvae of *Leptinotarsa decemlineata* (the Colorado potato beetle) and nymphs of *Myzus persicae* (the green peach aphid). None of the modifications were as active as the original neoclerodane diterpenes (klein Gebbinck, 1999). Introduction of an α-hydroxyl group, or in general oxidized or unsaturated structures could improve the capability to form covalent bonds. In this way an enal moiety might act as a -SH acceptor, reacting with sulfhydryl groups of a receptor protein. It would seem, however, that the principal mode of action of drimanes is binding to primary amine groups of a receptor: terpenoids, after all, do not possess N atoms which can be involved in S—C=N—O SO_3^- binding. This does not mean that unsaturated dialdehyde moieties do not contribute to the pungent taste, but that is a separate property.

The great specificity of olfactory receptor cells to discriminate between isomers and enantiomers of an odorant makes that volatile terpenes have an almost infinite number of possible additive, synergistic and inhibiting effects on attraction and repellence (in gustation on deterrence). This will be illustrated in chapter 5.1.

4.1.3 Homeostasis of cell numbers

The famous words of Goethe "In der Beschränkung zeigt sich der Meister" were addressed to artists and philosophers but could have been imposed on him by his studies of Nature: indeed, both in ontogeny and in fully developed organisms, the decision on what shall *not* live is continuously made. Differentiation of tissues is associated with programmed cell death; even a steady state is impossible without limitation of cell numbers. Every minute millions of cells die in our body, but because of homeostasis of cell numbers, we remain in balance: cell division is in equilibrium with cell death.

One may question why cells do not multiply more slowly and live longer. Situations exist in which a rapid supply of cells is live-saving as in the production of neutrophils, special white blood cells, the majority of which die within a few days. In case of an invasion by pathological microorganisms there is a rapid mobilization of neutrophils to combat the infection.

In the 1970s the idea that normal cell death is based on a well-defined and programmed cellular programme, completely different from what happens in necrosis, was new and took several years to become generally accepted. Raff (1998) describes how the first genes required for programmed cell death were identified and how indications were obtained that apoptosis depends on specific proteolysis. These proteases become functional by activation of inactive procaspases, resulting in a cleavage that produces active subunits (not unlike active G-protein subunits) and a prodomain that is discarded.

Apoptosis probably occurs in the whole living world, although caspase cascades may have developed later in evolution. Upon infection with a virus, *Escherichia coli* viral protein binds to a protease that becomes activated, cleaves, and inhibits protein synthesis and consequently virus multiplication, though killing the bacterium itself. In plants caspases may be involved in apoptosis, which is essential in e.g. senescence of flowers and leaves. The word "apoptosis", of Greek origin, means "falling leaf"!

The suicide activators may act via a proteolytic network supported by caspases, a family of cysteine proteases, some of which are involved in apoptosis by acting on specific inter-domain linkers, comparable to the mode of action of apical caspases that transmit signals to executioners by cleaving other zymogen inter-domain linkers (Salvesen and Dixit, 1997). This is a process of limited proteolysis in which a few cuts, usually only one, are made in inter-domain regions. This may result in either activation or inactivation of a protein, such as a suicide protein. Although other enzymes like nucleases and protein kinases are involved in cell death, caspases may be regarded to be a refinement of well-timed and limited proteolytic events in programmed cell death (Askenazi and Dixit, 1998; Thornberry and Lazebnik, 1998).

What triggers the suicide machinery in a living cell? That depends on whether a healthy cell needs to terminate its life e.g. in tissue differentiation, where many cells are selectively removed, or on their condition, e.g. when the cell is ill or damaged in some way. With some simplification apoptosis is triggered in the following ways:
- Upon damage of mitochondria
- Upon damage of DNA in the nucleus (activation of p53)
- By thyroid hormones in the blood
- A lack of apoptosis-suppressing signals (due to Bcl-2 binding)
- Inactivation of genes encoding proteins, which protect against apoptosis.

The last two ways in which apoptosis is triggered are indirect ones, by inactivating cell-saving systems. Programmed cell death seems to be a balance between two opposite processes that could be controlled by two independent genetic pathways. One system serves to protect cells from being driven into the apoptotic route. The membrane protein Bcl-2 inhibits apoptosis at a critical point of the mitotic G1 phase, when cytokines act as progression factors. Bcl-2 seems to inhibit transcription factors required for transcription of cytokines like IL-2, that acts on the primary progression factor for T cell production (Linette *et al.*, 1996). A shortage of Bcl-2 leads to immune-deficiency because T cells run into a shorted cell cycle and in contrast Bcl-2 leads to hyperplasia and could keep cells with pathological mutations alive, that should be destined to die. Overproduction of Bcl-2 occurs frequently in human cancers and contributes to refractory (therapy-resistant) cell lines.

Other proteins like Bag-1 enhance the anti-apoptotic effect of Bcl-2 and the same holds for serine/threonine protein kinase Raf-1. Probably Bag-1 stabilises Raf-1 in an active conformation once activated by other mechanisms or protects it from inactivation by phosphatase because it interacts with the membrane protein Bcl-2 (Wang *et al.*, 1996). Maybe even trimolecular complexes are formed, operating at the mitochondrial membrane. Ras proteins (in healthy cells the good ones) may activate Raf-1 at least in part by targeting it to membranes, as could do Bag-1 to complex formation (Wang *et al.*, 1996b). Fig. 4.5 depicts our view on interactions of Bcl-2 protein and caspases in apoptosis.

More direct systems lead to cell death. "Suicide" proteins are present in all living cells, hidden in the inter-membrane space of mitochondria. They are released when holes (gaps) are generated in the membranes by specific signalling systems, that on their turn are activated by DNA damage. In fact, any agent causing pore opening in mitochondria induces mitochondrial swelling and the release of a cell suicide protein. If the number of such mitochondria increases, the concentration of suicide protein in the cytosol reaches a threshold value required to initiate the apoptotic changes in the nucleus (Green and Reed, 1998; Skulachev, 1996).

Skulachev, of whom we give a translated reference, came very close. It appears that there is a double role for cyt. c. Normally it functions in respiratory-chain phosphorylation, generating ATP. Section 2.1.6 discusses a second symbiosis of eukaryotes (the first was the nucleus) with an eubacterium transforming into a mitochondrion. With the mitochondrion, the cell introduced a spy – it controls life and death by releasing apoptosis factors into the cytosol, to the benefit of the organism. (The reader is requested to refrain from making a connection between Skulachev's mitochondrion and a personage in "A spy who loved me", one of the 007 – James Bond movies). When mitochondria are damaged they can release cyt. c. Once in the cytoplasm, it binds to a protein and this complex activates procaspase-9, which initiates the caspase cascade (Fig. 4.6). Most of our knowledge of cyt. c-induced apoptosis is based on studies of vertebrates.

The release of cyt. c is not a one-way process; it is controlled by proteins of the Bcl-2 family, of which there are two main groups: those that promote apoptosis (Bax and Bak) by inducing its release, and those that inhibit it. The Bcl-2s seem to guard the key to the mitochondrial gate (Martinou, 1999). Precluding cyt. c release is not believed to be the sole function of Bcl-2, because Bcl-2 can still protect cells after release of cyt. c. Members of the Bcl-2 family, such as Bcl-x$_L$ may bind to Apaf-1, a mediator between cyt. c and procaspase-9, in this way inhibiting the death signal. This bond could be

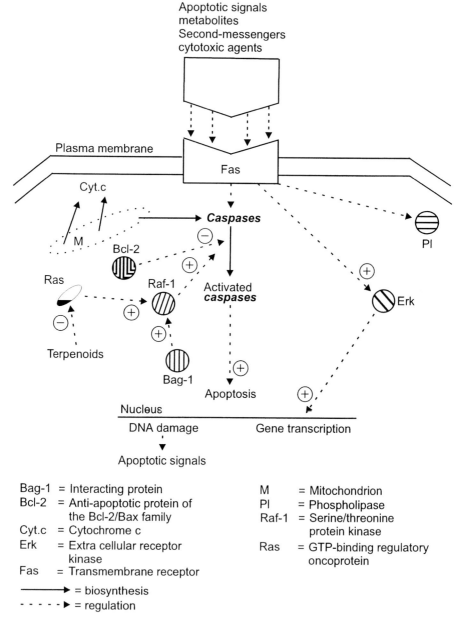

Fig. 4.5. Interactions of the Bcl-2 protein, caspases and other regulating proteins in apoptotic processes.

Chapter 4. Specific properties of terpenoids

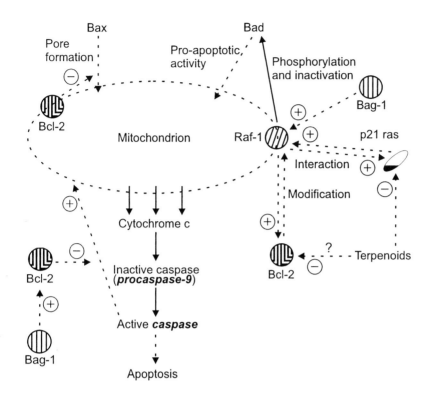

Fig. 4.6. The role of the mitochondrion in activation of caspases and in several regulating and oncoproteins involved in apoptosis.

prevented by activity of a member of the pro-apoptosis Bax family. Adams and Cory (1998) discuss models for regulation of the Bcl-2 family and the balance between pro-survival and pro-apoptosis genes. It should be realized that *all* Bcl-2-like genes are potentially oncogenic, especially upon mutation.

Recently, evidence has been obtained that in plants programmed cell death can occur independently of caspase activity. In this type of degenerative cell death there is not necessarily nuclear condensation, but activity of hydrolases from the lysosomes. Plant pathologists denominate this type of cell death oncosis. Plant tissues may not suffer from leakage of cellular contents that could result in tissue inflammation in animals. According to Jones (2000) plants and animals may share the mitochondrial cyt. c component of programmed cell death, based on an ancient mechanism by which the mitochondrium could kill its host. Jones (2000) points to evidence that Bcl-2 and Bax fulfill similar roles in plant cells and yeast as in animal cells, to which he adds that this evidence is indirect and needs confirmation by elucidation of Bcl-like plant proteins. In contrast to animal cells the plant cell vacuole can sequester hydrolases. This is of interest in relation to storage of terpenes in oil glands (Fig. 2.9C and D).

Damaged DNA in the nucleus can activate apoptosis and the same is true for accumulation of misfolded proteins in the endoplasmatic reticulum. A widely studied protein, p53, becomes activated in response to DNA damage. By various routes it can either cause apoptosis or stop cell division. Unfortunately the gene encoding p53 can be inactivated by mutation, resulting in cell proliferation (Raff, 1998; see also 4.1.4 and 4.1.5). DNA damage is only one factor that activates p53. According to Oren (1999) this can be achieved to some extend by increased translation of the p53 mRNA, although it is now generally accepted than an increase in active p53 occurs mainly through post-translational mechanisms. Oren (1999) mentions a number of stress factors that activate p53: DNA damage, ribonucleotide depletion, mitotic spindle damage, deregulated growth signals, even NO and hypoxia.

The thyroid hormones form a classic example of inducers of apoptosis, as correctly stated by Raff (1998). The larval stages of frogs have tails that disappear at metamorphosis and humans have an uneven number of fingers, not four or six, suggesting that cells have been eliminated between developing fingers and toes. In interference of thyroid hormones with retinoids, although not yet unraveled, terpenoids arrive at the scene.

Terpenoid hormones can stimulate apoptosis by a different mode of action. Ecdysone induced p53 expression in a p53-null human cancer cell line, resulting in irreversible cell growth arrest. In addition ecdysone enhanced the cytotoxic effect of the anti-tumour drug cisplatin. However, ecdysone-inducible p21 waf1 became also expressed and this conferred increased resistance to both cisplatin and etopoxide (Wang *et al.*, 1998), suggesting that enhanced apoptosis by p53 does not always increase susceptibility to chemotherapy (see 4.1.5).

When the gene expressing a Bcl-2 protein is not expressed, or when Bcl-2 is bound to a protein, which disables it to bind to the outside surface of mitochondria or other membranes, apoptosis can be strongly enhanced. The opposite is possible as well; e.g. an excessive production of Bcl-2 caused by chromosome abnormality (Adams and Cory, 1998). This occurs for instance in a form of lymphocyte cancer. The pathologically high Bcl-2 levels inhibit apoptosis of these cancer cells.

In plants, terpenoids can induce apoptosis. Izumi *et al.* (1999) found that geraniol was the most potent apoptosis-inducing terpenoid in shoot primordia of *Matricaria chamomilla*. The nuclei of cells treated with 5 mM geraniol were condensed and fragmented. From these results we cannot conclude if geraniol interferes with a caspase cascade in this composite plant.

In human leukemia U937 cells geranylgeraniol potently induces caspase-like activity, but geraniol itself did not (Masuda *et al.*, 1997). Apoptosis induced by farnesol, geranylgeraniol and several cytotoxic drugs with different modes of action was accompanied by accumulation of CDP-choline in two mammalian cell lines (Williams *et al.*, 1998). CDP-choline is an intermediate in phosphatidylcholine biosynthesis. The resonances of this compound that can be detected in a non-invasive way occur in a relatively well-resolved region of tissue spectra, which could make it a marker for apoptosis.

Invaders that need living cells for their own reproduction should prevent the initiation of early cell death and Salvesen and Dixit (1997) refer to inhibition of caspases by viruses as a strategy to limit cell response upon infection. An example is the baculovirus protein p35 that has the ability to terminate activity of almost all caspases. As baculoviruses are effective biocontrol agents cell death can be delayed by inhibition of caspases, but eventually necrosis leads to death. The baculovirus protein p35 could encode a homologue of Bag-1 that facilitates interactions of Bcl-2 with Raf-1. No homologues of p35 are known in mammals, but caution is justified with respect to possible mutants of baculoviruses with the ability to infect mammals. If caspases are involved in proteolysis, the basic process of the blood coagulation cascade (in addition to cell disruption (Ebola virus)), little imagination is needed to cast a horror scenario in which a novel generation of pathological viruses cause severe hemorrhages or stop programmed cell death in developing embryos, killing either mother or child.

4.1.4 Effects of terpenoids on Ras proteins

As discussed in section 3.5.2, Ras proteins are dependent on prenylation to interact with protein acceptors in the cytoplasmatic membrane. Box 4.2 is a (simplified) scheme of the Ras-Raf signalling pathway. Monoterpenes can decrease the levels of farnesylated Ras proteins, which rises the question if they do so by inhibiting FTase or different gene expression of Ras proteins, thereby inhibiting their biosynthesis. Other scenarios are thinkable (Fig. 4.7).

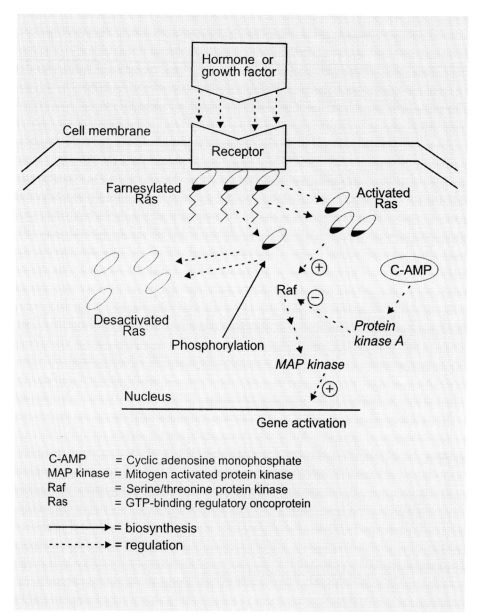

Box 4.2. Simplified scheme of the Raf/Ras signalling pathway with possible connections to the cyclic AMP pathway.

Chapter 4. Specific properties of terpenoids

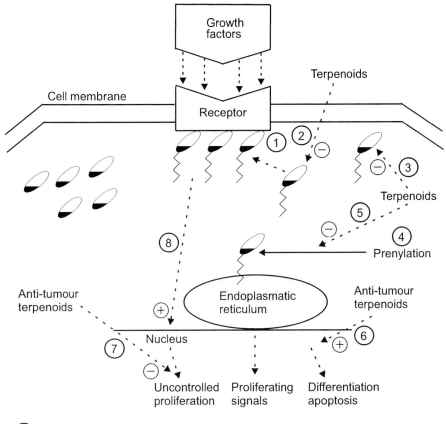

① = Prenylated Ras attached to the membrane for signal transduction
② = Inhibition and modification of prenylated Ras
③ = Interference with Ras membrane attachment
④ = Prenylation of Ras proteins
⑤ = Inhibition or interaction with Ras prenylation
⑥ = Interaction with cyclins and genes/caspase cascade
⑦ = Interaction with nuclear receptors
⑧ = Proliferating signals of the signal transduction cascade
⎯⎯→ = biosynthesis; biochemical processes
- - - -▶ = regulation

Fig. 4.7. Role of terpenoids in Ras activation/inactivation.

FTase inhibitors selectively block protein farnesylation and reduce the growth of many Ras-transformed cells *in vitro* and *in vivo*. They may prevent prenylation of other proteins than Ras proteins or interfere with Ras-induced ligand production.

Perillyl alcohol decreases the levels of antigenic Ras in human myeloid THP-1 and lymphoid RPMI-8402 leukemia cell lines. Neither perillyl alcohol nor its precursor limonene were found to act by depleting cells of farnesyl-PP or inhibition of FTase (Hohl, 1996), suggesting that the monoterpenes decrease farnesylated Ras by a different mechanism.

Limonene, perillic acid, perillyl alcohol, menthol and pinene inhibited growth of viral Ha-ras-transformed rat liver epithelial cells at concentrations of 0.25-2.5 mM. The major effect was not that of inhibition of HMG-CoA reductase, as 2 mM MVA only partly counteracted the growth-inhibiting effects of the terpenes. Western blot analyses of cytosolic and membraneous fractions of WB-ras cells treated with the monoterpenes indicated no change in Ras distribution, which means that they did not alter Ras plasma membrane association, suggesting no effect on prenylation of Ha-ras. Karlson *et al.* (1996) came to the same conclusion with regard to the effect of (+)-limonene and related monoterpenes on PANC-1 pancreas carcinoma (carrying a K-ras mutation), and 12V-H-ras-transformed rat fibroblasts. The monoterpenes perillyl alcohol and (+)-perillic acid methyl ester efficiently inhibited cell growth at 1 mM, but did not induce morphological reversion of 12-H-ras-transformed cells and did not decrease MAP kinase enzyme activity or collagenase promoter activity in PANC-1 cells, making it unlikely that they inhibit Ras function.

More studies can be quoted indicating that monoterpenes do not inhibit prenylation of Ras, but affect Ras more indirectly. However, Stayrook *et al.* (1998) come to the conclusion that perillyl alcohol does inhibit farnesylation of H-ras in pancreatic tumour cells and inhibits MAP kinase phosphorylation in H-ras. In pancreatic tumour cells carrying a K-ras oncogene there was decreased Ras farnesylation, but not sufficient to affect Ras GTP/GDP ratios or MAP kinase phosphorylation. Thus, the degree of inhibition of Ras prenylation may be dependent on experimental conditions, showing that *in vivo* experiments are crucial before Phase I studies on humans or domestics make sense. In addition, Hardcastle *et al.* (1999) conclude that these terpenes do inhibit FTase (see 3.5.3). Limonene and perillyl alcohol are now clinically evaluated in cancer patients (see 4.1.5 and 7.1.8).

Whether or not a particular monoterpene will inhibit Ras proteins would seem to depend on the type of tumour (Gille and Downward, 1999; Okada *et al.*, 1999; Ren *et al.*, 1997; Rubio *et al.*, 1999). One variable in this process may be the way Ras proteins travel from the endoplasmatic reticulum (ER) to the plasma membrane. Farnesylated Ras proteins first associate with ER and than with the Golgi apparatus, at least H-ras does. K-ras could pass the Golgi in a shortcut (Magee and Marshall, 1999). Fig. 4.8 represents our view of the possible sites of interference of terpenoids with signal transduction in cells with limonene as an example.

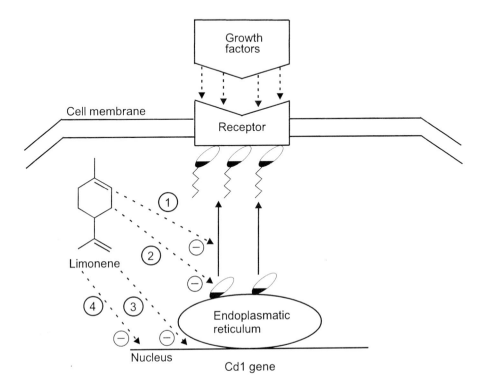

① = Inhibition and modification of activated Ras-proteins
② = Inhibition of farnesyl and geranylgeranyl transferase. Inhibition of Ras-protein prenylation
③ = Inhibition of cell cycle progression by inhibition of the cyclin D1 gene
④ = Inhibition of cell growth and proliferation by interfering with signal transduction
⟶ = biosynthesis; biochemical processes
- - - ▶ = regulation

Fig. 4.8. Effects of limonene on Ras.

A detailed discussion of these (and more) studies would take too many pages. The picture arises that when Ras protein prenylation is only partly inhibited, the effect on apoptosis is difficult to predict. Multiple Ras effector pathways contribute to G1 cell cycle progression, so it depends on how effectively a terpene can rule out an essential effector pathway.

The role of Ras in malignant development is not necessarily based on mutations. In human mammary cancer overexpression of normal c-Ha-ras is commonly observed. Steroid hormones can stimulate Ras gene transcription. Pethe and Shekhar (1999) demonstrated an interaction of c-Ha-ras sequence with estrogen receptor proteins together with a stimulation of estrogen on c-Ha-ras gene transcription. The authors suggest that alteration in transcriptional regulation of c-Ha-ras gene by estrogen may be an important factor in mammary cancer progression.

Ras proteins are not only important in case of malignancies. Normal, unmutated Ras proteins contribute to the life span of many cell types. In sympathic neurones Ras suppresses the p53-mediated pro-apoptotic pathway. It probably does so by activating signalling cascades, which suppress a p53 pathway (Mazzoni et al., 1999). In addition, Ras-activated MAPK that travels to the nucleus increased transcription of pro-survival genes. In plants, MAPK might be involved in transcription of pro-survival genes that become activated in different types of stress.

Normal Ras activity is needed for the growth and development of fungi. In *Colletotrichum trifolii* CT-ras is induced by a nutrient signal. In the absence of this stimulus, Ras is not induced, resulting in termination of growth and production of conidia. Truesdell et al. (1999) introduced a mutation in CT-ras by replacing a codon (GGT) by one with a single amino acid change. This mutated CT-ras induced abnormal, strong filamentous growth even under nutrient limitation, as well as defects in polarized growth with reduced differentiation. The same mutated CT-ras induced cellular transformation of mouse fibroblasts and tumour formation in mice! Cell polarity is important for many processes in eukaryotes. Truesdell et al. (1999) make a valuable remark in quoting previous work stating that hyphal growth shares elements with neurite extension in developing nervous systems. As discussed in section 2.1.2, fungi resemble more of vertebrates than of plants and could help to our understanding of the normal as well as pathological signals of the Ras cascade.

There is a continuous search for Ras FTase inhibitors other than terpenoids. Different classes of FTase inhibitors have been identified with little cytotoxic effects in normal tissues (Rowinsky et al., 1999). Theoretically they are beyond the scope of this book, but we should realize that combinations of terpenoid prenylase inhibitors with other FTase inhibitors may have additive if not synergistic anti-tumour potencies. This will be discussed in coming sections, a.o. 4.1.5. There is another argument for a broad spectrum of possible FTase inhibitors. The "bad" Ras proteins are mostly mutated Ras, and mutations may result in resistance to FTase

Chapter 4. Specific properties of terpenoids

inhibitors. Del Villar *et al.* (1999) reported on increased resistance of human FTase y361L to a tricyclic compound, acting on the essential CAAX group. This mutant exhibits affinity for a motif which is recognized by GGTase I. Such alterations in orientation are a threat to therapies based on prenylation inhibitors.

4.1.5 Cancer research

Sections 4.1.5 and 4.1.6 treat the effects of terpenoids on gene expression in malignant development of cell cycle progression and apoptosis, while 4.1.7 is dedicated to other effects of terpenoids, such as on the cell skeleton. Section 4.1.5 discusses the role of (exogenous) terpenoids on cells, which have already taken a malignant course of development, section 4.1.6 deals with redifferentiation and prevention. In papers on cancer medication and prevention the latter subject usually comes first. To our opinion the reader of this book is better prepared to follow the discussion of 4.1.6 after some information on the state of the art with respect to the effect of terpenoids on malignancies.

MONOTERPENES AND INTERMEDIATES

Documented effects of relatively simple terpenes on tumours are not of recent date only (Barton, 1959; Gould *et al.*, 1987; Moon *et al.*, 1983; Slaga *et al.*, 1982). Studies on the possible utilization of monoterpenes and isoprenoid intermediates against cancer are a mixture of cell and tissue culture, intact organisms, Phase I and II trials, dietary inclusions etc. The most often reported effects on cancer cells are inhibition of DNA synthesis, inhibition of prenyltransferases, suppression of HMG-CoA reductase, inhibition of phosphatidylcholine biosynthesis, decrease in cyclin D1 mRNA levels. Box 4.3 shows the activation of caspases with known and possible roles of terpenoids.

Evidence has been obtained that the effect of farnesol on the protein kinase C signalling system by inhibition of phosphatidylcholine synthesis, may be the cause of apoptosis in human acute leukemia CEM-C1 cells. Farnesol inhibited phosphatidylcholine-specific phospholipase C in CEM-C1 cells (Voziyan *et al.*, 1995). It decreased synthesis of diacylglycerol (DAG), an activator of protein kinase C (often referred to as PKC). The same research group demonstrated an inhibition of PKC activity in HeLa cells after incubation with farnesol. The authors mention an additional mechanism that can be stimulated by farnesol: degradation of HMG-CoA reductase, thus decreasing synthesis of farnesyl-PP, necessary for activation of Ras and other G-proteins, and for induction of cell proliferation (see also 4.1.4). It is possible that geranylgeraniol and farnesol also inhibit essential prenyltransferases of the cell systems. A more precise mode of action of farnesol on HeLa cells is its inhibition of CDP-choline:1,2-diacylglycerol choline phosphotransferase, leading to an accumulation of CDP choline and inhibition of phosphatidylcholine biosynthesis (Anthony *et al.*, 1999).

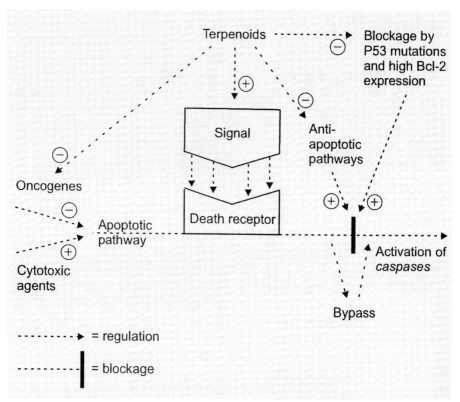

Box 4.3. Activation of caspases in tumour cells. Blockage of the caspase cascade by e.g. Bcl-2 activity can be overcome by terpenoids, creating a bypass resulting in apoptosis.

These facts may underlay the reported inhibition of pancreatic cancer growth by dietary farnesol and geraniol (Burke *et al.*, 1997). They demonstrated that these monoterpenes not only inhibit hamster pancreatic tumour growth *in vitro*, but also that they exert *in vivo* anti-tumour activity upon oral intake (20 g/kg diet). The specific inhibition of prenyltransferases could become a welcome tool in the treatment of pancreatic tumours that often have a bad prognosis.

The processes mentioned above are possibly under influence of the p53 gene and protein kinase C signalling systems. This suggests a relationship between these mechanisms and terpenes (Ohizumi et al., 1995; Voziyan et al., 1995). Farnesol inhibits also the growth of leukemia CEM-C1 cells (Melnykovych et al., 1992), and recently geranylgeraniol and farnesol appeared to induce apoptosis in A549 lung adenocarcinoma cells (Miquel et al., 1996).

PKC may not operate on its own. Protein kinase D (PKD) that is rapidly activated by tumour-promoting phorbol esters, is probably activated in living cells through a PKC-dependent signal transduction pathway. PKC inhibitors such as GF I completely abrogate PKD activation (Zuzaga et al., 1996).

The effects of monoterpenes on Ras proteins have been included in section 4.1.4. Limonene (Fig. 2.7) stops cells from installing Ras in their outer membranes. This breaks the flow of commands through Ras. Limonene interferes with the Ras cascade by preventing Ras proteins to influence the cells by reaching the cell membrane. Unfarnesylated Ras proteins do not associate with the plasma membrane and are incapable of cellular transformation; limonene inhibits the post-translational isoprenylation of p21 ras and other small G-proteins (Fig. 4.8). It should be noted that lymphocyte proliferation can be inhibited by D-limonene, making a sustained application of limonene in therapies at least complicated (regular check of lymphocyte levels and activity (Schulz et al., 1994)). The same type of interference can prevent other proteins to approach the cell membrane.

Perillyl alcohol is one of the most potent inhibitors of FTase and GGTase (Table 4.1), but its effect depends on the cell type. Ren et al. (1997) found perillyl alcohol to inhibit the *in vivo* prenylation of specific proteins in NIH3T3 cells by GGTase but not FTase. The inhibition of post-translational isoprenylation of p21 ras and other small G-proteins by monoterpenes has received considerable attention. Oral perillyl alcohol therapy (200 mg/kg/day) reduced vein graft intimal hyperplasia in rabbits (Fulton et al., 1997). For humans we estimate a maximal daily dose of 4800 mg/m^2.

The sensitivity of cell types, which seems to be a drawback of therapies based on monoterpenes alone, needs further investigation. Bardon et al. (1998) compared the effects of limonene, perillyl alcohol and perillic acid on cell cycle progression of human breast cancer lines. Perillyl alcohol was the most potent inhibitor of cyclin D1 gene expression and of cell growth. There was a distinct following order of sensitivity of the cell lines to the terpenes although visual effects were the same: growth inhibition was associated with accumulation of cells in the G1 phase with reduced transition to the S phase. The difference in sensitivity of the cell lines was independent of the enantiomeric composition of limonene and perillyl alcohol.

Table 4.1. Terpenoids with anti-tumour effects with known interference of signal transduction in tumour cells. PKC = protein kinase C; RAR = retinoid receptor

Terpenoid	PKC	CD1 gene	p21	p53	RAR	protein prenyl.	any caspase
ecdysone				X			
farnesol	X			X		X	
geraniol							X
geranylgeraniol				X		X	
geranyl geranoic acid							X
geranylgeranyl acetone							X
limonene		X	X	X		X	
myrcene				X		X	
perillic acid		X		X		X	
perillyl alcohol		X		X		X	
ginsenoside							X
gossypol		X					
ochraceolides	X					X	
retinoids				X	X		
vitamin D				X			
vitamin K2							X

Ohizumi et al. (1995) found geranylgeraniol to be a potent inducer of apoptosis in human tumour cell lines, including myeloid leukemia (K562), lymphoblastic leukemia (Molt3) and colon adenocarcinoma (COLO320 DM). These authors supposed that the action of farnesol and geranylgeraniol as apoptosis-inducing activity on human leukemia cell lines was triggered by inhibition of DNA synthesis. The main effect was DNA fragmentation with condensation and fragmentation of the cell nuclei. The authors suggest that inhibition of DNA synthesis triggers the induction of apoptosis. As prenylation of proteins is an essential step in signal transduction systems, the question arises if geranylgeranyl-PP is involved in the activity of tumour suppresser genes by prenylation of cell death-commanding proteins. To our opinion the synthesis of new proteins is not necessarily essential to activate cell death signal transduction systems. This view is illustrated by the fact that Ohizumi et al. (1995) also found that cycloheximide, an inhibitor of protein synthesis had no effect on the induction of apoptosis by geranylgeraniol in HL-60 cells.

Okada et al. (1999) in explaining the effect of geranylgeranyl acetone on human leukemia HL-60 cells stress that prior to DNA fragmentation, caspase-3(-like) proteases were markedly activated, and that this process might be responsible for the observed apoptosis. In addition, treatment with geranylgeranyl acetone interfered with processing and membrane localization of Rap1 and Ras and this is supposed to result from apoptosis.

As mentioned at the beginning of this section, monoterpenes may have additional targets (Table 4.1), among which is the nucleus. Monoterpenes can pass cell membranes in an easy way, which makes it understandable that moderate doses of limonene and its metabolites inhibit the expression of the cyclin D1 cycle-regulated gene quoted above (Bardon et al., 1998).

Retinoblastoma tumour-suppressor protein (Rb) is an essential G1-specific mediator that links Ras-dependent mitogenic signalling to cell cycle regulation. Inactivation of Ras in cycling cells caused a decline in cyclic D1 protein levels, accumulation of the hypophosphorylated, growth-suppressive form of Rb, and G1 arrest (Peeper et al., 1997).

Relatively few attempts have been made to utilize the penetrative properties of monoterpenes by combining them with other anti-tumour chemotherapeutics. The same is true for other biocides such as insecticides and herbicides (see chapter 6.2). Zhao and Singh (1998) could significantly increase the permeability coefficient of tamoxifen through porcine epidermis by combining it with eugenol and D-limonene in 50% ethanol. Partitioning of tamoxifen to the stratum corneum was markedly improved. Menthone enhanced the permeability of this drug by increasing extraction of the stratum corneum lipids.

Perillyl alcohol would seem to be rather selective with respect to carcinoma cells, leaving healthy cells in peace. In advanced mammary carcinomas cell cycle- and apoptosis-related genes were differentially expressed within 48 h of treatment. Although Bcl-2 and p53 remained unchanged, TGF-β-related genes were induced including their receptors, c-jun and c-fos were transiently induced within 12 h. Protein expression studies based on fluorescence immunohistochemistry together with confocal microscopy revealed co-localization of TGF-β1, the mannose 6-phosphate insulin-like growth factor II receptor, the TGF-β type I and II receptors *and* Smad2/Smad3 in epithelial cells (Arasi et al., 1999). Smad3 was induced during active carcinoma regression. The authors state that Smads may serve as biomarkers for anticancer activity. It is of importance to note that none of the perillyl alcohol-mediated anticancer activities were observed in healthy mammary tissues.

From these and other studies it would seem that metabolites of limonene are more potent than limonene itself. Especially S- and R-perillic acid and perillyl alcohol are effective inhibitors of protein prenylation with a major target in GGTase type I (Hardcastle et al., 1999). Monoterpenes that are less polar as perillyl alcohol (Fig. 7.4), such as limonene and myrcene, or more polar, such as perillic acid, diols or triols were less effective inhibitors of protein prenylation and cell proliferation. This may be due to the lower solubility in aqueous environments, such as the cytosol. Therefore, the *in vitro* effects (about 1 mM) may not be of the same magnitude, as those needed for an anti-tumour agent *in vivo*. Still, orally administered limonene (7.5% in the diet) or perillyl alcohol (2% in the diet) cause complete regression of more than 80% of rat mammary carcinomas with no apparent toxicity to normal host tissues. The effects, however, may be reversible because approximately half of the limonene-treated tumours reappeared upon withdrawal of the limonene treatment (Haag *et al.*, 1992).

To a biologist limonene (mostly D-limonene) is an exiting monoterpene with a range of functions in various organisms. It has bactericidal properties, it is an attractant for some, and a repellent to other insect species, it is produced by many plants (a.o. citrus fruits), it has anticancer effects, and it is well tolerated in cancer patients at doses of supposed clinical activity (Vigushin *et al.*, 1998). In the human body it is degraded into isoprenoids with their own anti-tumour action. In plants it can be converted to carvone (Fig. 2.8) that inhibits seed germination, a function that at first sight seems unrelated to the attraction of the aphid *Cavariella aegopodii* to caraway, one of its hosts. However, when we think of G-protein cascades, these facts are not so strange, and this is why we put it forward in this section. A closer look on the receptors for and impact of limonene metabolites on quite different organisms could greatly contribute to our understanding of their mode of action on cell cycle progression. This is also true for more complex terpenoids that have unexpected effects on mammalian cells, as well as on other organisms.

A synthetic geranylgeranoic acid (GGA) induced apoptotic cell death in a human hepatoma line (HuH-7), but not in mouse primary hepatocytes. This process was initiated by a rapid loss of the mitochondrial membrane potential in a medium with 10 µM GGA (Shidoji *et al.*, 1997). This does not occur with geranylgeraniol, indicating that a terminal carboxylic group of GGA is essential to get the membrane potential down. According to the authors, the derangement of the mitochondrion may be a primary but not a final target of GGA to induce apoptosis, and a subsequent activation of an ICE family protease cascade could be a prerequisite to GGA-induced apoptotic cell death. If only a massive derangement of the mitochondrial function occurs, cell death may result from mere necrosis due to depletion of cellular energy.

At present epithelial cancers of the lung (especially small-cell types), pancreas, colon and breast are difficult to treat with existing therapeutic modalities. Burke *et al.* (1997) found farnesol, geraniol and perillyl alcohol to suppress pancreatic adenocarcinomas of Syrian golden hamsters *in vivo*

Chapter 4. Specific properties of terpenoids

Table 4.2. Terpenoids with less defined modes of action against tumour cells. AG = angiogenesis; cytos. = cytoskeleton; CA = carcinogenes; HMGR = HMG-CoA reductase; diff. = (re)differention; ? = unknown

Terpenoid	AG	DNA	cytos.	G1 arrest	CA	HMGR	diff.	?
aphidicolin		X					X	
asprellic acids								X
betulinic acid derivatives								X
carotenes						X		
corosolic acid	X							X
curcumins							X	
dehydrothyrsiferol								X
farnesol		X	X	X		X		
geranylgeraniol		X	X	X		X		
geranylstilbenes								X
ginsenoides	X							
gossypol		X						
β-ionone				X				
kansuiphorin								X
limonene				X	X	X		
limonoids								X
menthol						X		
perillyl alcohol				X	X	X		
protolichesterinic acid		X						
oleanolic acid	X							
pinene						X		
remangilones								X
retinoids							X	
taraxasterol					X			
taraxerane					X			
taraxerol					X			
taxamairins			X					
taxol			X					
tingenone								X
tocotrienols							X	
ursolic acid	X							
vitamin K2					X		X	

with dietary treatments. The effects may be partly transcriptional and partly based on stimulation of apoptosis. The effective dose of farnesol and geraniol was 20 g/kg diet, without significant toxicity to the animals. This is another example of promising dietary supplements to be used in combination therapies (see Table 4.2). It is also a moment to remember the limonene-derived monoterpenes of Crowell *et al.* (1994) (see 4.1.1). Gould *et al.* (1987, 1990) found limonene among the anti-carcinogenic terpenoids in orange peel oil.

SESQUITERPENES, DITERPENES AND TRITERPENES

The low toxicity and the specific effects on apoptosis of monoterpenes are less common among more complex terpenoids, starting with sesquiterpenes and diterpenes. They tend to have a higher specificity and toxicity, and this holds for many organisms and not for mammalian cells alone. This does not mean that terpenoids from C-15 upward do not exert selective cytotoxicity. *Chiloscyphus rivularis* contains several sesquiterpenes. Of these, 12-hydroxychiloscyphone showed selective bioactivity in a yeast-based DNA-damaging assay and cytotoxicity to human lung carcinoma cells (Wu *et al.*, 1997). In an earlier study, Woerdenbag *et al.* (1987) found a dose-dependent growth reduction of fibrosarcoma FIO 26 in mice upon intraperitoneal injection (minimal dose 20 mg/kg). However, after an initial delay period, the tumour growth rate accelerated. Moreover, although these trials were done *in vivo*, FIO 26 is a non-metastasizing tumour, which has experimental advantages but is less realistic with respect to most human lung carcinomas. Metastases undoubtedly complicate the step from *in vitro* to *in vivo* studies, not to speak of Phase I and II trials. Metastasizing sub-populations of the original tumour may react differently (or not at all) to treatment with the same terpenoids, as is know for cytostatic drugs (see 7.1.8). We should realize that much of the research quoted in the present chapter is done with specific cell lines on culture media.

Several *neo*-clerodane diterpenes have physiological activities on mammalian tissues. Some are anti-inflammatory, some are cytotoxic and some combine a degree of cytotoxicity with anti-tumour effects. Their activities range from anti-microbial to plant and insect growth regulators. Hundreds of these compounds are mentioned by klein Gebbinck (1999) in his study of the synthesis of model compounds from natural clerodane insect antifeedants. A screening of clerodane diterpenes for medical purposes is time-consuming and is in its beginning. The wide range of effects of these compounds is a challenge to pharmacological research and should be stimulated. Labdane type diterpenes have promising effects on DNA synthesis of leukemic cell lines, but again there are great differences in activity already at the first screening.

A review by Lee (1999) describes research on cytotoxic natural plant substances. More than 100 new compounds and their synthetic analogues showed activity *in vitro* to human tumour cell lines and are of interest for

Chapter 4. Specific properties of terpenoids

further *in vivo* evaluation. Sesquiterpene lactones and triterpene glucosides are among these compounds. Triterpenes exhibit a wide range of pharmacological activities. Anti-tumour effects are often quite unexpectedly encountered in general screening programmes on ancient medicinal plants. Taraxasterol and taraxerol (Fig. 4.9) appeared to exert potent anti-tumour promoting activity in carcinogenesis tests on mouse skin and showed a noticeable inhibitory effect on spontaneous mammary tumours in mice (Takasaki *et al*., 1999). Taraxasterol is one of the triterpenes occurring in *Achillea millefolium*, an old herbal remedy of the Maritime Indians (Chandler *et al*., 1982). There are so many reports on triterpenes and triterpenic acids that it seems best to include some of this work in Tables 4.1 and 4.2 that give the main targets of the terpenes, rather than an attempt to discuss them all in this section.

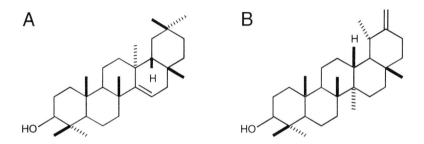

Fig. 4.9. The structures of **A**: taraxerol; **B**: taraxasterol.

A cell cycle arrest in G1 not unlike the effects caused by monoterpenes has been reported for triterpenic acids, such as ursolic acid. Ursolic acid is a potent inhibitor of the B16 mouse melanoma cell line (es-Saady *et al*., 1996). Defensive properties of ursolic acid (and of terpenes in general) may have been selected early in angiospermic development. Ursolic acid, isolated from *Crataegus pinnatifida* leaves inhibits chitin synthase II from *Saccharomyces cerevisiae* with an IC50 value of less than 0.9 µg/ml. Another triterpene acid, oleanolic acid, needed a concentration of 5.6 µg/ml under the same conditions (Jeong *et al*., 1999). Ursolic acid (most effective) and oleanolic acid inhibited tumour growth in *in vivo* effects of irradiation of a hematopoietic system in mice and enhanced post-irradiation recovery (Hsu *et al*., 1997).

Derivatives of natural steroid hormones e.g. of estrogens are frequently prescribed in cancer treatment. Their main targets are genes involved in the cell cycle as discussed in section 4.1.4. This book is on natural terpenoids, but as combinations of natural terpenoids with analogues or cytostatics are a promising development, they are discussed in section 7.1.8. Whereas tumour cells often exhibit insensitivity to anti-growth signals in the body, including hormones, those of other organisms still may induce a response in cancer cells. The insect hormone ecdysone induced p53 expression in the p53-null human lung cancer cell line H1299, resulting in irreversible cell growth arrest. In contrast to p53, p21 waf1 overexpression induced by ecdysone conferred increased resistance to the cytotoxic drugs etopoxide and cisplatin (Wang *et al.*, 1998). The authors conclude that the impact of p53 on susceptibility to chemotherapy may depend on the particular drug and type of DNA damage. Fig. 4.11 is a tentative scheme of terpenoid interaction with p53 mutation and expression.

Some triterpenes have been reported with a mode of action similar to that of monoterpenes such as perillyl alcohol, discussed above. Chloroform extracts of stems and bark of *Lophopetalum wallichii* were found to inhibit FTase. The extracts contained a number of known and new lupane-type triterpenes, of which ochraceolide A and ochraceolide B inhibited FTase with IC50 values of 1.0 and 0.7 µg/ml respectively (Sturm *et al.*, 1996). When evaluated against a panel of human cancer cells, ochraceolide A induced apoptosis, but ochraceolide B did not.

Terpenoids from ginseng such as ginseng saponins that have a dammarane skeleton can induce apoptosis in human cancer cells. Ginsenoside Rh2 induces apoptosis of human hepatoma SK-HEP-1 cells. The mode of action of this terpenoid is Bcl-2-insensitive activation of caspase-3 (Park *et al.*, 1997). Table 4.2 gives an overview of the effects of a selection of terpenoids on gene expression and the machinery of apoptosis. Natural terpenoids, which affect the cell cycle, can be found in unexpected sources, from marine organisms to newly discovered plant species. It is a hopeless task to present all the available references here; every week several interesting studies are published and at the moment the reader opens this book we are hundreds of possible references behind. Nevertheless, the material presented in this chapter may help to select the most apropriate type of terpenoid to introduce in a treatment against a particular type of cancer, especially in *combination therapies* as suggested in section 7.1.8.

We like to mention one other example: gossypol and its enantiomers. Gossypol (Fig. 4.10) is a well-known allomone from cotton, the synthesis of which is stimulated upon attack of a pathogen. In addition it is a defensive terpenoid against herbivorous insects. Probably its anti-microbial effect was selected earlier than that of an insecticide (Harrewijn, 1993). One could expect that allomones effective against a host of pathogens share an essential target of cell growth. The L-enantiomer of gossypol induced a dose-dependent mortality in a number of human tumour cell lines in a dose of about 20 µM. It was even more active than cisplatin, melphalan and

Chapter 4. Specific properties of terpenoids

dacabazine in two melanoma lines, than cisplatin and daunorubicin in a lung carcinoma line, and than hydroxyurea and busulphan in a leukemia line (Shelley et al., 1999). The authors mention DNA fragmentation as a major effect, a view that is supported by Ligueros et al. (1997) in their study of gossypol on human mammary cancer and fibrosarcoma cell lines. These authors found that the expression levels of Rb and cyclin D1 were decreased in a dose-dependent manner, resulting in inhibited DNA synthesis and G1/S block. An allomone, which attacks micro-organisms in their basic functions, seems to exert a similar effect on cancer cells.

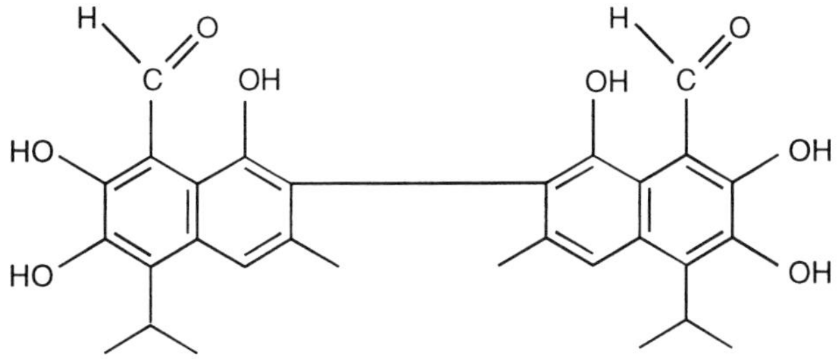

Fig. 4.10. The structure of gossypol, a plant allomone of mevalonic origin that has anti-tumour properties.

Retinoids, as we have seen in section 4.1.2, are involved in differentiation of many tissues and organs, processes that can develop pathologically when the concentration of the retinoids is not appropriate. As discussed in *VISION*, retinoic acid resembles JH of insects and JHs are sometimes included in studies of the effects of retinoids on tumours. Therapies based on retinoids have their drawbacks because of the potential overexpression of (mutated) p53 genes (Fig. 4.11) and possible side effects in tissues with rapid cell division such as in hematopoetic systems. However, although an overdose of vitamin A can stimulate skin cancer, metabolites of retinoic acid can counteract skin tumour promotion. Both retinoic acid and 5,6-epoxyretinoic acid inhibited the formation of induced mouse skin tumours based on the induction of ornithine decarboxylase (Verma et al., 1980).

Verma et al. (1980) could not establish an effect of JH3 on this type of decarboxylase. JHs could be kept in mind, however, in combination trials, as retinoids seem valuable terpenoids to include in combination therapies. As an example we mention the synergistic effects of 13-*cis*-retinoic acid and cisplatin in the treatment of squamous cell cancers, where the dose of retinoic acid could be kept ≤ 20 mg/day (Weisman et al., 1998). Trials have

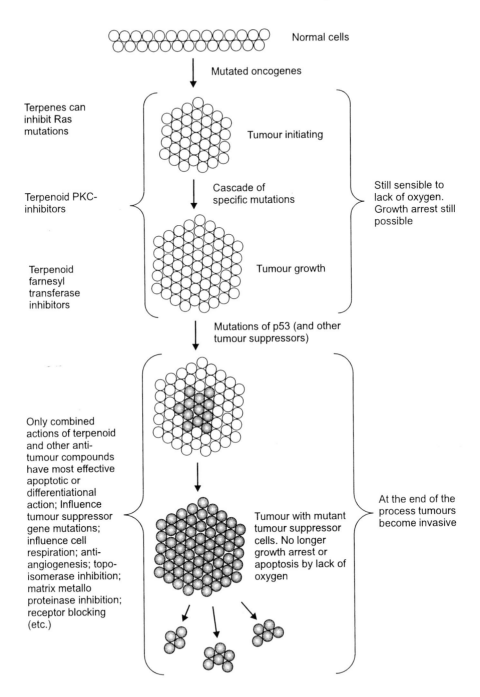

Fig. 4.11. Tentative scheme of terpenoid interaction with p53 mutation and expression.

Chapter 4. Specific properties of terpenoids

been done with combinations of retinoic acid and interferon-α on patients with metastatic renal cell carcinoma and advanced pancreatic adenocarcinoma, with promising results although no complete regression was observed in the advanced stages (Brembeck *et al.*, 1998). A combination of 13-*cis*-retinoic acid with interferon-α-2a resulted in a progression-free response in 7 out of 43 patients with renal tumours. Motzer *et al.* (1995) report that responding sites included bone metastases, a promising result although far from perfection.

In patients with acute promyelocytic leukemia a chromosomal translocation involving the RAR-α gene occurs. Smith *et al.* (1992) have made a review of clinical studies of retinoids with respect to their utilization in oncology. The major effects of retinoids seem to be on the RAR-α receptor, which might explain the high clinical response rate of promyelocytic leukemia to retinoic acid. In addition, positive responses have been observed with dermatologic malignancies including cutaneous T-cell malignancies and chronic juvenile myelogenous leukemia. Most interesting is that Smith *et al.* (1992) conclude that retinoids can induce differentiation in cell lines which exhibit properties of tumour cells. When cells, which have taken a malignant course of development, can be induced to differentiate normally, there is new hope for the treatment of cancer types for which eradication of tumour cells is not enough. This may not be a major problem for local skin cancers (except some melanomas), for carcinoma in the tail of the pancreas in which islet cells are involved, the situation is more severe, not to speak of types of leukemia. Even if all cancer cells can be eliminated, bone marrow transplantation is still essential. Retinoids are among the terpenoid messengers, which could shift the balance from stabilization of the disease (at best!) towards complete recovery. Many natural compounds have been found to be potentially valuable therapeutics in oncology, of which terpenoids comprise only one group. The question is: can terpenoids be only utilized to eradicate cancer cells, or do they have potencies to cure them?

4.1.6 "To kill or cure, ..."

As discussed in section 4.1.5, the opposite of inhibition of caspase activity, selective stimulation, but then in mammalian cells, could be a valuable therapy in addition to interference with the oncogene cascade. Given the high number of cell deaths in a healthy human adult (about 10^{11} per day), caspase activators should be selective for cancer cells. Probably the major site of action of terpenoids is inhibition of protein synthesis, indicating that enzymes are potential targets of terpenoids in programmed cell death cascades.

A different approach to treat malignancies would be to restore normal differentiation and to protect valuable cells to die prematurely. In fact, life or death of a cell is a matter of balance. The complex of Bcl-2, Bag-1, Ras and Raf-1 forms the yin side of a presumed yin-yang regulation of cell progression and death. Inactivation of Bcl-2, Raf-1 or Ras proteins in malignant cells can stimulate the yang side (apoptosis), but in many situations inactivation of these proteins alone may not lead to successful treatment of cancers. Not only should cells be arrested, ideally at least part of them should be cured and resume the normal course of development e.g. in hematopoetic systems. Terpenoids may assist in recovery of cancer patients by acting on:
- Inactivation of growth factors and triggering of pro-apoptotic proteins
- Differentiation and redifferentiation
- Stimulation of the autodefensive system
- Mevalonate depletion (in combination with toxicology profiles of regular chemotherapeutics)
- Inhibition of angiogenesis
- Interference with carcinogenic factors.

INACTIVATION OF GROWTH FACTORS AND TRIGGERING OF PRO-APOPTOTIC PROTEINS

Signalling cascades involved in apoptosis have been discussed in 4.1.5. Cancer cells often escape from the activity of effector caspases because their receptors do not become trimerized or dimerized (in the case of APAF-1), or because members of the Bcl-2 family block the release of cyt. c from the mitochondria. The pro-apoptotic proteins of the Bax group are related to those of Bcl-2, but in contrast to Bcl-2 members, insertion of a Bax member into the mitochondrial membrane induces the release of cyt. c and induces apoptosis. According to Pawlowski and Kraft (2000), the regulation of the pro-apoptotic Bax family is quite complex and may response to specific sets of multiple weak signals rather than to strong individual stimuli. One such factor is the increase in intracellular pH that allows the unfolding of Bax protein and insertion of its C-terminal hydrophobic domain into the mitochondrial membrane. An increase in pH can occur upon prolonged withdrawal of growth factor, which results in dephosphorylation of the N and C termini of the phosphorylated Bax molecule. This allows the two hydrophobic ends of Bax to conjugate, enabling insertion into the mitochondrial membrane. Terpenoids could interfere with the activity of members of the Bcl-2 family and the balance between Bcl-2 and Bax proteins (Figs 4.5 and 4.6). Members of the Bax family often lack amino acid domains present in those of the Bcl-2 family. It will be a long way before we know the effect of different terpenes on gene expression of all these proteins. The opposite of increased expression of Bax members (and caspase activity) may be desirable in other diseases such as Parkinson and Alzheimer, in which pathological activation of apoptosis is an object of study.

DIFFERENTIATION AND REDIFFERENTIATION

There may be a role for terpenoids in activity of Ras proteins that regulate cell growth as mentioned in 3.5. This is an almost open area for research, but it does not seem to be a hopeless enterprise. Differentiation therapy is already being conducted in some cases as a treatment for leukemia. All-*trans*-retinoic acid can be highly effective, but does not induce differentiation in all types of leukemia cells. Combination therapies could be more effective. Sakai *et al*. (1994) found that vitamin K2 (menaquinone) and not vitamin K1 (phylloquinone) is a potent inducer of differentiation of various human myeloid leukemia cell lines at different stages of maturation. The difarnesyl group of the side chain of vitamin K2 (Fig. 4.1) may be essential to these effects, as retinoic acid is constructed from isoprene units. In tissue cultures 1 μM was effective, suggesting that vitamin K2 could be safely used in combination therapies. A possible combination with vitamin K2 is geranylgeranyl acetone, which is used as an antiulcer agent. It induces differentiation of various human myeloid leukemia cell lines (Sakai *et al.*, 1993), as does vitamin K2. Doses and ratios should be experimentally established.

Metabolic labelling experiments have provided evidence that mammalian cells can utilize free geranylgeraniol (GG-OH) for protein prenylation and free farnesol (F-OH) for sterol biosynthesis. According to Crick *et al*. (1997) mammalian cells may be able of synthesizing farnesyl-PP and geranylgeranyl-PP by a "salvage pathway" in addition to the conventional *de novo* biosynthetic route via mevalonate. This salvage pathway provided an alternate route for the biosynthesis of physiological levels of GG-PP, and probably F-PP. To our opinion a possible salvage pathway should be investigated in mutated cell lines in case differentiation therapy with terpenoids is considered. Once GG-OH is produced, it depends on the specificity of alcohol/fatty aldehyde dehydrogenase system whether GGA is produced that may lead to apoptosis instead of differentiation, dependent on the cell type (see above). GG-OH and F-OH can induce actin cytoskeleton disorganisation and subsequently apoptosis in lung adenocarcinoma cells.

The question is if this mode of action remains *in vivo* in many types of lung carcinoma cells, as activation of protein kinase C prevents induction of apoptosis by GG-OH. As has been mentioned for the retinoids in 4.1.5, some terpenoids could have a double function, not only leading to tumour regression, but also to (re)differentiation.

A striking remark in a review by Gould (1997) on cancer therapy by monoterpenes is that they do not only inhibit prenylation of G-proteins but in addition up-regulate a gene in mammary carcinomas which degrades the mammary tumour mitogen IGF II and activate the cytostatic factor TGF-β. These and other alterations in gene expression lead not only to G1 arrest, followed by apoptosis, but in addition to *redifferentiation* and complete tumour regression with replacement of tumour cells by stromal elements.

One should be aware that redifferentiation of malignant cells does not necessarily make them to function as normal cells. However, it may cause them to listen to the signals, which initiate the apoptosis cascade. The chemopreventive effects of limonene and other monoterpenes during the initiation phase of e.g. mammary carcinogenesis could be due to the induction of Phase II carcinogen-metabolizing enzyme, resulting in carcinogen detoxification (Crowell, 1999).

Shi and Gould (1995) stress that limonene and perillyl alcohol have been shown to induce complete regression of rat mammary carcinomas due to effects on cytostasis and differentiation. They tested the ability of perillyl alcohol to induce differentiation in a well-characterized neuroblastoma cell model and found it to be a potent inducer of the neuroblastoma-derived cell line Neuro-2A. Development into normal cells was not based on cytostatic effects or inhibition of ubiquinone synthesis (compare with 3.4.1).

The rhizome of *Curcuma domestica* (c.f. *C. longa*) (Zingiberaceae) is a basic ingredient of curry powders. It contains curcumins, phenolic deketonics that apart from cholinergic effects, induce differentiation in promyelocytic leukemia cells, in particular in combination with low levels of vitamin D3 (Sokoloski and Sartorelli, 1998).

STIMULATION OF THE AUTODEFENSIVE SYSTEM

Hsu *et al.* (1997) report that the triterpene acid ursolic acid not only exerts *in vivo* anti-tumour activity upon irradiation, but also increased immunocompetence of leukocytes. The authors suggest that ursolic acid (and oleanolic acid) could help to reduce radiation damage to the hematopoetic tissue after radiotherapy.

As mentioned above, vitamin K2 and its derivatives enhance the effect of all-*trans*-retinoic acid (Yaguchi *et al.*, 1997). To a certain extent there is some form of autoprotection against leukemia, as vitamin K2 is produced by the human intestinal microflora (Conly *et al.*, 1994). Although pathotypes of *Escherichia coli* exist, the useful ones, which are intestinal symbionts, synthesize their own vitamin K2. This vitamin is the major electron carrier during anaerobic growth with various electron acceptors. The *men* A gene encodes the enzyme DHNA-octaprenyltransferase and its location to a 2.0-kb region of the *E. coli* genome have been established (Smitt *et al.*, 1998).

These facts underline the importance of an active intestinal flora, which, as every experienced physician knows, is dependent on a balanced diet, preferably rich in dextrarotating lactic acid. Chapter 7 gives dietary advises for this purpose. Destruction of the intestinal flora with antibiotics, although their use can be imperative, can disrupt the discussed form of autoprotection. Measures to rapidly restore this flora after treatment with antibiotics, as is common practice in France, should be recommended.

Prospects of a completely new method to stimulate the immune system come from the work of Jomaa et al. (1999). This group of the University of Würzburg in Germany has discovered that IPP, the important precursor of isoprenoid biosynthesis, activates Vγ9/Vδ2T cells, the major subset of human $\gamma\delta$T cells. IPP is produced in the symbiotic *E. coli* of humans. The authors found that DOXP addition potentiates the ability of *E. coli* extracts to activate the T cells. As is discussed in section 2.1.1, many microorganisms with their MAI metabolism produce DOXP and 2C-methyl-D-erythritol-4-phosphate. It is of great importance to determine the amounts of these MAI metabolites in other intestinal bacteria such as *Bifidobacterium* and *Lactobacillus* species.

As we have discussed before IPP can activate T cells of the immune system in mammals. In principle the mechanism of signal transduction and the coupling of the receptor to affector protein, is the same in yeast and mammalian cells. However, yeast and mammalian cell systems do not always use Ras proteins, but also protein kinase C (PKC). The mitogen-activated protein kinase cascade (MAP kinase) is responsible for bringing growth and differentiation signals into the cell nucleus. This is the case in many eukaryotic cell systems from yeast to man (Egan and Weinberg, 1993). Protein kinase A is also known to be related to cyclic AMP and adenylyl cyclase. PKC is one of the most important signal transduction systems in all kind of cell systems (Box 4.4). It is activated by the receptor-mediated hydrolysis of inositolphospholipids (Nishizuka, 1986). It is involved in the transport of extra-cellular messengers across cell membranes. This regulates many Ca^{2+}-dependent processes. Diacylglycerol can activate the enzyme, but the exact mechanism of activation of PKC is subject to continued research. PKC is supposed to be the receptor of tumour promoters such as phorbolesters and other diterpenes.

Proto-oncogenes such as c-fos and c-myc are targets of PKC. The enzyme can phosphorylate membrane proteins. It regulates the feedback control of cell surface receptors (down-regulation). Terpenoids can inhibit the enzyme. The biochemical mechanism of the negative feedback control system of PKC exercised on cell surface receptors needs further attention. A more detailed insight in this system is of great importance. Recently it was found

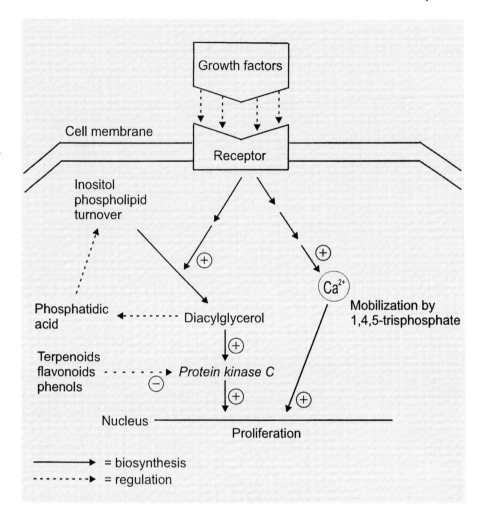

Box 4.4. Scheme of the protein kinase C signal transduction pathway.

that also G-proteins play roles in these systems (Liu and Simon, 1996). G-protein βγ-subunits can also regulate Ca^{2+} channels in the cyclic AMP system.

MEVALONATE DEPLETION

An effect of monoterpenes additional to those mentioned in the first part of this section is their suppression of HMG-CoA reductase. Complete inhibition leads to cell failure, so for the continued health of normal tissues one has to be careful with combinations of monoterpenes which have potent effects on this key enzyme. Mo and Elson (1999) surprisingly found that the growth of human mammary adenocarcinoma (MCF-7), human leukemic (HL-60) cells and murine B16(F10) melanoma cells was highly suppressed with β-ionone, a pure isoprenoid. The isoprenoid did not affect p53 functions. Another mixed isoprenoid, γ-tocotrienol, exerted similar effects. Both compounds initiated apoptosis and arrested cells in the G1 phase and suppressed HMG-CoA reductase. Both β-ionone and lovastatin interfered with the post-translational processing of lamin B, an activity essential to assembly of daughter nuclei. Mo and Elson (1999) suggest upon these (and other) results that the isoprenoids and lovastatin have additive and potentially synergistic inhibiting effects on tumour cell proliferation and initiation of apoptosis. A separate discussion of suppression of HMG-CoA reductase in cancer treatment is in section 4.1.8.

INHIBITION OF ANGIOGENESIS

A role of terpenoids in angiogenesis has not yet been discussed in this chapter. Vascularization of tumours is an important aspect of malignancies; when blood vessels grow into tumours their growth is not limited by lack of oxygen and nutrients. Moreover, metastasizing cells can leave the tumour and travel along blood vessels, a reason for intensive search for anti-angiogenic drugs. Ursolic acid and oleanolic acid inhibited angiogenesis in the chicken embryo chorioallantoic membrane, and the proliferation of bovine aortic endothelial cells. Again ursolic acid was 4-8 times more effective than oleanolic acid (Sohn *et al.*, 1995). The invasion of endothelial lung cells into the reconstituted basement membrane (the matrigel) is considered to be an essential event in tumour neovascularization. In rats, this process was inhibited by ginsenoside-Rb2 in a dose-dependent fashion. Intravenous administration of ginsenoside-Rb2 resulted in a remarkable reduction in the number of vessels oriented towards the tumour mass, although a significant inhibition of tumour growth was not achieved (Sato *et al.*, 1994). In contrast, intra-tumoral and oral administration of the terpenoid inhibited tumour growth and neovascularization. The authors discuss the contribution of ginsenoside-Rb2 in the inhibition of lung tumour metastasis.

INTERFERENCE WITH CARCINOGENIC FACTORS

A study of Giri *et al.* (1999) revealed that D-limonene has chemopreventive effects against hepatocarcinogenesis induced with N-nitrosodiethylamine and phenobarbital. The carcinogenic compounds increased transcription of c-jun and c-myc in nitrosodiethylamine-induced carcinogenesis. D-limonene completely inhibited such overexpression. Crowell (1999) in her review of monoterpenes in prevention and therapy of cancer mentions that the chemopreventive effect of limonene and other monoterpenes during the initiation phase of mammary carcinogenesis are likely based on the induction of Phase II carcinogen-metabolizing enzymes, resulting in carcinogen detoxification. The various effects of terpenes mentioned in section 4.1.6 could be additive, and, according to their mode of action, seldomly interfering in a negative way.

Perillyl alcohol is not only efficacious against progression of tumours, but also against their formation. Topical application of perillyl alcohol (10 mM) inhibited UVB-induced activator protein 1 transactivation in mouse skin and human keratinocytes. At this dose there was no apparent toxicity, suggesting that perillyl alcohol might be applied for chemoprevention of skin cancer (Barthelman *et al.*, 1998).

Valuable cancer chemopreventives may be developed from the roots of *Taraxacum japonicum* (Asteraceae). These contain the triterpenoids taraxasterol and taraxerol (Fig. 4.9), that exhibited potent anti-tumour-promoting activity in carcinogenesis tests of mouse skin (induced by a chemical initiator and a promoter). In addition they had an inhibitory effect on mouse spontaneous mammary tumours (Takasaki *et al.*, 1999), indicating that terpenoids with anti-tumour effects can be potentially anti-carcinogenic.

4.1.7   Terpenoids with different effects on cancer cells

Animal cells grow, multiply and function in an environment that is loaded with messages from hormones, cytokines and a host of other molecules in addition to growth factors. These messages need intermediates to be transmitted to the cell's interior. Moreover, the multiple signal cascades must be coordinated to result in an effective biological "cross-talk". Besides signalling pathways gating pathways exist, that have a function in regulation of cell systems. Both types of pathways are known to operate with cyclic AMP systems. The transmitted pathway receives information at the cell surface. A gating pathway can regulate information flow at any point of the transmitted pathway and can be activated or deactivated by cellular signals (Iyengar, 1996) (Boxes 4.5 and 4.6). This may explain why cyclic AMP stimulates cell growth in some cells, while inhibiting it in others. Work from several research groups demonstrates that any factor that increases cyclic AMP concentrations in the cells they are studying inhibits transmission of growth stimulatory signals through the Ras pathway. Such factors can be hormones that increase cyclic AMP concentrations or inhibitors that prevent its breakdown.

More complicated terpenoids may have more specialized properties in affecting specific cell systems and organisms. As an example may serve the action of the diterpenoids taxol (Fig. 4.12) and aphidicolin (Fig. 4.13), and limonoids. They cannot only inhibit cell growth, but in special cases they

Chapter 4. Specific properties of terpenoids

also induce differentiation. This is because they activate Ras genes that do not only provoke cell division, but also have a role in differentiation.

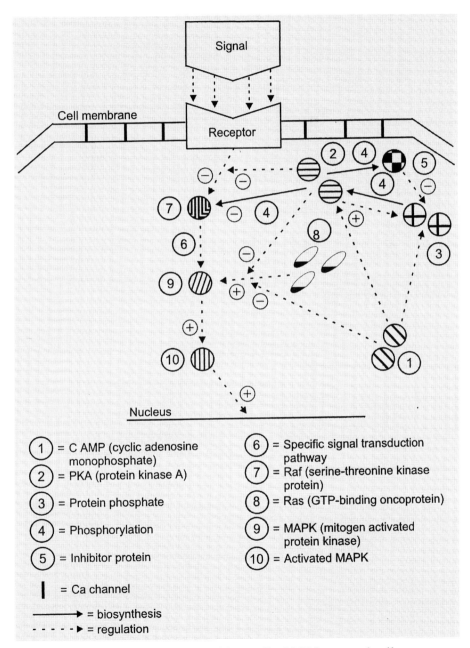

Box 4.5. Gating processes regulated by cyclic AMP in several cell systems.

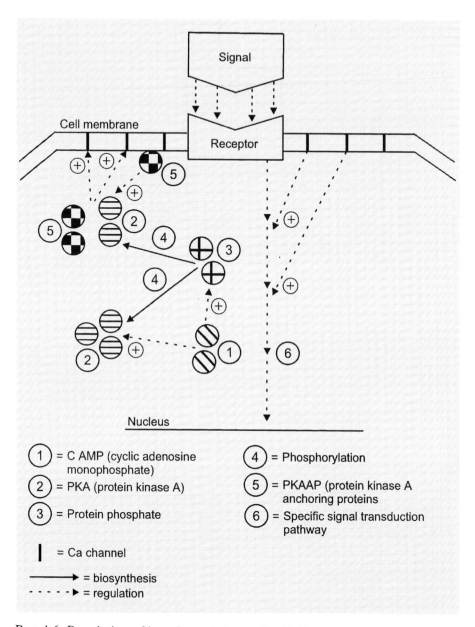

Box 4.6. Regulation of ion-channels by cyclic AMP via protein kinase A. Activation of Ras genes can even stop cell division and result in more specialized cell types.

On administering terpenoids to complete organisms the overall effect has to be considered. Not only the key enzyme HMG-CoA reductase can be affected, but also various other enzyme systems. Cyp450 can be stimulated and glutathion S-transferase can be raised (Aruna and Siuaramakrishnan, 1990). Such events can activate defensive mechanisms of cell systems. Compounds with a guaiazulene-type of structure, such as chamazulene (Fig. 2.10) and guaiazulene act as anti-tumour agents in mice as already shown by Barton in 1959, Graffi et al. in 1959 and Kraul and Schmidt in 1957. The mode of action appears to be based on inhibition of the respiration of the tumour cells. The terpenoids inhibited the succinyldehydrogenase activity in the cells and are relatively non-toxic to normal cells.

The utilization of lichen species is an old tradition in medicine. The triterpenic acid protolichesterinic acid has inhibitory effects on arachidonate 5-lipoxygenase. Ogmundsdottir et al. (1998) found protolichesterinic acid isolated from *Cetraria islandica* to inhibit DNA synthesis in three malignant human cell lines at concentrations above 20 µg/ml. The authors (from the Icelandic Cancer Society in Reykjavik) relate the induced cell death of the cancer cells to the 5-lipoxygenase inhibitory activity of the triterpenic acid. This lichen species (Iceland moss) has been imported into the Netherlands via the whale-hunting route of the Dutch, and the famous physician Hermanus Boerhaave (1668-1738) successfully prescribed extracts of this lichen to treat degenerative ulcers of the duodenum.

Taxol (Fig. 4.12) is an anticancer agent that can be isolated from *Taxus baccata* and from the stem of *Taxus brevifolii*, exerting activity against a broad range of tumours (Wani et al., 1971). Taxol can bind to cellular microtubules (Manfredi et al., 1982) and acts as a mitotic spindle poison, a mode of action already elucidated by Fuchs and Johnson in 1978. Taxol is unique in its inhibition of tubulin, a protein involved in mitosis (Wall, 1998). Preparations of taxol, registered as Taxotere® and Paclitaxel® are prescribed against breast and ovarian cancer. One should know that effective doses can cause severe irritation of the mucosa of the digestive tract, beginning with the mouth (Houtsmuller, pers. comm.).

Ohta et al. (1994) have established a taxol-resistant human small-cell lung cancer cell line by exposing H69 cells to stepwise increases in taxol concentration. The resistance of H69/Txl cells to taxol was 4.7-fold that of the original H69 cells, estimated with the tetrazolium dye method. Removal of the drug from the medium resulted in a 38% decrease in the growth rate of H69/Txl as compared with that in the presence of 330 nM taxol, suggesting that the taxol-resistant cell line was partially dependent on taxol! However, H69/Txl showed higher sensitivity to *Vinca* alkaloids such as vindesine, vincristine and vinblastine than the parental H69. An increased acetylation of α-tubulin was observed in H69/Txl cells compared with H69 cells. The acetylation of α-tubulin may be responsible for the taxol resistance and/or taxol-dependent growth of H69/Txl. Other members of the Taxaceae produce comparable terpenoids, some of which have anti-tumour activity, such as taxamairins from *Taxus mairei*.

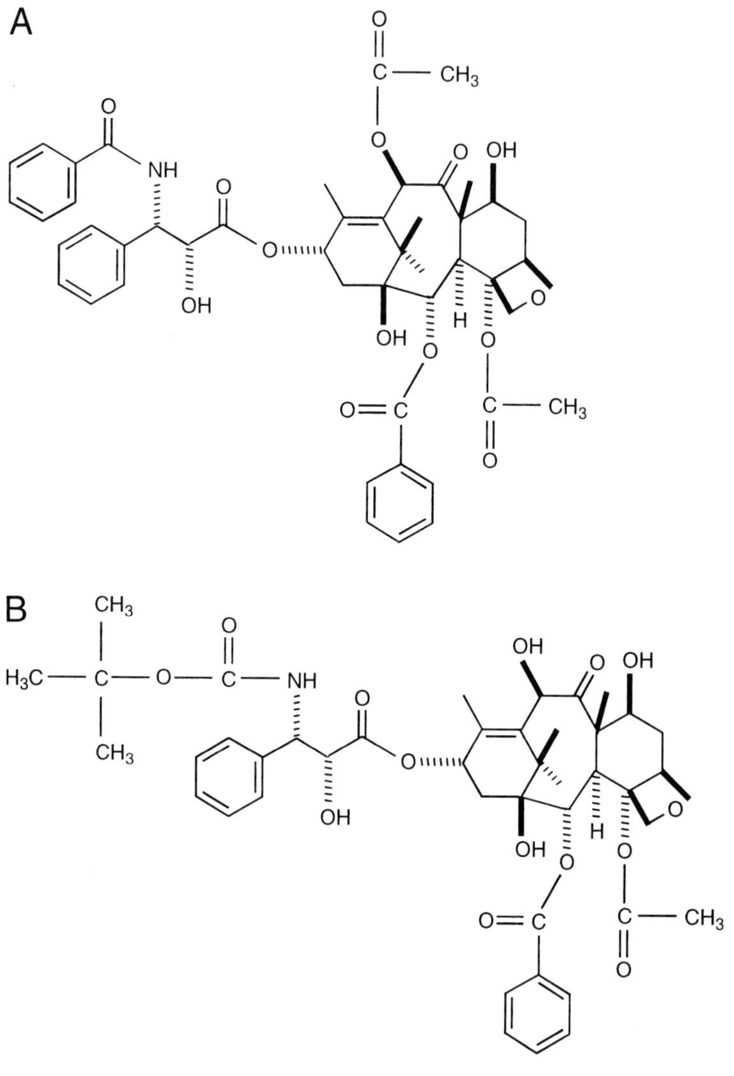

Fig. 4.12. Structures of (**A**) taxol and (**B**) taxotere.

Taxol can markedly increase the effectivity of cytotoxic drugs. The non-thiol-containing FTI-2148 is a highly cytotoxic agent for FTase (IC50: 1.4

Chapter 4. Specific properties of terpenoids

nM) and GGTI-2154 is selective for GGTase (21 nM). Combination with taxol resulted in enhanced anti-tumour activity (Sun et al., 1999). Continued research on the combination of taxol with cytostatics or apoptosis-stimulating agents is recommended, and this holds for many therapies in oncology.

Corosolic acid isolated from fruits of *Crataegus pinnatifida* var. *psilosa* (Rosaceae) displays potent cytotoxic activity against several human cancer cell lines. According to Ahn et al. (1998) the cytotoxicity is comparable with that of ursolic acid (discussed in 4.1.5), although a different mode of action cannot be excluded. The terpenoid can suppress PKC activity *in vitro*. Cytotoxic terpenoids do not necessarily act via the caspase cascade. Oncogenetically mutated Ras gene induces cellular degeneration accompanied by cytoplasmic vacuoles in human glioma and gastric cancer cell lines. The cytoplasmic vacuoles are mainly derived from lysosomes. This is an oncogenic Ras-induced cell death, which occurs in the absence of caspase activation, and is not inhibited by overexpression of anti-apoptotic Bcl-2 protein. Mutation of Ras genes might activate a cell death programme that cannot be overcome by a common anti-apoptosis mechanism (Chi et al., 1999). The authors suggest that studies of non-apoptotic cell death programmes could help to find genetically regulated cell suicide programmes that cannot be counteracted by Bcl-2, Bag-1, Ras or Raf-1 anti-apoptotic mechanisms.

Receptors for the steroid hormone vitamin D are expressed in many types of malignant cells. Vitamin D analogues can suppress the development and inhibit growth of human and animal tumours, including colon and prostate cancer. One of the drawbacks of high doses of vitamin D (calcitriol) is the induction of hypercalcemia. Smith et al. (1999) conducted a Phase I trial with doses of 2-10 µg calcitriol in patients with advanced solid tumours. Although hypercalciuria was seen in all patients, only those of the 10 µg regime developed hypercalcemia. This study indicates that 2-5 µg calcitriol can be administered with acceptable risk of toxicity. Again, calcitriol may be most effective in combination with other anti-tumour compounds. Einhorn et al. (1994) included calcitriol in their existing combination therapy of vinblastine, ifosfamide and gallium nitrate (coded: VIG). The combination was very effective against metastatic urothelial carcinoma, but seemed to be contra-indicated by its toxic side effects, particularly the development of granulocytopenia. To our opinion granulocytopenia can be regarded to be a response to the VIG contribuants, probably not to calcitriol. Therefore we suggest trials in which calcitriol be included in a higher dose (3-5 µg/day) in VIG-based chemotherapy, in which VIG is reduced by 20-50%.

4.1.8 Impairment of mevalonic acid synthesis in tumours

Inhibition of respiration of tumour cells or reduction of high turnover of metabolites needed for proliferation of malignant cells may be considered when apoptosis or cure of oncogene expression is not (yet) fully effective. An example is chronic lymphocytic leukemia (CLL) that mostly affects elderly people in which B-lymphocytes progressively proliferate. The CLL cells have an increased demand for cholesterol (membrane synthesis). The requirements are met by high HMG-CoA reductase activity together with receptor-mediated uptake of low-density lipoprotein (LDL), which is the major cholesterol-carrying lipoprotein in human plasma. The increased uptake of LDL is a dangerous situation, as LDL, once bound to a specific LDL receptor, is degraded in lysosomes, and releases its cholesterol that becomes available to the CLL cell. In addition, Siperstein stressed as early as in 1984 that impairment of negative cholesterol feedback is already present in the precancerous state of developing tumours. Therefore, pharmacological reduction of cholesterol levels in leukemic cells could help to prevent proliferation of malignant B-lymphocytes (see also chapter 7). Vitols *et al.*, (1997) tested the effect of the cholesterol synthesis inhibitors lovastatin (mevinolin), simvastatin and pravastatin on cellular thymidin uptake of CLL cells and immunoblastic B-cell lymphoma cells. Pravastatin was effective in lipoprotein-deficient serum only, indicating that the leukemic cells still could obtain cholesterol from LDL. Simvastatin, however, was a potent inhibitor of thymidine uptake in all culture conditions (20 µg/ml).

The possible effect of simvastatin on progression of CLL was tested in patients with a wide range of initial plasma cholesterol, each receiving 40 mg simvastatin daily for 12 weeks. Compared with culture medium conditions, this is probably not an optimal dose. No patient showed progressive symptoms during treatment; four patients developed progressive CLL upon cessation of simvastatin intake. Reduction of HMG-CoA reductase activity could be a valuable addition to combination therapies, including (in the case of CLL) e.g. chlorambucil, fludarabine and terpenoids (see chapters 6 and 7). Extended studies on tumour cell systems are essential, as there are indications that inhibition of HMG-CoA reductase can even *stimulate* DNA synthesis *in vitro*! Hohl *et al.* (1991) found a paradoxical increase in DNA synthesis in myeloid leukemia cells upon inhibition of HMG-CoA reductase.

Monoterpenes could both deplete isoprenoids by HMG-CoA reductase inhibition, and prevent prenylation of proteins such as derived from Ras and Ras-related oncogene families. In this way dietary inclusion of monoterpenes could alter malignant cell proliferation by acting on different processes in cell proliferation and differentiation.

Chapter 4. Specific properties of terpenoids

According to Bassa *et al.* (1999) the major effect of HMG-CoA reductase inactivation is inhibition of farnesylation and plasma anchorage of Ras protein. Such a type of inhibition is prevented by addition of MVA and F-PP, but not by cholesterol or low-density lipoprotein. Bifulco *et al.* (1999) come to a similar conclusion.

Elson *et al.* (1999) have made a review of the relative importance of HMG-CoA reductase in tumour cells. They arrived at the important conclusion that HMG-CoA reductase in malignant cells is often resistant to the endproduct feedback system operating in normal cells. As a consequence a pool of sterologenic pathway intermediates becomes available, intermediates that provide lipophilic anchors for membrane attachment and activity of growth hormone receptors, nuclear lamins A and B, and oncogenic Ras. Whether or not HMG-CoA reductase in tumour cells is less sensitive to normal feedback regulation, its abnormally high activity has been reported before. Bennis *et al.* (1993) measured a higher activity of this key enzyme in an A549 tumour cell line than in normal human fibroblasts. However, HMG-CoA reductase in tumour cells retains a high sensitivity to non-sterol isoprenoids. The available levels of a particular terpenoid in a diet are usually not enough to have a sufficient impact on tumour growth. According to Elson *et al.* (1999) the lowered cancer risk associated with diets rich in secondary plant metabolites could be due to combined actions of these compounds. This is exactly why we are strongly in favour of combination therapies (see 7.1.8).

The additive and potentially synergistic effects of dietary terpenoids on suppression of tumour cell proliferation and initiation of apoptosis could be based for a considerable part on their suppression of HMG-CoA reductase, independent of mutated Ras and p53 functions (Mo and Elson, 1999). To our opinion, different effects of terpenoids on cancer cells are important just as well. Moreover, where some terpenoids suppress HMG-CoA reductase, others can up-regulate its activity in a way insensitive to lovastatin, although they still may affect protein prenylation (Cerda *et al.*, 1999). Broitman *et al.* (1996) obtained similar effects of up-regulation of HMG-CoA reductase in tumours of the colonic epithelium by perillyl alcohol, but conclude that although the individual action of perillyl alcohol and limonene appear antagonistic *in vitro*, their overall effect, even if combined with HMG-CoA reductase inhibitors appears promising for the possible management of hepatic metastases from colon cancer.

An effective suppression of HMG-CoA reductase, useful in treating hepatoma, may be obtained with 25-hydroxycholesterol. Rat AH 136B ascites hepatoma cells treated with this sterol (4 µg/ml) responded with an irreversible entry into the sub-G1 phase. Complete inhibition of tumour development was attained with a combination of 4 µg/ml 25-hydroxycholesterol and 1 μg/ml all-*trans*-retinoic acid (ATRA) (Yokoyama *et al.*, 1999).

Treatment with 25-hydroxycholesterol efficiently blocked the proliferation of both human normal mammary epithelial cells and breast cancer cells. Upon removal of the inhibitor HMG-CoA reductase increased, followed by a rapid initiation of DNA synthesis in the tumour cells. According to Cerda *et al.* (1995) colon adenocarcinoma cell lines retain their capacity for repression by steroids, in particular 25-hydroxycholesterol, and are sensitive to lovastatin, which makes a combination therapy feasible.

Other natural terpenoids that could be effective in post-transcriptional suppression of HMG-CoA reductase are the tocotrienols, farnesylated analogues of tocopherols. This means that dietary vitamin E analogues may become a valuable addition to other treatments, aimed to suppress this key enzyme. The mode of action of tocotrienols seems to be mainly on degradation of the reductase protein, and to a lesser degree on its rate of synthesis (Parker *et al.*, 1993). It is of interest that 25-hydroxycholesterol strongly co-suppressed HMG-CoA reductase and mRNA levels together with the low-density lipoprotein (LDL) receptor protein.

Other plant-derived terpenoids could inhibit the transcript level of HMG-CoA reductase. In rats this enzyme increased after partial (2/3) hepatectomy, a process that was inhibited by β-carotene (70 mg/kg/day). This inhibition appeared to be post-transcriptional, so it did not affect the reductase gene.

More references could be quoted to support the conclusions and suggestions made in this section. The studies discussed in this section may serve to illustrate the importance of key enzyme suppression as one of the possible factors in combination therapies in oncology. Terpenoids of interest in cancer research acting on key enzyme suppression are included in Table 4.2. More detailed practical suggestions are in section 7.1.8.

4.1.9   Terpenoids with unexpected effects

Section 4.1.2 begins with insect JHs and their mode of action. They are internal messengers, although their production and activity can be regulated by external factors. There is evidence that in aphids JH has a function in the occurrence of parthenogenetic generations. Short photoperiods induce the production of sexuals and both the level of JH and of three monoterpenes are considerably lower under short day conditions in the aphid *Megoura viciae* (Hardie *et al.*, 1985; Harrewijn *et al.*, 1991). As yet little is known of an effect on the level of any HMG-CoA reductase. The MAI metabolism of the symbionts could be involved just as well (Table 3.1 and Fig. 3.7). *M. viciae* is known to produce monoterpenes: nepetalactol and nepetalactone (Fig. 5.3) are sex pheromones of this aphid (Dawson *et al.*, 1987). Related *Nepeta*

Chapter 4. Specific properties of terpenoids

terpenoids have an anti-fungal activity (Saxena and Mathela, 1996), illustrating the effect on fundamental mechanisms of living cells. It is not yet known if derivatives of these compounds have anti-tumour properties. These facts make it interesting to test compounds with proven pheromonal and hormonal action on insects on growth and cell division in vertebrates (tumour cells).

In writing this book we were surprised to encounter many more effects of terpenoids than we had dreamt of, as they have an impact on biological functions of organisms studied by completely different disciplines. Chapters 4 and 5 treat many of these; this section is intended to rouse the spirit of the reader and to prepare him (and her) for the second part of this book, which deals with the wealth of terpenoid messages dispatched in this world and covering a certain distance. This section illustrates that a particular terpenoid exerts effects on different organisms, effects which on first sight seem unrelated. However, this is mainly due to our lack of understanding of their mode of action.

An example is the range of effects of oleanolic acid and ursolic acid. In section 4.1.5 their protection against irradiation, and in 4.1.6 their inhibitory effects on angiogenesis of tumours has been mentioned, but these are only a few of their properties. Unfortunately, the information on these terpenoids is scattered over many disciplines. Oleanolic acid is an old oral drug, marketed in China against liver disorders. Its effect could be based on inhibition of toxicant activation. A review by Liu (1995) mentions their anti-inflammatory and anti-hyperlipidemic properties, and points to the stimulation of defensive mechanisms in the immune system. The triterpenoids oleanolic acid and ursolic acid exist widely in plants and food, as well as in medicinal herbs, of which the pharmacological utilization is mostly empirical.

Quite unexpectedly, ursolic acid exhibited anti-malarial effects with an *in vitro* anti-plasmodial IC50 value of 37 and 28 ng/ml with respect to a chloroquine resistant and a sensitive strain of *Plasmodium falciparum* (Steele *et al.*, 1999). Terpenoids of this kind affecting parasites, pathological micro-organisms with MAI metabolism should be investigated on MAI-metabolism inhibiting effects, as little is known on regulation of the MAI metabolism.

Aphidicolin (Fig. 4.13) is a diterpenoid that not only inhibits cancer cells, but in addition has an antiviral action. It is produced by the mould *Cephalosporium aphidicola* (Bucknall *et al.*, 1973). It is an inhibitor of cellular deoxyribonucleic acid and can inhibit the growth of Herpes simplex virus; cephalosporin P1 is another product of this genus of fungi. It is a diterpenoid of the nordammaranic type (Loder, 1972).

A triterpene isolated from the roots of *Maprounea africana* (Euphorbiaceae) appeared to be a potent inhibitor of the DNA polymerase activity of human immunodeficiency virus-1 (HIV-1) reverse transcriptase, with IC50 of 3.7 µM. The compound is not cytotoxic to cultured mammalian cells (Pengsuparp *et al.*, 1995). The mode of action of this triterpene (1-β-hydroxyaleuritolic acid 3-p-hydroxybenzoate) as inhibitor of AIDS virus involves non-specific binding to the transcriptase enzyme at non-substrate binding sites. These are *in vitro* experiments, but they demonstrate that a triterpene with almost unknown properties may contribute to the development of novel treatments against AIDS.

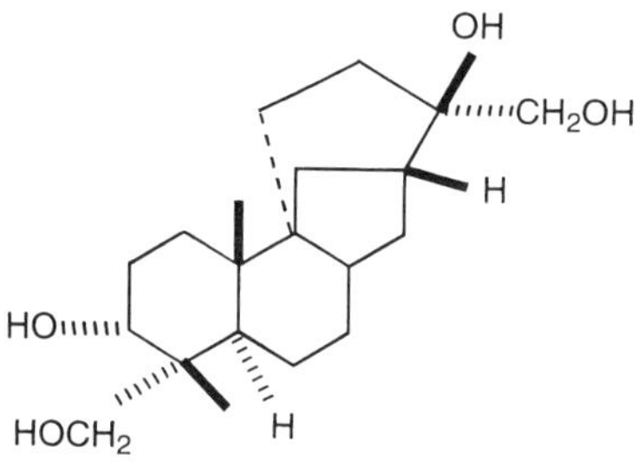

Fig. 4.13. Structure of aphidicolin.

Some properties of known terpenoids may be unexpected because nobody ever studied their effects. There are many studies on limonene, a major monoterpene in citrus fruit. The peel of *Citrus natsudaidai* (Rutaceae) contains auraptene, a coumarin-related compound with a geranyl-oxy group that enhances macrophage and lymphocyte functions in mice, so it can be regarded to be a stimulator of the immune system (Tanaka *et al.*, 1999). Interleukin 1-β production of peritoneal macrophages in auraptene-treated mice was significantly higher than that in the control group, and the same holds for tumour necrosis factor α production. One could think of dose levels of 100-200 mg/kg, which lays below the effective doses of limonene when applied as a single compound to combat tumours (200-400 mg/kg, see section 7.1.8). In addition, it might be combined with antibiotics in cases of peritonitis.

The monoterpene limonene has been discussed several times: its interference with protein prenylation and G-protein cascades is most striking. Limonene itself has many messenger functions between organisms, treated in the second part of this book, and its metabolite carvone has a number of effects that would seem to be unrelated. Potato growers use it to inhibit germination of tubers; plants concentrate it in their seeds to inhibit

Chapter 4. Specific properties of terpenoids

germination of competing plant species; the aphid *Cavariella aegopodii* is attracted to carvone as a major volatile of caraway; other insects, however, run for their life upon perception of carvone in an odour blend, and we use it in perfumes and liquors. Paracelsus (Theophrastus von Hohenheim (1493-1541) already prescribed seed extracts of caraway in the first half of the 16th century to ease crampy abdominal pain due to gas formation in the intestinal tractus. How have these functions been selected in the course of evolution? Is it possible to relate the effects to what is known of protein binding and messenger cascades? Before entering a discussion on biochemical selection processes and behavioural adaptation, we should look at the direct toxic effects of terpenoids that have no messenger functions *per se*, but that may have contributed to acquire functions as semiochemicals.

4.2 Direct toxic effects

4.2.1 Anti-microbial effects

On April 24, 1676, Antoni van Leeuwenhoek (1632-1723) described bacteria in an infuse of macerated pepper. Probably this is the beginning of practical bacteriology, because a few days later he observed that an extract of clove immediately immobilizes and kills the micro-organisms. Clove oil contains eugenol and other terpenoids (Table 6.1). This is the first scientific description of the anti-microbial effect of monoterpenes. The primary site of action of terpenes, aromatics and cyclo-alkanes is probably the cytoplasmic membrane, although the mechanisms are poorly understood.

The toxins of pathologic bacteria are mostly proteins of different classes, such as surface-acting toxins, some of which form pores in the cell membrane, toxins delivered into the cytoplasm by enzyme activity of a subunit or upon contact with the host cell ("injected toxins"). Members of the last two classes can interfere with different cell activities: protein synthesis, cAMP levels, G-proteins, protein kinases, actin polymerization, GTPase, dephosphorylation of kinase substrates, etc. The first class has a remote resemblance to the toxic effect of monoterpenes. The toxin of *Vibrio cholerae*, for instance, activates guanylyl cyclase type C receptor to elevate cGMP levels and activate cGMP protein kinase II, stimulate Cl$^-$ secretion and inhibit Na$^+$ absorption, resulting in diarrhea. The toxin, however, is a protein, with a different mode of action than that of monoterpenes.

Geraniol interferes with the membrane of yeast (Bard *et al.*, 1988). Hydrocarbon molecules accumulate in the membranes of cell organelles and micro-organisms, resulting in swelling of the membrane bilayer and increase in membrane fluidity. Geraniol was found to inhibit growth of the fungi *Candida albicans* and *Saccharomyces cerevisiae* strains. The results of Bard *et al.* (1988) demonstrate that the increase in membrane fluidity is greater in the more fluid bilayer, that is, at higher temperatures. The effect is minimal at 20°C and about 7 times higher at 37°C. The increased permeability to protons and ions leads to dissipation of the proton motive force (K^+ leakage) and impairment of intracellular pH homeostasis (Sikkema *et al.*, 1995). The symptomatic loss of cellular potassium may contribute to the anti-fungal effects of geraniol. At moderate ambient temperatures (soil, etc.) the effects on membrane permeability may not be sufficient to act as an anti-fungal compound. Partly disruption of membrane structure, however, opens interesting possibilities for combined treatments. Terpenes, as an example, can increase the effectiveness of antibiotics. They enhance the permeability of bacterial and mammalian cells to tetracycline, as already demonstrated by Piccinini *et al.* in 1972. The terpenes may have inserted themselves into the lipid layer of the cell membrane, thus influencing the selective permeability of the cell to foreign substances. The authors reported not only an increase in the anti-bacterial potency of the antibiotic (both *in vitro* and in experimental infections), but even a recovery of the anti-bacterial activity in resistant strains.

Monoterpenes specifically inhibit ammonia oxidation and may be responsible for the slow rates of nitrification in coniferous forests. The similarity in kinetic behaviour between monoterpenes and metal chelators make monoterpenes a candidate for chelating copper at the active site of ammonia mono-oxygenase. According to Ward *et al.* (1997) this may be the reason why monoterpenes in redwood needles and litter effectively inhibit ammonia oxidation and growth of *Nitrosomonas europaea*. The major compounds are limonene and α-pinene. All monoterpenes tested (except myrcene) were biologically active at concentrations well below the limits of their aqueous solubility (in the order of 10 µg/ml). Monoterpenes present in the water solution could be a factor in regulating nitrification rates in some forest soils.

Many monoterpenes produced by aromatic plants have effects on micro-organisms (Cowan, 1999; Rios *et al.*, 1987 and other authors in this section (see also 7.1.1)). A few examples are given in Table 4.3. They cannot only be used in pharmacology and in the kitchen, but they may also be useful in processed fresh vegetables products, which is a growing market segment. Wan *et al.* (1998) investigated the effect of basil essential oils with a high content of linalool and methyl chavicol on a range of Gram-positive and Gram-negative bacteria as well as yeast and moulds using an agar well diffusion method. Both monoterpenes showed anti-microbial activity against

Chapter 4. Specific properties of terpenoids

25 out of 35 strains with greatest sensitivity of *Penicillium expansum* and *Mucor piriformis*. Three *Pseudomonas* species, however, did not react and the same holds for *Clostridium sporogenes* and *Flavimonas oryzihabitans*.

Table 4.3. Effect of volatile terpenoids on bacteria (ph.t. stands for phenolic terpenoids)

Species	Gram-	Gram+	Inhibiting terpenoids
Bacillus subtilis		X	acorenone, rimulene
Citrobacter freundii (symbionts in gut of termites)	X		thujone
Enterobacter spp. (symbionts in gut of termites)	X		linalool
Escherichia coli	X		carvacrol (ph.t.), citronellal, citronellyl acetate, geraniol, linalool, neral, pulegone, terpinen-4-ol, α-pinene, α-terpineol, δ-3-carene
Flavobacterium suaveolens	X		germacrene, sabinene
Klebsiella oxytoca	X		cineole, thujone
Klebsiella pneumoniae	X		thujone, 1,8-cineole
Mycobacterium smegmatis			alantolactone, isoalantolactone
Proteus vulgaris	X		limonene (toxic to catflea), 1,8-cineole
Pseudomonas aeruginosa	X		linalool, pulegone
Salmonella spp.	X		β-caryophyllene, β-caryophyllene oxide
Shigella shiga	X		β-caryophyllene, β-caryophyllene oxide
Staphylococcus aureus		X	carvacrol (ph.t.), citronellal, citronellol, manool, pulegone, β-caryophyllene
Streptococcus faecalis	X		carvacrol (ph.t.), citronellal, citronellol, thymol
Vibrio cholerae		X	carvacrol, thymol (ph.t.), β-caryophyllene, β-caryophyllene oxide

Basil essential oil was active against all eight strains of yeast and the three strains of moulds tested. The results suggest that the oil could be a beneficial component of salad dressings or be used in preservation of vegetable products. A concentration between 0.1 and 1% is as effective as washing with 125 ppm chlorine. An excellent minireview on bacterial membranes, especially on the structure of Gram-negative cell walls is by Beveridge (1999). The relative importance of Gram-positive and -negative bacteria as pathogenic micro-organisms is discussed by Sriskandan and Cohen (1999).

The oil of *Artemisia annua* is toxic to malarial parasites (Rao et al., 1999). This seems a new development in the treatment of malaria. Paracelsus (see 4.1.9) applied *Artemisia* oil already in the first part of the 16th century against malaria (marshland-associated fevers).

The essential oil of the tea tree, *Melaleuca alternifolia* (Myrtaceae), has anti-microbial properties against a wide range of Gram-negative and Gram-positive micro-organisms as well as yeast. The major components of the oil are α-pinene, terpinen-4-ol, linalool and α-terpineol. Gustafson et al. (1998) studied the mode of action and found that the oil stimulates autolysis in exponential and stationary phase cells of *Escherichia coli*. Electron micrographs showed loss of electron dense material, coagulation of cell cytoplasm and formation of extra-cellular blebs. The coagulated material could represent denatured membranes and proteins. These phenomena have also been reported of membrane-active disinfectants like chlorhexidine and quaternary ammonium compounds. These disinfectants are believed to disrupt membrane structure; chlorhexidine causes the formation of electron dense blebs on the cell surface of *E. coli*, suggesting that tea tree oil causes similar collections of coagulated membrane and cytoplasmic constituents, which have been pushed through holes, produced in the cell wall. Lipophilic monoterpenes such as linalool, α-terpineol and 1,8-cineole can penetrate and disrupt lipid structures, explaining why the mode of action of tea tree oil closely resembles that of membrane-active desinfectants.

The main components of different essential oils, even from species of one family, have great variations and are dependent on the habitat. According to Manou et al. (1998) commercial samples of *Thymus vulgaris* essential oil are reasonably consistent, probably because they are mixtures of oils from different growing sites. The main component of this oil is thymol (about 40%), followed by *p*-cymene (about 25%), γ-terpinene (up to 10%), and a number of terpenes in lower concentrations, such as carvacrol, linalool, α-pinene, myrcene and γ-caryophyllene. Linalool, which is so important in tea tree oil and also in green tea leaves, is definitely not the major terpene in thyme oil. These big differences in the composition of essential oils explain why such a wide range of aromatic plants are used in medicine and dishes of Asia and the Indonesian Archipel. The practical use of essential oils and their combinations is more extensively treated in chapter 6.

Chapter 4. Specific properties of terpenoids

As mentioned above, monoterpenes in tea tree oil induce autolysis in *E. coli*. The oil can kill a wide variety of micro-organisms (Gram-negative and -positive bacteria, as well as yeast). The main components of the tea tree oil in the experiments of Gustafson *et al.* (1998) were linalool, α-pinene, terpinen-4-ol and α-terpineol; experiments on individual compounds were not included. Nevertheless this example illustrates that a few monoterpenes are toxic to many micro-organisms, some of which are symbionts of herbivorous insects. This likewise applies to farnesol of which the toxic effects on vertebrate cells have been discussed in this section. Farnesol causes a significant loss of intra-cellular diacylglycerol (DAG) in *Saccharomyces cerevisiae*, budding yeast. Although farnesol readily permeated the yeast cell membrane from a culture medium, its major effect was repression of cell cycle genes encoding DNA ligase and histone acetyltransferase, indicating interference with a phosphatidylinositol-type signalling (Machida *et al.*, 1999). The yeast cell growth was arrested at a much lower concentration of farnesol than can affect the fluidity of the cytoplasmic membrane due to its lipophylic property. In a medium with 400 μM linalool there was no inhibition in cell growth. The authors suggest a correlation between farnesol-induced growth inhibition in yeast cells and farnesol-mediated apoptosis in mammalian cells (compare with 4.1.5).

A toxic effect alone does not make a monoterpene a messenger between organisms, but for insects which avoid these compounds because they can affect their symbionts the messenger function becomes obvious (Harrewijn *et al.*, 1995). This is where the material of section 4.1.2 (odour perception), this section, section 4.2.2 and chapter 5 come together.

Micro-organisms themselves can produce several odorous volatile compounds. Alcohols, phenols, esters and other aroma substances that are common in higher plants are produced, often with peak concentrations during sporulation. Collins and Halim (1970) analysed the odour spectrum of a number of fungi and found that terpenes dominate in the essential oil of *Ceratocystis* spp. Although qualitative and quantitative differences were observed between species, the following terpenes were generally present: citronellol, citronellyl acetate, geranial, geranyl acetate, linalool, neral, nerol, neryl acetate and α-terpineol. The authors already suggested that the volatiles have an allelopathic function and enhance their own survival by an inhibitory effect on other micro-organisms.

For mammalian cell systems toxic effects are often remarkably low, except with regard to some groups of terpenoids (sesquiterpene lactones) and specific terpenoids. Sesquiterpene lactones inhibit essential enzyme systems, but are generally toxic. The terpenoid ester linalyl oleate is a cocarcinogen and is involved in the inhibition of DNA repair (van Duuren *et al.*, 1971). Special cell lines and cell lines derived from tumour systems are more vulnerable. Examples are HeLa cells, Hep G_2 cells and more in general, tumour cells developed from oncogenes (cells of mammary tumours, colon, bladder, etc.). Nerol, linalool, rhodinol, geraniol and farnesol are toxic to HeLa cells (100 μg/l), but without a thorough study terpenoids cannot simply be used in additive or alternative treatments, as many of them are not harmless. There are examples of selective toxicity on cell systems (Adany et al., 1994; Yazlovitskaya and Melnykovych, 1995 and section 4.1.7) which, however, are of a different order than those of the specific targets of terpenoids discussed in sections 4.1.5 and 4.1.6.

4.2.2 Toxic effects on insects and other animals

This section restricts itself to the toxic effects of terpenes on animals. Behavioural aspects are discussed in chapter 5. Roughly speaking, the relatively simple monoterpenoids are often toxic or repellent to insects, whereas sesquiterpenoids, diterpenoids and triterpenoids have more specific effects. The toxic effect of monoterpenes on insects may (at least partly) be based upon their inhibition of acetylcholinesterase (AchE), present in the neuromuscular junction, and with additional functions, such as in the peripheral sensory nervous system (Rice and Coats, 1994). Five monoterpenes representing distinct functional groups appeared to be effective competitive inhibitors of cholinesterase (Ryan and Byrne, 1988; Miyazawa et al., 1997), probably by hydrophobic binding to the substrate (Fig. 4.14). The authors found gossypol to be an uncompetitive inhibitor of AchE; that is it binds not to the enzyme but to the enzyme-substrate complex, thus preventing product formation.

As mentioned in section 2.1.4, many species of herbivorous insects harbour symbiotic micro-organisms with many functions related to their fitness. This makes them vulnerable to the antibiotic effects of terpenoids. Toxic terpenoids, which are volatile and can be percepted, could have made an important contribution to the development of plant-herbivore interactions as we know them today (Harrewijn et al., 1995).

The aphid alarm pheromone (E)-β-farnesene appeared to be toxic to Hep G_2-cells in vitro in relatively high concentrations, up to 0.1 mM (Cohen et al., pers. comm.). This, however, is the natural concentration in which (E)-β-farnesene occurs in aphid hemolymph. The compound appeared to disrupt the cell membranes resulting in increased fluidity, as mentioned above for geraniol. As some aphid species, which also contain farnesenes in their hemolymph, do not react to these sesquiterpenes as alarm substances, it could be that farnesenes have originally been selected for their defensive properties, and that the role of (E)-β-farnesene as an alarm pheromone is

Chapter 4. Specific properties of terpenoids 177

secondary. (E)-β-farnesene is toxic to aphids when applied topically at a dose of c. 1/3000 of the body weight (see 5.1.3).

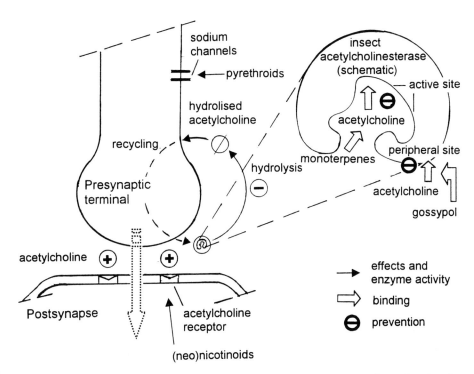

Fig. 4.14. The effect of monoterpenes and gossypol on signal conduction in an insect nervous system. Dotted vertical arrow = direction of conduction. Monoterpenes do not act on the acetylcholine receptor like nicotinoids but can bind to acetylcholine as competitive inhibitors; gossypol can bind to the acetylcholinesterase/acetylcholine complex in an uncompetitive way.

More complex terpenoids can have a more specific effect. Azadirachtin, a tetranor-triterpenoid is the major insecticidal component of neem oil (Fig. 2.16B). Doses smaller than 0.01% strongly reduce esterase activities in adults of the obliquebanded leafroller, *Choristoneura rosaceana* (Smirle *et al.*, 1996). As specific esterases are responsible for the development of insecticide resistance, neem could be a valuable tool in the management of resistant populations where esterase-based metabolism is involved. Neem oil did not affect gluthathione transferase activities. Azadirachtin affects activity of endosymbionts of aphids, reducing the production of *Buchnera* GroEL proteins (van den Heuvel *et al.*, 1998). This may be an indirect effect of azadirachtin, as it can negatively affect feeding activities.

Eucalyptus species contain mainly 1,8-cineole and jensenone, the latter being a phenolic substance. These compounds are moderately toxic to *Pseudocheirus peregrinus* (ringtail possum) and *Trichosurus vulpecula* (brushtail possum), causing post-ingestive emesis. The possums show an initial aversion to dietary cineole, although they can habituate to cineole and become conditioned to feed on low doses (Lawler *et al.*, 1999).

The reaction to toxic terpenoids of herbivorous animals is either a passive or an active response. In a passive response the animal ingests a certain amount of the terpene that is not perceived as a deterrent, until the toxic effects manifest themselves and feeding is terminated, not necessarily irreversibly. Feeding behaviour is modulated in a dose-dependent manner via the pharmacological effect of the ingested toxicant.

In an active response there is a reaction to the activity of chemoreceptors; the terpenoid is detected and at a certain threshold concentration the animal decides to stop feeding. Of course it is sensible for the herbivore to react at a concentration lower than that leading to intoxication. Section 4.2.2 is an introduction to the complex relationships between different organisms in which toxicity of terpenoids is only part of their properties, relationships that are discussed in more detail in chapter 5.

4.2.3 How to overcome autotoxicity

In higher plants terpenes are commonly stored in specific organs and within the secretory structures themselves, terpenes are accumulated in extra-cellular compartments away from living cytoplasm (Fig. 2.9). This is a protection against autotoxicity.

For small, medium-bound organisms there is the question of autotoxicity, connected with the production of toxic volatiles. One solution may be the release during sporulation, avoiding contamination of the substrate and at the same time defending the spores and the organs producing them. It has also been suggested that the formation of simple esters by fungi may be a mechanism for removing both alcohols and acids that would become toxic to the organism if allowed to accumulate in the medium.

Terpenes can be stored as essential oils in glands, ducts or other compartments. This reduces the risk of autotoxicity (Brown *et al.*, 1987) and this may have been the original function of oil glands and peltate trichomes in Angiospermae. An evolutionary advantage is an immediate release of allelopathic volatiles by young vegetative parts and reproductive organs in effective concentrations upon attack by micro-organisms and herbivores (Harrewijn *et al.*, 1995). The effective dose of terpenoids needed to affect micro-organisms and to disturb feeding behaviour or development of herbivores is often remarkably low.

Croteau (1988) has provided compelling evidence that monoterpenes turnover to catabolites, which can be re-utilized in lipid biosynthesis in the developing root or rhizome or become further oxidized in energy production. This process can start in undamaged oil cells, which are emptied of their contents.

Steroid hormones in plants could reach concentrations that are auto-destructive by interfering with normal development. In angiosperms, these are complex dynamics of these terpenoids of which defensive ones such as the phytoecdysteroids reach high concentrations in generative parts when they need a high level of protection (Zibareva, 2000). Production, storage and release of toxic terpenoids in plants is a well-timed process that can be induced and regulated by internal and external factors and which is the result of a dynamic balance between anabolic and catabolic processes.

Chapter 5

Functions of natural terpenoids in the interrelationships between organisms

Sir Vincent B. Wigglesworth says at the end of the last page of his book Insects and the life of Man (1976): "Science, however, is founded on faith and lives by faith. Without a deeply rooted instinctive belief in the existence of laws which reign throughout nature, the incredible labours of scientists would be without hope. We must just soldier on in that faith to whatever end it may ultimately lead".

With this in mind the material of chapter 5 is addressed to the reader. It has been written in the hope that some general rules concerning the functions of terpenoids acting as signalling compounds *between* organisms stand out and can be related to what happens *within* an organism as discussed in chapter 4.

When it comes to signalling compounds the relationships between plants and insects are by far the best studied ones; for that reason this chapter focusses on the relationship between plants and insects, although these are not the only organisms discussed.

When dealing with insects, at first sight the surveyability may seem to be served best by discussing different orders in the same sequence in all subsections. In doing so, there would hardly be any arrangement in the terpenoid structures, as the same compound can have different functions in various insect orders or families. Therefore each subsection is based on main classes of terpenoids: monoterpenes, farnesene isomers, sesquiterpenes and diterpenes, followed by more complex terpenoids. Within this classification the insect orders in which terpenoids have known messenger functions will be discussed in the same sequence: Orthoptera (crickets and grasshoppers), Dictyoptera (cockroaches and mantids), Isoptera (termites), Hemiptera (true bugs) with special emphasis on Homoptera (particularly aphids),

Thysanoptera (thrips), Lepidoptera (butterflies and moths), Diptera (true flies), Siphonaptera (fleas), Hymenoptera (bees, wasps, ants, sawflies, etc.) and Coleoptera (beetles).

Additional reading on perception mechanisms and plant-herbivore relationships is in "Chemical ecology of insects" (eds W.J. Bell and R.T. Cardé, Chapman and Hall, London), 1984 and in "Chemical ecology of insects II" (eds R.T. Cardé and W.J. Bell, Chapman and Hall, New York, 1995). These books are referred to in "General reading", positioned right after the References.

5.1 Semiochemicals

Terpenoids have an important role in signal transmission between organisms (Ourisson, 1990). Terpenoids form one of the most versatile chemical languages in the network of communication between plants and other organisms. Penuelas *et al.* (1995) even speak of a "plant language".

In Angiospermae, terpene synthesis is usually associated with glandular structures, with a peak activity during leaf expansion and a substantial production of essential oils (Fig. 2.9). Selective pressures by the climatic and edaphic phenomena of the Mediterranean regions of the world have favoured this aspect of secondary metabolism. They are characterized by long, hot and dry summers and the summer-associated leaf stress together with a high level of radiation are major factors that increase essential oil content (Ross and Sombrero, 1991). When stressful conditions cause the enhanced allocation of fixed carbon into the biosynthesis of essential oils, the availability of especially monoterpenes will not be the same throughout the seasons unless specifically induced by other external factors and this also holds for their use as kairomones by both beneficial and herbivorous insects.

There is general agreement that humans have sex pheromones, although attraction to the other sex is primarily based on optical stimuli and linguistic communication. This is not so in life forms which inhabit this planet for a much longer period than humans. Invertebrates, particularly insects, live in an environment characterized by chemical stimuli: odour and gustation. Finding a partner is a prerequisite for sexual reproduction and terpenoids have a dominant role in this form of chemical language. Section 5.1 could have started with the highly evolved relationship between flowers and pollinators, on which there is a wealth of literature. However, attraction of pollinators is only one function of plant volatiles and the result of a long process of specific adaptation. Besides, pollinating insects are not attracted

Chapter 5. Functions of natural terpenoids in the interrelationships between organisms   183

to plants by kairomones alone, which is illustrated a.o. by the sexual mimicry in orchids: flower morphology and colour, together with the production of *sex pheromones* attract male bees and wasps (Vogel, 1993).

From an evolutionary point of view, the sequence of terpenoid classes utilized in this section is based upon the relative importance of these compounds in attraction and avoidance, referring to the material of chapter 4.2. Sections 5.1.3, 5.1.4 and 5.1.5 deal with plant and animal coevolution, that has a long history of mutual interactions.

5.1.1   Sex pheromones and other semiochemicals with impact on behaviour

Of all main taxa discussed in this book: bacteria, fungi, plants, insects and vertebrates sex pheromones are known in which terpenoids have a considerable share. By far the best studied are insects, because they are the chief competitors with man for the domination of this planet. Once the chemical composition of a sex pheromone or an aggregation pheromone is elucidated, we can try to lure the insect to an odour source and subsequently kill it with an insecticide (that does no longer have to be applied on big areas) or by some trapping system. Do not think that "attract and kill" is a human invention! Many male insects do not survive attraction by a female: they may perish for a number of reasons, being consumed after mating is just one possibility.

The oogonium cells of the green alga *Oedogonium* spp. secrete the terpene sirenin (Fig. 5.1) that attracts male gametes, the androspores. This chemotactile stimulus is sufficient to attract the gametes (Machlis and Rawitscher-Kunkel, 1967). Many micro-organisms produce terpenoids (see 4.2.1) with different functions: yeast utilizes sterols and steroid lipids in sexual reproduction. Fungi are not so immobile as they would seem to be and resemble more to vertebrates than to plants (as discussed in chapter 2). In fungi with their early MAD metabolism steroid hormones known from vertebrates often take part in the complex processes involved in sexual reproduction. As an example, steroid pheromones belonging to the antheridiols are attractive substances to antheridial hyphes of *Achlya* spp. In fungi, terpenoids are often only one link in the chain of chemical messengers involved in reproduction and it is a trifle beyond the scope of this book to treat the major non-terpenoid cascades.

Terpenoids have sex-related signal functions in vertebrates. Squalene is one of 42 components identified in the preorbital secretion of *Damaliscus dorcas dorcas*, the bontebok, and *D. d. phillipsi*, the blesbok, both antelope species of South Africa (Burger *et al.*, 1999). Except for cholesterol that is also excreted, this is the only terpenoid, making it impossible to decide if squalene is essential for the messenger function of the blend.

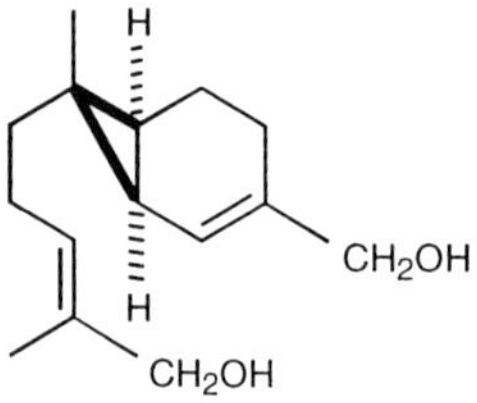

Fig. 5.1. Sirenin, a terpenoid sex pheromone of a green alga.

It is surprising that the same terpenoid can induce quite different types of excitation, depending on the organism. This is best illustrated by discussing the functions of members of one particular class of terpenoids at a time.

MONOTERPENES, SESQUITERPENES AND DITERPENES
After the first successful identification of bombykol, the sex pheromone of the silkworm by Butenandt and his co-workers in 1959 (this is a 16-C unsaturated aliphatic alcohol), it took another 12 years before a main component of the sex pheromone of *Periplaneta americana* (Dictyoptera) was elucidated. This component (periplanone B) is a sesquiterpene derivative (Fig. 5.2A). This was done by a combined research group in the Netherlands (Persoons, 1977; Ritter *et al.*, 1981). Interestingly, germacrene A (Fig. 5.2B) is a sex mimetic of *P. americana*. This terpene is one of the volatile metabolites excreted by Actinomycetes (Tsuchiya *et al.*, 1980). The periplanones are extremely active. Only a few molecules acting on the sensilla of males result in behavioural effects, and a little more brings them into sexual arousal. They seek to find the pheromone source and try to copulate with each other or even with objects in the neighbourhood.

In Coreidae species (Hemiptera) the complete sex pheromone blend comprises not only farnesenes, but also (3R)-nerolidol, β-ocimene and linalool (McBrien and Millar, 1999).

The sex pheromones of aphids (Homoptera, Hemiptera) are particularly well studied. Aphids can synthesize a range of monoterpenes and sesquiterpenes (Harrewijn *et al.*, 1991). In 1987 Dawson *et al.* published the identification of an aphid sex pheromone. Actually, it comprised two monoterpenes released from the hind legs of female *Megoura viciae*,

Chapter 5. Functions of natural terpenoids in the interrelationships between organisms

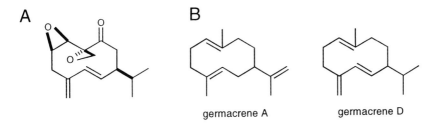

Fig. 5.2. **A**: periplanone B, a component of the sex pheromone complex of *Periplaneta americana*; **B**: two important germacrene isomers.

nepetalactol and nepetalactone (Fig. 5.3). Single-cell EAG recordings of olfactory cells associated with the secondary rhinaria of males (see 4.1.2) showed that both components were electrophysiologically active, however, the males did not react to a single component. Both components were needed to attract males. Nepetalactone is not specific for aphids. It is known from *Nepeta* spp. (Lamiaceae). Nepetalactone is also used in defensive secretions, like that of the coconut stick insect, *Graeffea crouani* (Smith et al., 1979). The strange reactions (catnip response) of members of the Pantherinae and Felidae to *Nepeta cataria* are based upon its smell, and not its taste. Probably the *Nepeta* monoterpenes cross-react with a naturally occurring social odour in these animals (Tucker and Tucker, 1988).

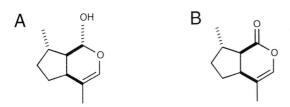

Fig. 5.3. The two components of the sex pheromone of the aphid *Megoura viciae*. **A**: nepetalactol; **B**: nepetalactone.

Because aphids are associated with agricultural plants, the search for aphid sex pheromones was continued, as they might be utilized in IPM programmes. Not that this would likely happen in summer, because aphids reproduce asexually on their host plants during summer. Host alternating species migrate to the primary (winter) host where sexual reproduction occurs, and interference with this phenomenon could substantially reduce aphid problems in the following season. In laboratories all over the world behavioural studies using olfactometers combined with EAG/GC technique were set up to find novel sex pheromones of aphids. Other monoterpenes, e.g. α-pinene and limonene were found to be active but surprisingly, several aphid species appeared to utilize nepetalactone and nepetalactol. Like in pheromones of Lepidoptera, the ratio between the two compounds was shown to be species-specific and reaction of the aphids can depend on the enantiomer. The sex pheromone of *Phorodon humuli*, for instance, was shown to be (4aR,7S,7aS)-nepetalactol or nepetalactol III that could be synthesized from the corresponding nepetalactone in the labiate plant *Nepeta mussinii* (Campbell *et al.*, 1990). Nepetalactones accumulate within the subcuticular cavity of peltate glandular trichomes of *Nepeta racemosa* (Clark *et al.*, 1997). There is an additional effect of volatiles of the primary host (*Prunus* spp.) on the attraction of male aphids, although the sexual forms can find each other without the apparent need for olfactory cues of the primary host. Different ratios of nepetalactol I and nepetalactone II are sex pheromones of *Aphis fabae*, *Acyrthosiphon pisum*, *Megoura viciae* and *Myzus persicae* (Dawson *et al.*, 1990; Hardie *et al.*, 1990). In *M. viciae*, the most effective ratio of nepetalactol/nepetalactone is 1:5, in *M. persicae* 2:1, in *A. fabae* 1:29 and in *Rhopalosiphum padi* 1:10. This suggests that aphids will not easily be mistaken in finding the right sexual partner.

When EAG recordings of complete antennae are made, the response to a volatile compound is dependent on both the sensitivity of the sensilla and the number of sensilla on the antennae. The male antennae of *A. fabae* are 1000-10,000 times more sensitive to nepetalactol and nepetalactone than to (E)-2-hexenal (Hardie *et al.*, 1994b). Males can locate a sex pheromone source in the field in the absence of host plant cues.

Lilley and Hardie (1996) did not find a significant difference in the response of male *Sitobion fragariae* to nepetalactone combined with plant material or to nepetalactone alone, although for the gynoparae a combination of both stimuli was more attractive than plant odour alone. The biological significance of this phenomenon is not quite clear. How do gynoparae aggregate in the neighbourhood of males, when the males are not yet there because they have to react to the sex pheromone of the gynoparae?

The major sex pheromone component of this aphid is (4aS,7S,7aR)-nepetalactone that can be prepared synthetically. When this enantiomer was extracted from plants it was less effective than the 99% (7S)-lactone (Hardie *et al.*, 1997). Trace amounts of other enantiomers reduce activity of the real sex pheromone components, and the same holds for (4aS,7S,7aR)-nepetalactol, that is specific for *R. padi*. Even when a specific pheromone is present as a minor component in a synthetic product, it can attract aphids,

Chapter 5. Functions of natural terpenoids in the interrelationships between organisms

e.g. *P. humuli* by traces of (4aR,7S,7aS)-nepetalactol in the synthetic pheromone of *R. padi*. So it appears that some aphid species react very sensitively to one specific enantiomer, whereas others use blends of enantiomers of nepetalactol and nepetalactone.

Females of the Mediterranean fruit fly, *Ceratitis capitata*, the medfly (Diptera), show antennal responses to the male-produced volatiles ethyl-(E)-3-octanoate, geranyl acetate, (E,E)-α-farnesene, linalool and indole. The main volatiles from ripe fruits of their host plant mango were identified as (1S)-(-)-β-pinene, β-caryophyllene and ethyl octanoate. There is some overlap with respect to ethyl octanoate, but the antennal reaction to this compound was weaker than to β-caryophyllene (Cosse *et al.*, 1995). Still, this may be another example of an additional effect of host plant odours on sex pheromones. This view is supported by the fact that EAG response amplitudes from both male and female antennae to the three mango volatiles were significantly greater than to a hexanol control. In addition, α-copaene that is a potent attractant for males of *C. capitata*, is present as a minor component of the essential oils of hosts such as orange, guava and mango (Nishida *et al.*, 2000). The hemolymph of males of the Caribbean fruit fly, *Anastrepha suspensa*, contains terpenoids among which ocimene that is supposed to play a role in sexual signalling (Teal *et al.*, 1999).

In Coleoptera, some weevils e.g. *Anthonomus eugenii* (Curculionidae) utilize geraniol (Fig. 2.5) and geranic acid as components of sex pheromones containing other hydrocarbons. The sex pheromone of *Tomicus minor* is (-)-*trans*-verbenol. Verbenone is a natural oxidation product of α-pinene and this compound is a repellent interfering with the attraction to pheromones and kairomones in bark beetles (Miller *et al.*, 1995). Males of the California five-spined ips, *Ips paraconfusus* produce ipsenol and ipsdienol (Fig. 5.4) as components of their aggregation pheromone. A related species, *Ips pini*, uses ipsdienol as an attractant, but ipsenol as an interruptant of the aggregation pheromone. The pheromone components are synthesized by the beetles via the isoprenoid pathway (Seybold *et al.*, 1995). JH has been shown to induce ipsdienol production *in vivo* (Ivarsson *et al.*, 1996). JH may be involved because it induces sexual maturation, and in this way a sesquiterpene induces the production of monoterpene alcohols that are the main components of the pheromone on the proper time. Recent studies of Seybold (1998) have revealed that JH3 probably stimulates the level of transcription of HMG-CoA reductase. In addition, the regulation of the isoprenoid pathway and pheromone biosynthesis by JH3 is linked to the physiological act of feeding in the phloem during host colonization. Not only does feeding stimulate *de novo* biosynthesis of ipsdienol by male *Ips pini*, but feeding also stimulates *de novo* biosynthesis of JH3 in this insect. See also chapter 2.

Possible terpenoid precursors of aggregation pheromone components of bark beetles are discussed by Byers (1995).

Picea abies is a host of the double spined bark beetle, *Ips duplicatus* that uses ipsdienol and (E)-myrcenol (Fig. 5.4) in a 5:1 ratio as an aggregation pheromone. Myrcene, that is present in the trees, could be a precursor of the pheromone components. Ivarsson and Birgersson (1995) however, found that both ipsdienol and (E)-myrcenol are produced *de novo* and not from myrcene (Fig. 5.4). It was possible, though, to block terpene biosynthesis of the insects with compactin, indicating that the key enzyme of mevalonic acid synthesis is involved in pheromone production. This can be seen as another example of herbivores, which mimic the odour of a host plant to stimulate aggregation (Harrewijn *et al.*, 1995).

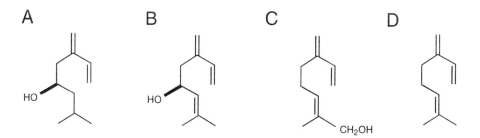

Fig. 5.4. (S)-ipsenol (**A**) and (S)-ipsdienol (**B**), components of the aggregation pheromone of *Ips paraconfusus*. **C**: (E)-myrcenol; **D**: β-myrcene. For explanation see text.

Many species of bark and ambrosia beetles utilize volatiles as cues for breeding site location, among which Flechtmann *et al.* (1999) found 18 oxygenated monoterpenes released by *Pinus taeda*.

FARNESENE ISOMERS

The sesquiterpene α-farnesene elicits a variety of responses in insects (see 5.1.2 and 5.1.3). In mice farnesenes are involved in sexual preference of females. Both α- and β-farnesene, probably originating in the accessory sex glands, are absent in the urine of immature males and significantly reduced in that of castrated males. In dominant males the concentrations show a dramatic increase and the weight of the preputial glands is higher (Harvey *et al.*, 1989). The urine not only induces female attraction and intermale aggression, but also causes social subordination resulting in reduced gonadal function.

Chapter 5. Functions of natural terpenoids in the interrelationships between organisms

If mice were insects, α- and β-farnesene could be denominated to be aggregation pheromones. Farnesene isomers are part of a more complex sex pheromone blend in *Anastrepha suspensa*, the Caribbean fruit fly. Farnesene (isomer not established), bisabolene, anastrephin and epianastrephin were identified from extracts of hemolymph of males and are compounds of a sex pheromone that in addition contains ocimene, nonenols and suspensolide (Teal et al., 1999).

In the insect order of Hemiptera farnesenes are constituents of aggregation pheromones. In Coreidae male *Amblypeeta lutescens lutescens* secrete (E,E)-α-farnesene, (E)-β-farnesene together with linalool and β-ocimene. According to McBrien and Millar (1999) these terpenes could be components of an aggregation pheromone. This, however, is not a rule in other Hemiptera. In Heteroptera (suborder of Hemiptera) many studies have been made on aphids. Neither α- nor β-farnesene are sex pheromones in these insects: they are toxic or have functions as alarm pheromones and hormones (sections 5.1.2 and 5.1.3). Alarm pheromones are perceived with the primary rhinaria and sex pheromones mostly with secondary rhinaria, if present (see 4.1.2). This means that in Hemiptera farnesenes can be perceived within primary or secondary rhinaria.

Farnesol isomers (Fig. 5.9, Box 2.3) are components of the aggregation pheromone of *Biprorulus bibax* (the spined citrus bug) that also contains linalool and nerolidol. Females of Tephritidae (fruit flies), belonging to the Diptera, are attracted to a three-component blend, one of which is (E,E)-α-farnesene. The other two are the monoterpene ester geranyl acetate and ethyl-3-octenoate (Landolt and Averill, 1999).

Males of the Scandinavian bumble-bee species *Bombus pratorum* mark small areas of the edges of birch leaves with compounds from the cephalic portion of their labial glands. These marks are used for a male-specific flight route and are patrolled by the males; they are supposed to attract conspecific unmated, young queens. The marking substance contains (Z)-11-octadecanol, hexadecanol and a pentacosadiene together with three terpenes: farnesol, geranylgeranyl acetate and farnesyl acetate (Bergman and Bergström, 1996).

In Coleoptera a range of terpenes have functions as sex pheromones. Members of some families utilize farnesenes, e.g. *Maladera matrida* (Scarabaeidae), where (Z,E)-α-farnesene (not (E,E) is a sex pheromone. Scarab beetles seem to have developed their own solution to recognize their sex pheromones from background noise. In contrast to Lepidoptera, where the exact ratio of the components of a sex pheromone blend does the job, scarab beetles utilize single constituents, but react to isomers or even enantiomers. This is possible because the pheromone molecule has a chiral centre that plays a determining role in pheromone perception (Leal, 1999). Enantiomers produced by another beetle species can act as a behavioural antagonist.

It appears that farnesenes are constituents of sex pheromone blends in a number of insect orders and seldomly operate on their own. If they do, the receptors seem to be specifically tuned to a particular enantiomer.

COMPLEX TERPENOID SEX PHEROMONES

Complex, less volatile sex pheromones could probably only act on a short distance or when released on a preferred mating place, e.g. on flowers. Little is known of these terpenoids that may be more effective when transported by a fluid than by air. Complex terpenoids such as steroid hormones are mostly internal messengers produced by endocrine glands, but distinction between internal and external messengers is not always clear. An example is JH or derivatives of this sesquiterpene that is produced by insects and that can act as a pheromone. Ants and termites mark their route between nest and food supply with a trail pheromone. The poison gland of the Pharaoh's ant, *Monomorium pharaonis*, contains (apart from some alkaloids) the trail pheromone faranal (Fig. 5.5A) that closely resembles JH2 (Fig. 5.5B). One can make artificial trails with this sesquiterpene that will be followed by the insects. A concentration in the magnitude of 10^{-2} µg/cm of trail is sufficient.

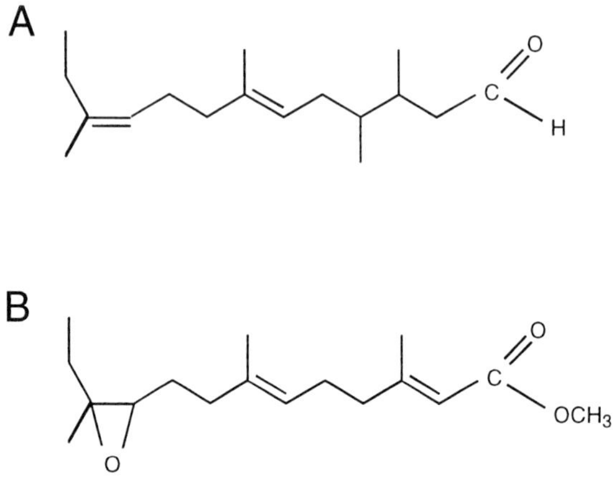

Fig. 5.5. Structures of **A**: faranal, a trail-following pheromone of the ant *Monomorium pharaonis* and **B**: the insect hormone juvenile hormone 2.

Chapter 5. Functions of natural terpenoids in the interrelationships between organisms

A number of steroids produced by mammals have a musky smell. Musk odour has been a component of perfumes for more than 3000 years, which may be partly based on non-olfactory sensory reception (Ritter *et al.*, 1981). Steroids may be components of a more complex blend, as cholesterol is in the preorbital secretions of the bontebok and the blesbok (Burger *et al.*, 1999). In mice a preputial-associated urinary sex factor is androgen-independent, although urinary factors are androgen-dependent (Ninomiya and Kimura, 1988).

Liquid excretions of some stinkbugs (Pentatomidae, Hemiptera) contain bisabolene epoxide and (Z)-α-bisabolene. In other species β-sesquiphellandrene is a component.

In conclusion, terpenoid sex pheromones are predominantly highly volatile monoterpenes and intermediates, and sesquiterpenes. Many of them are also produced by plants or by different insect species and have acquired functions as allomones and alarm pheromones. The same compounds can be constituents of sex pheromones and of kairomones. This is not surprising because finding a sexual partner, a mating place or a host plant is often based on long-range orientation.

5.1.2 Alarm pheromones and defensive secretions

Sex pheromones serve to arouse conspecifics; alarm pheromones have a similar role that can be extended to self-defense or defense of a colony. In this section we discuss terpenoid secretions of animals, in 5.1.3 that of plants. Because the role of farnesenes and complex terpenoids is limited terpenoids of this section are not subdivided into classes.

Very little is known of a role of terpenoids in arousal or defense of Orthoptera and Dictyoptera. In termites (Isoptera) that have a high social structure considerable literature exists on the composition of defensive secretions. Monoterpenes, sesquiterpenes and diterpenes are found, often in the oxygenated form (Prestwich, 1984). The soldiers of *Nasutitermes* spp. and of *Hospitalitermes* spp. secrete trinervitenes (diterpenes). In 1982 Baker *et al.* described the acyclic diterpene alcohol (R)-(-)-(E,E)-geranyllinalool in the defensive secretion of soldiers of *Reticulitermes lucifugus*, a compound which they had detected in 1981. At that time the important function of geranylgeraniol in gene expression and induction of apoptosis (chapter 4.1) was not yet known, but it is intriguing that a diterpene involved in basic cell functions may have been selected early in evolution as a defensive compound by members of a successful insect order. Diterpenes are also present in the defensive secretions of soldiers of the genus *Hospitalitermes*

(Chuah *et al.*, 1986), and can be present in the form of propionate esters (Prestwich *et al.*, 1980).

There is quite a variation of terpenoids in defensive secretions of termites. Eudesmanes are sesquiterpene derivatives; amitol from the soldiers of *Amitermes excellens* is a 5β,7β,10β-eudesmane (Naya and Prestwich, 1982). Much less variation is found in literature with respect to repellent terpenoids to termites themselves: pulegone, that is repellent to the German cockroach, is also repellent to *R. lucifugus* (Floyd *et al.*, 1976).

In 1972 Bowers *et al.* published the isolation, identification and synthesis of an aphid alarm pheromone, a publication of only two pages, but with a great impact on aphid biology and agriculture (see 6.2.4). The senior author of this book happened to visit the New York State Agricultural Station at Geneva at that time and was addressed by W.L. Roelofs, who said "we solved your problem!" That problem was virus transmission by aphids in seed potatoes. The alarm pheromone, (E)-β-farnesene, is present in the glandular hairs of some wild potato species, a.o. *Solanum berthaultii*. The idea was breeding of potato cultivars with glandular hairs excreting the alarm pheromone. It is still a sound idea, but after 30 years it appears difficult to breed potato lines with high yield combined with effective glandular hairs, although there is progress.

In aphids, the alarm pheromone is secreted from the siphunculi (Fig. 5.6). A muscle runs down the siphunculus to the abdominal sternite. When it contracts, a slit opens and a droplet of waxy exudate emerges. The fluid rapidly solidifies in the air and can play a defensive role by gumming up the cephalic parts of predators. In addition the alarm pheromone is released with the fluid and causes restlessness of nearby aphids that often disperse or drop down from the plant. Fig 5.6 shows that in contrast to an opposite opinion of some insect physiologists, there is no specific gland that produces (E)-β-farnesene. Actually a droplet of hemolymph is excreted, that also contains isomers of (E)-β-farnesene such as (Z,E)-α- and (E,E)-α-farnesene, so the alarm pheromone is not necessarily a single compound (Gut and van Oosten, 1985). These compounds may be released by the degenerating fat cells (C in Fig. 5.6).

That there exist at least three farnesene isomers has been known for some time: Bowers *et al.*, 1972; Bowers *et al.*, 1977; Nishino *et al.*, 1977; Pickett and Griffiths, 1980 (Fig. 5.9). In industrial production of (E)-β-farnesene a certain proportion of the isomers is also obtained, and the same is obviously true for biosynthesis. Still, the presence of these isomers in hemolymph of aphids is not just determined by accidental chemical processes, as particular morphs contain specific ratios of the isomers (Gut and van Oosten, 1985)

*Chapter 5. Functions of natural terpenoids in the interrelationships   193
          between organisms*

and sesquiterpenoids including farnesene isomers have specific effects on aphid physiology (Gut *et al.*, 1991).

Alarm pheromones in general often have a broad interspecificity and (E)-β-farnesene is no exception: it serves as an alarm pheromone for several, but certainly not all aphid species studied. In *Megoura viciae* the alarm pheromone consists of α- and β-pinene, together with limonene (Pickett and Griffiths, 1980) and in *Therioaphis maculata* it is germacrene A (Fig. 5.2B).

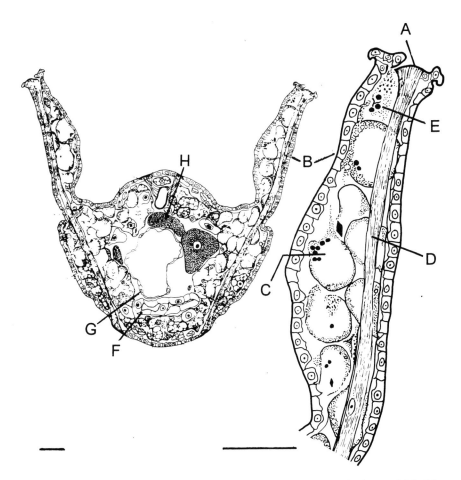

Fig. 5.6. Cross-section through the siphon region of the aphid *Myzus persicae*. A = siphon (or siphuncular) valve; B = siphon (siphunculus); C = degenerating fat cell; D = siphon valve retractor muscle; E = waxy droplet released into the hemolymph; F = intestinal tractus; G = hindgut; H = endosymbionts. Bar = 25 μm (with courtesy of M.B. Ponsen).

(E)-β-farnesene is a constituent of essential oils of many other plant species. One might expect that alighting of aphids on such plants be inhibited. However, this depends on the presence of additional volatiles interfering with the alarm response (which was not yet known in 1972). The essential oil of the hop plant contains (E)-β-farnesene. Nevertheless hops are readily colonized by *Phorodon humuli*, that normally reacts to (E)-β-farnesene. Dawson *et al.* (1984) demonstrated that (-)-β-caryophyllene and to a lesser extent α-humulene, both components of hop oil, strongly inhibit activity of (E)-β-farnesene. In our own experiments we have found that the degree of inhibition depends on the aphid species.

The monoterpene perillene (Fig. 5.7A), produced by a number of thrips species (Thysanoptera) is a repellent to workers of two ant species. It is often accompanied by rose furan (Fig. 5.7B) (Blum *et al.*, 1991). The components are present in anal exudates. *Thlibothrips isunoki* secretes β-myrcene as a defensive compound (Haga *et al.*, 1990). Probably this monoterpene hydrocarbon also serves as an alarm pheromone, because upon disturbance the aggregating individuals raise their abdomen, secrete a drop of anal fluid and rapidly move about. This species does not produce unsaturated carboxylic acids that are more commonly secreted by Thysanoptera. *Haplothrips leucanthemi* produces the terpene meillon, and the gall-forming thrips *Gynaikothrips uzeli* and *Liothrips kuwanai* secrete β-acaridial, but their function is not quite clear. Some terpenes may have other functions, such as sex pheromones, but even for Thysanoptera very little is known of sex pheromones (Terry, 1997).

The aromatic terpenoid juglone (Fig. 5.7C), a highly phytotoxic quinone produced by *Juglans nigra*, the black walnut, is a repellent to ants and is also synthesized by *Elaphrothrips tuberculatus* (Blum *et al.*, 1991). This brings us to ants (Hymenoptera) with their complex social structures like Isoptera.

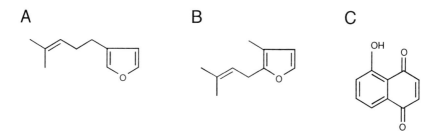

Fig. 5.7. Defensive terpenoids produced by both plants and insects. **A**: perillene; **B**: rose furan; **C**: juglone.

Chapter 5. Functions of natural terpenoids in the interrelationships between organisms

The Dufour glands of the army ant *Aenictus rotundatus* excrete a complex mixture of terpenoids, consisting more than 50% of terpenoids not generally present in Dufour glands. An example is (E)-β-ocimene in the glands of *Labidus predator* and *L. coecus* (Keegans *et al.*, 1993).

The poison gland of the ant *Myrmicaria eumenoides* secretes α- and β-pinene, limonene and myrcene. Separate colonies release different proportions of these monoterpenes. In higher concentrations these compounds evoke alarm behaviour and are virtually toxic to the ants, but in low concentrations the pinenes and limonene act as an attractant semiochemical (Howse *et al.*, 1977). The terpenes are used by other organisms, including plants, as defensive compounds (Table 6.3) and toxicity of higher concentrations of a semiochemical to the same species is no exception. (E)-β-farnesene, an alarm substance of aphids, is also toxic to aphids in moderate concentrations (see 5.1.3). Attraction and repellence of a volatile compound within the same species may be an equivalent to calling and shouting in vertebrates, albeit with a simple vocabulary. When individuals of different colonies meet each other they can discriminate between a calling signal of their own group and another one. It would be interesting to know whether the alarm response to a certain mixture is also restricted to members of the same colony or not, as an alarm response by members of the other colony could result in avoidance or flight behaviour.

Ants make their own antibiotics to protect their colonies from pathogenic fungi. The mandibular gland of *Lasius fuliginosis* contains several terpenoids. The main compound is also an uncommon one: 3-formyl-7,11-dimethyl-(2E,6Z)-dodecatrienal (original title: there must be some error here). It inhibited germination of spores of four species of insect pathogenic fungi in concentrations of 1-10 ppm (Akino *et al.*, 1995). More well-known terpenoids in the mandibular gland are limonene and geranylgeraniol, that can comprise over 50% of the secretion (Oldham *et al.*, 1994). Terpenoids were the main substances identified from cephalic secretions of formicine worker ants by Bergström and Löfqvist in 1970. Many of them elicit alarm behaviour of the workers. The dominating ones in the mandibular glands of *Lasius carniolicus* are citronella (oil) and geranylcitronellal.

The anal glands of ants of the subfamily of Dolichoderinae contain terpenoids with insecticidal properties. The principal repellent compounds of *Ochetellus glaber* secretions were determined to be (E,Z)- and (Z,E)-isomers of dolichodial in a 3:1 ratio (Cornelius *et al.*, 1997).

Leaf-cutting ants harvest specific plants as a substrate for their fungus gardens. Hubell *et al.* (1983) demonstrated that a terpenoid in the leaves of a neotropical tree, *Hymenaea courbaril* (Fabaceae), repels workers of *Atta cephalotes* and is toxic to the attine fungus.

Intraspecific terpene variation may play an important role in pine selectivity by leaf-cutting ants. Concentrations of myrcene (Fig. 5.4) and caryophyllene (Fig. 5.9) in *Pinus caribaea* were higher in needles of non-foraged trees than in those newly defoliated by *Atta laevigata* (Barnola *et al.*, 1994).

Among animals, arthropods are distinguished by their dominance in terms of numbers of species and individuals. The ubiquity of these populous organisms guarantees that they will be subject to great predatory pressure from both invertebrates and vertebrates. Arthropods themselves constitute a major group of predators. Ants, for instance, with their carnivorous propensity and an efficient system of prey acquisition, together with a richness in species (almost 15,000) form a serious threat to non-predatory insects that do not possess defensive jaws or legs. Undoubtedly the marine biology will reveal many terpenoid defensive compounds such as those produced by a Pacific nudibranch (Kubanek *et al.*, 2000, discussed in section 3.3.2). The present section focusses on insects.

Sometimes herbivorous insects react both to volatile host substances and to their own semiochemicals and the two can have an additive effect. The Mediterranean fruit fly, *Ceratitis capitata*, has calling males. The females react besides to ethyl-(E)-3-octenoate, to the terpenes geranyl acetate, (E,E)-α-farnesene and linalool. A strong antennal response was also recorded from two mango terpenes: β-caryophyllene and β-pinene (Cosse *et al.*, 1995).

Whether a terpene acts as a repellent or an attractant is not only species-specific but can also depend on the local concentration. Workers of *Trigona subterranea* (stingless bees) generate chemical trails with a secretion, dominated by citral. In the vicinity of the nest, however, sustained alarm behaviour is observed resulting from increased excretion of the same compound upon approach by enemies (Blum *et al.*, 1990). One should realize that citral is an effective fungicide, illustrating the multipurpose significance of monoterpenes. As mentioned before, citral is not a single component, but a mixture of two citral isomers, the acyclic monoterpenes geranial (citral A) and neral (citral B) (Fig. 4.2). The alcoholic versions exist named geraniol and nerol and have CH_2OH groups instead of CHO. Another example is the special composition and specific concentration of the separate volatile compounds of the tree *Fraxinus pennsylvanica* that exert an extraordinary repellency towards insects (Markovic *et al.*, 1996).

When applied to leaves of *Capsicum annuum* farnesol, 2,10-bisaboladien-1-one and geraniol inhibited settling of the aphid *Myzus persicae* in doses <100 µg/cm^2 (Gutiérrez *et al.*, 1997). We found farnesol (Fig. 5.9), a precursor of JH3 (Fig. 2.14), to cause a high mortality in nymphs of *Aphis fabae* in very low doses (van Oosten *et al.*, 1990).

Chapter 5. Functions of natural terpenoids in the interrelationships between organisms   197

Terpenes can mask the presence of terpenoid semiochemicals. Limonene, for example, inhibits attraction of *Hylobius abietis* to pitfall traps baited with α-pinene or to pieces of Scots pine (Nordlander, 1991). When only one part of caryophyllene (Fig. 5.9) is present on 300 parts of (E)-β-farnesene, the alarm reaction in the aphid *Myzus persicae* is not evoked, and the aphids will not disperse upon arrival of natural enemies. This phenomenon may be the result of a selection process that gives neighbouring conspecifics optimal chances for survival. As the aphids do not disperse they will not attack other plants and remain easy food for natural enemies.

This section has been focussed on insects as they are part of the pentagram depicting interrelationships that form the basis of this volume. Nevertheless terpenoids are important defensive compounds in other taxa. As an example Tomaschko (1994) reports that ecdysteroids from *Pycnogonum litorale* (Pantopoda, Arthropoda) act as a chemical defense against *Carcinus maenas* (Decapoda, Crustacea) in a marine predator-prey relationship.

Our last remark is on miriamin, a defensive diterpene from the eggs of a land slug (*Arion* sp.). It is an antifeedant to a coccinellid and probably to other predators. Schroeder *et al.* (1999) have named this compound miriamin as a tribute to Miriam Rothschild, a superb naturalist and unrelenting conservationist.

5.1.3   Repellents and deterrents: defensive metabolites by plants

There is a wealth of information on natural terpenoids with defensive functions in plants. Their ecological roles have a.o. been discussed by Langenheim (1994). It is easy to fill a big book with terpenoid allomones. Many of them are antibiotics, but do not have a messenger function *per se* and their effects on behaviour are limited. Many classes of volatile monoterpenes inhibit plant growth and can be regarded to have allelopathic functions (Romagni *et al.*, 2000). Some of these are utilized in agriculture and discussed in section 6.2.4. A selection has been made of material that facilitates discussions in chapter 5.2. The present section focusses on repellents and deterrents as defensive metabolites (allomones) of plants, directed towards insects (Fig. 5.8). Some of these terpenoids may not be perceived at all by the herbivores, but act as internal messengers after ingestion of plant material, e.g. JHs or their derivatives. They belong to the juvenoids, interfering with normal insect development and anti-juvenoids that prevent normal secretion of JH. Together with hydroxycortisones they form a class of toxic terpenoids interfering with normal messenger functions.

This class of plant allomones is excellently discussed in a book by J.B. Harborne (1993).

Normally speaking the sky is blue and the world around us is green. If there were too many herbivores our environment would look quite different. About half of the world's insect species are associated with plants. In the course of evolution both plants and insects have developed intriguing strategies to exploit each other at acceptable risk levels. Plants have to defend themselves against much more than herbivores alone, and host plant resistance is often partial.

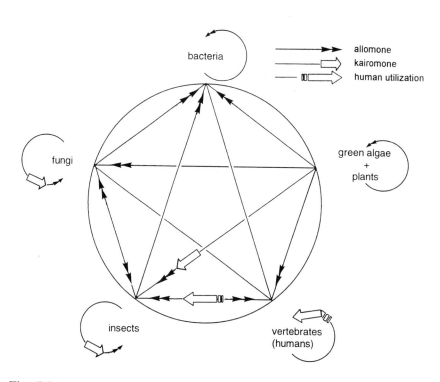

Fig. 5.8. Major semiochemical functions of natural terpenoids as allomones (repellents and deterrents) and kairomones. Humans utilize essential oils as semiochemicals.

When we watch our busy bees visiting flower after flower, collecting nectar and dispersing pollen we are fascinated by what seems to be an efficient relationship that benefits both plants and insects. A closer look at insect-plant relations, however, reveals that there has been a continuous war

Chapter 5. Functions of natural terpenoids in the interrelationships between organisms

without armistice, during which defense systems were developed to be broken by adaptations of herbivores. The utilization of volatile plant metabolites e.g. to find a host is only one aspect of the dynamic equilibrium between plants and insect species (see 5.1.4). Many terpenoid allomones have a degree of toxicity to herbivores that can be based on different modes of action (see 4.2.2). Herbivores try to avoid plants or plant parts that affect their metabolism and may react to relatively low concentrations of volatiles.

At higher concentrations many repellents and deterrents become toxic to the target insects, and that is probably the reason why they *are* repellents. The same holds for combinations of terpenoids or structurally joined terpenes. Linalool/limonene mixtures are allelopathic to fruit flies. They make citrus fruits resistant to these insects, but are not present in grapefruit (Greany *et al.*, 1983). Geranyllinalool is a natural insecticide in *Pinus pinaster*. It is toxic to subterranean termites of the genus *Reticulitermes* (Nagnan and Clement, 1990), but it is also present in termites living on *Pinus* spp. (Lemaire *et al.*, 1990). Twelve species of predatory or competing ants appeared to be very sensitive to this compound (Lemaire *et al.*, 1986).

Several terpenoid feeding inhibitors such as drimanes generate post-ingestive sublethal toxic effects, resulting in decreased food intake (Gols *et al.*, 1996; Ortego *et al.*, 1995). Messchendorp *et al.* (1998) were first in reporting a deterrent receptor of *Leptinotarsa decemlineata* larvae (Coleoptera). An epipharyngeal sensillum with chemosensory function was shown to contain five sensory cells. Electrical recording of these cells elucidated that one cell responded to water, a second to sucrose and a third one to a sesquiterpene drimane that acts as a feeding deterrent. The response of the "deterrent cell" increased with increasing doses of drimane, with a maximum response at 1 mM/l (an almost saturated solution). Moreover, the response of the sucrose cell was reduced when the sensillum was stimulated with a combination of sucrose and drimane, suggesting that drimane inhibits the response of the sucrose cell. From these results four main modes of action of (terpenoid) deterrents can be proposed:
- Competitive binding to receptors
- A signal from a deterrent cell to the CNS directly influencing feeding behaviour
- Interference with the response of cells sensitive to feeding stimulants
- A "subtraction" of stimulant and deterrent signals in the CNS interfering with the neural code.

The first two possibilities have been discussed in section 4.1.2, in odour perception and gustation. The third possibility has been proven to operate in many feeding deterrents (e.g. Schoonhoven and Luo, 1994). There may be some analogy with the phenomenon of interference as in light reflection and transmission; the same type of receptor cell can be sensitive to two different terpenes, but rise and decay of the response may be asynchronous, resulting in a (partly) extinguished signal. The fourth option has been discussed as a realistic phenomenon before (e.g. Schoonhoven and Blom, 1988).

Megastigmus pinus and *M. rafni* seed wasps (Hymenoptera) are repelled by limonene in the odour blend of cones of potential *Abies* host species and attracted to the blend of their real hosts containing other terpenes including α-pinene (Luik *et al.*, 1999). The inhibitory effect of limonene could be based on a similar mode of action as discussed above. With this in mind we compare the repellent and deterrent effect of different terpenoids on a number of insect herbivores.

MONOTERPENES, SESQUITERPENES AND DITERPENES

Geraniol (Fig. 2.5) is a repellent for *Schistocerca americana* (Orthoptera), whereas it is a well-known kairomone for honeybees (Hymenoptera (Dobson *et al.*, 1990)). Mulkern and Toczek (1972) have studied the toxic effects of plant extracts containing sesquiterpene lactones on *Melanoplus differentialis* and *Melanoplus sanguinipes*. Information on effective repellents to Orthoptera is scarce, but a considerable amount of work has been done on antifeedants. Research has been focussed on deterrent, toxic and growth-regulating effects of azadirachtin and other triterpene derivatives of the neem tree, *Azadirachta indica*. Mordue *et al.* (1993) have made a review on the antifeedant effect of azadirachtin and on neem extracts in general on feeding behaviour, morphogenetic peptide hormone release and loss of fitness of *Schistocerca gregaria* and *Locusta migratoria*. Biochemical effects on the cellular level are still poorly understood and this hampers the research on simpler, mimetic substances. See also Gols *et al.*, 1996.

Triterpene derivatives as present in neem seeds do not primarily exert a repellent effect on grasshoppers, although they have antifeedant properties. Aqueous neem extracts can be used in the field, a.o. against *Kraussaria angulifera*, the Sahelian grasshopper, and protect the crop from substantial feeding. First to third instar nymphs may additionally suffer from inability to moult or from incomplete sclerotization (Passerini and Hill, 1993).

A number of repellents to insects of other taxa are more or less effective against cockroaches. The group of Jain in India found zizanal and epizizanal, present in vetiver oil, to be repellent to cockroaches and to have low toxicity to mammals. An extended review including other classes of natural antifeedants is by Jain and Tripathi (1993). Limonene (Fig. 2.7) as a constituent of citrus oil has insecticidal properties against *Theganopheteryx* spp. (Taylor and Vickery, 1974).

Chapter 5. Functions of natural terpenoids in the interrelationships between organisms

Feeding deterrents against Isoptera can be isolated from *Pinus* needles, e.g. against *Odontotermes obesus*. Terpenes of another coniferous tree, *Taxodium distichum*, the bald cypress, are involved in resistance to *Coptotermes formosanus*, a Formosan subterranean termite (Scheffrahn *et al.*, 1988). Monoterpenoid alcohols, especially eugenol and geraniol were toxic to *C. formosanus* (Cornelius *et al.*, 1997). When termites were confined in containers with a treated and untreated layer of sand, eugenol had a detrimental effect on mortality and feeding at a concentration of 200 nl/g of sand. This, however, does not make eugenol a selective feeding deterrent to the termites. Its main effect is presumably because of the toxicity of the vapour. Because of its volatility and low toxicity to mammals, eugenol may be a candidate for use as a fumigant against subterranean termites.

Clerodane diterpenes from *Detarium microcarpum* leaves exhibited strong feeding deterrent activity against workers of the subterranean termite, *Reticulitermes speratus* at about 1% (Lajide *et al.*, 1995). As terpenes with comparable structures are produced by plants of different origin, it is not so strange that terpenes from the maritime pine *Pinus pinaster* are toxic to subterranean termites of the genus *Reticulitermes* (Nagnan and Clement, 1990).

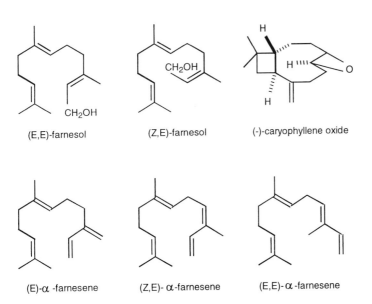

Fig. 5.9. Two farnesol isomers, caryophyllene oxide and examples of farnesene isomers. For functions see text.

According to Uchiyama *et al.* (1991) natural extracts of *Citrus* and *Phellodendron* can be combined with low doses of insecticides, such as pyrethroids, to obtain an effective control of termites.

(E)-β-farnesene is a well-known alarm pheromone of aphids. The isomer (E,E)-α-farnesene and a mixture of farnesol isomers (Fig. 5.9) cause a high mortality among nymphs of *Aphis fabae* and *Myzus persicae* when topically applied in a dosage equivalent to the amount of (E)-β-farnesene present in the hemolymph of one individual (van Oosten *et al.*, 1990) (Table 5.1). (E,E)-α-farnesene induces sexual characteristics, and the hemolymph of winged males and gynoparae contains an extra-proportional amount of (E,E)-α-farnesene (Gut *et al.*, 1991).

Four species of aphids that live on Sitka spruce, *Picea sitchensis*, appeared to have a different level of tolerance of two monoterpenes that are present in stems and needles: myrcene (Fig. 5.4D) and piperitone (Jackson *et al.*, 1996). The aphids were exposed to 3 µl of the terpenes in a glass vessel of 25 ml. This means that data are presented on the effect of terpene vapour on survival of the aphids, not on possible antifeedant effects. The authors conclude that myrcene and piperitone play a role in the resource partitioning shown by the four aphid species. The aphids also have specific morphological adaptations, such as the length of their proboscis, and appear to be physiologically adapted to exploit a particular feeding site. Adelgids, which are closely related to aphids, are known to have a limited tolerance of monoterpenes. High levels of santalene and camphor in red spruce (*Picea rubens*) inhibit colonization by *Pineus floccus* (Alexander, 1987) and high concentrations of limonene and myrcene in Douglas fir deter the adelgid *Adelges cooleyi* (syn. *Gilletteella cooleyi*) (Stephan, 1987).

Table 5.1. Percentage mortality (after one week) of L_1 nymphs of long-day-reared *Aphis fabae* treated with farnesol and (E)-β-farnesene (25 ng/mg). Solvent is isopropylalcohol

Morphs	Treatment		
	isopropylalcohol	farnesol	(E)-β-farnesene
L_1 stage from untreated virginoparous apterae	0	87	
L_1 stage from untreated virginoparous alatae	0	75	64
Desc. of treated virginoparous apterae	0	0	
Desc. of treated virginoparous alatae	0	0	43

Methyl salicylate and (1R,5S)-(-)-myrtenal that also stimulate olfactory cells in the primary rhinaria of *Aphis fabae*, are repellent to this aphid and in addition inhibit attraction to volatiles from broad bean, a natural host (Hardie et al., 1994). An abundant component of defensive resins produced by gymnosperms is (1S,5S)-(-)-α-pinene, which is metabolically related to (1R,5S)-(-)-myrtenal. The structure of these terpenes comes close to that of the phenolic allelopathic component methyl salicylate. The masking effects of these components on the odour blend of a natural host plant could serve as another example that the decision to accept or reject the odour of a plant is not made on the level of the sensilla.

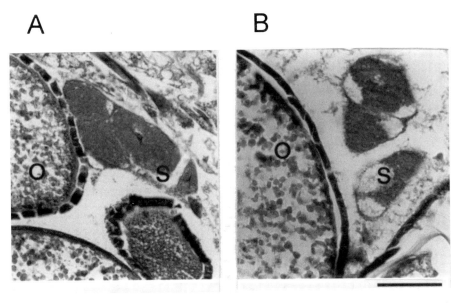

Fig. 5.10. Sections of oviparae of *Acyrthosiphon pisum*. **A**: mothers treated with solvent only; **B**: mothers treated with farnesene/pulegone. O = oocyte; S = mycetocyte with endosymbionts. Note enlarged oocyte in B. Bar = 50 μm.

The aphid *Acyrthosiphon pisum* is reluctant to penetrate into leaves of host plants treated with myrcene. Treatment of *Myzus persicae* and *Megoura viciae* with menthol results in disturbed feeding behaviour and even in cannibalism, which is a well-known phenomenon among Thysanoptera. Probing behaviour of *M. persicae* on leaf discs treated with geraniol, farnesol (Fig. 5.9) and bisabolene was strongly inhibited during 10 min of electrical recording of stylet penetration (Gutiérrez *et al.*, 1997). Pulegone (Fig. 2.5) has no effect on host plant acceptation but reduces reproduction in parthenogenetic individuals, affects wing dimorphism by stimulating the alate course of development, and inhibits oviposition in sexuals of *A. pisum* (Fig. 5.10). This may be caused by malfunction of the trophocytes that are dependent on amino acid homeostasis realized by the endosymbionts of the genus *Buchnera* (Douglas, 1992; Douglas and Prosser, 1992). Pulegone has anti-microbial properties and destroys the symbionts in a dosage of 300 ng/mg of aphid (Harrewijn *et al.*, 1994).

The vapour of (E)-β-farnesene in concentrations of 500 mg/m^3 and higher causes high mortality in the whitefly species *Trialeurodes vaporariorum* and *Bemisia tabaci* (Klijnstra *et al.*, 1992) (Table 5.2). The experiments were done in perspex cages. One should realize that when working with controlled release dispensers in glasshouses, the vapour pressure of (E)-β-farnesene must be kept high enough to obtain a sufficiently lethal effect. Although these experiments illustrate why (E)-β-farnesene is a repellent terpene, it acts as a toxic allomone in this situation. The same holds for plant-derived JHs. Their pathological effects on insects a.o. discussed by Bowers (1985), Harborne (1993) and Slàma (1987) is different from that of semiochemicals.

Table 5.2. Mortality of adult whiteflies on tobacco plants in 50 l perspex cages with filter paper discs containing 15 mg of (E)-β-farnesene

Treatment	Mortality (%)		
	24 h	48 h	72 h
Control	30	42	44
10 mg	28	73	75
30 mg	80	88	90
100 mg	92	92	92

It is possible that the biosynthesis of mevalonic acid takes more than one pathway in insects and the regulatory mechanism is not necessarily the same as in mammals (Havel *et al.*, 1986). This would explain the additive or even synergistic effect of monoterpenes and sesquiterpenes on vital functions in aphids. In this respect the MAI metabolism of the *Buchnera* symbionts should be kept in mind (see section 2.1.4).

The oil of *Artemisia annua* (Asteraceae), containing pinenes, cineole, camphor and santonin inhibits ovary development of *Dysdercus koenigii* (Hemiptera) when applied topically (Rao *et al.*, 1999). An explanation could be that the oil interferes with the reproduction system that is hormonally regulated. The authors found a greater than normal neurosecretory activity in the brain of the insects.

Aphids are equipped with a number of detoxifying enzymes such as esterases and oxidases that become active upon salivation or are secreted into the lumen of the intestinal tract. Aphids are not used to acquisition of high dosages of terpenoids because they feed on phloem sap. Volatile terpenoids, which are toxic to herbivores of various insect taxa, have reported anti-microbial properties and act as allomones and repellents against aphids (Harrewijn et al., 1995). The olfactory receptor neurones of aphids certainly respond to terpenes. Visser et al. (1996) compared the peripheral perception of plant volatiles by aphids with the electroantennogram (EAG) technique in four species: A. fabae, Brevicoryne brassicae, M. persicae and M. viciae. A. fabae responds relatively strongly to (E,E)-α-farnesene, B. brassicae to carvone, especially to (R)-(-)-carvone, and to (E)-β-farnesene. Besides to the green leaf volatiles, nitriles and isothiocyanates, the antennal preparations reacted to the following terpenes: (1S)-(-)-α-pinene, (1S)-(-)-β-pinene, (S)-(+)-carvone, (R)-(-)-carvone, α-terpineol, linalool, geraniol, nerol, citronellal and (+)-citronellol. One should realize that a response of antennal sensillae is not the same as the response of an individual aphid, and that the experiments were done with apterous aphids. Nevertheless these results suggest that aphids can respond differently to plant volatiles among which terpenes.

Aphids can smell terpenes with other functions than sex pheromones or allomones. A range of plant volatiles is perceived that can function as kairomones, synergistics to sex pheromones, or as allomones. (S)-(-)-β-pinene, (S)-(+)-carvone, (R)-(-)-carvone, α-terpineol, linalool and citronellal evoke strong EAG responses in Megoura viciae (Visser and Piron, 1995). EAG responses from winged virginoparae (summer form) and gynoparae (autumn form) of A. fabae revealed that peripheral olfactory perception of most volatiles is not significantly different between these forms. In addition, the responses of adult wingless virginoparae to volatiles of two alternative hosts showed similar perception (Hardie et al., 1995). This suggests that any odour discrimination between these host plants be not perceived at the level of the peripheral olfactory receptors.

The psyllid Trioza apicalis feeds on carrots, parsley and caraway, but dislikes dill, on which oviposition is poor. All belong to the Apiaceae. Nehlin et al. (1996) have compared oviposition on the plants with the spectrum of volatile compounds released from the different species. Subspecies of carrots (Daucus carota) were characterized by high contents of sabinene (mostly the (+)-enantiomer) and α-pinene (both enantiomers). In dill (Anethum graveolens) myrcene was dominating, followed by other monoterpenes, among which β-pinene. Analyses by other groups rank myrcene as only a trace compound and D-carvone as the major volatile of dill (Table 6.1). Some females will sit on dill, but do not deposit eggs. Myrcene may act as an oviposition inhibitor, although on *Carum carvi*, which can be fed upon, there was no oviposition either. Carvone is a major terpene of *C. carvi* (Table 6.1). The volatile spectrum of this species is strongly dominated by carvone, followed by limonene and α-terpinene. This leaves the possibility that oviposition stimulants (such as from carrots) are more important in host plant acceptance than inhibitors.

The mint monoterpene pulegone (Fig. 2.5) has a repellent action on larvae of *Spodoptera eridania* when included in the diet (Gunderson *et al.*, 1985). Diterpene acids of *Larix laricina* are potent antifeedants to *Lymantria dispar*, the gypsy moth, that can recover from and compensate for short-term exposure to monoterpenes (Powell and Raffa, 1999). Little is known of terpenoid repellents to Thysanoptera. Higher concentrations of the terpenes may have a repellent action at close range, a well-known phenomenon with kairomones that originally had a function as allomones (Harrewijn *et al.*, 1995). However, little is known of repellents to thrips: butyric acid additions seem to exert such effects, but possible combinations with terpenes have yet to be investigated. Some thrips species seem to be repelled by crushed basil leaves, suggesting that linalool, eugenol, caryophyllene (Fig. 5.9) or 1,8-cineole are candidates (see Table 6.1). With respect to host plant resistance a number of plant factors are known, ranging from glandular hairs to a low amount of aromatic amino acids, but at present resistance based upon secondary metabolites, let alone terpenoids, is hidden in a "black box".

Both camphor and 1,4-cineole (Fig. 2.7) are effective repellents to *Delia radicum*, and both compounds are produced by the camphor tree, *Cinnamomum camphora*. These terpenes can be extracted or synthesized, but in effectivity they are surpassed by the simple phenolic salicylaldehyde (den Ouden *et al.*, 1996).

The vapour of pine oil inhibits the oviposition response in the onion maggot, *Delia antiqua*. The principal deterrent property of this oil (Norpine 65, Northwest Petrochemicals, Anacortes, Washington) was found to reside in three monoterpenes, 3-carene, limonene and *p*-cymene (Ntiamoah *et al.*, 1996). Inhibition, however, was not complete and the authors advise a monoterpene-based deterrent to be combined with other deterrents or with attractants in another area. A cheap source of pine oil is a by-product from wood pulp mills. It usually contains more than the three monoterpenes mentioned. Oils containing myrcene, α-phellandrene and β-phellandrene could also be effective, depending on the ratio of the constituting monoterpenes.

Pine oil also inhibits feeding and oviposition of *Musca domestica*, the housefly. For this insect the most active substance appeared to be linalool (Magange *et al.*, 1996). The authors state "because fly maggots naturally develop in and rely on microbe-rich organic sources, gravid females may perceive and avoid oviposition sites that are rich in anti-microbial compounds such as linalool". Linalool is not mentioned as a main substance in pine oil by Ntiamoah *et al.* (1996). Magange *et al.* (1996) suggest that further improvement of the effect of pine oil may be obtained when it is enriched with Z-menth-2-en-ol and E-piperitol. Larval development of *M. domestica* is inhibited by farnesol (Sharma and Saxena, 1974). When topically applied, citronellic acid and thymol were most toxic among several monoterpenes tested. LD50 was reached with less than 40 µg per insect. Pulegone (40 µg) and (S)-limonene with 50 µg came close (Lee *et al.*, 1997).

For those dipterans, which are not regarded to be very noxious or harmful, not many studies have been made on toxic or repellent phytopharmaceuticals. Still, investigations on terpenes have recently been made for *Drosophila melanogaster*, the model insect of those working on gene expression. Franzios *et al.* (1997) tested essential oils extracted from the mint species *Mentha pulegium* and *M. spicata* for insecticidal and genotoxic activities on this laboratory insect. They also tested the main constituents of the oils: pulegone, menthone and carvone. Pulegone was found to be the most toxic, causing death of the insects in extremely small amounts, followed by carvone and menthone. Surprisingly, the essential oil of *M. pulegium*, consisting for more than 75% of pulegone, was 9 times less effective than would be expected from its pulegone content alone. The reverse was true for menthone: the oil was 6 times more effective than could be expected when its lethal effects were linearly dependent on menthone. Tests with mixtures of the three consisting terpenes strongly suggested that synergistic/antagonistic phenomena exist with respect to the terpenes in the oils. Somatic mutation and recombination tests with the oils and their constituents revealed that *M. pulegium* does not exert any mutagenic or recombination effects, but *M. spicata* shows strong mutagenic activity, and no recombinogenic activity. Menthone appeared to be a potent mutagenic but not recombinogenic inducer. Again, the toxic and genotoxic properties of the essential oils were not in accordance with those of their major constituents. *M. spicata* is rich in perillyl alcohol, which has strong anti-tumour properties (see 4.1.5 and 7.1.8). This example illustrates that a natural product, e.g. an essential oil, may exert a different effect than its contributing terpenes.

We like to stay for a moment with the same Greek research group who's members in addition to the above mentioned investigation, tested the effect of essential oils obtained from oregano plants on *Drosophila*. Oregano is not a specific species, but a favourite spice in the Italian and Mediterranean kitchen, containing the aromatic Lamiaceae *Origanum vulgare*, *Coridothymus capitatus* and *Satureja thymbra*. In addition to the general information in Table 6.1 the main constituents of *O. vulgare* and *C. capitatus* are carvacrol (75-82%) and *p*-cymene (6-10%), *O. vulgare* being characterized by a noticeable content of γ-terpinene (6%), whereas *S. thymbra* is rich in thymol (35%), *p*-cymene (27%), γ-terpinene (14%) and α-pinene (4%). The essential oil of *S. thymbra* was found to be the most effective as an insecticide, followed by *O. vulgare* and *C. capitatus*. The same unexpected antagonistic effects as found with the mint species occurred with the oregano oils. Thymol, with a LD50 of 2.6 µl is less effective than carvacrol, with a LD50 of 1.6. Still, the oil of *S. thymbra* was more effective than that of the other species with their much higher content of carvacrol. Mixtures of carvacrol and thymol were less toxic than each of the compounds itself, suggesting antagonistic phenomena. The overall effects of the oils may be more complex, as more than 40 other minor volatiles are present in each of the oils, and the picture could be different with respect to other insects. It certainly is to micro-organisms: thymol is synergistic to the anti-microbial effect of carvacrol.

The oils of the three oregano plants did not exert any mutagenic or recombinagenic activity. Thymol alone was found to be a potent mutagenic but not a recombinagenic inducer. Cooparative screening for spontaneous or induced mutagenesis even showed that the oil of *O. vulgare* subsp. *hirtum* reduces mutagenic processes. This is in accordance with the proposed anticancer potential of oregano oil with its effect on glutathion *S*-transferase (Lam and Zheng, 1991; see also 4.1.7).

The number of cases of head and body lice infestation throughout the world should not be underestimated as it seems to run into hundreds of millions a year. Lice quickly develop resistance to existing pediculicides which increases the need for effective lice repellents that should be non-toxic and non-irritating. Although the repellent activity of essential oils and their components against blood-sucking arthropods has been known for a long time, their efficacy against human lice is poorly documented. Mumcuoglu *et al*. (1996) compared five essential oils (nine compounds) to N,N-diethyl toluamide (DEET) for their repellent activity against the human body louse, *Pediculus humanus humanus*. Tested on corduroy patches apart form the human body the most effective repellents were DEET and citronella. Practical applications with serial dilutions on human hairs revealed that citronella was the most potent repellent for lice, followed by citronellal, rosemary, geraniol and DEET. The differences were not big, providing that

Chapter 5. Functions of natural terpenoids in the interrelationships between organisms

proper dilutions were used. DEET remained effective at a dilution of 1:32, geraniol at 1:8, citronella at 1:4 and rosemary and citronellal at 1:1.

Most blood-sucking insects react positively to CO_2 produced by their hosts. Tests neglecting this phenomenon could produce less reliable data. Mumcuoglu *et al.* (1996) have elegantly solved this by treating the hairs with ammonium bicarbonate, which is also attractive to lice.

D-limonene is derived from citrus peels and as such registered, and approved by the Food and Drug Administration for flea and tick control. For special purposes one could think of combinations of these monoterpenes, either with each other or with low doses of insecticides acting on cholinesterase or the central nervous system.

Sadof and Grant (1997) found that varieties of *Abies grandis* (Scotch pine), which are resistant to *Dioryctria zimmermanni* (Hymenoptera), emitted relatively high levels of limonene compared with susceptible ones. A deterrent effect of toxic terpenes is not uncommon. The Sitka spruce weevil, *Pissodes strobi*, oviposits in early summer on one-year old terminals of young Sitka spruce. Although limonene acts as an attractant, myrcene and α-pinene are repellents. The preferred material for oviposition contained significantly less myrcene than slow growing terminals and current growth (Hrutfiord and Gara, 1989). Monoterpenes in the oleoresin of *Picea glauca* and *Larix laricina* are toxic to spruce beetles, *Dendroctonus rufipennis* and to bark beetles, *D. simplex*. Besides, the monoterpenes seem to act as deterrents, as limonene, 4-allylanisole (methyl chavicol), myrcene and β-phellandrene inhibited the response of the spruce beetle to frontalin, whereas myrcene and limonene inhibited the response of the eastern larch beetle to seudenol (Werner, 1995). Limonene, myrcene and β-phellandrene were found in greater quantities in the wound tissues than in undamaged tissues, indicating the presence of an induced defensive system. Defoliated *L. laricina* trees responded with increased levels of limonene in the oleoresins during the first 2 years of successive defoliation, but these levels decreased during the 3rd and 4th year of successive defoliation.

Leaves of two highly aromatic plants, *Artemisia tridentata* (Asteraceae) and *Monarda fistulosa* (Lamiaceae) inhibited oviposition by *Zabrotes subfasciatus,* the Mexican bean weevil, in beans in concentrations less than 1% w/w. Most of the volatile components from the dried leaf material were terpenoids, with camphor and 1,8-cineole (4.0 mg/g) being most abundant in *A. tridentata* and carvacrol being most abundant in *M. fistulosa* (Weaver *et al.*, 1995).

COMPLEX TERPENOIDS

Azadirachtin is not the only effective deterrent in neem. Kearney *et al.* (1994) tested a variety of whole plant and *in vitro* cell cultures from neem seedlings of Ghanaian origin on feeding behaviour of *Schistocerca gregaria* (Orthoptera). Azadirachtin was present in seed extract, but was not detected in extracts of other plant parts including callus.

The triterpenoid azadirachtin is a feeding deterrent to *M. persicae*. It reduces the period of sustained phloem feeding from plants treated systematically with the compound. Thus, it decreases the acquisition and transmission of phloem-limited and persistently transmitted viruses. Besides, studies on *Nilaparvata lugens* (the brown planthopper) have shown that the symbiont population declined after treatment with azadirachtin-containing extracts (Raguraman and Saxena, 1994). Azadirachtin presented in artificial diets to *M. persicae* hardly acts as a feeding deterrent but clearly influences the endosymbiont population. At 2600 ppm azadirachtin displayed an aphid mortality of 80-90%. At dosages ranging from 100 to 2500 ppm the endosymbiotic bacteria become degenerate and development of the aphids is strongly inhibited. Moreover, the synthesis of symbionin, a protein abundantly produced by the endosymbiont, and released into the hemolymph, is affected. As a consequence, the transmission of luteoviruses, like potato leafroll polerovirus (PLRV) is strongly inhibited since symbionin is the key component in determining the persistent nature of these viruses in aphids (van den Heuvel *et al.*, 1994, 1997). Morphological aberrations on the bacterial endosymbionts were observed in aphids which fed on 2560 ppm of azadirachtin, and between 320 and 2560 ppm the release of *Buchnera* GroEL into the hemocoel was reduced (van den Heuvel *et al.*, 1998) (Fig. 5.11). There may also be an effect on the recognition by receptors of the salivary glands, resulting in inability of the virus to pass the membranes of these glands. At dosages lower than 100 ppm, azadirachtin still causes a considerable reduction in PLRV transmission, although no allelopathic effects on the aphids were observed and the endosymbiont population was similar as in untreated aphids. These results are somewhat in contrast to those of Mordue *et al.* (1996), who found that 10-20 ppm azadirachtin in the diet of *M. persicae* caused a 20-fold reduction in the number of nymphs together with a high proportion of non-viable offspring. This suggests that there is a direct effect on the reproductive system prior to an antibiotic effect on the endosymbionts.

Treatment of leaf discs with 1% neem seed oil, originally containing about 4000 ppm of azadirachtin had the same effect on reproduction of two out of three aphid species as 40 ppm of pure azadirachtin (Lowery and

Chapter 5. Functions of natural terpenoids in the interrelationships between organisms   211

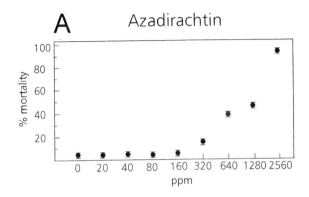

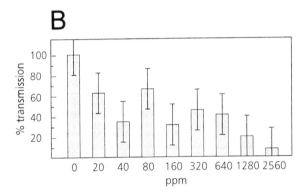

Fig. 5.11. The effect of azadirachtin on **A**: mortality of *Myzus persicae* and **B**: the aphid's ability to transmit PLRV. One-day-old nymphs were allowed to feed for 4 days on chemically defined diets containing the indicated concentrations of azadirachtin. Error bars represent the 95% confidence intervals for the means (with courtesy of J.F.J.M. van den Heuvel).

Isman, 1996). In *M. persicae* and *Nasonovia ribisnigri* this is probably caused because fewer embryos develop and in addition large numbers of dead offspring are produced. However, the reduction in reproductive rate is not primarily caused by increased embryo mortality. Azadirachtin may cause reduced oocyte maturation with the result that fewer embryos develop and growth of embryos is delayed. According to Schmutterer (1987), the sterilizing effect of azadirachtin results primary from inhibition of oogenesis and vitellogenesis. Trophic cells or young oocytes may be more sensitive to neem than are developing embryos. As in the case of pulegone, quoted above, azadirachtin could affect the trophocytes by acting upon the symbionts. Probably there is more to it, as aphids exposed to neem seed oil or azadirachtin produced large numbers of dead offspring, a pathological process which we have observed several times in aphids during our studies with other terpenoids. The effective dose mentioned by Lowery and Isman (1996) is lower than found by van den Heuvel et al. (1994). Moreover, the latter group used a chemically defined diet whereas the study by Lowery and Isman was with leaf discs. Besides, this low dose did not have an effect upon *Chaetosiphon fragaefolii*, but as leaf discs were used it is difficult to say whether the differences are based on the aphid species or on uptake of azadirachtin by their host plants.

Neem seed oil at a concentration of 1.0% or the equivalent amount of pure azadirachtin (about 40 ppm) applied to leaf discs also resulted in a high mortality of second instar nymphs of both *M. persicae* and *N. ribisnigri* (Lowery and Isman, 1994). It was observed that in this stage the aphids failed to moult or could not escape the exuvium. When directly applied to the insects either by application to the dorsal abdomen or by spraying, the dose needed for LC50 was substantially higher than the calculated dose effectuated after leaf disc treatment, indicating that the active components of neem exert their effects mainly after ingestion. A strong antifeedant effect is less probable, as it does not explain the pathological effects on embryonal development described above. Starved aphids do hardly produce offspring, but do not show these pathological effects.

Nisbet et al. (1993) using the electrical penetration graph (EPG) technique (Fig. 5.12) to analyse feeding behaviour of adult *M. persicae* on tobacco systemically treated with azadirachtin, found a reduction in the mean duration of individual periods of E(pd) patterns. They did not discriminate between E1 (salivation) and E2 (ingestion) waveforms, so a conclusion that a reduction in phloem sap intake occurred can only follow after a substantial reduction in the proportion of E patterns, as part of the time E1 will be produced, not contributing to food intake. Their data seem to justify the statement that starting with 300 ppm of azadirachtin, phloem sap intake is reduced. The authors do not make such a precise statement, but refer to "shortened periods of phloem contact" with progressing periods of non-penetration activity. The question remains if this behaviour is caused by a direct antifeeding effect or by a physiological effect of azadirachtin *after* intake of a certain amount during phloem feeding.

Neem seed oil contains other triterpenoids, some of which are also toxic to aphids. Salannin, for instance, is toxic to *Lipaphis erysimi* with a LC50 of 0.06% when topically applied (Singh et al., 1988).

Chapter 5. Functions of natural terpenoids in the interrelationships between organisms

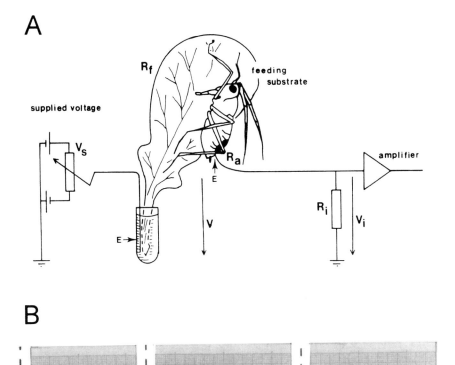

Fig. 5.12. The EPG technique applied to aphids. **A**: scheme of the circuit. V_s = adjustable supply of DC potential to plant; V = circuit potential (which is sum of V_s); E = electrode potentials (arrows); V_i = signal potential before amplification; R_a = aphid resistance; R_i = input resistance of amplifier; R_f = (negligible) resistance of feeding substrate. **B**: C and E patterns produced by *Myzus persicae* on a potato leaf, including a cell puncture (pd) and the onset of salivation (E1) and ingestion (E2) in a phloem vessel (with courtesy of W.F. Tjallingii).

Pistachio species, widely distributed in the Mediterranean and Middle-East areas are known for the oleoresins in their trunks. They mainly consist of euphane, dammarane and oleanane types. Gall-inducing aphids such as *Slavum wertheimae* and *Pemphigus utricularius* do not produce galls on all specimens of *Pistacia atlantica*, especially on plants growing in Iran. Monaco *et al.* (1982) found gall-free specimens to contain two pinane monoterpenes, (+)-9,10-cyclopropylterpinen-4-ol and (+)-9,10-cyclopropyl-terpinene-2,4,-diol and they suggest that these compounds play a specific repellent role against aphids and are responsible for the lack of galls.

A direct antifeedant effect of neem odour on *Nephotettix virescens* (green leafhopper) seems to be comparable with the effects observed by Nisbet *et al.* (1993) on aphids. Saxena and Khan (1986) using an electronic device monitored the feeding behaviour of *N. virescens* on rice plants kept in an area permeated with the odour of *Azadirachta indica* seed oil. Phloem feeding decreased progressively with increased concentration of neem oil. This was partly compensated by increased xylem drinking. The authors suppose that the increased intake of xylem fluid is due to the insect's effort to excrete harmful chemicals from the body. In the described experiments volatile neem oil substances may have entered the body during respiration.

Lepidoptera, among which *Spodoptera* species are particularly sensitive to azadirachtin. The complexity of the molecular structure of azadirachtin has precluded its synthesis for pesticide use, although various drimane structures can be synthesized (klein Gebbinck, 1999) and research into simpler mimetic substances is ongoing. Mordue and Blackwell (1993) have made a comparative review on the effects of pure azadirachtin on different insects and have indicated areas for further research.

Wanner (1997) found that 180 ppm azadirachtin applied to the crowns of black spruce trees caused a mortality of up to 93% of larvae of *Choristoneura fumiferana,* (the spruce budworm). However, as can be expected with feeding deterrents, some feeding occurs before the effects of azadirachtin or neem extracts (also tested) become manifest, resulting in not more than moderate protection of cones. Excellent protection of foliage was obtained with systemic treatment (injections) averaging 1.5 g azadirachtin per tree. The author concludes that neem seed extract, formulated for systemic application, could provide an alternative for control of forest pests in urban environments.

Simmonds *et al.* (1996) evaluated the antifeeding activity of some of the 60 *neo*-clerodanes that have been isolated during systematic studies of the genus *Salvia* (Lamiaceae) growing in Mexico. Eight of the compounds showed potent antifeedant activity in both choice and no-choice bioassays to larvae of *Spodoptera littoralis*. The same metabolites also stimulated dose-dependent responses from neurones in the lateral styloconic sensilla of this insect.

In Lepidoptera a relatively high proportion of complex terpenoids act as feeding deterrents. Messchendorp (1998) tested 15 drimane antifeedants on

Chapter 5. Functions of natural terpenoids in the interrelationships between organisms

larvae of *Pieris brassicae* (Lepidoptera) applied to leaf material of their host plant *Brassica oleracea*. The most potent antifeedants were structures with a lactone group on the B ring (Fig. 5.13). The deterrent cell in the medial sensillum styloconicum was excited by drimane solutions, suggesting that there is a direct inhibiting effect on the centre of the CNS dealing with the neural code.

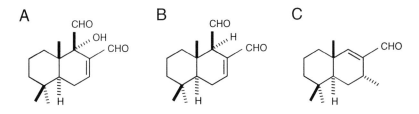

Fig. 5.13. **A**: warburganal; **B**: polygodial; **C**: polygonal.

In contrast to the effect of drimane compounds on *P. brassicae*, the most potent deterrents for the Colorado potato beetle were drimane aldehydes and not the lactones (compare 5.1.3). The natural drimane sesquiterpene dialdehydes polygodial and warburganal were the most active antifeedants (Gols et al., 1996). The deterrence of the lactones is highest when the lactone group is present at C8/C9. On potato plants effective concentrations of these compounds (about 5 $\mu g/cm^2$) induced phytotoxic effects, so additional research is necessary to evaluate the promising effect of these compounds.

Herbivorous insects are usually devided into generalists and specialists. This generally accepted terminology is somewhat misleading: even generalists do not thrive on many plant taxa, but may accept up to a few hundred of host species. Specialists are oligophagous or monophagous. The main theories of plant chemical defense agree on a tendency of herbivores to specialize on certain plant families or species, even parts. Aphids are specialist feeders on phloem tissues where they probably meet only low concentrations of terpenes (Harrewijn, 1993).

Another example is formed by the bark beetles, discussed above. They can cope with a number of terpenoids, while others are dangerous to the beetles and are avoided if possible. Many terpenoids contribute to the odour of conifers and increased release of volatiles by rupture of oil-containing tissues (as happens in wind-damaged trees) make pine trees easy to trace by insects. The high concentration of terpenes in the air acts as a defense, apart from some adapted species, which utilize defensive metabolites to find their host by breaking the defensive barrier and turning allomones into kairomones.

An excellent survey of plant hormones illustrating the astonishing activity of phytoecdysteroids has been made by J.B. Harborne in chapter 4 of his book "Introduction to ecological biochemistry" (1993). There is no need to repeat the effects on many insect species, as these are variations on the same type of defense. In addition, estrogenic hormones and phytoecdysteroids have been discussed by Slàma (1987). More on the interaction of phytoestrogens (not all of them are isoprenoids) with estrogen receptors (including vertebrates) is in Kuiper et al., 1998.

5.1.4 Kairomones

One of the best-studied relationships between plants and insects is that of the pollinators. The mutual benefit of this relationship seems to speak for itself – yet it is the result of a long-lasting contest between two different forms of life, be it only a short period on the astronomical time scale – angiosperms date from about the beginning of the Cretaceous. The transition from anemophily to entomophily in which pollen are dispersed by insects is not the "raison d'être" of kairomones. Long before insects started their useful work, volatile allomones, which were less toxic or could be detoxified had acquired functions as kairomones, e.g. in gymnosperms.

Plant volatiles can be used as kairomones to improve pheromonal cues to locate males, as pheromones can be used to facilitate the localization of a primary host plant on which sexual partners are present (Nottingham et al., 1991). Nepetalactol is a powerful pheromone of the aphid *Phorodon humuli* that not only attracts males, but also serves as an aggregation pheromone for gynoparae, that prefer to alight close to oviparae already established on primary hosts (Lösel et al., 1996). Gynoparae of this aphid species are particularly sensitive to plum (*Prunus domestica*) volatiles, which could improve their ability to locate suitable hosts to produce oviparae. Males also react to *Prunus* volatiles, but do so much stronger to a combination of the sex pheromone and *Prunus* odour. The mere presence of the ubiquitous *Prunus* plants indeed does not guarantee that oviparous females can be found on all of them, so a reaction to a pheromonal clue is of an additional advantage. Although a mixture of isomers of (Z,Z)-nepetalactol is considerably less attractive then the pure sex pheromone, the isomers do not interfere with the behavioural activity of the (Z,Z)-compound (Lösel et al., 1996).

*Chapter 5. Functions of natural terpenoids in the interrelationships   217
          between organisms*

MONOTERPENES, SESQUITERPENES AND DITERPENES

Fungal odour components were tested on two sympatric species of fungivorous collembolans with the EAG technique (Hedlund *et al.*, 1995). It appeared that camphor might be used as a kairomone to find fungal food resources, as does dipentene, 1-octanol and 2-methyl-1-propanol. One of us, when studying trail compounds of termites, found interestingly that a Collembola species living in symbiosis with the termites, responded to sesquiterpenes from sandalwood, in this way tracing the termites. It also responds to the trail pheromone of the termites. We think that many of these advanced studies, although not completed, have at least been initiated.

Orthoptera feed on many monocots and often their feeding sites form a closed area of host plants, such as grassland. Insects, which feed on hosts, which are easy to trace do not need much kairomones. The same holds for Dictyoptera, many of which are predators, e.g. mantids. Nevertheless volatiles of barley, oat and wheat attract insects and apart from the "green" volatiles such as (Z)-3-hexenyl acetate terpenes like (E)-β-ocimene, caryophyllene, α-copaene, α-farnesene and linalool oxides are present in the leaves (Buttery *et al.*, 1985).

As mentioned in previous sections of this chapter, olfactory receptor cells of aphids respond to a range of terpenes (e.g. Visser *et al.*, 1996). However, without extended information on the behavioural reactions of the insects it is difficult to say which terpenes act as kairomones, although EAG response profiles of hosts and non-hosts may provide a key. Visser and Piron (1995) recorded EAG responses of *Megoura viciae* to general green leaf volatiles and more specific plant volatiles. Aldehydes like hexanal and (E)-2-hexenal elicited the strongest response, followed by alcohols, in which a double bond decreased the response. Although the response to terpenes was somewhat less, peaks were 30-80% of (E)-2-hexenal, in the following order: citronellal, β-pinene, linalool, geraniol + citronellol, and eucalyptol (= 1,8-cineole) (Box 5.1).

Apart from peak size two other events underlay the effectiveness of sensory transduction: rise of the waveform which is proportional to the velocity of odour transport to receptor sites (time needed to open ion channels and to depolarize the dendritic membrane, see 4.1.2), and speed of the decay, reflecting inactivation of odour molecules. By comparing these parameters of potential kairomones, one could make a first selection for later behavioural tests.

What can EAGs tell us when we compare a generalist aphid like *M. persicae* (> 300 host plants) with oligophagous aphid species such as *A. fabae, B. brassicae* and *M. viciae*? Visser *et al.* (1996) normalized absolute EAG responses of these aphid species and expressed them as percentages from that of 1% (E)-2-hexenal. With all four species 14 mono- and sesquiterpenes were tested. The authors were particularly interested in the response to non-host plants and emphasize that the sensilla of *A. fabae* and *M. viciae* (aphids which do not feed on cruciferous plants) show about the same response to isothyocyanates as those of *B. brassicae* that does feed on these plants. Within the context of this book it is of interest to know how the sensilla reacted to the terpenes. *B. brassicae* sensilla responded most to (S)-(+)-carvone and (R)-(-)-carvone followed by citronellal. Carvone is a major terpene of caraway, a host plant of *Cavariella aegopodii* that is attracted by carvone (Chapman *et al.*, 1981).

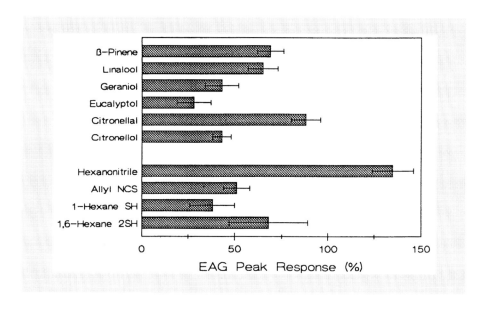

Box 5.1. Normalized EAG peak responses of apterous virginoparae of the aphid *Megoura viciae* to plant volatiles diluted in paraffin oil (at the source: 1% v/v). Bars show means and 95% confidence intervals (n=9-14). This response profile illustrates how small changes in the structure of volatile terpenes affect their perception by the olfactory sensilla. Hexanonitril probably acts as a strong repellent (with courtesy of J.H. Visser and P.G.M. Piron).

In contrast, the response of *A. fabae* sensilla to carvone was only half that of the mean response calculated for the four aphid species and the same holds for linalool. *M. persicae* and *M. viciae* had an average response to all

Chapter 5. Functions of natural terpenoids in the interrelationships between organisms

terpenes. All aphids responded well to (E)-β-farnesene. The sensilla of *A. fabae* were particularly sensible to (E,E)-α-farnesene. This farnesene isomer is rather toxic to this aphid species (van Oosten *et al.*, 1990).

Although several odours among which fruit odours such as benzaldehyde are attractive to Thysanoptera, reports on terpenoids are in minority. *Frankliniella occidentalis*, the western flower thrips, is attracted by anisaldehyde (Brodsgaard, 1990). The other terpenoid volatiles tested are eugenol and geraniol, that are attractive to the New Zealand flower thrips, *Thrips obscuratus* (Teulon, 1988). Both compounds are also attractive to *F. occidentalis*, but did not significantly increase trap catches in the greenhouse when applied to blue sticky traps (Frey *et al.*, 1994). Whatever volatiles are used, it is not easy to improve sticky traps in glasshouses with attractive odours, a problem well known from pheromone research. Unwanted changes in release rates can be caused by changes in temperature and humidity, and temperature-dependent air movement makes it difficult to sustain a proper odour gradient. Moreover, plant odours may interfere with the baits. Frey *et al.* (1994) found that geraniol attracted the thrips in *Chrysanthemum* but not in *Sinningia*, possibly because some odour was interfering with antennal receptors for geraniol. Comparable problems may be met with whiteflies and aphids.

In *Pieris brassicae* the response of the deterrent cell in the medial sensillum styloconicum to drimane compounds contributes significantly to inhibition of feeding behaviour of larvae (Messchendorp *et al.*, 1996b). Interference with cells sensitive to feeding stimulants seems less probable, because the large variability found in feeding inhibition by drimane structures corresponds with the variability in responses of the deterrent cell. The lactones were most active, whereas polygodial and warburganal were less effective.

Guerin and Visser (1980) recorded EAG responses of male and female *Psila rosae* (Diptera), the carrot rust fly, to odour compounds associated with umbelliferous host plants. Odour perception in this insect involves reception of green leaf volatiles, among which hexanal and *trans*-2-hexanol. Another group of receptors is particularly sensitive to terpenes. Strong EAGs were elicited by *trans*-methyl-iso-eugenol (Fig. 5.14), carvone, β-caryophyllene, linalool and *trans*-2-nonenal. Sensitivity to β-caryophyllene (Fig. 5.9) and iso-eugenol was outstanding, although there were consistent differences in the dose-response curves of individual antennae at higher test concentrations (Fig. 5.14). Both terpenes are present in the leaves of carrots.

Terpenes of natural food sources have a function in learning of ants (Hymenoptera). Recruited workers of the leaf-cutting ant *Acromyrmex lundi* learn the odour of the food fragment carried to the colony by a scout, and use this cue as a decision criterion when collecting food. Citral, a natural deterrent odour (Fig. 4.2), can also be associated as a food signal, showing how powerful olfactory learning is in this ant species (Roces, 1994).

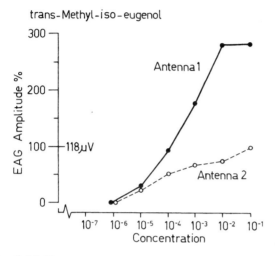

Fig. 5.14. Dose-response curves of EAGs of two antennae of two individuals of *Psila rosae* to *trans*-methyl-iso-eugenol. Concentrations refer to dilution in paraffine oil (v/v). Each point represents one stimulation (with courtesy of J.H. Visser).

For those hymenopterans which are regarded to be useful, such as pollinating species, research is more focussed on attraction than on repellence. Repellents to honeybees are not so interesting for the same reason, but attractive odours have long been studied. Honeybees are known to be attracted to flowers by a blend of volatiles, composed of a mixture of "green" volatiles and a choice of simple volatile metabolites including terpenes. Geraniol, a main component of rose odour, is an important terpenoid attractant and eugenol is a repellent, in contrast to coleopteran beetles that often are attracted by eugenol or methyl eugenol.

In 1629 John Parkinson, an English herbalist, wrote of the use of "balme" for attracting swarms of honeybees, which was probably *Melissa officinalis*. This herb is still called "balm" in the USA. Parkinson may have quoted Virgilius, who already mentioned the crushed plants for the same purpose. This is in contrast to the opinion that he actually meant *Hyssopus officinalis*, another Mediterranean labiate, with medicinal properties, but that is a shrub and not so easily crushed. Moreover, records show that melissa oil was used as a swarm attractant in the 1930s. At least 23 compounds have been identified as constituents of the oil of melissa (with the common name lemon balm or bee balm). Citral (Fig. 4.2) constitutes 61% of the oil, with traces of geraniol and nerol. It is interesting that terpenes synthesized in the Nasonov gland include geraniol as a major component, with citral and nerolic acid and the successful utilization of synthetic mixtures of these compounds in attracting swarms has been reported (Burgett, 1980).

Probably toxic effects and deterrence have preceded the use of terpenes as kairomones by herbivorous beetles (Coleoptera). Intermediate effects have also been found, but it is not always easy to find the missing link. The sesquiterpenes albopetasine and bakkenolide can be isolated from *Petasites albus* (Senecioneae), whereas furanopetasine and isopetasine are characteristic for *P. hybridus*. The leaf beetle *Oreina cacaliae*, which is seldom found on *P. albus*, ate less of leaf discs treated with albopetasine and bakkenolide than from untreated discs, but did not discriminate between untreated discs and those treated with sesquiterpenes from preferred hosts also belonging to the Senecioneae, and containing other sesquiterpenes such as adenostylone, deacyladenostylone and *neo*-denostylone (Hägele *et al.*, 1996). It also accepted discs treated with furanopetasine and isopetasine, although *P. hybridus* is not a host. Probably the beetles never encounter this plant because it is found at lower altitudes than the beetles, but it may be an oversimplification that the authors give this as an explanation to the phenomenon. Does this mean that the beetles never learned to recognize the sesquiterpenes from *P. hybridus* as an allomone? Either these compounds are not toxic, or they are, but in that case it is strange that the beetles keep feeding on the treated discs, at least for the experimental period of 24 h. Possibly the sesquiterpenes studied have long-term effects, which would explain sustained feeding for a number of hours, and non-acceptance of non-hosts which are frequently encountered.

There are many studies showing that monoterpenes present in conifers act as attractants to bark beetles. Often one particular terpene does the job, a reason why a combination of an attractant with an insecticide is tested as a promising "lure and kill" procedure. The insecticide or some other ingredient, however, can severely reduce the attractive effect of the terpene. The tile scarab beetle *Maladera matrida* is particularly attracted by eugenol and both sexes responded optimally with no further improvement upon the addition of other lures. The baits were no longer attractive, however, when eugenol was combined with the insecticide methamidophos, as its repellency overrode the attraction of the lure (Benyakir *et al.*, 1995).

Hylobius abietis, the pine weevil, is highly attracted to volatiles produced by *Pinus sylvestris* (pine) and *Picea abies* (spruce). Two of the 30 known receptor neurones of this beetle are specialized for pinene and limonene. Wibe *et al.* (1998) showed that one receptor neuron is specialized on α-pinene and the other for limonene. In addition, the pinene receptor appeared to response noticeably better to (+)-α-pinene than to the (-) enantiomer. The limonene receptor responded better to (-)- than to (+)-limonene. Enantiomeric ratios of many other volatiles may be important for the pine weevil in discriminating host and non-host odours. It would be particularly interesting to test enantiomers of terpenes of conifers which are regarded as non-hosts, such as *Juniperus communis*, that contains 100 times more (+)-sabinene than (-)-limonene.

Although much less intensively studied, defoliating insects also seem to locate their hosts by homing in on monoterpenes released by host trees (Leather, 1996).

The major component of rape flower odour is (E,E)-α-farnesene (>60% of the composition). Although seed weevils (*Ceutorrhynchus assimilis*) move upwind on the reception of rape flower odour, they were unresponsive to (E,E)-α-farnesene in an olfactometer (Evans, 1992). However, the attractive response to artificial flower odour was absent when the farnesene component was omitted from the odour, suggesting an additive or even synergistic effect of (E,E)-α-farnesene on rape flower odour. The author also found a reduction in the sensory response of olfactory receptors on the antennae when α-farnesene was omitted.

5.1.5 Synomones

Another intriguing strategy to explore a certain habitat is the production of bodyguard-attracting volatiles by plants with either a high risk to become attacked, or that have already been infested by herbivores. Lima bean plants downwind of nearby conspecifics infested with spider mites produced an extreme proportion of (E)-β-ocimene (Dicke *et al.*, 1990). Natural enemies can distinguish between attacked and intact plants. This, of course, does not help the plants already heavily infested, but increases the chances for survival of conspecifics at the location (Dicke *et al.*, 1990b).

Plants have a dynamic role in mediating the interaction between herbivores and natural enemies of herbivores. Whereas green leaf volatiles, butyrates and several cyclic terpenes are released from storage or are synthesized from stored intermediates upon attack by herbivores, acyclic terpenes seem to be synthesized *de novo* following insect damage. Paré and Tumlinson (1997) found a number of acyclic terpenes among which (E,E)-α-

Chapter 5. Functions of natural terpenoids in the interrelationships between organisms

farnesene, (E)-β-farnesene, (E)-β-ocimene and linalool to be biosynthesized in cotton when attacked by fourth instar caterpillars of *Spodoptera exigua*. The induction of synthesis of terpenoid volatiles is, at least partly, triggered by an elicitor from the oral secretions of the feeding caterpillars.

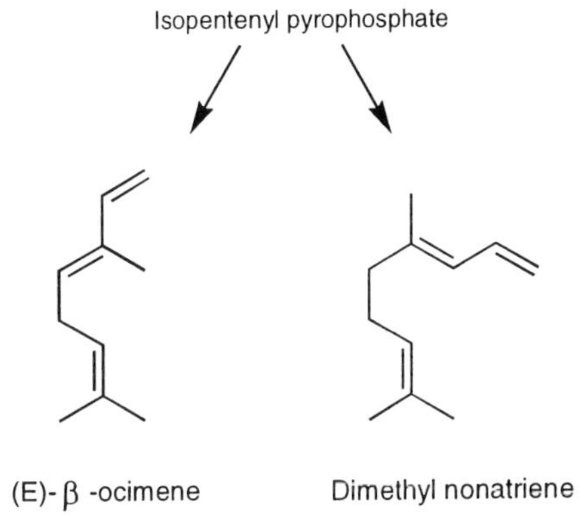

Fig. 5.15. Biosynthesis of (E)-β-ocimene and dimethyl nonatriene from isopentenyl pyrophosphate (IPP) by cucumber plants attacked by *Tetranychus urticae*, a herbivorous spider-mite. These terpenes attract another mite species predating on *T. urticae*.

The complexity of the interaction between plants and herbivores is well illustrated by the herbivore-induced synomone given off by cucumber leaves under attack of the spider mite *Tetranychus urticae*. The volatiles released include (E)-β-ocimene and the homoterpene (3E)-4,8-dimethyl-1,3,7-nonatriene (Fig. 5.15). The latter compound acts as an attractant to the predatory mite *Phytoseiulus persimilis*, which in turn reduces the herbivore mite population.

The nonatriene homoterpene is a common component of the volatile blend of angiospermic flowers and is known to be released from Lima beans as well as cucumber infested with *T. urticae*. Paré and Tumlinson (1997) demonstrated that (3E)-4,8-dimethyl-1,3,7-nonatriene is biosynthesized *de novo* upon attack by herbivores. Its precursor is (E)-3S-nerolidol. However, uninfested leaves and flowers of several plant species can convert exogenous nerolidol into the nonatriene, so in infested plants some elicitor should stimulate the step from nerolidol to the nonatriene (Fig. 5.15). Bouwmeester et al. (1999) demonstrated that the activity of (E)-nerolidol synthase increased by a factor 6 upon spider mite infestation. It comes to no surprise (see section 2.2.2) that the enzyme is Mg^{2+}-dependent, as is the case with other sesquiterpene synthases. Indeed the enzymatically-produced nerolidol was (E)-3S-nerolidol. The conversion of nerolidol to (3E)-4,8-dimethyl-1,3,7-nonatriene occurred without herbivory or elicitor treatment, suggesting that the enzymes for this conversion were already expressed. So what regulates the final release of the nonatriene? Nerolidol itself is not released from the cucumber and bean leaves. We could think of either a stimulation of nerolidol synthase or a deregulation of a possible feedback regulation of the nonatriene on its own synthesis that a.o. inhibits nerolidol synthase. Mechanical damage does not stimulate the synthesis of nerolidol from F-PP (Bouwmeester et al., 1999).

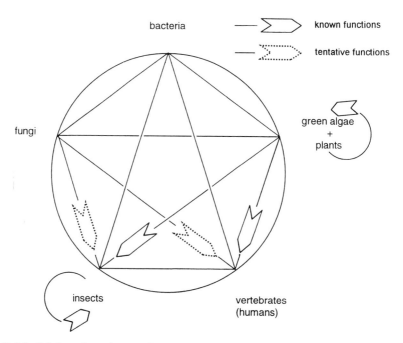

Fig. 5.16. Major functions of natural terpenoids as known and tentative synomones.

Comparisons of the composition of volatile blends of different plant-herbivore complexes show that specific terpenes as reported from herbivore-induced synomones are neither present in the volatile blends of mechanically damaged cabbage plants, nor were they recorded from cabbage plants infested with *Pieris rapae* or *P. brassicae* (Geervliet et al., 1994), suggesting that the specialist parasitoid *Cotesia rubecula* uses other information than odour sources when foraging on *Pieris* caterpillars. In general, predators seem to evoke a stronger "cry for help" from attacked plants than parasitoids do.

According to Geervliet *et al.* (1994) predators may depend less on visual information than parasitoids do, or they are able to detect much smaller deviations in a complex odour blend than we can even measure today. Anyhow, attraction of predators by herbivore-infested plants is a given fact. In olfactometer assays, the predatory mite *Amblyseius cucumeris* responded significantly more to volatiles from chrysanthemum plants infested with *Frankliniella occidentalis* than to volatiles from uninfested plants. The increased attraction was attributed to higher amounts of germacrene D in infested plants (Manjunatha *et al.*, 1998). Olfactometer tests with germacrene D were also positive. As mentioned in section 5.1.1, germacrene D is a sex attractant of *Periplaneta americana*, and germacrene A is an alarm pheromone of the aphid *Therioaphis maculata*. This again illustrates that structurally related terpenes have acquired different functions (Fig. 5.16). It also illustrates that the relation between a plant and its bodyguard can be rather specific. We mentioned above that the predatory mite *P. persimilis* is attracted by (E)-β-ocimene plus a nonatriene released from cucumber leaves.

The larvae of *Spodoptera* spp. are often chemically inconspicuous, but when they feed on plants alarm signals are produced by the attacked plants that are exploited by predators and parasitoids to locate their prey. Terpenes can be among these volatile compounds. When larvae of *Spodoptera exigua* start to feed on corn seedlings there is a rapid increase in the release of "green odours": (Z)-3-hexenal, (E)-2-hexenal, (Z)-3-hexen-1-ol and (Z)-3-hexen-1-yl acetate. When the larvae were removed from the plants, the release of the compounds decreased immediately. However, many hours later, the plants started to release indole, (E)-β-farnesene, (E)-nerolidol and a few other terpenes (Stowe *et al.*, 1995). Artificially damaged plants do not emit the terpenes and indole. As indole is a breakdown product of chlorophyll, these results indicate that the large amounts of terpenes are not just produced by a disorganized plant metabolism. Natural selection may be in favour of the production of a more specific blend of volatiles upon herbivore feeding, a blend that cannot be mistaken for the one released after mechanical damage, or activity of vertebrate herbivores. The latter are of no use to the parasitoids and predators. Still, plants of the same family or even genus probably emit different blends, which makes learning processes of the hunters essential for successful foraging.

Damage induces a rise in activity of lipogenase pathway products and the volatile aldehydes and alcohols may directly affect insect herbivores. After leaf damage young tomato leaves can reduce aphid populations, and the effect was brought about by both C-6 aldehydes and alcohols, but not by volatiles of the terpenoid pathway, such as 2-carene and limonene that were also released (Hildebrand *et al.*, 1993). Again this leaves the possibility of a more specific function (attraction of natural enemies) for the terpenes.

Terpenoids form an important group of herbivore-induced plant volatiles. If they are synthesized *de novo* upon herbivore attack, one might expect that their emissions follow the photoperiod. As discussed in chapter 2, monoterpenes are thought to be mainly derived from IPP via the DOXP pathway, a MAI route, with pyruvate + glyceraldehyde-3-P as substrates. Sesquiterpenes are thought to be mainly synthesized from MVA via a MAD route, although exchange of the intermediate IPP is possible. Both routes are based on glucose, the production of which is dependent on photosynthesis. Indeed linalool, (E)-β-ocimene (monoterpene synomones) and (E,E)-α- and (E)-β-farnesene show an emission profile similar to the light cycle (Paré and Tumlinson, 1999). This should be kept in mind when drawing an experimental scheme for studies of synomones.

Synomones are not necessarily released from damaged plant parts. Paré and Tumlinson (1998) demonstrated with chemical labelling experiments that a chemical messenger is transported from the damaged area to undamaged leaves to trigger synomone synthesis. What triggers the production of (terpenoid) synomones? Glucosidases secreted by chewing insects could cleave sugars bound to terpenes, but that would be a local reaction. An intermediate signalling molecule suggested to play a dominating role is jasmonic acid that can be synthesized from linolenic acid. Its function could be comparable to that of salicylic acid that triggers defensive reactions upon attack by micro-organisms. Linolenic acid can be hydroxylized to hydroperoxylinolenic acid. So far, this is a normal process leading to green leaf volatiles and to the octadecanoic pathway, the latter being involved in jasmonic acid synthesis. Feeding by e.g. caterpillars can accelerate hydroxylation of linolenic acid and may shift the balance towards jasmonic acid (Paré and Tumlinson, 1999). This leaves the questions if, and which secondary messengers are employed to sites distal to the feeding sites. What happens if plants are covered entirely with herbivores? As discussed in section 4.1.6, activation of protein kinase C is a very important process in various types of response to stress in eukaryotic cell systems. Experiments including the protein kinase C cascade could increase our understanding of the adjustment of the "plant language" to cope with an attack by herbivores in addition to other stress factors.

Chapter 5. Functions of natural terpenoids in the interrelationships 227
between organisms

Why are volatiles released from stressed plants more reliable for insects operating at the third trophic level to home in than those released from uninfested plants? It cannot simply be a matter of concentration. Although the emission of volatiles can increase upon herbivore attack (Turlings *et al.*, 1990) many plant species emit more volatiles when not attacked than other species do when they are under siege (Harrewijn *et al.*, 1995). More probable it is the specificity of the blend or the ratio between jasmonate-induced volatiles and the background of green volatiles plus basic fragrance (a shift towards jasmonate could reduce activity of the green volatile pathway).

Methyl jasmonate (Fig. 5.17) is a volatile compound that acts as a plant hormone. When exposed to bean plants it induces attraction of predatory mites (Dicke *et al.*, 1999), a process resembling volatile induction within an attacked plant. However, it is not very probable that methyl jasmonate is released in sufficiently high concentration to induce defense in conspecifics (Turlings *et al.*, 1990).

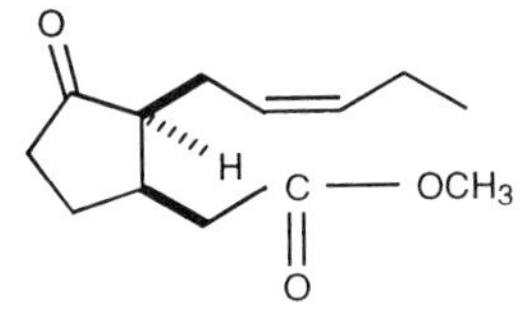

(3R,7S)-(+)-epi-methyl jasmonate

Fig. 5.17. Methyl jasmonate that induces (or increases) the production of synomones upon attack by herbivores. Synomones (see Fig. 5.15) recruit natural enemies of herbivores.

Plants reveal a variety of defensive strategies ranging from thorns, spikes, glandular hairs and cuticular waxes to secondary metabolites and inhibitory proteins and enzymes. Natural enemies form an army protecting against herbivores, an army consisting of hirelings rewarded by prey or oviposition sites. Investment in bodyguards may be seen as a consolidation of an altered production of volatiles induced by attack of different herbivores. From an evolutionary perspective, the cost of synomone production can be reduced if already existing enzyme systems are recruited for enhanced release and an altered ratio of odour components in infested plants. This subject is further discussed in section 5.2.2.

The communication between plants and bodyguards opens possibilities to stimulate the third trophic level in IPM programmes (see chapter 6).

5.2 Relationship between structure and function

The anatomical basis of receptors of terpenoid messengers shows more unity than diversity. As discussed in section 4.1.2, olfactory and gustatory receptor cells are closely related. Receptor cells of sensilla often have a limited degree of specialization. Receptor cells which are stimulated by pheromones as well as by host plant odours are no exception (Visser, 1986). Roughly speaking, the mode of action of terpenoid messengers on cell organelles is in interfering with protein prenylation or G-protein cascades, and in sensilla they are transported by binding proteins and involved in enzymatic cascades leading to ion channel opening. It is a challenging idea that understanding of the similarity in the mode of action of terpenoid messengers is of mutual interest to different disciplines. First, let us see where we are (section 5.2.1) and if we can climb high enough in the almost untraversable grounds to see how the formations below us have risen as they did (sections 5.2.2. and 5.2.3).

5.2.1 Similarities between receptors

When a volatile (terpenoid) messenger molecule enters a sensillum, it is carried by a fluid to a low molecular weight binding protein and this is the beginning of an enzymatic cascade with subsequent activity of heterotrimeric guanosinetriphosphate-binding proteins (G-proteins) and effector enzymes. Photons and odorants activate an inward current in receptor cells. The signal transduction pathway in eukaryotic cells not in direct contact with the outer world utilize a similar mechanism in which G-proteins are likewise involved (Box 3.2). These G-proteins release subunits that reach the nucleus or bind to regulators of the G-protein signalling pathway, eventually influencing gene expression. The various sections of chapter 4 have demonstrated that terpenoids can both induce and interfere with a messenger cascade. The first activity is based on binding to a surface protein, the second is an activity in protein binding again, in prenylation of proteins or just the opposite: inhibition of prenylation.

This is what makes terpenoids so intriguing: throughout eukaryotes their targets expose great similarities, but the course of following events can be anything from interference with the genetic code to directing cohorts of herbivores to our crops.

THE GENETIC CODE
Attraction of herbivores by natural kairomones is based on genetic codes. As Barbieri (1998) states: modern biology does not recognize any other code

Chapter 5. Functions of natural terpenoids in the interrelationships between organisms

in nature but the genetic code. This author stresses that pattern formation (remember the retinoids) and signal transduction (from monoterpenes to steroid hormones) can be accounted for precisely by the existence of organic codes. The origin of a new organic code is in fact the basic mechanism of macroevolution.

As we have said before, terpenoids cannot multiply themselves, they are biosynthesized via enzyme systems that in their turn are dependent on the expression of genes. Specialists of different disciplines reading this book (we hope they do) undoubtedly will have their own vision on the biological significance of the existence of similarity in terpenoid receptors. The cancer specialist may be interested in binding properties of terpenes in prenylation of proteins, the pharmacologist looks at limonene and its derivative perillyl alcohol with different eyes as the biologist does to perillene, as the latter terpene may serve to defend more than thrips colonies alone.

However, we agree with Barbieri (1998) that semantic transformations and biological significance do not exist in the organic world and should play no part in our reconstruction of development and evolution. Only recently the study of protein sequence differences between fungi, plants and animals have been applied to estimate divergence times between these phyla. Major questions as on the reliability of the molecular data and on the enormous temporal gap between the calculated origin of phyla and their appearance in the fossil record makes it difficult to achieve a consensus of key dates, e.g. of the evolutionary separation of plants and animals (Brooke, 1999). We could add: let alone, that the study of gene development has provided information substantial enough to reconstruct coevolution of messengers and their receptors.

The long-distance effects in complex feedback mechanisms as discussed in chapter 3 cannot operate without additional receptors, and novel targets of terpenoids are continuously found. Jiang and Koolman (1999) recently included moulting glands in the list of target tissues of ecdysone agonists. Ecdysteroid biosynthesis in insects is under complex control: endocrine, neurocrine and feedback systems are involved. The authors have added an additional feedback loop repressing steroid genesis in demonstrating that a 20-hydroxyecdysone agonist acts directly on the moulting gland of *Calliphora vicina*, the blue blowfly. This example illustrates that what to our opinion may seem a well-developed, finely tuned regulatory mechanism already is still more complex than originally thought and is apt to be further developed.

CONSOLIDATION OF MESSENGER/RECEPTOR RELATIONSHIPS

What is the selective force that makes some organic (terpenoid) novelties irreversible and result in a better fitness of the organism? Maybe it is the possibility to explore another habitat and to survive there. Take the study of Jiang and Koolman (1999) mentioned above. Many potential host plants produce powerful phytoecdysteroids which renders them non-hosts to herbivorous insects because of the supernumerous moults and other pathological effects brought about by high ecdysteroid titres (Harborne, 1993), which comes in addition to the ecdysteroid production of the insect itself. Therefore, it is of advantage when efficient feedback loops are selected that can reduce the pathological effect of phytoecdysteroids. Once new regulatory systems have proved to support the organisms in exploiting other hosts, they become obligate.

Likewise, new genes expressing receptors that can bind potential messenger molecules or are more specific e.g. with respect to chirality of volatile terpenes can improve orientation in an environment polluted as it were with nonsense information. Genes can duplicate and they can become deleted. Upon deletion of genes expressing novel receptor and signalling proteins the organisms may no longer be able to compete with organisms better equipped to read the complex chemical language. This is an old evolutionary principle.

Methanogenic Archaebacteria were viable in anaerobic environments where CO_2 and geological H_2 were abundant (see chapter 2.1). Martin and Müller (1998) discuss their partnership with free living Eubacteria that became the symbiont. The meeting had to occur under conditions that kept the host viable from the start, but once the joined organisms become deprived of H_2, the original host is dependent upon the heterotrophic eubacterial symbiont. According to Martin and Müller (1998) this is a strong selective force that irreversibly associates symbiont and host, but opens the way to exploit organic substrates.

In the last decade of the 20th century studies on the role of odour binding proteins (OBPs) in stimulus recognition have revealed a number of OBPs, albeit mainly in pheromone reception. Nevertheless it is clear that it will be difficult to understand how relatively few proteins can deal with the enormous variety of odorants. Steinbrecht *et al.* (1996) state: "Even considering the fact that the number of OBPs per species is steadily increasing with refined separation methods, it is hardly conceivable that OBPs will do more than crudely pre-select, while the ultimate recognition of the stimulus remains the task of the membrane-bound receptor molecule of the sensory dendrites". Still, the same type of sensillum responds to pheromone in male *Bombyx mori* and to the plant odours linalool and

Chapter 5. Functions of natural terpenoids in the interrelationships between organisms

benzoic acid in females. Steinbrecht *et al.* (1996) mention that male and female sensilla have different types of OBPs, but selectivity is limited.

In section 4.1.2 we have tried to explain how the neural code could cope with complex blends of volatiles. From that discussion and material of the present section it would seem that new combinations of OBPs, G-proteins and interactions in the neural network were more important in the interpretation of external messengers than the invention of completely different receptor proteins with their corresponding genetic code.

SAFE BIND, SAFE FIND

Technically speaking, the genetic code is a tool to make tools, and as Hogg (1961) stated on page 76 of his technical book on lockmaking: "in the 19th century there were tools to make tools; and tools to make tools to make tools". There is a similarity with the evolution of the messenger/receptor combinations: the genetic code provides enzymes and receptor proteins, and these are chemical tools to construct the unique relationship between messengers and the signalling cascade leading to a behavioural response. Allow us to compare the receptor cell with a fine detector lock as invented by Jeremiah Chubb in 1818 (Fig. 5.18). Most elements of this ingenious lock, however, were not new, but variations on an existing theme. There was a key; there were levers (invented by Barron in 1778), springs, and a bolt. The quality of the lock greatly depends on the accuracy with which key and levers are constructed. It is quite clear that the number of permutations depends on the number of steps carved into the key (corresponding with the number of steps with which the levers can be risen), and the total number of levers. The keyhole itself could provide a first barrier, as it had been for centuries; but the form of the keyhole can be seen at first inspection – the form of the keyhole could be compared with the anatomy of the sensilla – they afford a first selection of molecules that can reach the binding protein (OBP). Box 5.2 compares a detector lock with a messenger/receptor relation.

Long before Chubb made his locks locksmiths devised the principle of the "ward", a hidden device frustrating the movement of any key in a lock but the one that fitted to it. Chubb refrained from wards, as they are easy to overcome and his locks offered a much more advanced form of protection. It would seem that Nature has not invested too much in "wards", given the limited number of binding proteins. The next level of selection is in the different height of the notches in the levers with corresponding bits in the key. The key should lift them to *exactly* the right height to let the bolt stump travel through the notches. This can be compared with the receptor proteins, as many as the genetic code permits.

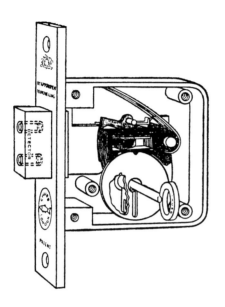

Fig. 5.18. A Chubb Detector lock. The detector spring has caught a lever which has been lifted out of position by the use of the wrong key, thus preventing the bolt from being withdrawn (after Hogg, 1961).

	Lever lock		Receptor cell	
Initiator	key		messenger molecule	
Entrance to the receptor	keyhole		anatomy of sensillum	
Access to receptor mechanism	ward		binding protein	
Permutations	{ notches		receptor protein	
	number of levers		number of receptor cells	
Main filter	"detector" lever		glomerulus	
The "owner"	resets the lock		CNS	

Box 5.2. Comparison of key and lock with messenger and receptor.

So far, a considerable number of keys can be made for one type of lock, but this is not enough to keep a good technician from picking the lock. In fine tumbler locks, there are as many as five or six levers, enormously increasing the number of permutations. This is realized in animal receptors with various receptor cells with their different sensitivity to messenger molecules (although there is some overlap). The comparison can be drawn into more detail. The 1818 six-lever lock had another essential feature: the "detector" lever. If (with a false key or tools) only *one* of the levers happened to be "over-lifted", the detector lever was lifted and immediately blocked in that position. Impossible to work on the lock any more; a hopeless position for the lockpicker. This is a kind of definite filter as found in the glomerulus (see 4.1.2). Of course the lock was not permanently blocked; the owner could reset the detector lever by inserting the proper key and turning it counter-clockwise. This is how the brain can overcome an inadequate message (an imperfect key). The system is reset to await the next messenger (key) to enter the sensillum (lock). Hogg (1961) seems to have foreseen the comparison we have just made by entitling his book "Safe bind, safe find". Indeed, upon adequate binding and processing of the messenger the herbivore will safely find its target.

In conclusion, what has been depicted in the establishment of a locksmith can be compared with convergent evolution and is found in the relation between terpenoid messengers and their receptors. The building elements in the lock were not new, but ingenious variations and combinations were. The same holds for the elements of receptor cells – it is the combination of previously developed elements that makes the relation between messenger and receptor unique. Which role this combination played in the evolution of behaviour is discussed in section 5.2.2.

5.2.2 The role of terpenoids in the evolution of insect behaviour

The material presented in chapters 4 and 5 has demonstrated that terpenoids in general have a great impact on growth and development of animals. The insatiable curiosity of entomologists has certainly contributed to our understanding of the basic principles of the production and function of semiochemicals and has opened new outlooks on the evolution of behaviour.

In the same chapters we have discussed the highly evolved effector network operating within cells in which terpenoids are involved, a role they played before organisms sieged the land and plants became consumed by herbivores. The terpenes and terpenoids were there for several reasons and we should be careful to impute insects a major role in the selection process of terpenoids in relationships between plants (or other producers) and their environment.

Did insects stimulate plants to produce the enormous variation in allomones, among which terpenoids? That would imply that their anti-microbial and fungistatic effects as discussed in section 4.2.1 are secondary. The question is: which type of defense is most important in a particular habitat? To understand the selection process of particular types of chemical defense, a number of main theories have been developed.

MAIN THEORIES OF PLANT CHEMICAL DEFENSE

Theories on chemical defense try to classify plant defense according to (1) cost of defense in relation to the necessary level of protection (*tissue value hypothesis*), (2) plants which are easy to find or are inapparent (*apparency hypothesis*), (3) plants in lush or poor habitats (*resource availability hypothesis*) and (4) an integration of (2) and (3), the *habitat templet model*. They are surveyed by Feeny (1991) and are mainly the work of entomologists. No need to say that (2), for instance, can hardly be applied to micro-organisms that do not go in search of plants as insects do.

All theories agree on that there exist different types and levels of defense, but seem to ignore the fact that plants need to defend themselves against much more than herbivores alone. The main features of the four classes just mentioned lead to contradicting predictions of the adaptations of the herbivores (Harrewijn, 1993). From a physiological point of view, there are drawbacks. One example: in hypothesis (2) apparent plants need continuous defense against many insects and in (3) slow-growing plants are also forced to invest in high levels of defense (fast-growing plants escape because of their biomass and abundant production of seeds). According to the *habitat templet model* apparent plants in poor environments need high levels of "quantitative" defense. Quantitative defensive compounds are present in relatively high concentrations, such as resins (terpenoids) and tannins (phenolics), or compounds generally toxic to many herbivores.

Alkaloids and cyanogenic glucosides are highly toxic, but they are nitrogen-dependent and better suited to be produced in lush habitats. Comparison of production costs of plant defense with risks of losses to herbivory have been made of specific relations (Schoonhoven, 1999), but the very high concentrations of resins and tannins present in a.o. ferns do not speak for the best possible defense in poor habitats (Harrewijn, 1993).

Terpenoids, the production of which does not withdraw nitrogen from the plants, could be produced under both lush and poor conditions, provided there is enough light to realize biosynthesis of geranyl-PP via MAI with its DOXP pathway generating isopentenyl-PP and dimethylallyl-PP (see chapter

Chapter 5. Functions of natural terpenoids in the interrelationships between organisms

2). This may be a reason why plants of the Mediterranean region produce so much volatile terpenes (Ross and Sombrero, 1991).

Strange enough, none of the theories trying to explain the combined stress of environmental factors and herbivores on the evolution of plant chemical defense has paid attention to the role of volatiles, let alone terpenes. Still, many allomones are volatiles. We have made an extended review on the evolution of plant volatile production in insect-plant relationships elsewhere (Harrewijn et al., 1995). In this review we agree on the fact that vulnerable plant parts developed a high degree of protection. In Senecionaceae a few major groups of secondary metabolites such as pyrrolizidine alkaloids and benzofurans frequently occur, but they also contain sesquiterpenes of the furanoeremophilane type. Most of these compounds occur only in the rhizomes and are either absent from the leaves or can only be traced in very low quantities (Hägele et al., 1996). So far only cacalol has been found in considerable amounts in the leaves of *Adenostyles alpina* (Asteraceae) (Harmatha, unpubl.) and petasine in the leaves of *Petasites hybridus* (Wildi, unpubl.). In all cases investigated, there was considerably more of the terpenoids in flower buds and rhizomes than in the leaves. The presence of the above mentioned terpenoids in valuable storage tissues such as rhizomes is seen by Hägele et al. (1996) as a protection against soil dwelling herbivores, among which are several snail and slug species, as all furanoeremophilane sesquiterpenes except albopetasine reduced food consumption of the generalist snail *Arianta arbustorum*.

We also agree with Feeny (1991) that the essence of chemical plant defense is to be found in tissue value and apparency or vulnerability. Unfortunately, the different hypotheses on ecological patterns of chemical resistance in plants give little emphasis on the possible role of volatile allomones.

Volatile terpenes as components of essential oils are known for their antiviral, bactericidal and fungicidal properties (see 4.2.1 and in addition Croft et al., 1993; Gangrade et al., 1990; Harrewijn et al., 1995; Hudson, 1990; Schilcher, 1987). The biosynthesis of an almost incredible variation of volatile terpenes by bacteria, plants and fungi may have been consolidated for mutual protection, quite indifferent from the appearance of insects on the scene that have their share in the process of selection of allomones. Many terpenoids that protect plants against microbial pathogens are toxic to insects and act as repellents or deterrents (see 5.1.3).

Fig. 5.19 depicts the possible evolutionary sequence of the functions of terpenoids in the relationships between bacteria, plants, fungi, insects and vertebrates, based on the pentagram we used before. Bacteria defend themselves against viruses and other bacteria, fungi against bacteria and other fungi, plants against all of them, including plants! The monoterpene carvone (Fig. 2.8) present in the seeds of a number of Apiaceae like caraway and dill (but also in citrus peels) inhibits germination of seeds of other species, even germination of potato tubers (Oosterhaven *et al.*, 1995), and thyme monoterpenes inhibit germination of seeds of other species (Taraye *et al.*, 1995). Terpenoids not only exert allelopathic effects on seeds of competing plants but also in the rhizosphere on soil micro-organisms (Langenheim, 1994). We appreciate this as a chemical weapon in the competition on resources in the habitat. The question arises: did insects adapt their behaviour to the specific ratio of terpenes produced by plants or did the plants adapt their biosynthetic systems upon herbivore pressure?

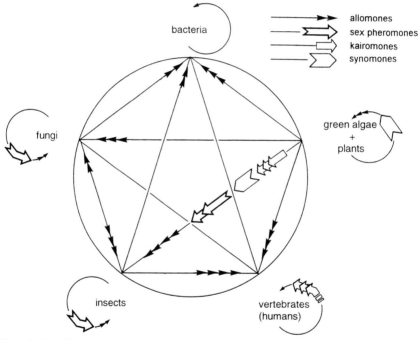

Fig. 5.19. Possible evolutionary sequence of the major functions of terpenoids. Single arrows symbolize early evolution, multiple arrows symbolize later developments. Number of arrows heads corresponds with the order of sequence *within* a class of pheromones, not exactly *between* classes.

PRODUCTION OF TERPENOIDS AND ADAPTIVE BEHAVIOUR

The specialization of the aphid *Cavariella aegopodii* on caraway, utilizing carvone as a kairomone (Chapman *et al.*, 1981), can be seen as a specific

Chapter 5. Functions of natural terpenoids in the interrelationships between organisms

adaptation rather than as a driving force to produce carvone. There is a wealth of literature on pollinating insects. The many volatile excretions by generative plant parts has had considerable appeal to those studying pollination and have been thought to be there to serve mutual interests of plants and insects (Bergström, 1987). To our opinion however, attraction of pollinators has developed secondary; the first interactions of insects with plant volatiles (particularly terpenes) developed because of the chemical identity of some insect pheromones with plant volatiles (Harrewijn et al., 1995). Chapter 5.1 provides many examples of identical terpenes with functions as sex pheromones or plant allomones. Consider the following scenario:

The first mutual interactions of insects with (terpenoid) plant volatiles developed because of chemical identity of some insect sex pheromones with plant volatiles. This made it easier to find a sexual partner. Section 5.1.3 gives examples of insects, which are assisted in locating their partners by odours of plants, e.g. aphids and winter hosts. Subsequently, the attraction of insects to plants was favoured by nutritional rewards and resulted in pollination. In the course of this process terpenoids with similar structure acquired different functions, that is: from allomones to kairomones and synomones.

We have argued above that defensive terpenoids were omnipresent in land plants. Gymnospermae do not attract many insects for pollination as dispersal of pollen is mostly by wind although dioeceous species with an effective insect pollination exist. Volatiles with a proven insecticidal activity were present in times that the Gymnospermae and more ancient plants were dominating the landscape and Angiospermae just appeared. According to Jones and Firn (1991) general metabolic traits that induce diversity may have been selected for very early in evolution and a continuing production of a diversity of terpenoids has proceeded *independent* of the adaptation of consumers (Fig. 5.20, compare with chapter 2.2).

Investment in defensive metabolites was a logic step in the development of highly specialized reproductive organs. In contrast to gymnosperms stamen and anthers of angiosperms stand out in the open, unprotected to infection by micro-organisms and damage by herbivores. Why is there such a high proportion of terpenes in the allomones of generative parts? There is little time to mobilize chemical defense, as is the case with phytoalexins in more expended plant parts. Young, developing vegetative parts are likewise at great risk and this may explain why Gershenzon et al. (1993) found little turnover of volatiles in young leaves of plants known for their high content of secondary compounds. Members of Ranunculaceae and Rosaceae, supposed to be early angiosperms, show remarkably low infestations with microfungi (Ellis and Ellis, 1985).

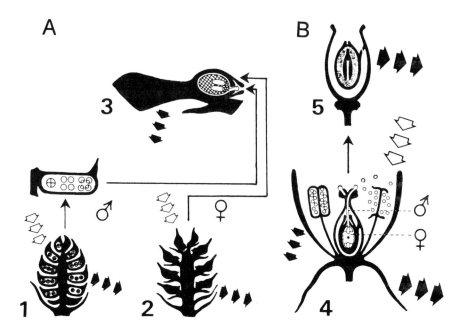

Fig. 5.20. Evolution of defense and attraction from gymnosperms (**A**) to angiosperms (**B**). → = defensive volatiles; ⇨ = attractive volatiles. 1: male cone; 2: female cone; 3: seed scale with developing embryo; 4: dioecious flower; 5: seed (after Harrewijn *et al.*, 1995).

Young, nutritious plant parts close to a mating place could increase the attraction of pollinators. However, the resulting herbivory is a drawback. Angiosperms have developed a novel strategy to that end: they made inflorescence more attractive as a feeding site. The introduction of effective pollination by insects in gymnosperms (particularly Diptera and Hymenoptera), as occurred in *Ephedra* spp. (Gnetinae), was associated with the production of nectar by specialized sterile flowers among the male gametophytes (Bino *et al.*, 1984). In *Gnetum* spp. of the Malayan region both ovules on female strobili and sterile ovules on male strobili secrete pollination droplets in the evening. The visitors of extant *Ephedra* species have been adapted to exploit highly concentrated sugary exudates and have licking or chewing mouthparts.

Chapter 5. Functions of natural terpenoids in the interrelationships between organisms

The hypothesis that sex pheromones were once the basis for a role of terpenes in the relationships between plants and insects is challenging, but what do we know of these relations today that speaks for an evolutionary scenario as depicted above? Orthoptera have strong relations with monocots, especially Gramineae, which may explain a dominating role of green leaf volatiles. The sex pheromones of Lepidoptera are mostly unsaturated straight-chain aldehydes, alcohols and acetates and gustation is an important factor in host plant acceptation, in which non-volatile terpenoids are involved. Moreover, adults do not partake of the same food as their larvae.

Diptera and Hymenoptera are credited for their early association with plants. Section 5.1.1 has shown that a.o. linalool and farnesenes are sex pheromones, the effect of which is increased by plant odours. The relation of Hymenoptera with e.g. Rosaceae have been extensively studied by Dobson *et al.* (1990), a relation which we have discussed with respect to the dominating role of sesquiterpenes in the sepals and gynoecium (Harrewijn *et al.*, 1995). Many Homoptera have terpenoid sex pheromones, as do the Coleoptera that utilize particularly pinenes and their oxidation products, and myrcene derivatives. All of these terpenes have antibiotic properties (Table 6.1). We hope that these facts will spark a somewhat different approach of the study of semiochemicals from an evolutionary point of view.

Many examples of anatomical adaptations that are the result of co-evolution of plants and pollinators can be given, such as sexual mimicry in orchids, which is an example of obligatory behaviour: flower morphology and colour together with the production of sex pheromones attract bees and wasps (Vogel, 1993).

To what extend had the pollinating insects to adapt to their potential hosts and which effect did they exert on the production of terpenoid allomones? We strongly favour the vision that the impact of herbivorous insects on (terpenoid) defense of plants has been limited (Harrewijn *et al.*, 1995). We are supported by Jermy (1991) who states that adaptation to host plants is more important than the evolution pressure of insects on plants. In a review on the evolution of plant-insect relationships (1993) he concludes that the radiation of insect species occurred much later than the evolution of plant traits, and that indications of an evolutionary feedback from insects to plants are poor. Our vision is also supported by Crawley (1989), who stresses that the role of insect herbivores in plant population dynamics is typically less pronounced than that of vertebrate herbivores. The plant-eating dinosaurs developed along with the angiosperms may have become much more threatening indeed.

DETOXIFICATION OR AVOIDANCE?

Following the evolutionary scenario of utilization of plant allomones with similar structures as those of sex pheromones of insects, resulting in herbivory, behavioural adaptations are a logic step. The many (terpenoid) allomones induced insects to develop strategies to cope with such substances, either by detoxification or by avoidance. Box 5.3 gives an overview of terpenoids of land plants involved in plant-animal relationships.

Insects are relatively insensitive to toxic substances for a number of reasons (Harrewijn, 1993). In short: (1) their intestines have peritrophic membranes that can selectively absorb macromolecules, (2) the midgut epithelium has a high potential of detoxifying enzymes. In mammals detoxification is mainly in the liver, so toxins are already circulating in the blood before they can be broken down; in insects detoxification starts before they are released into the hemolymph, (3) the insect blood/brain barrier is very effective, protecting the central nerve system, (4) the circulation system of the hemolymph (that does not need to transport oxygen) is very simple and intoxication is not immediately fatal, and (5) insects expend less energy to maintain body functions as they are poikilotherm.

The sensitivity to cardiac glycosides is remarkably low, but terpenoids can still be rather toxic. Although insects are less sensitive to toxins than vertebrates, they generally try to avoid terpenoids, except for specific relations in which allomones have acquired functions as kairomones (see 5.1.4). Why do herbivorous insects tend to become specialists that are less fitted to cope with a broad spectrum of allomones?

First, herbivorous insects may be poikilothermic and need less fuel than vertebrates, but they need a much higher protein level in their food than mammals do. The protein to carbohydrate ratio is somewhere between 1:1 and 1:3, whereas in mammals it is an average of 1:8. Population development of herbivorous insects is fully dependent on protein sources. Young plant parts and developing fruits are rich in protein, and these tissues generally have the highest levels of toxic metabolites.

Second, proteins themselves cannot be traced from a distance. Amino acids act as feeding stimulants, but their local presence is not necessarily related to host plant quality. When insects utilize allomones to that end, they must turn them into kairomones.

Third, plants are endowed with a variety of defense systems many of which contribute to a strong selective force on the insects and attain plants possibilities for escape in *time* and *space* (Harrewijn, 1993), of which allomones are only one system.

Fourth, there are the anti-microbial volatiles, among which many

*Chapter 5. Functions of natural terpenoids in the interrelationships   241
           between organisms*

Class	Especially in:	Nr. known (at least)	Functions *
Monoterpenes	Angiosperms (Lamiaceae) Gymnosperms	1000	Repellents Pheromones Kairomones Synomones
Sesquiterpenes	Angiosperms Gymnosperms	1000	Repellents Pheromones Kairomones Synomones
Sesquiterpene lactones	Angiosperms (Asteraceae)	600	Deterrents
Diterpenes	Angiosperms Gymnosperms	1200	Deterrents Barriers
Triterpenes	Angiosperms	700	Deterrents Antifeedants
Carotenoids	Green plants and animals	400	Attractants
Cardenolides	Angiosperms (Apocynaceae, Asclepiadaceae, Scrophulariaceae)	200	Deterrents (bitter)
Cucurbitacins	Angiosperms (Cucurbitaceae)	60	Deterrents (bitter)
Limonoids	Angiosperms (Meliaceae, Rutaceae, Simaroubaceae)	120	Deterrents (bitter) Antifeedants
Saponins	Angiosperms Caryophyllaceae	600	Antifeedants (hemolysis)

* NB: these are functions as semiochemicals only. Many have different biological activities; e.g. carotenoids are involved in energy transmission and protection against free radicals. Antifeedants are supposed to act more slowly than deterrents.

Box 5.3. Production of terpenoids by terrestrial plants with functions as semiochemicals (from different sources, a.o. Harborne, 1977; Wink, 1988).

terpenes. The anti-microbial effects have been mentioned above, but come into a different light when we realize that herbivorous insects that are no predators and lack essential nutrients such as sterols, are confronted with biochemical barriers. To overcome this problem insects evolving from saprovory, fungivory and carnivory to herbivory acquired various symbiotic relationships with micro-organisms (Barbosa *et al.*, 1991; Douglas, 1992; Schwemmler and Gassner, 1989; Southwood, 1996). These micro-organisms ranging from bacteria to yeast and fungi are particularly sensitive to volatile terpenes. Generally speaking antibiotics which inhibit synthesis of proteins or gene expression will more easily provoke the occurrence of resistant biotypes of micro-organisms than antibiotics which affect membrane structure of cell walls. This may be the reason why essential oils already produced by gymnosperms and early angiosperms have reported allelopathic effects on a range of micro-organisms (Croft *et al.*, 1993; Kubo, 1992; Linskens and Jackson, 1991) Table 4.3 gives an overview of anti-microbial terpenes.

Their mode of action has been described before (section 4.2.1). Many volatile allomones may be toxic to herbivores because they disrupt their symbionts. Mollema and Harrewijn introduced the term Volatile Anti Symbiont Allomone (VASA) at the 9th Symposium on Insect-Plant Relationships (1996). Southwood (1996) who agreed upon the potential significance of VASAs stated "perhaps the most dramatic aspect of symbiotic micro-organisms is the recognition that plants may produce volatile antisymbiont allomones". When monoterpenes of *Abies grandis* (grand fir) were tested for their effect on *Scolytus ventralis* it appeared that vapours of all monoterpenes caused beetle mortality within 4 h at doses normally found in the trees. Incorporation of any of these monoterpenes into a growth medium substantially reduced growth of the symbiotic fungus *Trichosporium symbioticum*. Limonene was the most inhibitory terpene (Raffa *et al.*, 1985). We have mentioned our own experience with the detrimental effect of pulegone on *Buchnera* symbionts of aphids before.

Terpenoids have had a profound impact on the behaviour of herbivores in the coevolution of plants and insects. Avoidance behaviour with respect to repellents and deterrents and detoxification of particular allomones was essential to utilize allomones as kairomones recruiting all the flexibility the genetic code could dispose of. An intriguing phenomenon is pharmacophagy, in which allomones are incorporated for defense or even are utilized as sex pheromones. As an example, adults of *Athalia rosae ruficornis* feed on trichomes of *Clerodendron trichotomum* (Verbenaceae) and sequester a series of *neo*-clerodane diterpenoids. The plants serve as a rendez-vous site for the sawflies. The clerodendrins seem to be utilized in sexual

Chapter 5. Functions of natural terpenoids in the interrelationships between organisms

communication and may have an additional function in defense against natural enemies (Amano *et al.*, 1999).

From mating to foraging is quite a step. A transgression into herbivory is necessarily associated with orientation to different plant parts, different odours, and to be aware of other enemies. Risks are often highest when searching for a feeding or oviposition site; hence there are advantages of an accurate and rapid search image and of the various neural mechanisms to limit the input of unnecessary information (Bernays, 1996; Southwood, 1999).

In conclusion, herbivorous insects that substantially depend on their microbial symbionts can be expected to have evolved strategies to protect these organisms. Incorporation of symbionts in specialized host cells, the endocytes, has metabolic advantages and could provide a degree of protection, although the example of pulegone illustrates that this is not always sufficient. The existence of plant volatiles, affecting the insect's firmness for reproduction, subsequently causing avoidance behaviour, is highly advantageous (Gould, 1984, 1991). Berenbaum (1988) agrees with our vision that plant substances toxic to the insect's symbionts can lead to avoidance behaviour.

The relationships between plants and herbivores as we know them today are the result of a continuous process of selection of defense systems and adaptation by the insects, of which the behavioural adaptations have resulted in a dynamic equilibrium of shared chemistry, or at best in mutual benefit, such as in pollination.

5.2.3 Signal transduction and neurological adaptations

In section 5.2.2 we discussed some major principles operating in the interactions between two biological kingdoms: plants and insects. One principle is that nothing *is* what it *seems*, and that nothing *stays* what it *is*. Probably early pollinators were attracted to flowers because of chemical cues for their sexual life. The pollinators were not (yet) pollinators, the flowers were not (yet) the flowers as we know them today, and some of the allomones acquired different functions by turning into kairomones and synomones. This process is still ongoing and the insects are restlessly adapting.

Interpreting and acting adequately to the incoming stream of sensory information is an enormous and complex task. Making a mistake in the process of selecting a host, a partner or shelter can be fatal. Not attending external information (increasing the efficiency of the response dynamics) is known in humans, although there are still caveats in our understanding of information processing of gustatory and olfactory systems in the CNS, the brain stem and even in thalamic projections (Dulac, 2000).

But where in insects is the decision made to *not* attend the information? Bernays (1996, 1999) stresses that for a herbivorous insect it is not only a matter of appropriate actions, like feeding when the insect needs energy, or avoiding allomones. Rapidly choosing an appropriate behaviour is also valuable in a dangerous world, where a prompt response may be just as important as accuracy. To solve this problem in insects one of the following strategies could have been selected:
- A strong reduction in the types or responses of receptors, resulting in a silencing of the neural network
- A filter system that suppresses unnecessarily information in an early stage of processing
- A CNS that can cope with all incoming information and make an immediate ranking of the visual, olfactory and gustatory signals, adjust the output of motor neurones and adequately modify behaviour.

REDUCTION IN TYPES OF RESPONSES

Let us attend the first point (Fig. 5.21A). To arrive at a limited number of very specific signals the combination of odorant binding proteins (OBPs) and receptor cells should be highly selective with respect to a certain semiochemical. Much work on the selective properties of OBPs has to be done and we should realize that the present knowledge of OBPs, which are supposed to be specific, such as the pheromone binding proteins (PBPs) indicates that their variations may be limited, and that OBPs may not always be needed as a carrier molecule for the stimulus (Steinbrecht *et al.*, 1996).

The input of irrelevant information can be temporary limited. The thresholds of taste perceptors to stimulating semiochemicals can be higher in satiated individuals; among grasshoppers the tips of the sensilla (Fig. 4.4B) can be closed in individuals which are not in search of food (Bernays, 1996).

Electrophysiological recordings have shown that single receptor neurones do not have a high degree of specialization and are not particularly sensitive, although there are exceptions (Visser, 1986). Specialist receptor cells exist that are narrowly tuned to one or a few compounds, such as those sensitive to the sex pheromone nepetalactol/nepetalactone in aphids. *Pieris* butterflies are specialist herbivores of cruciferous plants. *Pieris* caterpillars (and many other lepidopterans) are endowed with generalist deterrent receptors that respond to a broad spectrum of secondary plant substances. So far, nothing new. Recently van Loon and Schoonhoven (1999) found a second type of deterrent chemoreceptor in the sensilla styloconica of the maxillae, which is very sensitive to cardenolides (Box 5.3). The generalist deterrent receptor

Chapter 5. Functions of natural terpenoids in the interrelationships between organisms

has a threshold that lies 50–100 times higher. The lateral deterrent chemoreceptor responds only to cardenolides and is completely insensitive to other secondary metabolites. In contrast, the medial deterrent receptor cell is a true generalistic neuron, responding to over 50 different feeding deterrents among which terpenes such as drimane antifeedants (Messchendorp, 1998; Messchendorp et al., 1996+b).

With respect to the first level of "signal simplification", receptor cells for repellents or deterrents are relatively limited in number, and specific. They may be even sensitive to a specific enantiomer of a terpene only (see 4.1.2 and 5.1.3). The balance of activity elicited by general repellent (or deterrent)

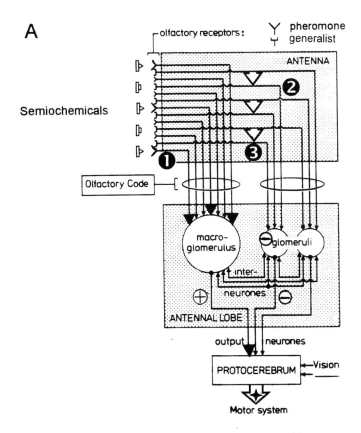

Fig. 5.21A. Reduction in types of responses in an olfactory network based on (1) highly sensitive specialist receptors and (2) generalists with a relatively high threshold, (3) mutual inhibition of generalists at the level of the sensillum. In this example the insect decides to follow an odour plume leading to a sexual partner. Compare with Box 4.1.

receptors and specific ones is important for the behavioural decision. In *Pieris* larvae the system facilitates discrimination between allomones of cruciferous plants and those of other chemical categories. According to van Loon and Schoonhoven (1999) specialist neurones could have evolved from generalist neurones through gradual loss of some receptor sites, or from token stimulus receptors, in which the synaptic connection changed into the opposite direction. To our opinion the latter possibility implies that these insects were already adapted to hosts of which they could detoxify some allomones and utilize them as kairomones, a process that became refined by specific repellent or deterrent receptors.

Generalist receptor cells detect a broad variety of odour components. Still, some noise reduction can be obtained because firing of one receptor cell can inhibit the response of another in the same sensillum (Lemon and Getz, 1999; Messchendorp *et al.*, 1996; Ochieng and Hansson, 1996; Schoonhoven and Luo, 1994).

SECOND-LEVEL FILTERING

The spectra of generalistic receptor cells often overlap although their response profiles can be different. There is a functional convergence as the axons of the receptor neurones terminate in a glomerulus from which a limited number of output neurones reach the protocerebrum. This system is further improved by the presence of other glomeruli in the same lobe and inhibitory effects of interneurones, the basics of which are discussed in section 4.1.2. This leads to an improvement of the signal-to-noise ratio. The neural activities in the output axons of the generalist receptors in the antennal lobe are affected by quantities of components in an odour blend. The quality of the blend can enhance this signal via specialist receptors of kairomones or suppress it via specialist receptors of repellents (Box 4.1). In this way, a combination of specialist and generalist interneuron types leads to a separation of pattern activity. Specialist interneuron types react to "quality", the generalistic types to "quantity".

When an insect flies in search for a host plant, the activities in the array of peripheral neurones (the across-fibre pattern) are influenced by interactions between the components of odour blends. The insect subsequently adjusts its speed and direction until the fingerprint of volatiles corresponds with its internal standard. During evolution, this standard is modified by reduced sensitivity to particular allomones and by recognition of others as kairomones.

Another phenomenon could improve second-level filtering. There is stimulus-evoked oscillatory synchronization of neural assemblies in visual and olfactory systems; a subject already touched in section 4.1.2. In insects,

Chapter 5. Functions of natural terpenoids in the interrelationships between organisms

the information on odour identity seems to be contained in spatial and temporal aspects of the oscillatory population response. Experiments with locust synapses downstream from the antennal lobe (so, close to the sensilla) indicate that their responses to odours become less specific when antennal lobe neurones were desynchronized by picrotoxin injection (Laurent et al., 1999). These results suggest that spike tuning and synchronization are important parameters for odour perception or sensation. Picrotoxin is a receptor agonist.

If repellent terpenes have similar antagonistic effects they might desynchronize neuronal ensembles tuned to kairomones or sex pheromones, and frustrate complete interneuron types (Fig. 5.21B). If so, we could think of terpenes interfering with prenylation of proteins. Fig. 5.21 gives only two examples of many possible variations. In as yet unpublished experiments with aphids we found remarkable interactions of terpenoids. Combinations of geraniol and myrcene evoked alarm reactions comparable to the effect of (E)-β-farnesene. We note that an important part of the geraniol molecule is identical with the farnesene-tail. The same is true for the myrcene molecule and the farnesene-head. The combination nerol and myrcene does not evoke any reaction.

Obviously, isomerism is important in connection to the farnesene receptor and its signal transduction system. It is significant that α-farnesene does not evoke any reaction. However, combination of α-farnesene with myrcene gives a similar reaction as does the natural alarm pheromone (E)-β-farnesene. It is intriguing that two common plant odours if combined evoke an alarm reaction in aphids. An explanation could be that an overlap in the odour perception of the different terpenes occurs, caused by a minimal time-lapse in ion channel opening induced by the terpenes. Specific terpene-induced modifications of G-proteins could play a role. Another explanation is an interaction of the generalist receptors with the specialist farnesene receptor, in which geraniol and myrcene are general repellents. This leaves a third possibility: binding to farnesene-OBP with that part of the molecule which is identical to farnesene.

In this respect it is intriguing that Kurahashi et al. (1994) showed that odorants can activate and also suppress the inward current in vertebrate olfactory receptor cells. Amyl acetate suppressed a response to D-limonene. The latency of suppression was shorter than that of the inward transduction, suggesting that suppression is due to a direct effect of odorants on ion channels or on a second messenger system rather than to an effect on olfactory receptor proteins. Although these are vertebrate receptors, the inhibition of terpenoid-sensitive receptor cells would seem to operate via the G-protein cascade.

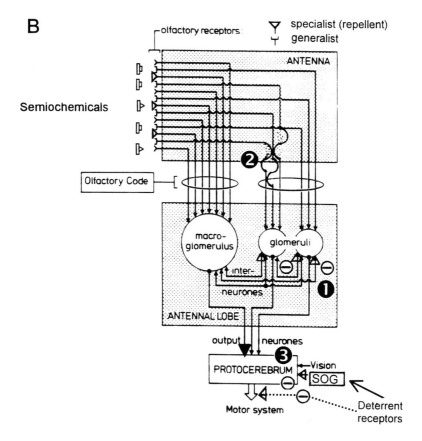

Fig. 5.21B. Second-level filtering and integration of sensory information by (1) inhibition of interneuron types, (2) desynchronization of the signals reaching the glomerulus and (3) integration in the protocerebrum – not always possible – with a gustatory signal. In this example the insect decides *not* to accept a plant as a host. SOG = suboesophageal ganglion. Compare with Box 4.1 and Fig. 5.21A.

INTEGRATION BY THE CNS AND MOTOR OUTPUT

In the protocerebrum of insects, the olfactory information is integrated with other sensory modalities, and output elements are connected to the motor system controlling behaviour. An excellent review of neural coding of general odours in insects that includes a discussion of olfactory coding in the antennal lobe projection neurones (transmitting signals from the glomeruli to the mushroom bodies) has been made by Lemon and Getz (1999). In the

Chapter 5. Functions of natural terpenoids in the interrelationships between organisms

insect brain there is not such a restricted area known where information from gustatory receptors is processed as is known for olfactory information. Most axons from gustatory receptor cells project to the suboesophageal ganglion (SOG), that is connected to the brain, and major processing of taste information is thought to occur in the SOG (Fig. 5.21B). It is beyond the scope of this book to discuss processing models in detail, but we note that they should incorporate sensilla with broad response profiles, e.g. to sugars, salts, and amino acids. Moreover, the axons may arrive from different sites, including the legs. The axons from gustatory receptors often terminate in the ganglion of the segment in which they occur and there is no known gustatory centre.

Sensitivity to antifeedants can decline in time (habituation) and it can increase (sensitization). Based on genetic codes, the neural mechanisms offer a form of protection against unwanted decisions, and this may overrule hunger or search for a sexual partner. The aphid *Myzus persicae* can be given access to nutritionally suboptimal chemically defined diets, which the aphid will ingest, supposed there is no deterrent incorporated. However, while reduction of the total amino acid concentration to 50% hardly reduced diet intake, omission of methionine resulted in ingestion of only 40% of that on the full diet (Harrewijn and Noordink, 1971), Obviously, the CNS of this aphid species refuses to accept a food source which lacks the essential amino acid methionine.

In the course of this discussion we may have become convinced that for insects odours are important to find their mates and host, but we should not overlook the "last moment" decision of the CNS, which can again be based on odour perception. Both feeding and oviposition can be set to "go" in a very short period. It is possible to think of feeding, and perhaps also oviposition, as being determined by the switching on or off of a pattern generator (Chapman, 1999).

NEUROLOGICAL ADAPTATIONS

In insects, neurones from the olfactory lobe are directly connected with the mushroom bodies that are the principal site of the brain for learning and memory. Just as vertebrates make associations between particular events, insects could do this with respect to e.g. feeding and odours, which they experienced at the time. The relation between certain terpenes and receptors is often rather specific. A reduced sensitivity to a particular repellent terpene, or of the corresponding synapse may result in the acceptance of a different range of plants.

The presence of different cells responsive to a wide range of different metabolites is typical for herbivorous insects and adaptation to a new host would require the loss of its sensitivity to deterrents in the potential host. This process would render a specialist feeder more of a generalist unless it acquires specific tuning to recognize new allomones as kairomones. According to Jermy (1993), this must have happened often during evolution.

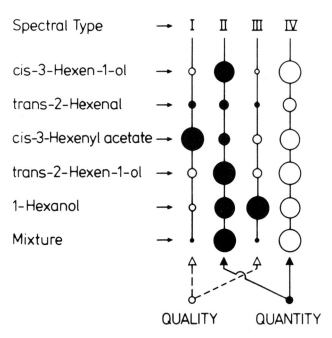

Fig. 5.22. Mean relative response spectra of interneuron types (I-IV) in the antennal lobe of the Colorado potato beetle, visualized in the areas of circles. Excitation shown as solid circles, inhibition as open circles (with courtesy of J.H. Visser).

To realize adaptation to new hosts it would be advantageous when (a) specialist receptor cells could detect plant volatiles with a degree of similarity to semiochemicals of conspecifics, (b) odour specificity of a receptor cell type can change, and (c) the neurological system can discriminate between the mere presence and ratio of odour components. With respect to (a), receptor cells which are stimulated by pheromones as well as by host plant odours indeed have been found (Visser, 1986). Concerning (b), Frey and Bush (1990) in their study of *Rhagoletis* sibling species concluded that genetically based differences in antennal sensitivity contribute to different host preferences. And to illustrate (c), response centres in a receptor population of many neurones of the Colorado potato beetle to different volatiles show a gradual differentiation of individual receptors from generalists to specialists.

That different compounds in a mixture do not simply contribute to the neural response in an additive manner is already shown in Box 4.1. It appears that the pattern of neural activity across the axons towards the olfactory brain already shows separation (Fig. 5.22). One channel (e.g. II or IV), consisting of generalist receptors, mainly reacts to a number of components (quantity), either by excitation or by inhibition. The eventual change in neural activities in that channel depends on the ratio between volatiles in an odour blend. Another channel (e.g. I or III), consisting of more specialized receptors mainly responds to one particular odour component. In the Colorado potato beetle activities of specialist interneurones of such channels were not affected by a mixture of green leaf odour components (Visser and de Jong, 1988). Such a pattern of neural organization discriminates between concentration and ratio of odour components and greatly contributes to trace a source emitting a number of terpenes at a time in an environment rich in background odours (some of which are also terpenes).

IT'S IN THE AIR!

In conclusion, terpenoids have an important role in all types of semiochemicals. They have acquired specific relationships with the receptors involved in perception of semiochemical messengers, a specificity that goes as far as the level of enantiomers. Understanding of the role of binding proteins in the perception of terpenoids is in its infancy, but the material presented in chapter 4 on messenger functions on the cellular level and the studies of different types of semiochemicals as discussed in this chapter underline their significance in G-protein cascades and their role in neurological adaptations.

In his final lecture at the 10th Symposium on Insect-Plant Relationships (1999), Chapman focussed on the neural perspective of insect-plant relations. The conclusions of this lecture, entitled "It's all in the neurones" opened with: "I have tried to show how changes in insect behaviour that may lead to host plant switches and ultimately to the evolution of new species, may be related to single changes in the nervous system". We think that chapter 5 of this book has demonstrated that terpenoids, especially volatile terpenes constitute an ancient class of carbon compounds that have had, and still have a profound impact on the evolution of insect behaviour and in the consolidation of biotic pollination in preangiosperms. To our opinion studies on the role of volatile terpenes in the transition of allomones to sex pheromones and kairomones should be advocated. To state it differently: much happens in the neurones when terpenes are in the air, which is a tribute to the actor George Formby who hazarded to fly in fragile airplanes in a retake of a 1935 movie, similarly entitled as its main song "It's in the air". Perhaps, this song says it all.

Chapter 6

Terpenoids in practice

6.1 Natural terpenoids and their sources

There are several good handbooks on terpenoid chemistry that describe isolation and purification procedures, mainly for scientific purpose. Titles are quoted under the additional heading "General reading". When isolation and purification techniques are of importance for a better understanding of a particular study, they have been included in the discussion on that study. The reader will be referred to such techniques by looking up an appropriate solvent (such as methanol) in the Index. During preparation of this book, potential readers requested us to include some information on essential oils and simple isolation procedures. This is presented in sections 6.1.1 and 6.1.2.

6.1.1 Terpenoids in land plants and simple isolation procedures

At the end of the 18th century it became known that essential oils consist of volatile components, in contrast to the previous opinion that odorants are dissolved in fatty oils, as is common practice in perfumery. Another denomination is *etheric oils*, derived from the Greek word "aither", which means "air from the heavens", the lower regions of the cosmos.

Chapter 2.2 describes the biochemical background of the production of terpenes as present in essential oils together with organs adapted for storage. Flowers with thick petals, such as roses, magnolias and clove contain essential oils in their epidermal cells. Gardeners use to say that white flowers generally smell stronger than coloured ones (jasmine!), but that may be particularly true for cultivars of domesticated plants – there exist strongly smelling species among wild Ranunculaceae, Rosaceae and Magnoliaceae with coloured sepals or petals. These families are regarded to be more primitive angiosperms, indicating that essential oils were produced early in angiospermic evolution (see 5.2.2).

Angiosperms of Mediterranean origin, among which many labiates are a rich source of essential oils containing an array of monoterpenes and higher terpenes. Some species can be easily grown at higher altitudes, provided there is enough light (DOXP pathway!) and a soil type relatively poor in nitrogen. In medieval times medicinal herbs were brought over the Alps and grown in monastic and cloister gardens, especially with the Benedictines, Dominicans and Carmelites. One of the secret recipes of cloister pharmacology was "eau des Carmes", indicating from which order it originated (it was a.o. based on the oil of marjoram). The advantage of cloister gardens is that brick walls, storing heat and protecting the Mediterranean plants from wind and frost usually surround them.

Probably the best known genus of Asteraceae (Compositae) with respect to the use of essential oil in pharmacology is *Matricaria*. It is rich in α-bisabolol and matricine (Table 6.1), which is formed from bisabolol via a photochemical reaction in glandular hairs (see 2.2.2). Together with chamazulene these sesquiterpenes have potent anti-microbial effects, and have been utilized since ancient times to treat ulcers, dermatological and intestinal disorders. The spasmolytic effects of extracts, however, are due to flavonoids. Species of this genus also produce farnesene isomers, which we used in our study on their effects on aphids (Harrewijn *et al.*, 1993, 1994).

A thorough study on the pharmacology of *M. chamomilla* is presented in a book by Schilcher (1987), in which we come across that the blue oil fraction containing chamazulene was already distilled in 1488 by Hieronymus Brunschwig. Derivatives of azulene have promising effects as specific antibiotics and have been screened; especially in Germany in the 1950s for their anti-tumour effects (Barton, 1959; Kraul and Schmidt, 1957).

A more modern way to obtain plants rich in terpenoids is to grow plantlets or shoots on a growing medium. Fulzele *et al.* (1995) established multiple shoot cultures of *Artemisia annua* on a medium supplemented with N(6)-benzyladenine and indole-3-acetic acid. Plantlet formation occurred on a medium containing naphthaleneacetic acid and kinetin. Addition of gibberellic acid and ethephone to the culture medium in a bioreactor stimulated the synthesis of terpenoids. A production of more than 1 mg% fresh weight camphor, 0.15 mg% 1,8-cineole and 0.65 mg% β-caryophyllene is possible. Table 6.1 gives an overview of terpenes in essential oil of herbs and trees. It cannot be complete but shows which terpenes are most commonly found. An extended work on volatile compounds in foods and beverages is by H. Maarse (ed., 1991). Chapters 12 and 13 of that book deal

Chapter 6. Terpenoids in practice

with spices and condiments (resp. by M.J. Richards and M.H. Boelens) and give a wealth of information on essential oils.

An interesting genus is *Artemisia*. Paracelsus already prescribed extracts of members of this genus against intermitting fevers, caused by malaria parasites and matricin has found to be an active anti-malaria compound. *Matricaria* is closely related to *Tanacetum*. Apart from the fact that there is some confusion about the taxonomy of this genus, chemosystematic approach of the Asteraceae has given little attention to *Tanacetum*. Thomas (1989) has made a study of the phytochemistry of leaf and flower oils of *T. cilicium*, a perennial plant widely distributed in Turkey, Syria and Lebanon. The flowers produce twice as much oil as the leaves, but there is no significant variation in the composition of leaf and flower oils. This makes the plant interesting as a source of (E)-β-farnesene, as the major constituents of the oil are γ-cadinene and (E)-β-farnesene with only a trace of α-farnesene. In addition, the oil is rich in α-pinene, β-pinene, sabinene, 1,8-cineole and chrysanthenol (an oxygenated monoterpene). In total 32 mono- and sesquiterpenes were found, with 1,8-cineole in common with *Artemisia* spp.

Adekenov *et al.* (1996b) obtained four sesquiterpene lactones from areal parts of *T. vulgare* in Kazakhstan. A combination of these compounds, called "Tavurae" was highly active against the indoor pests *Tetranychus urticae* and *Trialeurodes vaporariorum*. The same group isolated 10 different sesquiterpene lactones from plants of the family Asteraceae with weak to strong antifeedant effects on different insect species (Adekenov *et al.*, 1996).

Trees and shrubs of the Burseraceae are known for their aromatic resins. An example is *Commiphora mukul* (c.f.: *Balsamodendron mukul*), which is found on the slopes of hills in India. Incisions in the bark cause the resin to run out, which hardens to tears of "guggul". In this form it is harvested and purified by boiling it in a decoctum of other herbs. It is utilized against chronical disorders because of its antiseptic, adstringent and carminative properties. It is reputed for its antirheumatic effects (see 7.1.5).

Not only green plants, but also with the exception of some Myxomycetes all classes of fungi produce volatile metabolites. Many are found among the Basidiomycetes, in particular the wood-rotting fungi. Collins (1976) gives a small review of the isolation and characterization of volatile odorous substances produced by micro-organisms together with their possible use as sources of flavour compounds. The essential oils of *Ceratocystis virescens* and *C. variospora* are particularly rich in the monoterpenes geraniol, geranyl acetate, citronellol and linalool. These fungi grow readily in submerged culture and do not require unusual conditions in terms of light, temperature

Table 6.1. Terpenoids in essential oils of some herbs (**A**) and trees (**B**). ✤ = anti-tumour terpene; ● = strong anti-microbial terpene

Plant species	Common name (or plant type)	Main compounds
A		
Agrostachys hookeri	(Euphorbiaceae)	14-dehydroagrostistachin✤, agroskerin✤, agrostistachin✤, 17-hydroxy-agrostistachin✤
Anethum graveolens	dill	myrcene?, α-phellandrene, D-limonene ✤ ●, (S)+carvone, *cis*-ocimene, β-phellandrene, β-pinene, γ-terpinene ●
Anethum sowa	Indian dill	carvone, D-carvone derivatives, eugenol, limonene ●, methyl cinnamate
Anisomeles indica (*A. ovata*)	fang feng cao (flowering shrub)	anisomelic acid✤, ovatodiolide✤, 4,7-oxycycloanisomelic acid✤
Artemisia absinthium	wormwood	camphor, α-phellandrene, β-caryophyllene ●, (+)-isothujone, 1,8-cineole ●, (-)-thujone ●
Artemisia annua	(Asteraceae)	artemisinic acid, artemisinin ✤, bisabolene derivatives, deoxyartemisinin ✤, fenchene, E-β-farnesene, α-pinene ●
Artemisia argyi	(Asteraceae)	borneol, camphene, camphor, carvone, isoborneol, E-carveol, α-phellandrene, α-terpineol
Artemisia vulgaris	mug wort	thujone ●, 1,8-cineole ●
Baccharis megapotanica	(Asteraceae) flowering herb	baccharinoids B1, B2, B3, B7✤
Carum carvi	caraway	carvone, D-limonene✤ ●
Caryophyllus aromaticus	clove	eugenol
Citrus	citrus peels	D-limonene✤ ●
Coriandrum sativum	coriander	camphor, geraniol (acetate) ●, linalool ● and enantiomers, α-pinene ●, γ-terpinene ●, *p*-cymene ●
Coridothymus capitatus	ingredient of "oregano"	carvacrol ●, *p*-cymene ●
Cuminum cyminum	cumin	cumin aldehyde (phenolic), β-pinene, γ-terpinene ●, *p*-cymene ●
Elephantopus coroliniamus	(Asteraceae)	isodeoxyelephantopin✤,
Elephantopus tomentosus	hairy elephant's foot, ku di dan	tomenphantopin A, B✤
Elettaria cardomonum	cardamome	borneol, sabinene, terpinen-4-ol ●,

Chapter 6. Terpenoids in practice

Plant species	Common name (or plant type)	Main compounds
		α-terpinene, α-terpineol ●, 1,8-cineole ●
Eucalyptus	(Euphorbiaceae)	1,8-cineole ●
Foeniculum capidaceum	fench	fenchol
Foeniculum vulgare	fennel fruit	anethole, fenchone, limonene❖ ●, phellandrene
Geum urbanum	avens	eugenol
Hedychium coronarium	ginger lily, tu qiang huo	coronarins A, B, C and D❖
Hyptis capita	spikenard	hyptatic acid❖, ursolic acid❖, 2-α-hydroxyursolic acid❖
Iris missouriensis	Missouri iris	iso-iridogermanal❖, zeorin❖
Juniperus communis	(Cupressaceae)	cadinene
Matricaria chamomilla (= *recutita*)	chamomile	chamazulene ●, bisabololoxides, matricine ●, α-bisabolol ●, (E,E)-α-farnesene, (Z,E)-α-farnesene, (E)-β-farnesene
Matricaria matricarioides	(Asteraceae)	(E)-β-farnesene
Melaleuca alternifolia	tea tree	linalool ●, terpinen-4-ol ●, α-pinene ●, α-terpineol ●, 1,8-cineole ●
Melissa officinalis	balm	citronellal ●, geranial ●, geraniol ●, linalool ●, neral ●
Mentha piperita	peppermint	limonene❖ ●, menthol, menthofuran, menthone, sabinene ●, β-phellandrene, β-pinene, 1,8-cineole ●
Mentha pulegium (c.f.: *Pulegium vulgare*)	mint, penny royal	menthol, menthone, pulegone ●
Mentha spicata (c.f.: *M. viridis*)	spearmint (green mint)	carvone, limonene❖ ●, menthol, perillyl alcohol❖, terpinen-4-ol ●
Nepeta cataria	catnip	nepetalactone
Nepeta racemosa	(Lamiaceae)	nepetalactone
Ocimum basilicum	sweet basil	caryophyllene ●, eugenol, linalool ●, methyl chavicol, (E)-β-ocimene, (Z)-methyl cinnamate, 1,8-cineole ●
Ocimum suave	(Lamiaceae)	eugenol, linalool ●, methyl chavicol, (E)-methyl cinnamate
Origanum majorana	marjoram	carvacrol ●, terpinen-4-ol ●, thymol ●, (E)-sabinene hydrate ●, (Z)-sabinene hydrate ●
Origanum vulgare	"oregano", wild marjoram	carvacrol ●, γ-terpinene ●, *p*-cymene ●

Plant species	Common name (or plant type)	Main compounds
Pimpinella anisum	anise	anisyl derivatives, methyl chavicol, (E)-anethole, mostly phenols
Rhabdosia effusa	(Lamiaceae) flowering herb	effusantin A❖
Rhabdosia excisa	flowering herb	excisanin A, B❖
Rhabdosia longikaurin	flowering herb	longikaurin B❖
Rhabdosia sculponeata	flowering herb	sculponeatin C❖
Rhabdosia umbrosa	flowering herb	kamebakaurin❖, kamebanin❖, leukaminins A, B❖, umbrosin A❖
Rosmarinus officinalis	rosemary	borneol, borneol acetate, camphor, carvacrol ●, eugenol, thymol ●, verbenone, α-pinene ●, 1,8-cineole ●
Ruta graveolens	herb of grace	limonene❖ ●, α-pinene ●, 1,8-cineole ●
Salvia spp.	sage	camphor, isothujone ●, thujone ●, 1,8-cineole ●
Salvia spp.	Spanish sage	sabinene
Salvia spp.	clavy sage	linalool ●, linalool acetate ●
Satureja thymbra	*Origanum vulgare* + *Coridothymus capitus* = oregano	thymol ●, α-pinene ●, β-cymene, γ-terpinene ●
Thymus capitatus	Spanish oregano	carvacrol ●, thymol ●
Thymus vulgaris	thyme	camphor, carvacrol ●, carvone, geranial ●, linalool ●, myrcene, neral ●, thymol ●, α-pinene ●, β-caroyphyllene, 1,8-cineole ●
Thymus zygis	(Lamiaceae)	carvacrol ●, thymol ●

B

Aesculus hippocastanum	horse chestnut (tree)	barringtogenol-C21-angelate❖, hippocaesculin❖
Bolbostemma paniculatum	tu bei mu (climbing vine)	tubeimoside❖
Brucea antidysenterica	bitter-barked tree	bruceanols A, B and C❖, bruceantinoside C❖, yadanzioside N❖, 1,11-dimethoxy-canthin-6-one❖, 1-hydroxy-11-methoxy-canthin-6-one❖
Brucea javanica	ya dan zi, Kosam seed (bitter-barked tree)	yadanzioside N❖, yadanzioside O aglycone❖, yadanzioside P❖
Ginkgo biloba	maidenhair tree	ginkgolides❖, ginsenosides

Plant species	Common name (or plant type)	Main compounds
Euptelea polyandra	euptela (forest tree)	3-O-acetyloleanolic acid❖
Maytenus spp.	(Celastraceae) (tropical trees)	maytenfoliol❖
Simaba multiflora	(Simaroubaceae) bitter-barked tree	senecioyloxychaparrin❖, 13,18-dehydro-6-α
Taxus baccata	(Taxaceae) (evergreen shrub)	taxol❖
Taxus mairei	yew	taxamairins A, B❖
Triperygium regelii	yellow vine	regelin❖, regelinol❖, triptolide❖

or pH. The anti-tumour effects of geraniol and geranyl acetate (Table 4.1) make culture of these fungi particularly interesting.

Many aromatic plants can be grown in gardens of moderate regions. A sunny site should be chosen, protected against strong wind. Use well-drained soil, relatively poor in nitrogen (50 kg N/ha is raining down in Western Europe every year). Directives for cultivating (from seeds or nurseries) are mostly on the packing; for those who want to prepare a growing site Table 6.2 gives a list of aromatic plants rich in terpenoids, which are easy to grow. For that reason Table 6.2 does not provide data on e.g. clove or cardamom (the latter being a perennial plant in India and the West Archipel, more than 4 m high with optimal seed production in years 3-6). These seeds are better bought from a reliable company. Pharmacological properties of the plants in Table 6.2 have been checked in several American, European and Indian sources. It should be noted that their qualities as herbal remedies are not always based on the terpenoid content alone. As an example, *Hypericum perforatum* (Fig. 2.9) has antidepressive properties. According to Tyler (1994) this may be due to the metabolites hypericin, pseudohypericin and related naphtodianthracins, of which hypericin has proved to inhibit monoamino oxidase A and B *in vivo*. Apart from essential oils, *H. perforatum* also contains quercetin, a phenolic flavonoid with a.o. protein kinase inhibiting properties. More information on aromatic plants is found in Hay and Waterman (1993).

Some aromatic plants can be grown in hydroponics, making the grower independent of soil quality. Composition of the essential oils and yield can be kept under control via the mineral nutrition. Examples are *Anethum graveolens* (Udagawa, 1995), *Angelica archangelica* (Letchamo *et al.*, 1995), *Mentha piperita* (Srivastava *et al.*, 1997; Takano and Yamamoto, 1996), *Ocimum basilicum* (Smith *et al.*, 1997), *Perilla frutescens* (Takano and Ohgi, 1994) and *Thymus vulgaris* (Udagawa, 1995). Extended

Table 6.2. Easy-to-grow aromatic plants, rich in terpenes. R = rich soil type, M = medium, S = sandy soil, H = much humus. s= stimulating, i = inhibiting. ✳ sunny site, ◗ can be partly shadowed, °C = higher temperature

Species	Soil type	Light	Main applications
Acorus calamus (Araceae)	Wetland	✳	Peeled rootstocks – in aperitivs; nervous intestinal disorders
Agrimonia eupatoria (Rosaceae)	R/M	✳	Flowers and leaves, gall stones, gall colics
Alchemilla vulgaris (Rosaceae)	RH	◗	Leaves and roots – adstringent, lotions, hemorrhages
Anethum graveolens (Apiaceae)	M	✳	Seeds – culinary, anti-emeticum, carminativum, milk secretion (s)
Artemisia absinthium (Asteraceae)	MS	✳◗	Leaves – in vermouth, gall colics
Artemisia vulgaris (Asteraceae)	MS	✳◗	Leaves – culinary, gall production, intestinal disorders, blood circulation (s), hemorrhoids
Conyza canadensis (Asteraceae)	MS	✳	antidiarrhoeic, diaretic, arthostis
Coriandrum sativum (Apiaceae)	M/P	✳	Seeds – culinary, carminativum, digestivum
Foeniculum vulgare (Apiaceae)	M	✳◗	Seeds – culinary, carminativum, digestivum, conjunctivites, milk secretion (s)
Geum urbanum (Rosaceae)	R	✳	Roots and complete plants – dermatitis, gingivitis, sedative, tonicum
Humulus lupulus (Urticaceae)	M	✳◗	Flowers/strobuli – adjuvants (beer), gallenicum, incontinence, sedativum, diurethicum
Hypericum perforatum (Hyperiaceae)	MH	◗	Flowers and leaves - anti-depressivum, blood circulation (s), gall production, vegetative dystonia, tonicum, retroviruses (?)
Lamium album (Lamiaceae)	P/M (Ca^{2+})	◗	Flowers + young leaves – menstruation problems, panariticum, soporific
Lamium galeobdolon (Lamiaceae)	P/M	◗	Flowers – cosmetics
Lavandula officinalis (c.f.: *L. angustifolia*) *L. spica* (Lamiaceae)	M (Ca^{2+}) °C	✳	Flowers – cosmetics, vegetative dystonia

Chapter 6. Terpenoids in practice

Species	Soil type	Light	Main applications
Matricaria chamomilla (c.f.: *M. recutita Chamomilla officinalis*) (Asteraceae)	M	✳◗	Flowers, roots – anti-inflammatory, anti-spasmodic, bursitis, gastro-intestinal spasm, stomach ulcers *Helicobacter* (?)
Melissa officinalis (Lamiaceae)	M/P H	✳	Leaves – culinary, attracts honey bees, antibacterial, gastic catarrh, sedativum, spasmolyticum, menstruation problems
Mentha piperita (= *M. aquatica* x *M. viridis*) (Lamiaceae)	R/M H	✳	Leaves – in mints, digestivum, desinfectans, sedativum, spasmolyticum
Mentha spicata (c.f.: *M. viridis*) (Lamiaceae)	M	✳	Leaves – anti-tumour
Ocimum basilicum (Lamiaceae)	M	✳	Leaves – culinary, in Chartreuse liquor, periodontitis, insommia
Origanum majorana (Lamiaceae)	RH	✳	Complete plants – culinary, carminativum, digestivum
Rosmarinus officinalis (Lamiaceae)	M °C	✳	Leaves – activating (bath) cosmetics, acne, lumbago, diurethicum, blood circulation (s), hair
Ruta graveolens (Rutaceae)	Rocky, °C	✳	Complete plants – diurethicum, sedativum, spasmolyticum, rectum prolaps
Thymus vulgaris (Lamiaceae)	R/M	✳	Flowers + leaves – culinary, cosmetics, anti-microbial *V. cholerae*, bronchitis, acne, intestinal parasites

information on the application of hydroponics is presented in the bibliographies of ISOSC (International Society for Soilless Culture), e.g. in the 7th bibliography (1999, ed. A.A. Steiner).

To obtain essential oils in some cases complete plant parts can be utilized in dried form, without any isolation procedure. According to Johnson *et al.* (1985) capsules filled with freeze-dried leaves of *Tanacetum parthenium* (about 50 mg daily) were effective as prophyllaxis against migraine, probably due to the sesquiterpene lactone parthenolide. In England the herb has been taken to this purpose by large numbers of people continuously and for many years, without apparent side effects. However, these are relatively small doses. It should be noted that in complete plant parts the whole range of natural terpenes is present, together with other secondary metabolites. This, of course, can be an objection for selective application of specific terpenes. This holds even more for air-dried plants, be it outdoors or indoors.

Monoterpenes are rather volatile and their content may be considerably reduced during drying. Air temperature should preferably not become above 35°C. Nevertheless; the use of dried plant parts such as flowers has been common practice over centuries. The advantage of course is a simple and cheap procedure and can result in a good product, especially when higher terpenes and terpenoids are involved. Fig. 6.1 shows an easy-to-construct outdoor drying system and Box 6.1 gives an overview of isolation procedures.

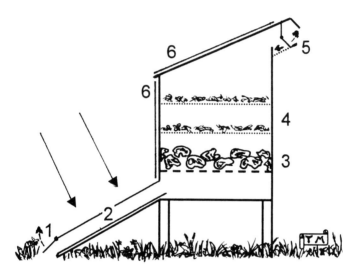

Fig. 6.1. A simple outdoor drying system for aromatic plants. 1 = adjustable air inlet; 2 = glass panels and heat absorbing black plate; 3 = grate with stones preventing condensation after sunset; 4 = grids with plant material; 5 = adjustable air outlet (operated with a mechanical thermostat); 6 = heat-reflecting shields.

Because of the hydrophobic character of most terpene structures, water extraction is of limited use, but when extracts have to be made which should be relatively cheap, crushed plants are mixed with (warm) water or methanol/water. Of course these crude extracts contain several metabolites, but it is a regular procedure in developmental countries. Williams and Mansingh (1996) have published a critical review of the potential of neem compounds for their use in the tropics. Neem extracts are toxic to over 400 species of insect pests, some of which have developed resistance to conventional pesticides. The authors stress that although the composition of

Chapter 6. Terpenoids in practice 263

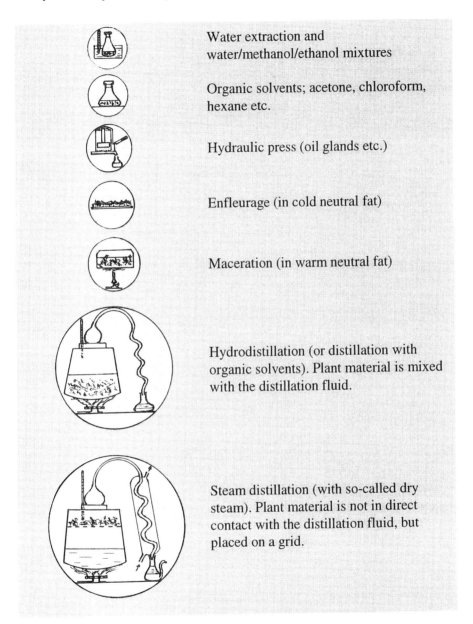

Water extraction and water/methanol/ethanol mixtures

Organic solvents; acetone, chloroform, hexane etc.

Hydraulic press (oil glands etc.)

Enfleurage (in cold neutral fat)

Maceration (in warm neutral fat)

Hydrodistillation (or distillation with organic solvents). Plant material is mixed with the distillation fluid.

Steam distillation (with so-called dry steam). Plant material is not in direct contact with the distillation fluid, but placed on a grid.

Box 6.1. Procedures to isolate terpenoids from plants.

the extracts is not always consistent, a big advantage could be a very slow development of resistance, if any. This is comparable to what happens in combination therapies (7.1.8). The crude extracts are composed of several compounds with a different mode of action and therefore interfere with olfaction, mating, reproduction, etc. often in the same insect species.

Moreover, the only cheap and feasible programme for developing countries at this time is the utilization of crude extracts or fractions of these extracts. As different neem ecotypes vary in biological activity selection processes utilizing biotechnology could be effective in developing superior neem varieties.

6.1.2 More complex isolation techniques

There are several good textbooks on isolation of terpenes. Titles are under the heading "General reading". As this is not a textbook on chemical techniques we keep this section concise. Terpenes from plants rich in oil glands can be removed with a hydraulic press. A few main techniques have been developed to isolate or extract essential oils, resulting in high yields:
- Enfleurage
- Maceration
- Distillation
- Extraction.

ENFLEURAGE - This is a delicate procedure to transfer essential oils of precious flowers into a fatty substance, such as lard. It is done at room temperature. Glass plates are covered with a layer of neutral fat of about two cm thickness. Intact flowers are put on this layer and kept there for 2-4 days, depending on the species. The flowers are then replaced by fresh ones, and this procedure is repeated 8-12 times. Therefore, this procedure can only serve for plants of which the flowering time is long enough. The lard is removed and covered with 96% ethanol for about two weeks. The filtered fluid is a basis for perfumes.

MACERATION - Again a neutral fatty substance is used, often lard, but in this procedure it is heated until it has just passed the melting point. The flowers or leaves are put in and carefully stirred. The mass is allowed to cool down to room temperature. After 3-4 h, heat again and replace the plant parts by fresh ones, and repeat this for a total of 10-12 times. The fat is melted for the last time, the plant parts are removed and the material is filtered through a sieve with coarse mesh. As in the enfleurage procedure, 96% ethanol is added (about 1:1) and the bottle put aside for several weeks. At 30°C this period can be reduced, but care should be taken not to overheat.

DISTILLATION - Hydrodistillation is a relatively simple technique. The distilling bulb is filled with plant material (roots are cut into pieces) and

Chapter 6. Terpenoids in practice

covered with distilled or demineralized water. Via a (water-generated) cooler the distilled liquor drops into the receiver, from which it is poured out into a low, cooled glass cylinder. The terpenes will collect at the surface, as water remains below, and should be immediately removed by suction. A few drops of the essential oil in ethanol release a strong odour. For greater quantities a stainless steel kettle can be used instead of a glass bulb. Steam distillation works with so-called dry steam. Plant material is placed on a grid through which the overheated steam is forced. When steam is circulating under low pressure in the extraction chamber, osmotic processes transfer essential oils into the aqueous phase. One can use a.o. a so-called Florentine bottle to collect the distilled mixture of water and essential oil (Box 6.1); a simple device to separate water from essential oil. Instead of water, mixtures of ethanol and water can be used (wine), resulting in a two-phase distillation. The majority of ethanol is distilled first and dissolves most lipophilic compounds, followed by an aqueous phase.

EXTRACTION – More complicated is extraction with organic solvents: methanol, ethanol, hexane, etc. Detailed procedures are described by a.o. Dobson, 1991; Gershenzon *et al.*, 1989; Maarse and Kepner, 1970; Meyer-Warnod, 1984; Picman *et al.*, 1993. It is not unusual to combine steam distillation and organic solvent extraction.

Limonene and menthone are the major monoterpenes present in the youngest leaves of *Mentha piperita* (peppermint). The proportion of limonene decreases rapidly with development, while menthone increases, to be converted into menthol at later stages (Gershenzon *et al.*, 2000). A similar process is found in *Carum carvi* (caraway), where limonene is converted into carvone in later stages, resulting in accumulation of carvone in the seed (Bouwmeester *et al.*, 1998). Although the enantiomeric compositions of dill and annual and biennial caraway seed essential oils were similar, hexane extraction is to be preferred for a precise determination of monoterpene composition and contents in caraway and dill seed. When isolated via hydrodistillation the more polar carvone was extracted more efficiently than the apolar limonene (Bouwmeester *et al.*, 1995). Gershenzon *et al.* (2000) who isolated the monoterpenes with diethyl ether, followed by analyses with GC/MS, concluded that biosynthesis and emission of terpenes occurs in different compartments and that the rate of rise and decline of monoterpene biosynthesis, based on the expression of the corresponding genes is a useful object of study to improve terpene production in this species.

We isolated pure (E)-β-farnesene from chamomile oil by a simple column-chromatographic technique. Small amounts were obtained by eluting the oil dissolved in hexane on a Florisil column (40-60 mesh). Higher amounts were obtained with basic aluminium oxide columns. The first fraction contains apolar monoterpenes. The head fraction contains (E)-β-farnesene; the next fraction with a blue colour is chamazulene. (E,E)-α-farnesene could be obtained in the same way by purification of hexane extracts of apple peels. The farnesenes are highly unstable and working with hexane preserves the sesquiterpenes from oxidation. The hexane can be removed in a little stream of nitrogen gas prior to experimentation.

A discussion of biological activity together with simple extraction and isolation techniques of sesquiterpene lactones of members of the Asteraceae is provided by Rodriguez (1976). He draws special attention to the glandular trichomes of *Parthenium hysterophorus* that contain large quantities of parthenin, the cause of serious contact dermatitis e.g. in India, where the herb has been introduced from the Americas. Parthenin is produced in seedlings older than 10 days, where it is restricted to the trichomes on stem and leaves. Parthenin, and some closely related lactones, forms two unstable adducts with L-cysteine. The two active sites on the molecule may contribute to its strong sensitizing properties (Picman *et al.*, 1979).

The centre of pharmacognosy of the University of Groningen, the Netherlands, has made an intensive study of terpenoids in the genus *Eupatorium* (Woerdenbag *et al.*, 1991). Besides alkaloids of the pyrrolizidine type, Asteraceae are known for their widespread and generally occurring sesquiterpene lactones. In *E. cannabinum* the major sesquiterpene lactone is eupatoriopicrin (Fig. 6.2B), derived from the basic sesquiterpene lactone structure, the germacranolide skeleton (Fig. 6.2A) and its isolation and purification is described in a review (Woerdenbag, 1986). Extracts are usually made with methanol or ethanol. The extracts are concentrated by evaporation of the organic solvent, diluted with water and subsequently cleared with a solution of lead acetate. The supernatant is then extracted with chloroform. After evaporation of the solvent the crude extract is obtained.

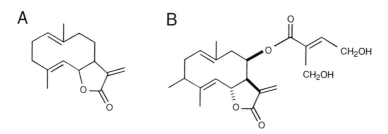

Fig. 6.2. **A**: germacranolide skeleton; **B**: eupatoriopicrin, a sesquiterpene lactone from *Eupatorium* spp.

Chapter 6. Terpenoids in practice

Mono- and diterpenes of gymnosperms were isolated by Gershenzon *et al.* (1993) by soaking samples in *tert*-butyl methyl ether for two days at room temperature (gentle shaking). They partitioned the extract against a solution of 100 mM ammonium carbonate (pH 8.0). An organic phase containing monoterpenes should not be heated and can be dried over anhydrous magnesium sulfate. Monoterpenes can be separated from diterpenes by adding diethyl ether to a sample, washing with water and passing over a column of silica gel (100-200 mesh). Samples for diterpene analyses were treated with diazomethane, resulting in methylation of diterpene acids, and then evaporated until dried.

Luis *et al.* (1992) used air-dried (room temperature) material of *Salvia canariensis* (Lamiaceae) to isolate diterpenes. The plantlets were macerated with distilled acetone at room temperature for two days, followed by the silica gel procedure as described above.

The second chapter of the valuable standard work on the genus *Azadirachta* by Schmutterer (ed., 1995) gives an extensive overview of diterpenoids, triterpenoids and tetranortriterpenoids of *A. indica*. Chapter 8 of that book is devoted to other *Azadirachta* species and to species of *Melia* with their chemistry and biological activity.

Besides isolation, proper determination and quantitation of terpenoids in plant extracts often has a priority, especially when relatively simple methods can be developed. As an example we mention the search for alternative methods for the quantitation of ginkgolides in extracts of *Ginkgo biloba* by van Beek *et al.* (1991). Ginkgolides are very potent and selective platelet-activator-factor (PAF) antagonists. Phytopharmaceuticals based on these highly oxidized diterpenes belong to the best selling drugs in Europe. The metabolites are unique for *Ginkgo*, the leaves of which have been utilized since ancient times to treat vascular disorders in elder people, in particular those related to blood clotting and infarct phenomenons.

With the recent development of new LC/MS interfaces on-line coupling of HPLC and MS together with ionization techniques higher selectivity in analyses of unknown plant components is possible, covering terpenoids of MW 200-800.

SOLUBILIZATION AND STABILIZATION

For practical applications e.g. in insect control and domestic use it is of importance to prepare cheap and reliable formulations of terpenoids. Most of them, starting with essential oils, dissolve poorly in aqueous medium. Ethanol is expensive in many countries, methanol is a cheaper alternative, but rather toxic. A further simplification is transfer of a methanolic solution into a mixture of water + methanol (about 95/5), in which procedure warm water is sometimes used. Bacterial contamination and other decomposition processes may seriously reduce the reliability of these methods when the product is used in field conditions.

The application of organic co-solvents to increase solubility of essential oils in water is reasonably satisfactory but has its drawbacks for large-scale production. Dispersion can be improved with neutral detergents such as Tween 80, Triton X-100 and lauryl sulphate, but the solutions often loose their transparency. For industrial processes detergents of the anion type are an economic option, in particular metal salts of sulphate and sulphonates, or fatty acids of natural origin. An existing patent is EP 92-304923 (1992, Martin *et al.*). Another possibility is the use of polyethylene glycols (patent US 93-99879, 1993, Slavticheff and Barrow). It should be noted that polyethylene glycols are not suitable for human consumption.

Complexing agents such as cyclodextrins or polysaccharides are other possibilities to increase solubility of terpenes in water. Patents on these techniques are pending or have been confermed (e.g. US patent 94-2331173, 1994, Eskins *et al.*). As discussed in chapter 2, a glycoside bond is a common way along which nature facilitates transport of terpenoids in aqueous media. In addition, this is a protection against autotoxicity (see 4.2.3), indicating that complexing terpenes with low molecular weight-carbohydrates can reduce their biological activity. Glucoside bonds are not always easy to obtain in the laboratory, as is a.o. the case with carvone (Oosterhaven, pers. comm.). As a rule glucoside bonds are possible with hydroxyl groups and heterocyclic nitrogen compounds. In spite of their low complexing affinity monosaccharides in reduced form (e.g. sorbitol) can be applied to solubilize essential oils, a procedure which in addition makes use of an anionic detergent in water a bit above room temperature. A patent on this process is pending and may provide detailed information on the ratios of detergents, monosaccharides and essential oils. It is easy to test this simple procedure on a small scale, but care should be taken to check the chemical reactivity of the terpenes upon binding to a saccharide.

According to Isman (1999) formulations of essential oils with proper surfactants and humectants could mitigate risks of phytotoxicity on most horticultural and ornamental plants, retaining sufficient activity against pests. A practical remark: as in distillation procedures, reaction vessels containing essential oils should never be of copper, aluminium or even iron. Use glass, ceramics or stainless steel instead.

Jarvis *et al.* (1998) studied the stability of azadirachtin in aqueous and organic solvents to improve the quality of this powerful natural pesticide. They reported that azadirachtin solutions in organic solvents such as methanol are stable at room temperature, at higher temperatures it is decomposed. Organic solvents like chloroform and dichloromethane produce

Chapter 6. Terpenoids in practice

acidic compounds if radiated with sunlight, so these solutions must be kept in the dark.

However, while azadirachtin (Fig. 2.16B) is relatively stable to heating when not dissolved, it is rapidly destroyed or altered by heating in aqueous solution and methanol. At 50°C the half-life in methanol was only 7 days and at 90°C not more than 7 h. In distilled water the decomposition is even more rapid: 15 and 18 min respectively. The major decomposition product is 3-acetyl-1-tigloylazadirachtin. Two other triterpenoids of neem, nimbin and salannin are more stable, particularly nimbin (Fig. 6.3B). Another problem with azadirachtin is its limited pH stability. Azadirachtin is most stable in mildly acidic medium. According to Jarvis et al. (1998) and others, stability is greatest between pH 4 and 5 and declines rapidly at pH 7. The authors advice to avoid storage of aqueous solutions of azadirachtin unless they are deep-frozen.

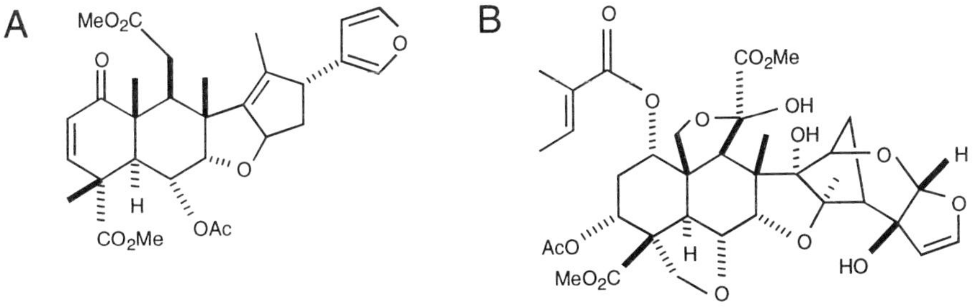

Fig. 6.3A: 3-acetyl-1-tigloylazadirachtin, the main decomposition product of azadirachtin; B: nimbin, a more stable decomposition product. Compare with Fig. 2.16B.

All over the world essential oils can be obtained in local shops although the degree of purity is not always guaranteed. Information on the distributors can be found a.o. on the electronic world wide web (WWW).

6.2 Utilization of terpenoids in biology and agriculture

This section is closely related to chapter 5.1 that discusses the functions of natural terpenoids in a more theoretical context. The coming sections focus on the utilization of natural terpenoids for beneficial insects, in insect control, resistance breeding and a few selected applications in biology and agriculture.

6.2.1 Beneficial insects

Section 5.1.4 describes quite a number of terpenes, which are attractive to insects, among which those in *Melissa officinalis* that attracts honeybees. Selection procedures could increase their activity with respect to pollinating insects. The use of terpenes is not limited to their functions as kairomones (Box 6.2). Within the context of this book this is a marginal topic, because the mode of action of many terpenoids effective in insect control has not always been investigated and information on repellent or deterrent action is lacking, leaving their functions as messengers open. With this in mind, we discuss a few important studies that may contribute to the application of terpenoids.

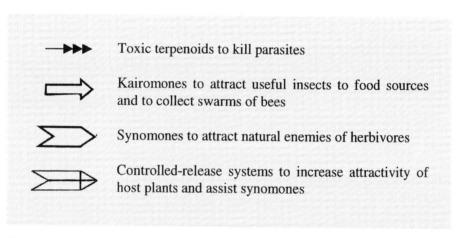

Box 6.2. Usage of terpenoids in the management of beneficial insects.

According to Wilson *et al.* (1990) menthol fumigation in bee-hives is effective in controlling tracheal mites as parasites of honeybees. The mites (*Acarapis woodi*) enter the prothoracic tracheae and feed through the walls on hemolymph. In moderate climates control of the mites is possible by applying 50 g of menthol crystals per hive in a perforated bag. To our knowledge, nobody ever saw colonies of mites running out of the treated bee-hives, so menthol most probably kills them in the honeybees. The bees do not seem to be disturbed by the menthol vapour. This induced Kevan *et al.* (1999) to investigate the effect of feeding menthol to honeybees in order

Chapter 6. Terpenoids in practice

to obtain a systemic medication. Honeybees can be fed menthol by feeding them sucrose syrup (50%) containing menthol dissolved in absolute ethanol (320 mg/ml). Menthol became detectable in the hemolymph after about 2 h and could reach a dose of 100 µg per honeybee before becoming toxic.

It is questionable if feeding of menthol to honeybees is of practical importance to control of tracheal mites, but the method is interesting with respect to the combination of terpenes as an alternative to synthetic acaricides for the control of a major problem in honeybee keeping, *Varroa jacobsoni*. Unfortunately, very few terpenes have proven to be successful in combating this parasitic mite in field trials. Imdorf *et al.* (1999) have made an extensive review of the potential use of essential oils for the control of *V. jacobsoni*. The authors conclude that thymol causes the highest mite mortality; menthol, which is promising against tracheal mites, is much less effective. Commercial products are usually composed of a mixture of terpenes. Api Life VAR® (LAIF Chemicals, Italy) is a porous ceramic carrier impregnated with a mixture of thymol (76%), eucalyptol (1,8-cineole) (16%), menthol (4%) and camphor (4%), but according to Imdorf *et al.* (1999) not all components contribute to the varroacidal activity, which apart from their selective activity can be due to inadequate vapour pressure. This does not mean that unlimited additions of essential oils are harmless, as the bees may meet difficulties in accepting their feed, a reason why it is advised to provide winter-feed before fumigation treatment starts.

In their own experiments the group of Imdorf obtained promising results with a combination of menthol and camphor. In Europe, thymol and blends with thymol are widely used to control *V. jacobsoni*. Because of possible residues of essential oils in honey, maximal residue limits (MRL) have been fixed. This is because terpenes can dominate the odour of any products, not for safety regulations, as according to the EU regulation Nr 2377/90 thymol belongs to the non-toxic veterinary drugs (group II) for which there is no MRL.

Honeybees encounter essential oils during foraging and this could be a reason why they endure substantial concentrations of terpenes. Beekeepers in Western Europe feed their colonies in early spring and frequently include aromatic plants such as *Mentha spicata* cv. *crispa*. In this period the breeding chambers are disinfected with essential oils. In northern regions where nectar cannot continuously be collected one could try to feed menthol, thymol or terpene mixtures in sugar solutions before and after the collection period.

The attraction of useful predators to plants infested with herbivores is discussed in section 5.1.5. A few terpenes would seem to recruit natural enemies in a number of situations: (E)-β-farnesene, (E,E)-α-farnesene, germacrene D, linalool. The tachinid parasitoid *Cyzenis albicans* of

Operophtera brumata, the winter moth (Lepidoptera), is attracted by borneol, that is released by leaves of *Quercus garvyana* (Roland *et al.*, 1995). The terpenes just mentioned could increase the attractivity of infested plants when applied in slow-release formulations in IPM systems. Examples are cyclodextrin inclusion complexes, micro-particles and degradable, more compact polymers.

In other situations slow-release formulations with terpenes can be used to assist existing repellent plant metabolites. Examples are UV-cured combinations of pre- and copolymers of acrylates. From these classes of matrices zero order or first order-controlled release of terpenes can be obtained (TNO patent Nr 0410530, Appl. Nr 90202002.3, 1995: Dispenser for insect semiochemicals, method for the preparation thereof and method for the control of insects).

6.2.2 Insect control

The material presented in section 5.1.3 comes from two different sources: mere academic studies of insect-plants relationships and case studies on the control of herbivorous insects. Many suggestions for practical application of our knowledge of messenger functions of terpenoids can be derived from that section, e.g. sesquiterpene dialdehydes as antifeedants for the Colorado potato beetle (Gols *et al.*, 1996) and the numerous pieces of advice in the papers and books of H. Schmutterer. In addition, the titles of papers quoted in the References at the end of the present volume may serve to select appropriate procedures to control insects with terpenoids.

In this section we evaluate terpenoids with messenger functions in practice, devided into three groups:
- Repellents, deterrents and attractants
- Those which interfere with internal messengers and development
- Those with unknown messenger functions (classified as "toxic" terpenoids).

REPELLENTS, DETERRENTS AND ATTRACTANTS

Table 6.3 gives a number of terpenoids that are known to be repellent to insects. After reading the elaborate section 5.1.3 one may expect an extended list of repellent terpenes and deterrent terpenoids. However, there is a remarkable phenomenon: in many studies of plant parts with repellent properties there is more than one odour component, making it impossible to say which volatile terpene is *the* repellent constituent, not to speak of additive or synergistic effects with volatile compounds of different classes. This may not be so surprising. We have demonstrated (see 5.1.3) that combinations of monoterpenes are effective alarm and repellent semio-

Table 6.3. Single natural terpenoids with proven repellent or deterrent action against herbivorous insects

Monoterpenes	Insect groups and species
camphor	aphids, *Zabrotes subfasciatus*, *Delia radicum*
D-carvone	aphids, flour beetles
caryophyllene	thrips spp. (?)
carvacrol	*Zabrotes subfasciatus*, *Spodoptera littoralis*
cineole	*Delia radicum*, *Zabrotes subfasciatus*
citronellal	*Aedes aegyptii* (larvae), body lice
p-cymene	*Delia antiqua*
eugenol *	*Spodoptera littoralis*, termites, (thrips spp. ?)
(E)-β-farnesene	aphids
geraniol	aphids, body lice, *Aedes aegyptii*, locusts, planthoppers
geranyllinalool	termites
juglone $^\oplus$	ants, bark beetles, *Leptinotarsa decemlineata*
limonene	aphids (Adelgidae), ants, *Delia antiqua*, larch beetles, spruce beetles, seed wasps
linalool	aphids, houseflies, (thrips spp. ?)
perillene	ants
β-phellandrene	spruce beetles
α-pinene	ants, aphids, psyllids
β-pinene	aphids
pulegone	*Blattella germanica*, termites
myrcene	ants, aphids (Aphididae + Adelgidae), larch beetles, psyllids, spruce weevils
sabinene	aphids
thujone	*Aedes aegyptii* (larvae)
thymol	houseflies (?)
verbenone	bark beetles, pine beetles, pine weevils
More complex terpenes	
azadirachtin	aphids, locusts, planthoppers, spruce budworm, *Spodoptera* spp.
clerodane diterpenes	termites
drimanes	*Leptinotarsa decemlineata*, *Pieris* caterpillars
polygodial	aphids, *Leptinotarsa decemlineata*, *Pieris* caterpillars
warburganal	*Leptinotarsa decemlineata*, *Pieris* caterpillars

* Phenolic terpenoid; $^\oplus$ related naphthoquinone; ? = merely indicative

semiochemicals in aphids, and sections 5.1.3 and 6.2.2 illustrate that in consolidated insect-plant relationships more selective and powerful signals can be expected from multicomponent semiochemicals and multi-stage defense (Box 6.3). We therefore strongly suggest incorporating comparative experiments with multicomponent blends versus single components in studies of repellent and deterrent properties of plant parts. This provides indispensable information in selecting resistant plant genotypes.

A mixture of synthetically produced farnesenes, containing a high proportion of (E)-β-farnesene, has been combined with a low dose of an insecticide to obtain an effective aphicide. It is registered as Panic® and aimed to kill the aphids because mobility of the insects increases upon perception of (E)-β-farnesene, thus enhancing contact with the insecticide. This will work for aphids, which are sensitive to (E)-β-farnesene, but it should be noted that there is a wide variation between species, and probably even between biotypes. In experiments with an (E)-β-farnesene preparation containing minor amounts of isomers such as (E,E)-α-farnesene that can have unexpected side effects (van Oosten *et al.*, 1990) we found an order of magnitude difference between the response of biotypes of *Myzus persicae* to the alarm compound. Montgomery and Nault reported already in 1977 that although (E)-β-farnesene is an alarm pheromone for many aphid species, there is a 500-fold difference in response threshold between aphid species. *M. persicae* is one of the most sensitive species that reacts to only 0.1-1 ng, whereas *Aphis fabae* needs 50-100 ng. A relatively high amount of (E)-β-farnesene in a combination aphicide would result in more species to react, but has two drawbacks: it is more expensive and it causes sensitive species to overreact and drop down immediately, thus retarding further contact with the insecticide. When the aphids recolonize the plants, the volatile farnesene fraction has become less effective.

Hori (1998, 1999) tested the repellence of 13 labiate essential oils against the aphid *M. persicae*. Two of these, rosemary and thyme were strongly repellent in olfactometer tests. The main component of rosemary oil is 1,8-cineole, but this terpene alone did not repel *M. persicae*, but a different aphid species: *Neotoxoptera formosana*. Fortunately Hori and Komatsu (1997) tested other components of rosemary oil and found linalool, camphor and α-terpineol to be repellent terpenes. However, the effect of single components was an order of magnitude less than that of the complete oil. The authors concluded that synergism of the components of rosemary oil is responsible for the repellent action. We hope that someone else will do comparable studies with thyme oil or other plants from Table 6.1.

The results of choice tests can greatly differ from those with no-choice tests. All 10 oils of labiate plants inhibited *M. persicae* from settling on a meridic artificial diet in a choice situation, but in no-choice tests only thyme, spearmint and pennyroyal oils affected settling (Hori, 1999). Thyme oil emanates through a Parafilm membrane and exerts both repellent and deterrent effects; spearmint caused aphids to starve, as they did not feed on

Chapter 6. Terpenoids in practice

sachets containing the oil. Pennyroyal oil showed no repellence but inhibited feeding by an unknown mechanism. When the aphids do not smell rosemary oil, they will feed, illustrating its genuine repellent character.

Verbenone and ipsenol are anti-aggregation pheromones of *Ips pini*, the pine engraver. Verbenone- and ipsenol-impregnated beads (2.5 mg verbenone and 0.05 mg ipsenol/m^2) were very successful to prevent attacks of this insect on felled pines (Devlin and Borden, 1994). Verbenone is a well-known anti-aggregation pheromone of Scolytidae and there is substantial literature on the application of verbenone to protect *Pinus* spp. However, although verbenone alone has been reported to successfully reduce

Box 6.3. The role of terpenoids in multi-stage defense of plants to herbivorous insects. **A**: plant with high level of multi-stage defense; **B**: plant with a weak defense. 1 = repellent odours; 2 = attractive odours (kairomones); 3 = deterrents e.g. in glandular hairs; 4 = toxic terpenoids (allomones); 5 = ratio of allomones (*) / nutrients (●); 6 = escape from; 7 = attraction of natural enemies by synomones. Several scenarios of attack and defense can be depicted dependend on the activity of terpenes in each stage.

colonization of *Dendroctonus frontalis* (the southern pine beetle) without a negative effect on its natural enemies (Salom *et al.*, 1995), and Clarke *et al.* (1999) mention the technique of placing verbenone pouches on freshly attacked trees plus a buffer of treated uninfested trees as a new tactic for controlling the southern pine beetle, the results with *Ips typographus* (the spruce bark beetle) are less promising. Von Niemeyer *et al.* (1995) found that while verbenone had proved to be an effective anti-aggregation terpene in trials with rotating slot traps, the compound was attractive in combination with host tree odours. Attacks of the bark beetles occurred close to and even under the verbenone dispenser. Photoisomerization can convert verbenone into chrysanthenone, but Von Niemeyer *et al.* (1995) demonstrated that after exposition to full sunlight verbenone dispensers within a slot trap were still highly attractive to the bark beetles. These trials again illustrate the importance of a proper translation of laboratory experiments to field trials with their often unexpected drawbacks.

Table 6.3 is constructed utilizing data from more studies than those presented in previous sections, but presents a limited number of terpenoids. Carveol, as an example, is a component of some repellents, but it is not the only one and for that reason not incorporated in the table, which shows that much remains to be done! In dose-response relations one should give attention to isomers and enantiomers. Even humans can discriminate between optical isomers of carvone; one with its typical smell of caraway, the other with its mint aroma (see 6.2.4).

After a closer look at the deterrent effect of more complex terpenoids we arrive at a similar conclusion as discussed with respect to monoterpenes. Powell *et al.* (1996) found the dialdehydic sesquiterpenoid polygodial to inhibit probing of aphids into tobacco leaves and to reduce transmission of potato Y potyvirus, however, this inhibition did not last very long. When aphids were confined to the leaves they resumed probing within a few minutes. Combination with other sesquiterpenoids could have improved this effect.

As discussed in section 5.1.3, Messchendorp (1998) tested 15 individual drimane antifeedants on lepidopteran larvae, resulting in a final selection of only a few active compounds (Fig. 5.13). However, combinations of these drimanes were not included in the experiments.

Combination of neem seed oil components may not be successful in inhibiting probing behaviour of *M. persicae*. Lowery *et al.* (1997) found that although neem seed oil inhibits transmission of non-persistent viruses by this aphid, oil-free neem seed extract did not, suggesting that it is the presence of the oil itself rather than the limonoids that interferes with virus transmission.

Comparison of various investigations of the deterrent and toxic effects of azadirachtin and neem extracts that contain more components leads to the

Chapter 6. Terpenoids in practice

conclusion that the neem extracts are more active than pure azadirachtin when applied at equivalent azadirachtin concentrations, indicating that azadirachtin is not the only active compound in neem. This may be due to other triterpenoids of neem, nimbin and salannin, but a definite conclusion must await the results of combination experiments.

Insects vary markedly in their behavioural sensitivity to terpenoids and this also holds for azadirachtin. Mordue *et al.* (1996) compared the sensitivity of two locust species to azadirachtin at the chemoreceptor level. *Schistocerca gregaria* appeared to be very sensitive (ED50 0.001 ppm), in contrast to *Locusta migratoria* (ED50 3.0 ppm). In the case of *S. gregaria* we may speak of a genuine feeding deterrent, whereas *L. migratoria* ingested a significant amount of azadirachtin, that acts as a weak antifeedant for this insect. Still, lower doses of antifeedant terpenoids can be effective against herbivores in situations where rapid disruption of virus transmission is not essential. The current recommendation that commercial neem preparations be applied at 50-125 ppm may strongly reduce their selectivity with respect to natural enemies (Schmutterer, 1990). Mordue *et al.* (1996) argue that reduced azadirachtin concentrations may weaken their antifeedant effects, but secondary effects on development and metabolism could still reduce pest populations and spare beneficials.

Van Leeuwenhoek (1632-1723) applied clove oil, rich in eugenol and β-caryophyllene against herbivorous insects. The clove product Syzar® successfully inhibits feeding and oviposition of Agromyzidae (Diptera) in chrysanthemums. Eugenol, that is mentioned in section 5.1.3 to exert high fumigant toxicity to subterranean termites, may be used to control above-ground and shipboard infestations of *Coptotermes formosanus* (Cornelius *et al.*, 1997).

Many Coleoptera are pests of stored products. Measures to protect stored cereals were already devised in the ancient Orient (from about 3000 BC), including volatilization of essential oils in special censers originating from dynastic Egypt. Probably these censers had originally been developed for sacral and ritual purposes but the vapours of oleoresin of a.o. frankincense (*Boswellia* spp.) and myrrh (*Commiphora* spp.) has bacteriostatic and insecticidal properties. Pyrolysed frankincense repels *Acanthoscelides obtectus* (bean weevil) and *Sitotroga cerealella* (Angoumois grain moth). Levinson and Levinson (1998) give a list of aromatic plants from which fumigants were prepared. These authors also quote more recent reports on the effectiveness of essential oils against coleopteran storage pests and mention that the insecticidal effect of *Artemisia absinthium* (wormwood), *Coriandrum sativum* (coriander) and *Inula conyza* (fleabane) were recognized and recommended by Varro (116-27 BC) and Plinius Secundus Major (23-79).

Dried leaves and essential oil of *Ocimum suave* (Lamiaceae), an aromatic perennial shrub native to Africa and India is used by local farmers to protect stored maize and sorghum. Obeng-Ofori and Reichmuth (1997) found that the essential oil of this plant is both an effective repellent and highly toxic to a number of stored-product Coleoptera. The oil consists for about 60% of eugenol, for 15% of (E)-β-ocimene, but has less than 2% linalool, that is the main terpene of the oil of *Ocimum cannum*, utilized in Rwanda (dried leaves mixed with edible beans). Eugenol alone is very effective when mixed with the stored product: only 1 μl/kg was sufficient to obtain more than 95% mortality within 24 h, and 3 μl/kg was 100% effective against four coleopteran species. Obviously, contact with the beetles is optimal when the grain is treated with solutions of eugenol in acetone. The mode of action is not yet known. Linalool is known to act as a competitive inhibitor of acetylcholinesterase, to our knowledge only in invertebrates (Ryan and Byrne, 1988), and eugenol has fungicidal and bactericidal properties. A disadvantage of pure eugenol is its low persistence in the stored product without a proper formulation, although that may be improved. Detailed toxicological studies are needed before eugenol treatment can become a safe stored product protection, but the effects are very promising.

Members of the genus *Ocimum* have essential oils rich in caryophyllene, eugenol, linalool, ocimene (various ratios of enantiomers) and β-pinene (Obeng-Ofori and Reichmuth, 1997). Different parts of holy basil, *O. suave* (c.f.: *O. sanctum*) have traditionally been used to control various ailments and as insect repellents. One could think of a combination of essential oils of a few species, in order to develop preparations containing both eugenol and linalool. These oils or extracts could be dissolved in other plant oils to obtain persistent mixtures, or coated to dried leaf material of other aromatic plants. Holy basil is also widespread in the tropics, which makes it an interesting plant for further research. A synopsis of the genus *Ocimum* has been made by Paton (1991).

When applied to rice or wheat the essential oil of *Elettaria cardamomum* (Zingiberaceae) completely suppressed reproduction of two stored-product insects, *Sitophilus zeamais* and *Tribolium castaneum* at a concentration of 5.3×10^3 ppm (Huang *et al.*, 2000).

Attractive odours, be it sex pheromones, aggregation pheromones or kairomones, can in principle be utilized to protect plants or products by the technique of "attract-and-kill" or "lure-and-kill", in which the insects are attracted by a trapping system containing a semiochemical and become subsequently killed. This is a common practice in the control of Lepidoptera, in which pheromone traps are utilized to monitor and combat herbivorous members of this order.

Monitoring and attract-and-kill of stored-product pests with pheromone traps has been developed for Coleoptera, but the latter technique is not yet common practice. Plarre and Vanderwel (1999) in their excellent review of pheromones of stored-product beetles mention trials with pheromone traps containing a pathogenic protozoan. Although the trials were highly successful they met resistance from public health departments, as happens in Europe with micro-organisms which are pathogenic to insects, such as preparations of *Bacillus thuringiensis*. Another possibility is the semiochemical-loaded funnel trap, devised to keep the insects in a container, from which escape is almost impossible. Development of these systems is a challenge to the inventive entomologist.

Monitoring to obtain information on pest density and distribution of Coleoptera is technically possible because many semiochemicals have been identified, but the traps often catch only the species for which the semiochemical is specific. Although it is evident that monoterpenes of the host tree stimulate so-called pioneer beetles to colonize their hosts, von Niemeyer and Watzek (1996) did not find any increase in catches of *Ips typographus* in Pheroprax© traps and *Pityogenes chalcographus* in Chalcoprax© traps when 2.0 ml dispensers with (-)-α-pinene and (+)-limonene were included, in spite of trials by other workers indicating at least additive effects. The authors suggest that other terpene combinations may improve the pheromone traps, although one should realize that other host odours might interfere with the added monoterpenes. Nevertheless, monitoring is successful for some species and a number of commercial products exist for monitoring of forest beetles, such as Pheroprax©, Linoprax© and Chalcoprax©. Protection of spruce with pheromone traps can be improved with repellent semiochemicals applied at a different site, such as polyethylene strips with verbenone in the case of *I. typographus* and polyethylene ampoules containing terpinen-4-ol in case of *P. chalcographus* (Vité and Baader, 1990). This bipolar strategy of behavioural manipulation, termed "stimulo-deterrent diversion" (SDD), could be applied to individual plants (Miller and Cowles, 1990) and to separate coniferous trees. For the integration into IPM systems we refer to Schlyter and Birgersson (1999).

INTERFERENCE WITH INTERNAL MESSENGERS AND DEVELOPMENT

As discussed in section 5.1.3, repellents and deterrents have acquired these functions because they exert pathological effects upon contact or ingestion, or because their molecular structure acts as an agonist of such compounds. Observations of insects acutely poisoned by components of essential oils refer to a neurotoxic mode of action (see 4.2.2, 5.1.3 and Fig. 4.14). Pulegone is probably the most potent inhibitor of acetylcholinesterase

(AchE) in invertebrates; in general ketones tend to be more potent than corresponding alcohols (Ryan and Byrne, 1988, Miyazawa et al., 1997). The degree of toxicity *in vivo* is not always correlated with the inhibition of AchE *in vitro*, which a.o. can depend on the penetrating properties of the essential oils.

According to Isman (1999 and Enan et al. in that paper) many monoterpenes are antagonists of octopamine receptors that are involved in neurotransmission in insects, but the receptors are different from those of vertebrates. This makes essential oils a source of potentially selective insecticides. The sublethal behavioural effects on insects and mites indicate that practical application via programmed-release techniques offers a high degree of safety with respect to vertebrates.

In field trials, thymol, carvacrol and pulegone appeared to be acutely toxic to *Spodoptera litura,* the tobacco cutworm (Isman, 1999). The group of Isman tested a prototype mixture of essential oil as an emulsifiable concentrate against cabbage pests. Their preparation provided good control of larvae of *Plutella xylostella* (diamondback moth) and *Trichoplusia ni* (cabbage looper). We used the solubilization techniques as mentioned in 6.1.2 to test the effect of mixtures of essential oils on aphids (not published). Solutions of essential oils, realized with a complexation technique, containing less than 2% active ingredients, were very effective against aphids (a.o. *Macrosiphum euphorbiae* and *Myzus persicae*) on a number of plants (including potato). The major components of the oils were eugenol, 1,8-cineole, menthol, α-pinene and thymol. Other combinations may also be effective, but one should be careful not to exceed phytotoxic concentrations. Isman (1999) correctly mentions that the most active terpenes are phytotoxic at lower concentrations (e.g. eugenol and thymol).

Ryan and Byrne (1988) refer to the possible development of resistance of AchE to essential oils as is known for insecticides and has been demonstrated for aphids (Devonshire, 1989). When used as mere repellents, resistant biotypes may not easily be selected, but it remains to be seen what happens in case of octopamine receptors. Bala et al. (1999) isolated two daphnane diterpenoids via methanol extraction from roots of *Lasiosiphon kraussianus* (Thymelaeaceae) with insecticidal effects on *Drosophila melanogaster* (Diptera), *Aphis gossypii* and *M. persicae* (Homoptera). The compounds did not exert an effect on AchE and according to the authors this may hold for all daphnane diterpenoids. The woody herb, common in Sudan and East Africa has antileukemic properties and the two active diterpenoids could be identified. An ingested dose of 2 mg/kg does not cause any toxic effect in mice, rats and rabbits. The anti-tumour properties suggest that daphnane diterpenoids stimulate apoptosis, so it would be interesting to investigate the prime target in insects.

Chapter 6. Terpenoids in practice

It has been mentioned before (see 5.1.3) that limonoids, in particular azadirachtin, affect endosymbionts of aphids. Other terpenes, such as pulegone have been proved to destroy symbionts of herbivores and can deprive insects of an important source of isoprenoids by inhibiting MAI metabolism (see 2.1.1 and 2.1.4), and disrupt the symbiotic relationships within the insect. Herbivorous insects deprived of their microbial brokers meet a biochemical barrier of host plant utilization (Douglas, 1992). In the big neem book of Schmutterer (1995) growth inhibition, sterility and reduced fitness are frequently mentioned. Lowery and Isman (1994) found neglectable effects of neem seed oil and pure azadirachtin when ingested via a diet on survival of adult aphids, but nymphs were subjected to apolysis (pathological swelling) in preparation for ecdysis and either failed to moult or could not escape the exuvium. In the summer season aphids are viviparous and in viviparous aphids. Lowery and Isman (1994) found a transgenerational effect of neem on survival, which could be explained by reduced viability of the embryos (peritrophic cells?), or damaged chorions.

Failure to moult is considered the most economically important effect of neem. There is substantial literature on the interference of azadirachtin on ecdysteroid metabolism (a.o. quoted in Lowery and Isman (1994) and the book by Schmutterer (1995)). It can cause both reduction of ecdysteroid titres by inhibiting release into the hemolymph and inhibit degradation of ecdysteroids, although it does not seem to affect the synthetic capacity of the CA. In their experiments with neem oil on *Choristoneura rosaceana* (Lepidoptera), the obliquebanded leafroller, Smirle *et al.* (1996) obtained indications that the oil inhibits esterase activities but not glutathion transferase activities. Pure azadirachtin was not included in the experiments, but the authors suggest that neem could be useful in situations where insecticide resistance has developed due to elevated esterase activity.

Phytoecdysones, quite common among ferns and gymnosperms can cause severe malformations that can be the result of premature moulting, puparium formation or supermoults. This particularly holds for holometabolic insects but as we have seen above, they can also affect hemimetabolics like aphids. According to Lafont *et al.* (1991) effects on reproduction requires relatively high doses, in the order of 500-3000 ppm in a diet. This is beyond the level expected for a natural interspecific messenger and commits them to the black box of "toxic terpenoids".

Already during the ACS symposium in 1984 (publ.: 1985) Bowers provided a list of phytoecdysteroids with similar structures as existing in Crustacea and pointed to the fact that the insect moulting hormone, 20-hydroxyecdysone, is the most commonly encountered phytoecdysteroid. It would seem that nature armed itself!

The concentration of ecdysteroids in plant parts can reach incredible values. Lafont *et al.* (1991) provide a table of presence of 20-hydroxyecdysone in several plant families. The highest concentrations are in flowers and fruits of *Serratula inermis* (Asteraceae): 2% of dry weight and in the stem of *Diploclisia glaucensis* (Menispermaceae): 3.2% of dry weight. This makes it improbable that it has a function as a hormone in these plants.

The fact that so many plant species harbour ecdysteroids may explain why many herbivorous insects posses rather efficient detoxification mechanisms, enabling them to cope with up to 5000 ppm of 20-hydroxyecdysone in their diet. This makes breeding of cultivars with even higher phytoecdysteroid concentrations questionable, as ingestion of higher doses by vertebrates can exert pathological effects on carbohydrate and MAD metabolism.

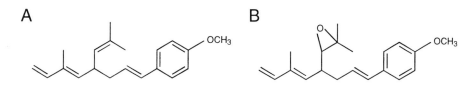

Fig. 6.4. **A**: juvocimene I and **B**: juvocimene II, originating from *Ocimum basilicum* (c.f. *Ocimum sanctum*), sweet basil.

Azadirachtin is more difficult to detoxify and this may explain its effectiveness as a feeding deterrent. This brings us to two other classes of isoprenoid insect phytohormones that can be utilized in insect control: phytojuvenoids and anti-JHs. Numerous compounds with low to moderate JH activity have been found in plants, among which the juvocimenes, present in the essential oil of sweet basil (Bowers, 1985) (Fig. 6.4). These induce strong morphogenetic effects a.o. against milkweed bugs in picogram quantities. The suppression of adult differentiation and the prolongation of immature stages by JHs make them interesting candidates for effective insecticides. However, natural JHs are neither stable enough under field conditions, nor active enough to be suitable for commercial use. A search was made for novel natural JHs of which analogues could be synthesized and programmes were started to synthesize possible juvenoids based on farnesol ethers (Henrick, 1991). A substantial part of section 3 of the proceedings of a conference on insect chemical ecology (ed. I. Hrdý, 1991) is devoted to juvenoids. Chapter 7.3 will provide more detailed information, but it should be noted in this section that JHs and anti-JHs have little effect on early instars. This is a drawback for the control of pests in crops, which are particularly exposed to early instars, as is the case with an infestation of *Heliothis virescens*, the cotton bollworm.

We refer only shortly to anti-JHs, which have substituted chromene structures and are no terpenoids. One might expect the anti-JHs to duplicate the surgical removal of the corpora allata (CA), resulting in precocious metamorphosis. The precocenes were the first natural anti-JHs isolated from *Ageratum houstonianum*. The list of sensitive insect species appeared to be limited (Bowers, 1985) so screening programmes for more broad-spectrum anti-JHs started whilst the mode of action of the precocenes was investigated. In the CA, monoxygenases induce formation of highly reactive derivatives that upon reaction with proteins induce necrotic cell death (Henrick, 1991). This, of course, is a drawback with respect to cytotoxic effects in mammals. Chapter 7.3 will give some comments to synthetic anti-JHs.

TOXIC TERPENOIDS

The precocenes mentioned above exert mainly toxic effects and the same holds for a number of essential oils in insecticidal plants. In fact, mere toxic effects are beyond the scope of this book that deals with natural messengers. Undoubtedly the mode of action of some terpenoids classified as "toxic" will be elucidated in the future, and therefore some key reports on toxicity of terpenoids are discussed.

Lee *et al.* (1997) tested several monoterpenes on their acaricidal and insecticidal activities. Monoterpenes exert no rapid contact activity on *Tetranychus urticae* but we mention the three most effective ones with their respective LC50s (ppm in solution applied on leaves): carvomenthol (59), terpinen-4-ol (96) and the phenolic terpenoid eugenol (219). Two insect species were subjected to topical applications as well as vapour of the monoterpenes, in the latter case mixed with soil. In topical applications the four monoterpenes most toxic to *Musca domestica* (the housefly) with their LD50s were: thymol (29), citronellic acid (32), pulegone (39) and limonene (50). Not all treatments of *Diabrotica virgifera virgifera* (the western corn rootworm) were reliable, but from those that were we mention thujone (12), citronellol (15), pulegone (38) and isopulegol (47). In soil tests the sequence of most active monoterpenes was different, with carveol (10), thymol (10), citronellol (21) and borneol (30) as the top four in their activity against the western corn rootworm.

Some monoterpenes were phytotoxic to corn roots and leaves, with expressed symptoms at 50 ppm of (R)-citronellol, (S)-eugenol and (R)-carvone. Pulegone with its high degree of insect toxicity did not affect the plants up to 500 ppm (highest dose tested) and the same holds for (S)-α-pinene ((R)-α-pinene was only slightly active). The bioactivity of the monoterpenes was less than that of conventional organic insecticides. For the best ones this was a factor 4 compared with pyrethrum against the housefly and a factor 10 compared with carbofuran of chlorpyrifos against the western corn rootworm. We notice that Lee *et al.* (1997) did not test combinations of monoterpenes.

Rice and Coats (1994) report that in topical and fumigant bioassays of *M. domestica* aldehydes of monoterpenes were more toxic than alcohols and vapours of bicyclic ketones were more toxic than monocyclic ketones. In the topical and larval bioassays monoterpenoid alcohols with 3 C=C double bonds were more effective than saturated alcohols.

Franzios *et al.* (1997) tested the essential oils of *Mentha pulegium* and *M. spicata* (see Table 6.1) on insecticidal and genotoxic activities on *Drosophila melanogaster*. Both oils are commercially exploited. Toxicity was expressed as the fraction of adult flies emerging from treated larvae and the number of larvae surviving to adulthood. The oil of *M. spicata* was superior to that of *M. pulegium*, probably due to its high pulegone content (76% of the oil used in these tests). The authors included separate compounds in their tests and concluded that both synergist and antagonist phenomenons exist that may define the toxicity of the oils depending on genetic and geographical parameters, Menthone reduces the effect of pulegone, but the remaining components of the oil of *M. pulegium* contribute to its final toxicity, a phenomenon reported for many secondary plant substances (Harborne, 1993). Quantifying synergy (and antagonism) can be based on isobolographic analysis that is also used in pharmacology (Nelson and Kursar, 1999).

Mosaic spots on the wings of F1 progeny of *D. melanogaster* were used to measure mutagenetic effects of the oils and their constituents carvone, menthone and pulegone. The effects on somatic mutation were tested with concentrations slightly above LD50. Nothing happened after treatment of larvae with the oil of *M. pulegium*. The oil of *M. spicata* did not increase recombination of subclones (expressed in twin wing spots), but it did increase point mutations and somatic recombination effects. The only monoterpene exerting similar effects was menthone, but *M. spicata* is not rich in menthone, so the *M. spicata* effect cannot simply be attributed to the composition of its oil. In addition, we refer to the anti-tumour effects of perillyl alcohol (see 4.1.5, 4.1.6 and 7.1.8), another component of *M. spicata* oil, which, however, was not tested in the experiments quoted.

The sesquiterpenes (E,E)-α-farnesene and (E)-β-farnesene can be toxic to insects at low doses (see 5.1.2 and 5.1.3). In our own experiments the alarm pheromone of aphids, (E)-β-farnesene, was lethal after application of 100 ng/aphid. Farnesene vapour is also toxic to *Trialeurodes vaporariorum*, the greenhouse whitefly. Farnesene vapour (with 50% of (E)-β-farnesene; 5 mg of the mixture in a box of 0.5 l) killed 100% adult whiteflies within 48 h, and 60% in slightly ventilated boxes (Klijnstra et al., 1992). When farnesene vapour was released from dispensers (1.5 mg/h) the total amount released in 72 h was about the same as one "shot" of 5 mg; nevertheless mortality reached only 50% in 72 h. It remains to be investigated if farnesene vapours can effectively be combined with monoterpenes.

It is not always easy to discriminate between mere antifeedant and toxic effects, as toxic compounds often have repellent or deterrent properties. González-Coloma et al. (1995) studied the effects of sesquiterpenes from *Senecio palmensi* on the Colorado potato beetle. Both bisabolene and silphinene were antifeedants to this insect, although less active than juglone. Exposure of fourth instars to the compounds over a 24-h period resulted in reduced feeding and growth rates. As the beetles were more sensitive to these sesquiterpenes in choice than in no-choice assays it was thought essential to distinguish between antifeedant and toxic effects. To that end growth efficiencies were calculated as the slope of the regression of relative growth rate on relative consumption rate.

Jimenez et al. (1997) compared the insecticidal effects of four limonoids with toosendanin, a commercial insecticide derived from the neem tree, on *Ostrinia nubilalis*, the European corn borer. The limonoids showed comparable activity to toosendanin and had low cytotoxicity to three human cell lines. A review of insecticidal and acaricidal properties of neem terpenoids and their usage in tropical pest management has been made by Williams and Mansingh (1996). This review includes extended tables on the effectiveness of neem extracts on different insect orders plus Acarina. Neem seed kernel extracts are particularly valuable because they hardly stimulate selection of resistant insect biotypes. The crude neem extracts are composed of several bioactive substances acting on various aspects of herbivore physiology, and the authors underline that it is unlikely that insects develop resistance within a few generations. The review refers to the rapid rate with which some herbivorous insects e.g. *Plutella xylostella* develop resistance against various insecticides and mention that two strains of this insect species exposed to neem extracts for more than 42 generations did not develop resistance. In fact, neem extracts are multicomponent insecticides corresponding with multicomponent defense systems (see 5.1.3) and not unlike multicomponent treatments of cancer (see 7.1.8). Insecticides based on azadirachtin alone may not be so durable, but have the advantage of a more consistent composition and better standardization.

According to Williams and Mansingh (1996) the effects of neem products on non-target organisms are limited, particularly on predators and parasitoids. Neem terpenoids must be ingested to become effective, and at the third trophic level (Fig. 6.5) lethal concentrations are rarely reached, although there may be some repellent or deterrent effects. Neem extracts seem to be remarkably benign to spiders, wasps, butterflies and honeybees. Effects of neem on aquatic fauna are discussed in 6.2.4.

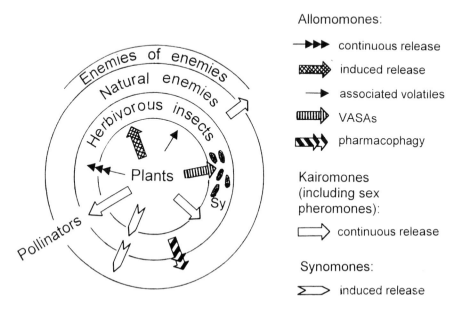

Fig. 6.5. Function of volatile terpenes on four trophic levels of plant-insect relationships. Sy = (endo)symbionts. VASAs = volatile antisymbiont allomones (see 5.2.2).

Singh *et al.* (1997) reported that ethanol has a better extracting capacity of neem compounds than petroleum ether from leaves, the epicarp and mesocarp of fruits, whereas petroleum ether scores better with fruits and seed kernels. When these extracts were diluted until the same concentration of active ingredients was reached, little differences existed with regard to insecticidal effect on *Dysdercus koenigii*, the red cotton bug.

6.2.3 Resistance breeding

According to Leather (1996) the ecological implications of resistance in conifers are rarely discussed. In particular the reduction in genetic material caused by clonal selection could have serious consequenses. Jermy (1990) in his review of the use of antifeedants in pest control states that genetic engineering studies should shed light on the possibility of transferring the

Chapter 6. Terpenoids in practice

genes responsible for the production of antifeedants from one plant into another. He also states that it is deplorable that in spite of the recognition of the importance of reliable large-scale field experiments not much has been done to prove the practical significance of the many data collected so far.

Leather (1996) states that utilization of resistant trees has five important advantages over selecting for silvicultural traits only:
- No additional pest control costs
- Operates also at low levels of pest incidence
- Reduces insect populations cumulatively
- Avoids toxic residues and environmental pollution
- Fits well in other IPM strategies in existence.

Some terpenes could be utilized as anti-attractants; they do not have an intrinsic repellence but can reduce the attractiveness of other compounds. A possible candidate is β-phellandrene. Terpenes that can act as repellents or deterrents are α-pinene, β-pinene, 3-thujone, 3-isothujone and juglone, a related aromatic compound. As the forest industry is of increasing importance on this world resistance breeding of gymnosperms with special attention to terpenoids should be stimulated.

Trioza apicalis (Homoptera), the carrot psyllid, is a pest that attacks carrots, especially in northern Europe. Nehlin *et al.* (1996) compared the attraction and oviposition on four members of the Apiaceae and found that two *Daucus carota* subsp. (carrots) were strongly preferred above *Anethum graveolens* (dill) and *Carum carvi* (caraway). The major volatiles of carrots were α-pinene (21-28%) and sabinene (29-44%). Dill, although rich in a.o. myrcene (27%), β-pinene (9%), β-phellandrene, ocimenes (11%) and γ-terpinene, hardly contains sabinene and much less α-pinene. Caraway is rich in limonene and carvone. The authors, who found a substantial part of α-terpinene in caraway only determined the limonene content (60%) of the oil of caraway and conclude that limonene could make this species unattractive to the carrot psyllid and that the high contents of α-pinene and sabinene could make the two carrot subspecies preferred by the psyllids. Depending on the growing phase of caraway, carvone plus limonene could act as anti-attractants. This type of knowledge is very useful if one does not want to change the original ratio of volatile terpenes in a crop, but chooses to include additional anti-attractants. Another possibility is intercropping with plants releasing repellents, which overrule the kairomones of the crop itself.

Although the so-called green leaf volatiles have a dominant position in the volatile spectrum of grasses and cereals, terpenoids can make a serious contribution to the presence of volatile metabolites in undamaged leaves. Buttery *et al.* (1985) found (E)-β-ocimene to be the major component in oat leaves and (E)-β-ocimene and caryophyllene to be main components of wheat leaf volatiles, together with less common terpenoids α-copaene, α-farnesene and linalool oxides. The authors stress that a possible role of these terpenoids in insect resistance should not be overlooked, especially in comparison to volatiles emitted by damaged plants, where oxidative enzyme activity causing breakdown of lipid and carotenoid components results in the release of aliphatic aldehydes and alcohols, which does not occur in the intact plants. When glandular structures are not abundant, as is often the case with vegetative plant parts, the defensive function of terpenes emitted by those plants should not be overestimated. Herbivorous insects may just as well use these odours as kairomones, as negative effects of terpenes upon eating these plants are limited. Buttery *et al.* (1984) also analyzed the volatile components of different parts of the red clover plant and compared the results with data on alfalfa volatiles. The amount of volatile oil obtained by Tenax trapping was less than 0.1 ppm for leaves and seedpods and about 0.5 ppm for flowers. (E)-β-ocimene, (E)-β-farnesene and caryophyllene were common to both alfalfa and clover (leaves and pods). Alfalfa sesquiterpenes that also include γ-muurolene and α-copaene were found to be attractive to the alfalfa seed chalcid, indicating a function as kairomone. Such terpenoid kairomones may be located in the cuticular waxy material and the first contact of the insects with the plant would be this material. Any deterrent effect of the terpenoids could therefore easily be established.

In section 5.1.3 we mentioned the repellent effect of (E)-β-farnesene on glandular hairs of wild potato species. This sesquiterpene is released in alarm-evoking quantities when the hairs are broken. A drawback in resistance breeding is that the tetralobulate glands on the epidermis of cultivars do not readily rupture upon contact. In addition to type A hairs, type B hairs of *Solanum berthaultii* exudate esters of short-chain branched carboxylic acids and these exudates could contribute to the defense of potato leaves, a reason why resistance breeding should focus on both phenomenons. It should be noted that in some countries (e.g. Peru) potatoes should preferably be resistant to *Phthorimaea operculella* (the potato tuber moth) (stored potatoes), for which selection techniques have been developed.

A special multi-author section of the 8th Symposium on Insect-Plant Relationships was dedicated to host plant resistance and application of transgenic plants (eds S.B.J. Menken, J.H. Visser and P. Harrewijn, 1992) and we refer to this section for the state of the art with respect to techniques and methodologies of introduction and establishment of plant resistance to herbivores. In addition, Tingey (1986) has made a valuable review of techniques for evaluating plant resistance to insects (chapter 9 in Insect-Plant Interactions, eds J.R. Miller and T.A. Miller).

Chapter 6. Terpenoids in practice

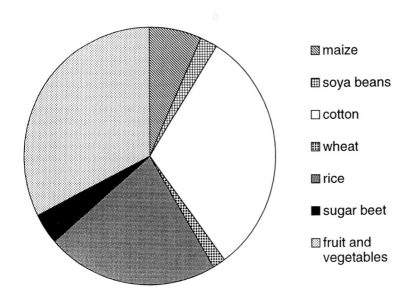

Fig. 6.6. Relative world insecticide usage in important crops (after County Natwest Woodmac).

Cotton, rice, vegetables and fruits, sugar-beet and maize are the crops with the biggest usage of pesticides (Fig. 6.6). Terpenoids are involved in resistance mechanisms in many of these crops (see 5.1.3 and 6.2.2).

The review by Tingey (1986) provides major collections of plant germplasms, several of which store sources of rice, beets, vegetables, maize and many other crops such as oilseed crops, beans, cassava, potatoes and sorghum. Special attention is given to sampling of insect populations and plant growth and damage assessment. The review ends in listing of research objectives appropriate to a multidisciplinary approach of the study of plant resistance that is fully adequate with respect to resistance mechanisms in which terpenoids are involved. Detailed descriptions of olfactometers to test repellent effects of terpenes are given in Smith *et al.*, 1994. An open Y-track olfactometer for recording of aphid behavioural responses to plant odours has been presented by Visser and Piron (1998).

In conclusion, terpenoids are involved in many resistance mechanisms of plants against herbivorous insects. Many studies have been made on their functions as repellents and deterrents, but substantial data on their mode of action (antisymbiont allomones!) and effectiveness in field trials are often lacking. A wealth of information has been collected on the sensory level, but many studies seem to be run out of financial support after concluding that "the mode of perception as well as the structure-activity relationship of terpenoid antifeedants differs considerably between insect species" (Messchendorp, 1998). This author states that "for practical implementation of terpenoid antifeedants in crop protection, this means that research should be focussed on selected pest insects". This is multidisciplinary research that includes more than structure-activity relations on the sensory level. Time has come to study the role of repellent and deterrent terpenoids in (host) plant acceptation in the following sequence:

1. Effects in choice situations
2. Effects in non-choice situations
3. Effects on the sensory level
4. Effects on behaviour – including multicomponent blends
5. Mode of action with respect to:
   - Nervous system
   - Intestinal tractus, digestion
   - Reproduction
   - Symbionts (VASAs)
6. Adaptation of the insects
7. Field experiments with combined usage of other (biological) control agents
8. Localization of responsive plant genes
9. Resistance breeding.

One may end up with a crop that is only partly resistant to the herbivorous insect(s) studied. Such crops can be valuable in IPM programmes and can be combined with semiochemicals of a different structure than produced by the plants.

6.2.4 Other applications

ANTI-MICROBIAL EFFECTS

The basics of anti-microbial effects of essential oils are explained in section 4.2.1. Wan *et al.* (1998) screened essential oils of basil against a range of Gram-positive and Gram-negative bacteria, yeast and moults. The main components linalool and methyl chavicol were particularly effective against *Bacillus cereus*, *Mucor piriformis*, *Penicillium expansum* and *Salmonella typhimurium*, based on the agar well differentiation method. A slightly less effect was obtained with *Aeromonas hydrophilla* that frequently occurs on lettuce, but according to Wan *et al.* (1998) the oil is effective enough to be a beneficial component of salad dressings or other vegetable products. In contrast to washing procedures with 200-300 ppm chlorine at

Chapter 6. Terpenoids in practice   291

low temperature basil essential oil does not produce chloramines and trihalomethanes, which are undesirable by-products.

Centaurea nicolai (Asteraceae) produces a number of sesquiterpene lactones. Vajs *et al.* (1999) discovered a new guaianolide in the aerial parts of this plant. This compound, 3-deacetyl-9-O-acetylsalogranolide A, is a derivative of salogranolide A and appeared to exert anti-fungal effects against *Aspergillus niger, A. ochraceus, Penicillium ochrochloron, Cladosporium cladosporoides, Fusarium tricinctum* and *Phomopsis helianthi.*

Carvacrol, the major component of essential oil of oregano and thyme, has been tested against various micro-organisms. We apologize for not quoting all studies separately. The effects on *Penicillium* spp. are limited, but in addition to *Aspergillus* and *Fusarium* spp. mentioned above, inhibiting effects on *Rhizobium* and *Rhodobacter* spp. have been reported, as well as on micro-organisms more pathogenic to vertebrates: *E. coli* (pathogenic types), *Pseudomonas aeruginosa, Salmonella typhimurium, Staphylococcus aureus* and *Vibrio vulnificus* (thymol is particularly effective against *V. cholerae*). This is of interest to food technologists as well as physicians, and inevitably there is some overlap with section 7.1.1.

Carvacrol is particularly effective against the food-borne pathogen *Bacillus cereus.* This spore-forming pathogen is associated with milk and other dairy products, but also with rice and meat products (soup). Ultee *et al.* (1998) found an exponential decrease in the number of viable *B. cereus* at concentrations above 1.0 mM/l. Addition of 3 mM/l resulted in a 10^4-fold reduction in the viable count in samples taken within 40 s after addition of carvacrol, suggesting a rapid interaction with the phospholipid bilayer of the cell membrane (see 4.2.1). Spores of *B. cereus* were less sensitive to carvacrol than the vegetative phase. Nevertheless, spores hardly survive 3 mM/l. Both fungicidal and bactericidal effects of carvacrol are lowest at neutral pH and considerably increase below pH 5.5 and above pH 8.0, although there is an upper limit due to the expected dissociation of the phenolic proton from the hydroxyl group of carvacrol.

The low active concentration of 1 mM/l or less makes carvacrol an interesting bacteriostatic for processed food, although Ultee *et al.* (1998) state that *B. cereus* grown and exposed at 8°C are less sensitive than those grown at 30°C.

Both limonene and carvone mask odour mixtures in rodents (Laing *et al.*, 1989). Humans that can discriminate between (+) and (-) enantiomers of terpenoid odours diverge widely in their relative sensitivities. According to Polak *et al.* (1989) odour discrimination of α-ionone enantiomers involves at least two receptor types of opposite chiral selectivity, and their distribution varies independently in the human population. Perfumers should take this in mind when selecting test persons to evaluate new blends of terpenes in cosmetics. Laska and Teubner (1999) found that humans as a group were only able to discriminate the optical isomers of α-pinene, carvone ((S)-(+)-carvone = caraway or "kummel" odour, (R)-(-)-carvone = mint odour) and limonene and failed to do so between the (+) and (-)-forms of α-terpineol, β-citronellol, camphor, fenchone, menthol and rose oxide. As in the experiments of Polak et al. (1989), error rates were quite stable and did not differ significantly between sessions, indicating that learning or training effects do not exist. See also Laska et al., 1999.

INHIBITION OF SEED GERMINATION

Carvone not only acts as a fungicide; in addition it inhibits seed germination of angiosperms (see 5.1.3). Other essential oils exert this effect, such as 1,8-cineole (Romagni et al., 2000). Terpenoids with a certain degree of phytotoxicity often act as inhibitors of seed germination. Some monoterpenes have a high degree of allelopathy against natural grasses (Fischer et al., 1994). This phenomenon is also known for more complex terpenoids. As an example we mention the limonoids, modified triterpenes, of which a wide range is produced by Meliaceae. Céspedes et al. (1999) tested a 23-R/23-S-mixture of photogedunin, the major nortriterpenoid of *Cedrela ciliolata* on its effects on germination, growth and seed respiration of a few monocots and dicots. At 500 µM the mixture caused complete seed germination inhibition in *Triticum vulgare* and *Lolium multiflorum*, but had much less effect on dicots. Acetylation of the C_{23} hydroxyl group with active anhydride/pyridine resulted in a ten times higher activity. These compounds also strongly inhibited stem elongation and weight increase of the monocots. Photogedunin that may act as an uncoupler of phosphorylation and in higher concentrations as an inhibitor of respiration redox enzyme, is much less active against dicots. Derivatives of photogedunin could be developed into selective growth inhibitors of monocots.

Carvone also acts as a natural potato sprout growth inhibitor. Originally its mode of action was thought to be on HMG-CoA reductase alone, but our present understanding of MAI metabolism makes this less probable. Interactions with membrane processes such as co-translation or modulation of the membrane environment are other possibilities. (S)-carvone could act as an intermediate leading to enhanced degradation of HMG-CoA reductase, analogous to the effect of farnesol on this key enzyme in animal tissues (Oosterhaven, 1995). In his thesis on the usage of (S)-carvone as an inhibitor of potato sprouting, this author concludes that (S)-carvone does not block protein synthesis in general. After prolonged exposure to (S)-carvone no cambium layers develop but the effect of this terpene is reversible. (S)-

Chapter 6. Terpenoids in practice

carvone can conjugate with glutathion, maybe also with saccharides and delays suberization (Oosterhaven, 1995).

After harvest, potato tubers are usually stored at 6-8°C with application of synthetic sprout inhibitors such as propham or chlorpropham. The "green" markets require potatoes free of ©IPC residues and application of carvone as a natural compound can meet these requirements, at the same time stimulating growing of caraway as a source of carvone. To this we add: be aware of the aphid *Cavariella aegopodii* that is not an invited guest in caraway. In Western Europe, caraway fields are often close to roads lined with poplars, and this is the primary host of this aphid. The problem can be avoided by planting other easy growing lane trees such as willows, birches, *Fraxinus* spp. etc.

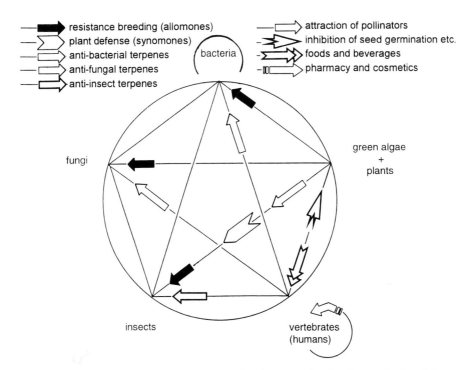

Fig. 6.7. The application of terpenoids by humans in the interrelationships between biological kingdoms.

Not that the natural sprout inhibitors are new: they were already utilized in ancient Inca cultures. This highly civilized population stored potato tubers in bins together with twigs of *Minthostachys* spp. (Muña), that controlled sprouting as well as attack of herbivorous insects. During potato storage, carvone is best applied in moderate weekly doses and not in high doses with great intervals, to avoid any deformation of aerial parts after planting. According to Stammati *et al.* (1999) neither (S)-carvone nor carvacrol, cinnamaldehyde and thymol has mutagenic effects in microbial and mammalian short-term assays. However, there was a marked dose-dependent differential toxicity of carvone on DNA repair. Several potato cultivars are reputed for small pseudo-mutagenic deviations. If carvone inhibits DNA repair it could exert some effect on gene expression, resulting in a small proportion of plants with deformation of leaves or other aerial parts. Fig. 6.7 depicts the application of terpenoids in human activities.

TOXICITY OF TERPENOIDS TO NON-TARGET ORGANISMS

According to Isman (1999) the widespread usage as flavouring agents in foods and beverages makes it not surprising that most terpenoids present in essential oils have minimal vertebrate toxicity. This author mentions that rat oral LD50 values range from 2-5 g/kg or more. The latter value is the upper limit for acute toxicity tests by the U.S. Environmental Protection Agency. Isman refers to the fact that most components of essential oils with proved repellent/insecticidal properties are on the GRAS (Generally Regarded As Safe) list of the U.S. Food and Drug Administration (FDA). According to Tyler (1994), some oils are included in FDA category III, which often is a result of (quote): "compartmentalized thinking" and "denying for whatever reason the effectiveness of products that everyone knows *are* effective, and which could be verified by simple experimentation". Chapter 7.1 gives a few salient examples, but it should be noted that potential toxicants in herbs must be carefully examined, not only *in vitro*, before declared to be safe. Pulegone is recognized in many countries as a potential dangerous terpene, and there are reasons to be careful with carvone. The monoterpene carvone has been discussed within various contexts in this book. As one of the main constituents of spearmint oil it is often included in toothpaste and chewing-gum. Paulsen *et al.* (1993) reported that (R)-carvone has occasionally caused allergy upon oral contact with dental products containing 5% (R)-carvone. Jager *et al.* (2000) studied the *in vitro* metabolism of (R)-(-)- and (S)-(+)-carvone in rat and human microsomes. It appeared that (4R,6S)-(-)-carveol was NADPH-dependently formed from (R)-(-)-carvone (Phase I) and that Phase II conjugation of carveol isomers by rat and human liver microsomes in the presence of uridine S'-diphosphoglucaronic acid only revealed glucuronidation of (4R,6S)-(-)-carveol. In contrast, (4S,6S)-(+)-carveol was produced from (S)-(+)-carvone. This points to stereoselectivity in Phase I and Phase II carvone metabolism, which can determine biotransformation of carvone in mammals.

Chapter 6. Terpenoids in practice

In caraway (S)-carvone is converted into (R)-limonene (syn.: D-limonene), that has an industrial usage for degreasing metal before coating and painting. Karlberg *et al.* (1992) point to the fact that these products can contain as much as 95% of (S)-limonene and studied the allergenic potential of this terpene together with its natural oxidation products. It appeared that (R)-carvone and a mixture of (E)- and (Z)-isomers of (+)-limonene oxide are potent sensitizers, whereas (S)-limonene itself is not an allergen. Therefore, care should be taken with prolonged air exposure of products containing high concentrations of S-limonene.

Isman (1999) states that given their relative lipophilicity and high vapour pressure, essential oils are not likely to leach in groundwater or to persist in soil or sediments. In tests with the rainbow trout 96-h LC50 values of 6.6 and 61 ppm for α-terpineol and eugenol were obtained. According to Williams and Mansingh (1999) the rainbow trout showed no toxic effects when exposed to concentrations of 5 mg/l of Margosan-O, a commercial neem seed product. These authors refer to trials with neem oil against larvae of *Culex pipiens*, in which 0.01% neem oil was effective, but had no effects on insectivorous fish, tadpoles and water fleas.

Corbet *et al.* (1995) investigated larvicidal properties of a few other terpenes on their effects as surfactant mosquito larvicides. They compared eucalyptus oil, turpentine (two grades) and 1,8-cineole as a component of eucalyptus oil with familiar larvicides like Arosurf and Golden Bear oil. All compounds were tested in a dose of 0.13 μl/cm^2. Refined turpentine acted best, even faster than Arosurf during the first 24 h, but was surpassed by crude (and cheaper) turpentine for a longer period. According to the authors plant essential oils merit further attention as widely available and environmentally benign mosquito larvicides.

The presence of limonene in flea-shampoos has been mentioned before. Other well-known recipes are combinations of N,N-diethyl-m-toluamide (DEET) with pyrethrum, or other mono-terpenoid repellents. Mumcuoglu *et al.* (1996) tested the repellence of essential oils containing other terpenes than limonene on *Pediculus humanus humanus*, the human body louse. Although DEET was a good repellent and remained effective at a dilution of 1:32, geraniol did so at 1:8, citronella at 1:4 and rosemary and citronellal at 1:1. In contrast to the other terpenes, citronella activity lasted for almost 30 days, ranking it close to DEET. Rosemary and citronellal lasted more than 14 days and may also successfully be used as a repellent.

In conclusion, knowledge of the mode of action of a terpenoid together with advanced knowledge of a specific insect species is essential to develop a method for optimal application in agriculture. There have been many disappointments with regard to the effects of specific terpenoids in practice after successful laboratory trials. Combinations are often more effective than single compounds, but these studies are time-consuming. Back to nature can be an option. Terpenoid mixtures with additive or synergistic effects can be isolated from micro-organisms, plants and insects, and their constituents can be isolated, fractionated, combined or recombined.

The field of terpenoids is enormously extensive resulting in a branched discipline. Ignorance of information, which is in fact available, may have frustrated the application of effective terpenoids. Electronic data systems could be of assistance to combine the proper compounds, isomers and enantiomers. The design of effective synthetic derivatives not existing in nature could help to limit the number of terpenoids in a combination or to select more effective molecules (see 7.3).

Chapter 7

Natural terpenoids to the benefit of human health

7.1 Applications in medicine

In chapter 4 messenger functions of terpenoids on the cellular level have been extensively discussed within a general biological context, aiming at a better understanding of the mode of action of terpenoids of different classes.

The present section provides more practical information on the usage of natural terpenoids in medicine and production of cosmetics. To that end this section has the following subheadings:
- Anti-microbials
- Analgesics
- Cholesterolemia, vascular problems including PVD
- Tracheal and bronchial disorders (COPD)
- Arthritis, rheumatism, inflammatory disorders
- Intestinal disorders
- Stress-related problems; sedatives
- Cancer therapies.

There are so many fascinating aspects of the pharmacological usage of natural terpenoids, that a selection must be made. We hope that the background information presented in the previous chapters together with the material presented below may serve to set a trail to be followed in further developments, and we will focus on studies heralding such courses of development.

7.1.1 Anti-microbials

The physiological effects of terpenoids (mostly monoterpenes) on microorganisms are discussed in section 4.2.1, and Table 6.1 gives herbal sources of these compounds; this section enters into more practical aspects.

Solutions for the treatment of disorders of nasal and oral mucosa containing terpenes often include the following ingredients: camphor (0.1-0.15%), 1,8-cineole (0.5-0.7%), fenchone (0.05-0.3%), menthol (0.2-0.7%) and linalool (0.5-0.8%). Classic nasal antiseptics may in addition contain chlorbutanol (0.15-0.25%), and nasal decongestants may include ephedrine (0.4-0.6%).

Dentifrices with adjuvants to treat inflammatory gum diseases can be based on extracts of caraway, chamomile, fennel, and St. John's wort, but one should be careful with caraway oil at concentrations above 0.2% because of its expected sensitizing effects. A Chinese recipe for a fluid in which a non-woven fabric is drenched (facial masks) to control a.o. stomatitis contains borneol, menthol and 1,8-cineole in a ratio of 5:10:2, together with extracts of flowers of *Lonicera japonica* (Caprifoliaceae) and roots of *Isatis tinctoria* (Brassicaceae).

The effectivity of chamazulene (from chamomile) and eugenol (from clove) as anti-microbials is well known and commonly applied. Thyme oil, with its major components thymol and *p*-cymene is widely utilized in pectoral remedies. The essential oil of the resin of *Commiphora mukul* (Burseraceae) has antiseptic properties; for long times tinctures have been used to treat gingivitis (Bhati, 1950).

An exiting development is based on the inhibition of the MAI metabolism of bacteria (see 2.1.1) with its key metabolite DOXP, a development which can also be important in insect control (see 6.2.2). Rohmer (1998) underlines this view in referring to the fact that in bacteria DOXP is subjected to intramolecular rearrangement, involved in the formation of the branched isoprenoid skeleton, with methyl-D-erythritol as a putative intermediate. The DOXP pathway represents a novel target for anti-bacterial drugs. In the malaria parasite *Plasmodium falciparum* two genes are known to encode the enzymes DOXP synthase and DOXP reductoisomerase, indicating that isoprenoid synthesis in *P. falciparum* depends on the DOXP pathway, probably located in the apicoplast. Fosmidomycin and a derivative inhibited DOXP reductoisomerase of this pathogen and suppressed the *in vitro* growth of multi-drug-resistant *P. falciparum* strains (Jomaa *et al.*, 1999b). In addition, mice infected with the rodent malaria parasite *P. vinckei* were used after therapy with the drugs. The effectiveness of terpenoids on the DOXP pathway in *Plasmodium* spp. may open new possibilities for the treatment of malaria. Steele *et al.* (1999) found the triterpenoids ursolic acid and

Chapter 7. Natural terpenoids to the benefit of human health

oleanolic acid to be effective against *P. falciparum*, including a chloroquine resistant biotype (dose ranging from 28-90 µl/ml). Ursolic acid can a.o. down-regulate (metallo)proteinase genes (Table 7.4). Within this context the oil of *Artemisia annua* (Asteraceae) (Table 6.1), with its toxicity to malaria parasites is interesting (Rao *et al.*, 1999).

7.1.2 Analgesics

Eugenol is an old analgesic used against toothache. To that end it is brought into the cavity. A cheap semi-permanent dental filling material with analgesic properties can be prepared by mixing (1) eugenol and SIO_2 (particle size about 15 µ), ratio 51:49 wt% and (2) a fixative comprising ZnO, olive oil, SIO_2 (same particle size, preferably methylized) and Zn acetate, ratio 80:14:5.2:0.8 wt%. Before filling, equal quantities of (1) and (2) are mixed for half a minute and can be used after 4 min. It has been developed by Futami and Shunichi (pat. dem. Fr. 2.605.882, 1988) and could be utilized for domestics as well as humans.

Many ointments and lotions exist to relieve pain in muscles, joints and underlying tissues, solutions of camphor (0.1-2.5%) or menthol (0.1-1.0%) in ethanol being the simplest. Ivy *et al.* developed an external analgesic lotion based on camphor, menthol, 1,8-cineole and ginseng extract with jojoba oil. Details for preparation are in Can. Pat. Appl. CA 2.025.078 (1991).

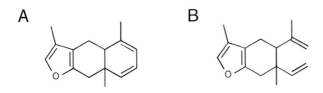

Fig. 7.1. Two analgesic terpenes from myrrh. **A**: furanoeudesma-1,3-diene; **B**: curzarene.

The monoterpenes mentioned in these examples and the isolation procedures presented in chapter 6.1 invite to some experimentation. One could also apply to the genus *Commiphora*, mentioned above. Among the resins secreted by members of this genus, myrrh is a natural compound common in northeast tropical Africa, and was used by the Jews as an ointment. Myrrh is supposed to have analgesic properties, which induced Dolara *et al.* (1996) to identify the constituents of hexane extracts of myrrh from *C. molmol*. Three sesquiterpenes were identified, which were tested intracerebroventricularly in mice (1.25 mg/kg). Two sesquiterpenes (Fig. 7.1) had significant analgesic effects and oral doses of 50 mg/kg had similar effects as morphine (5 mg/kg). The analgesic effects were blocked by naloxone, indicating interaction with brain opioid mechanisms. According to Dolara *et al.* (1996) usage of myrrh as an analgesic may have been replaced by opium derivatives, but the effect of these sesquiterpenes on central opioid pathways makes their structure interesting.

Tanacetum parthenium (Asteraceae), feverfew, is a medicinal herb used against arthritis and migraine. Johnson *et al.* (1985) recruited a number of migraine sufferers to test leaves of feverfew in a double blind trial and found that feverfew taken prophylactically prevents attacks of migraine. The authors prepared feverfew capsules from freeze-dried leaves, containing 25 mg dry material, rich in sesquiterpene lactones, the principal one being parthenolide. Commercial preparations of feverfew that can be standardized could provide an attractive alternative to e.g. ergotamine. According to Tyler (1994) 250 µg of parthenolide is now considered an adequate daily dose. However, not all feverfew products contain 0.2% parthenolide that would cover the daily dose of one tablet of 200-300 mg plant material. Moreover, a low percentage of the consumers develops some irritation of the buccal mucosa, lips or tongue. It should be mentioned that other *Tanacetum* species can be rather toxic, e.g. *T. vulgare*, of which the essential oil has been used against intestinal parasites such as ascarides: fatal cases have been reported.

7.1.3 Cholesterolemia, vascular problems, PVD

In mammals, endproduct regulation of cholesterol biosynthesis is an effective form of autoregulation, which is part of a multivalent feedback system operating in steroid production (see 3.2.2). Unfortunately, this system can be subjected to congenital defects resulting in a high LDL/HDL ratio (see Krieger, 1998). Besides cholesterol itself a number of isoprenoids are known to reduce activity of HMG-CoA reductase: farnesol, geranyl-PP, squalene, dolichol, oxysterols and steroid hormones (see 3.3.1, 3.2.2 and 3.4.2). However, these are isoprenoids produced within an organism and in their natural form they are not be applied externally, although the future may bring analogues that could be effective in *per os* medication. Other compounds, originally derivatives of terpenoid metabolites of fungi such as compactin, lovastatin and simvastatin inhibit HMG-CoA reductase much more efficiently. Garlic, of which the utility for the treatment of hyperlipoproteinemia in general is approved by the German Commission E should be mentioned, but the active components are not terpenoids.

Risks of high cholesterol levels can be reduced by some dietary measures (Box 7.1), in which terpenoids or at least isoprenoids play a role:

Chapter 7. Natural terpenoids to the benefit of human health 301

(1) Inhibition of HMG-CoA synthese
(2) Stimulation of HMG-CoA degradation
(3) Decreasing the ratio of LDL/HDL (lipoproteins)
(4) Inhibition of cholesterol uptake
(5) Reduction of the effect of blood plasma clotting factors
(6) Recognition and adequate treatment of peripheral vascular disorders (PVD)

Box 7.1. Dietary measures to reduce the risk of high cholesterol levels.

Ad (1). Effective pharmacological inhibitors of HMG-CoA reductase are available. As this book does not focus on commercial key enzyme inhibitors we restrict ourselves to one remark: lovastatin (mevacor) and simvastatin (zocor) have a side effect that can be beneficial to some patients: they stimulate bone morphogenetic protein-2 (BMP-2), that induces bone growth. In addition, oxysterols could be very effective, but an important drawback is their pathological effect on macrophages and their neurotoxicity. Clare *et al.* (1995) investigated the toxicity of oxysterols on human monocyte-macrophages *in vitro*. Of the oxysterols assessed, 26-hydroxysterol was most toxic; 7-position derivatives also induced cell damage, though at higher concentrations and 25-hydroxycholesterol was least toxic. These effects were counteracted by cholesterol, but that of course is only of scientific importance and has no practical meaning. Of all oxysterols tested, 25-hydroxycholesterol appeared to be the most toxic to sympathic neurones of the rat superior cervical ganglia (Chang *et al.*, 1998), 50% of the neurones being killed in 36 h with 10 µM. Most cells did not exhibit structural changes as observed in programmed cell death. The effects were counteracted by vitamin E and this also holds for the toxic effect of 7-β-hydroxycholesterol on rat cerebellar granular cells (Chang and Liu, 1998), but it remains to be seen if vitamin E (and other protectors e.g. methyl-β-cyclodextrin) can be combined with oxysterols in a pharmacologically active preparation.

An ayurvedic medicine reputed to reduce synthesis of cholesterol and lipoproteins and to stimulate fat combustion is Trifagul, that is composed of different herbs including *Commiphora mukul* with its high resin content (Tripahti *et al.*, 1968). The ayurvedic physician Sushruta (≈ 1000 BC) already proposed that the resin (guggul) reduces fat deposition.

Ad (2). Degradation of HMG-CoA reductase is regulated by signals generated downstream of MVA. Farnesyl-PP would seem to be the source of the positive signal for degradation of this key enzyme (Gardner and Hampton, 1999; see also 3.2.2 and 3.3.2). According to these authors who announce exploring clinical possibilities to stimulate key enzyme depletion, sterols (and farnesol) increase key enzyme degradation.

Ad (3). The ratio LDL/HDL can be reduced by regular exercise that stimulates production of prostaglandins active in LDL transport and deposition. In addition, oils rich in n-3 polyunsaturated fatty acids induce an increase in HDL with a corresponding decrease in LDL. A dietary supplement of 20 ml cod liver oil did not increase total cholesterol, but enhanced the level of HDL in serum (decreasing LDL) (Simonsen *et al.*, 1988). Other sources of polyunsaturated fatty acids are a.o. olive oil (especially Ω-3-type) (preferably cold pressings), sunflower oil and soya oil.

Ad (4). A reduced intake of cholesterol via our food is a classic measure to lower cholesterol levels, but this section discusses dietary additions to realize this effect. One possibility is to look at cholesterol-containing nutrients with a low LDL/HDL ratio. In contrast to the long-lasting opinion that consumption of eggs should be limited because of the high cholesterol content of egg yolk, the relatively high amount of HDL compared to LDL does not seem to catagorize eggs among high-risk nutrients (apart from contamination with e.g. *Salmonella* spp.). A recent development is to include phytosteroids into margarine, which changes them into nutriceuticals. Such phytosterols are 24-ethyl-5α-cholestan-3β-ol and sitostanol esters. They are claimed to substantially reduce (40%) intestinal cholesterol uptake, however, they reduce total serum cholesterol levels to a lesser extent. This could be explained by feedback of cholesterol on its own biosynthesis in healthy persons. When serum cholesterol values decrease, the feedback effect is also expected to be less. Two such products are marketed today. It should be noted that phytosteroids would be most valuable if their effect would be on the LDL/HDL ratio; a mere reduction of say 20% in serum cholesterol may be just as well obtained by eating less margarine.

Ad (5). Polyunsaturated fatty acids can prolong the prothrombin time, reducing aggregation of blood platelets. The effect of n-3 polyunsaturated fatty acids seems to be on vitamin K-dependent plasma clotting factors, as diets enriched in these fatty acids affect coagulation factors dependent on vitamin K (Andriamampandry *et al.*, 1998).

Ginkgo biloba is a plant with promising anti-clotting properties. This tree contains quite a number of secondary metabolites, including terpenoids, a flavone of the rutin type and other non-terpenoid substances. Extracts made with mixtures of acetone and water incorporate diterpene lactones, the ginkgolides that inhibit platelet activating factor (PAF). *Ginkgo* spp. are mysterious trees. Darwin called them "living fossils" that managed to survive among competitive angiosperms with a much higher fertility. Primitive ginkgos needed 25 years to enter the generative phase, and seeds did not survive for long. But ginkgos take their time: they can become more than 2000 years old. Van Beek (2000) who just finished editing a book on *G. biloba*, comparable to the neem book by Schmutterer (1995) mentions that a number of antioxidants in ginkgo could explain their long life. In addition, they have developed a "secret weapon": ginkgolides, poly-oxidated diterpenes with a bitter taste that act as defensive compounds to herbivores. The ginkgolides, in particular ginkgolide B (Fig. 7.2B) block the effects of PAFs that are produced in various tissues and that in addition to blood clotting release inflammatory substances and cause bronchiconstriction. Ginkgo products are a.o. marketed by Dr. Willmar Schwabe GmbH & Co and Ipsen Pharmaceuticals.

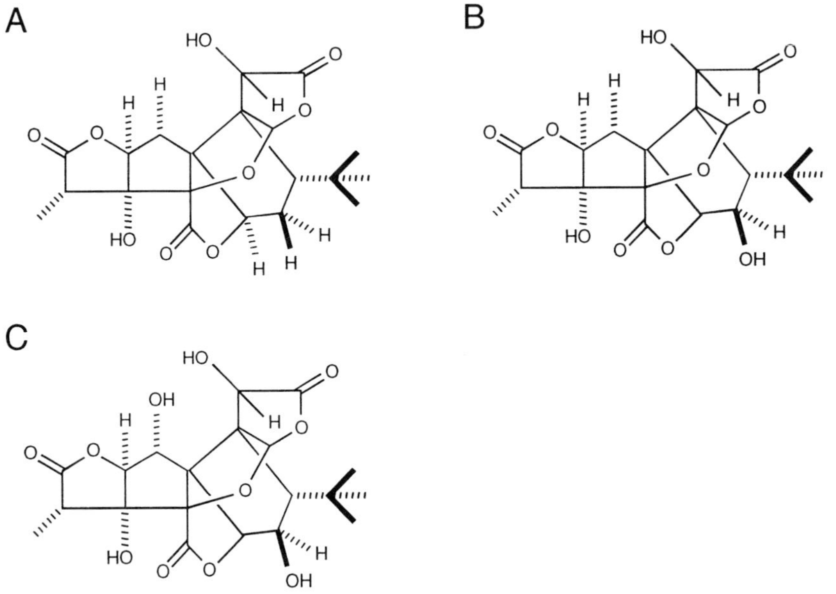

Fig. 7.2. The structures of: ginkgolide A (**A**); B (**B**) and C (**C**); members of a family of potent PAF (platelet activating factor) antagonists.

Ad (6). It is beyond the space allotted for this section to discuss all cardiovascular and peripheral vascular disorders (PVD). Some are based on systemic autoimmune diseases. Cardiovascular disorders are not necessarily based on cholesterolemia, but also on pre-existing conditions such as cardiac trauma, ischemic processes and myocarditis.

Flavonoids of ginkgo and not ginkgolides are responsible for a reduction in ischemic brain damage. The German Commission E put ginkgo extracts on their lists for treatment of cerebrovascular disorders. Herbs containing terpenoids to treat PVDs are in Table 6.2, the most common ones being *Artemisia vulgaris*, *Hypericum perforatum* and *Rosmarinus officinalis*. The essential oil of these herbs is applied externally, that of *R. officinalis* is reputed for its effect on blood circulation when added to a bath, although we find it difficult to understand how relatively low concentrations can act as a tonic. The oil is also taken as a drug.

Almost anecdotally is a paper by Nikolaevskii *et al.* (1990) claiming that inhalation of volatile oils of lavender and monarda reduce cholesterolemic lipoprotein depositions in the aorta of rabbits and reduce the risk of formation of atherosclerotic plaques. To that end, the oils should reach a concentration of 0.1-0.2 mg/m^3 of air. Maybe it is worth trying on humans, as volatile terpenes can be absorbed by the pulmonary mucosa.

Venal disorders are often due to a gradual destruction of the elastic fibres in the tunica adventitia, the major part of the wall of veins. Destabilization of cholesterol-containing membranes of lyposomes results in activity of lysosomal enzymes that alter the venal walls upon which electrolytes and proteins leach into the surrounding tissues. The venal walls become dilated and valvular insufficiency and reduced blood circulation belong to the syndrome that includes oedemas, thrombosis and thrombophlebitis. Few people know that the seeds of the common tree *Aesculus hippocastanum* (Hippocastanaceae), horse chestnut, and a few related species contain triterpenoid saponin glycosides, effective in the treatment of varicoles. Extracts of the big brown seed, usually based on a mixture of ethanol and water, can be dried and taken orally (doses 40-160 mg/day) or used in ointments. For external application one could combine the extracts with oil of *R. officinalis*. Preparations of horse chestnut should not be prescribed to treat arteries with their thick tunica media.

7.1.4 Tracheal and bronchial disorders (COPD)

A Chinese asthma centre investigated the essential oil of *Artemisia argyi* that has anti-asthmatic properties. Members of the genus *Artemisia* produce a range of different terpenes (Table 6.1). (E)-carveol and particularly α-terpineol appeared to be the most active terpenes. These results are confirmed by the Cooperative Group of Anti-asthmatica in Zhejang (Japan). This group also tested oil of *A. argyi* to conclude that both (E)-carveol and

L-α-terpineol are anti-asthmatic. Other investigations mention myrcenol and nerol to exert anthi-asthmatic activity.

In Japan several formulations of breath deodorants, nasal and oral sprays for treatment of bronchitis, rhinitis, etc. have been patented together with tubes, filters and nozzles for manual production of aerosols. Carriers are e.g. kaolin, talc and diatomaceous earth (in the Netherlands cigars containing the latter compound are positioned into the lowest quality!). In this way, menthol mixed with zeolite and ethanol is used against bronchitis. Juergens et al. (1998a) evaluated the potential role of L-menthol and mint oil as an anti-inflammatory drug, using paraffin as a solvent. The investigations were performed on LPS-stimulated monocytes from healthy volunteers. L-menthol considerably suppressed inflammation mediators by monocytes. L-menthol was preferable to mint oil and can be supplied in enteric coated capsules. The authors suggest further clinical trials of L-menthol for treatment of bronchial asthma and rhinitis. The same research group in Bonn, Germany reports that 1,8-cineole as present in eucalyptus oil inhibits both pathways of arachidonic acid metabolism (Juergens et al., 1998). Volunteers were subjected to in vivo trials and monocytes were isolated before therapy (3 x 200 mg 1,8-cineole/day), on day 4 and four days after discontinuation of the treatment (day 8). The production of inflammation mediators decreased with 30-40%, indicating that 1,8-cineole is an interesting terpene for further studies. To our opinion this certainly holds for combinations of L-menthol, 1,8-cineole and L-α-terpineol. In a later paper of that year Juergens et al. (1998b) reported a dose-dependent and highly significant inhibition of the production of tumour necrosis factor α, interleukin 1-β, leukotriene B4 and thromboxane B2 by 1,8-cineole. This is the first report on the action of a monoterpene as an inhibitor of cytokines, a property that makes this monoterpene of interest to incorporate in treatments of steroid-sensitive COPDs. The following recipe was passed to us by an experienced COPD specialist. Add the following quantities (ml) to 100 ml ethanol 96%: 1,8-cineole (40), oil of rosemary (20), pine oil (local, 20), menthol (20), and oil of lavender (5). Menthol can be dissolved into the oils of the other herbs, after which ethanol is added.

Many patients of pulmonary emphysema suffer from abnormally high populations of micro-organisms in their alveoli, among which *Pseudomonas aeruginosa*. As discussed in previous sections, compounds inhibiting MAI metabolism of these bacteria could help to keep these populations at a low level. Paradols from *Zingiber officinale* (Zingiberaceae) are effective against four bacteria tested: *Botryodiplodia theobroma*, *Proteus aeruginosa*, *P. vulgaris* and *Staphylococcus aureus* (Oloke et al., 1989). In 1989 the DOXP pathway was not yet elucidated, but with the present knowledge these trials merit continuation.

7.1.5 Arthritis, rheumatism, inflammatory disorders

There exist numerous recipes to compose ointments for the treatment of arthritis, rheumatoid arthritis and rheumatism, many of which contain terpenoids. This is not the place for extended lists with recipes. Moreover, as the personage Dr Astrov in a play by the Russian physician and dramatist Tsjechov (1860-1904) says: "If there are many remedies for a disease, probably *none* is actually effective". Treatments based on derivatives of corticosteroids are no subject of this volume.

Medieval recipes using herbs containing terpenoids are still accessible, e.g. the *Canon medicinae* by Ibn Sina, with many Indian and Arabic medicines. Inflammatory disorders were recognized as a separate category; recipes were often very complicated. We quote the Sloane manuscript 2463 for a remedy against "sores of the womb of women" (translated): "Take 4 drachms of gum tragacanth, 2 drachms of borax, 1 drachm each of camphor, armenian bol, alkannel, sandragon, half a drachm each of mastic, myrtle, white lead (!), olibanum, litharge, 4 ounces of oil from the stove, 8 drachms of rose water. The tragacanth and borax is dissolved in rose water and filtered, the rest is ground to a paste and blended with the previous mixture, and introduced into the vagina". This seems to be a multicomponent treatment *avant la lettre*.

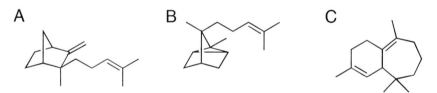

Fig. 7.3. The structures of **A**: α-santalene; **B**: β-santalene; **C**: (±)-β-himachalene.

Guggul (or guggulu), a preparation of the resin of *Commiphora mukul*, mentioned above, has anti-inflammatory properties, resembling those of hydrocortisone and is used in an ayurvedic medicine to treat arthritis. *Conyza canadensis* (c.f.: *Erigeron canadensis* (Asteraceae)) is an annual plant originating in North America, introduced into Europe in the 17th century. It can be easily grown on sandy soil and sunny sites. Its essential oil has been used as an antidiarrhoeic and diuretic medicine, but it is probably of greater interest because of its promising effect on spondylarthritis, chronic arthrosis of the spine and gout. Lenfeld *et al.* (1986) investigated the effects of extracts in ethanol, methanol and petroleum ether to find that the latter solvent is effective in isolating the sesquiterpene hydrocarbons β-santalene, cuparene, α-curcumene, γ-cadinene and β-himachalene (Fig. 7.3). The activity of the latter compound was studied and its anti-inflammatory activity was proved. A mixture of neutral oil and ethanol can be used as a vehicle. Extracts of *C. canadensis* stimulate migration of leukocytes and the formation of granulation tissues.

Most ointments to treat rheumatism contain salicylic acid, however, that is not a terpenoid but a related aromatic derivative. Some contain menthol as well as mint oil, but to our opinion one better makes a choice between these two, and could include thyme oil instead. A general recipe supposed to have additive effects may be as follows: camphor 2.0, menthol 1.0, thymol 0.5, methyl salicylate (or salicylic acid) 15, hydrocortisone 0.5, and excipients to make 100 parts. Salicylic acid can be dissolved in a little acetone prior to inclusion in an ethanolic solution that can be made up with chloroform or acetic ether.

7.1.6 Intestinal disorders

An unexpected application of terpenoids is in the prophylaxis of oral mucositis that may develop during repeated cycles of chemotherapy (taxol!) or after irradiation therapy. Carl and Emrich (1991) determined whether or not mucosites could be prevented by using extracts of chamomile (Kamillosan liquidum) as an oral rinse. The authors found that resolution of mucositis is accelerated by Kamillosan rinse and that prophylactic oral care is advised to minimize the effects of irradiation and chemotherapy.

The treatment of gastric ulcers with a triterpene glycoside has been developed in the Netherlands during the Second World War. This triterpene (glycyrrhizic acid) is present in the roots of *Glycyrrhiza glabra* (Fabaceae), licorice, and a preparation of licorice was made by a pharmacist in the province of Friesland, to be introduced on a bigger scale by Revers, a physician. The senior author remembers that his own father was successfully treated for peptic ulcers by his family physician in 1946, who called it an "astonishing development". An undesirable side effect is high tension; patients should therefore eat lots of fresh vegetables to restore their potassium balance, because licorice tends to retent sodium (and water) and to excrete potassium by inhibiting 11β-hydroxysterol dehydrogenase.

Glycyrrhizic acid probably inhibits prostaglandin dehydrogenase and reductase, increasing the protective effects of prostaglandins on the gastric mucosa. A dose of 200-600 mg glycyrrhizin/day is recommended for a limited period. Regular measures of blood pressure, especially in elder people, should accommodate this treatment.

It should be noted that many gastric ulcers are associated with duodenum ulcers, if these are not the primary cause of the complaints. Licorice is also effective to treat these disorders. In 1962 Adami *et al.* reported the effect of natural and synthetic terpenoids on gastric ulcers of guinea pigs. Squalene and β-sitosterol had curative properties, farnesenes were completely inactive, but farnesol was. Not surprisingly, β-carotene had protecting and curative properties – daily consumption of carrots was common practice in Folk medicine to treat gastritis.

Patients with the irritable colon syndrome (or irritable bowel syndrome, IBS) mostly show increased colon motility 60-90 min after food intake, often associated with a three-cycle-per-min increase in contractile activity. Those who suffer from alternating diarrhoea and constipation commonly have abdominal pain. According to Rhodes *et al.* (1987) carminative oils provide an effective treatment for IBS either delivered in coated hard gelatine capsules or as a rectal suppository. They recommend 0.15 ml peppermint oil 3 times/day before meals. Apart from *Mentha piperita*, *Chamomilla officinalis, Coriandrum sativum* and *Melissa officinalis* are candidates to support treatment of IBS, because of their spasmolytic properties (Table 6.2). A browse through other reports points to 1,8-cineole as a spasmolytic compound. There is no reason why these essential oils should not be combined with psyllium seed bulk preparations (e.g. Metamucil, Konsyl), that can be taken three times daily (1 tablespoon in any fluid), but *not* together with the essential oil.

Cholesterol-containing gallstones can be dissolved in mixtures of an essential oil (e.g. D-limonene or menthol) with fatty acids or glycerides. The combination of monoterpenes with non-ionic surfactants has been tested in hospitals in Japan and may be interesting as pre-operative treatment of gallstones with a perfusion technique. With respect to the portal vein system in the liver, one should be aware of oedemas caused by such mixtures.

7.1.7 Stress-related problems, sedatives

The tranquillizing effect of valerian is not based on the sesquiterpenes of its essential oil, valerenic acid and valeranone. Although the herb is effective, and many secondary metabolites have been identified, the chemical background of its tranquillizing properties is not exactly known. This holds for more remedies with sedative effect that are known to contain terpenoids, but as solid proof is missing that they are responsible for the physiological action, we do not include them in this section. Terpenoids are utilized in aromatherapy (Buchbauer *et al.*, 1991, 1993; Buckle, 1993).

The essential oil of *Foeniculum vulgare* (Apiaceae), fennel, obtained from the seeds, contains 50-60% anethol which helps do calm down nervous indigestion. The sedative and carminative properties of fennel have been used in Europe since the Physician of Court (the court of Emperor Ferdinand I), P.A. Matthiolus, published an essay on fennel in Prague in 1563.

Some nervous disorders are associated with digestive problems. *Lavandula officinalis* (Lamiaceae) is not only reputed for its sedative properties, it is particularly useful in vegetative dystonia. To treat this problem one makes a tea of two teaspoons of the flowers in 250 ml of water, which can be sweetened with honey. The same Dr. Matthiolus advised an extract of lavender in wine. In Russia lavender oil is administered by inhalation (≈ 1.0 g/m^3, 20-30 min daily).

Humulus lupulus (Cannabaceae), hop, is a good herbal support to a tranquil sleep. The strobiles (hops) have glandular trichomes and their essential oil (containing a.o. β-caryophyllene) and high resin content make it a producer of terpenoids. However, there are more secondary metabolites in hops and it is unclear on which ones the sedative properties are based.

The anti-depressive effect of hypericin derived from *Hypericum perforatum* (the oil glands are depicted in Fig. 2.9) is already mentioned in section 6.1.1 and could be based on inhibition of monoamine acid oxidase (MAO, type A and B). In Europe it is generally accepted as an anti-depressive herb and is marketed by many producers. In the Netherlands alone there are more than 15 different products on the market, with varying doses of hypericin in one unit. The amount of hypericin in one capsule or tablet ranges from 210-940 µg, depending on the producer.

A patent on administration of essential oils by inhalation or transdermically to treat stress-related problems is Eur. Pat. Appl. EP 183,436 (1986, Warren *et al.*) and includes nutmeg oil, mace extract, neroli oil, valerian oil, myristicin, isoelemicin and elemicin.

Melissa officinalis (Lamiaceae), balm, has long been recognized as a sedative herb, especially for stress-related palpations of the heart and to treat stomach problems. It was already prescribed by Paracelsus (1493-1541). Tea can be made from two teaspoons of dried leaves in 250 ml water and fresh leaves can be added to soup.

7.1.8 Cancer therapies

There is no doubt that prevention is better than the best possible treatment. Protection against mutagenic agents is a consolidated principle in nature. Those who have a fresh egg for breakfast may wonder why the eggyolk is yellow. Blount *et al.* (2000) put the same question to answer: because it has a high carotenoid content, protecting against free radicals.

A large body of epidemiological evidence, based on data from studies on animals and humans strongly supports relationships between dietary constituents and the risk of specific cancers. Many studies discussed in chapter 4.1 pay attention to (additional) protective effects of anti-tumour terpenoids. In many countries prospective cohort studies are made on the relationship between macro- and micronutrients and the development of tumours.

Although high vegetable, fruit and cereal intake in general tend to offer a better protection against cancers of the alimentary and respiratory tracts than to hormone-sensitive cancers (a.o. Block *et al.*, 1992; Jacobs *et al.*, 1995) the increased intake of specific nutrients could become a nutriceutical approach to prevent the development of malignancies in individuals with increased congenital risks (Singletary, 2000). This may hold for antioxidants such as ascorbic acid, selenium ions and isoprenoids like vitamin E, vitamin A and β-carotene (Wattenberg, 1992) as well as for more specific metabolites. Examples are lycopene, specific carotenoids and retinol in relation to prostate cancer (Giovannuci *et al.*, 1995; Khachik *et al.*, 1995).

According to Kusamran *et al.* (1998) some Thai bitter gourd fruits contain anti-carcinogens, that may inhibit activity of mutagen-stimulated enzymes in the liver. Many products based on natural terpenoids have already been marketed, such as Ginkgo biloba 50:1, produced by Jarrow Formulas Inc.

Many dietary factors with anti-tumour properties are terpenoids, of which the mode of action has been discussed in chapter 4.1. Several non-steroid and more complex terpenoids are at present incorporated into Phase II trials, alone or in combination with more classical cytostatic drugs for the treatment of cancers. Depending on the type of cancer (and the research group) different strategies are followed (Box 7.2).

(1) Inhibition of protein prenylation (gene expression)
(2) Selective induction of apoptosis
(3) (Re)differentiation of tumour cells
(4) MVA depletion
(5) Inhibition of angiogenesis
(6) Modulation of steroid receptors
(7) Fixation of the cytoskeleton
(8) Monoclonal antibodies

Box 7.2. Major strategies to treat malignancies.

The effects of different classes of terpenoids on cell cycle progression and apoptosis are treated in sections 4.1.4-4.1.8. In the discussion of the practical

Chapter 7. Natural terpenoids to the benefit of human health

application of natural terpenoids we follow the same sequence as presented in Box 7.2. When we finished this section in summer 2000, we noticed an increasing interest in inhibitors of protein prenylation, in particular of Ras proteins (see 4.1.4) and in selective initiation of protein cascades involved in apoptosis. Terpenoids exerting such effects on the cell cycle are highlighted in section 4.1.5. The numerous studies published in addition to those quoted in chapter 4.1 compel us to make a selection. Tables 7.1 and 7.2 give an overview of non-steroid terpenes and more complex terpenoids that are tested in Phase I or more advanced trials of anti-tumour compounds.

Ad (1). Hohl *et al.* (1998) elucidated that farnesyl-PP derivatives inhibit Ras farnesylation on a stereochemically-dependent way. Only (E,E)-α-hydroxyfarnesyl phosphonate inhibits Ras processing in human leukemic cells, while (Z,E)-α-farnesyl- and geranylphosphonates are inactive. The first compound does not affect squalene synthase or cholesterol synthesis. These results indicate that the length of the prenoid chain and stereochemistry of farnesyl-PP derivatives allow selective inhibition of prenylation of Ras proteins and could be developed into new anti-tumour drugs.

A novel inhibitor of FTase is a compound with a similar structure as limonene, encoded XR3054 that inhibits farnesylation of CAAX recognition peptides *in vitro* with an IC50 of 50 μM (Donaldson *et al.* 1999). It reduces farnesylation of p21 ras and inhibits proliferation of prostate and colon carcinoma cell lines, but to a much lesser extent mammary carcinoma cell lines. This is not a completely natural terpenoid, but this study illustrates that limonene and its derivatives are terpenes on which a new generation of anti-tumour drugs can be based. Treatment with limonene alone via a diet may approach the tolerable level; we estimate effective doses to be about 20 g/day for an adult person. This dose comes close to those proposed by Burke *et al.* (1997) for the treatment of pancreatic cancer with farnesol and geraniol, although that is based on diet weight.

According to End (1999) the year 2000 will be a significant date for Ras-related therapies, as the data obtained with numerous FTase inhibitors will be followed by Phase II trials. One of the promising terpenes is perillyl alcohol (compare 4.1.4, 4.1.5 and Table 7.1), that has now proven to selectively induce G0/G1 arrest and apoptosis in myeloid leukemic cells (Sahin *et al.*, 1999) and mammary tumour cells (Satomi *et al.*, 1999). Ripple *et al.* (1998) treated 18 patients with advanced malignancies with perillyl alcohol, in doses ranging from 800 mg/m^2-2400 mg/m^2. There were two heavily pre-treated ovarian cancer patients who experienced reversible ≥ grade 3 granulocytopenia. In all patients disease stabilization was obtained for ≥ 6 months. These clinical studies are a vanguard of trials in which derivatives of limonene are combined with other terpenoids and with synthetic cytostatics.

Table 7.1. Examples of the effect of non-steroid terpenoids on the cell cycle of normal and malignant cells and on complete organisms *

Terpenoids	Mode of action
Farnesene	- inhibition of its own biosynthesis in plant cells
	- insecticidal and hormonal actions in insects
Farnesol	- selective inhibition of N-type Ca^{2+} channels
	- interference with phosphatidyl-inositol-type signalling
	- regulated degradation of HMG-CoA reductase
	- post-transcriptional regulation of HMG-CoA reductase
	- induction of apoptosis (in human lung adenocarcinoma A549 cells)
Farnesyl acetate	- reduction of HMG-CoA reductase levels and sterol biosynthesis
	- stimulation of degradation of HMG-CoA reductase
	- blocking of protein prenylation
	- inhibition of DNA replication in hamster and human cells
Geraniol	- anti-bacterial action against *Salmonella*
	- inhibition of pancreatic tumour growth
	- apoptosis-like cell death in a plant (*Matricaria chamomilla*)
	- suppression of the growth of hepatomas and melanomas transplanted to rats and mice
Geranylfarnesol	- differentiation and apoptosis in leukemia cells (mouse myeloid leukemic M1 cells)
Geranylgeraniol	- inhibition of the orphan nuclear receptor LXR-α
	- differentiation and apoptosis in leukemia cells
	- induction of caspase-3 activity during apoptosis in human leukemia V937 cells
Geranyllinalool	- insecticidal action
	- defensive properties of pine wood and of termites producing this compound against other insects
Isopentenyl adenine	- formation of isopentenyl adenylated proteins with a function in cytokinin regulation of proliferation and differentiation in plants
Limonene	- inhibition of isoprenylation of Ras-proteins and other oncogenic proteins (c-jun and c-myc)
	- inhibition of cell growth, cell cycle progression and cyclin D_1 gene expression in human mammary cancer cell lines

Terpenoids	Mode of action
Linalool	- anti-convulsant properties
	- inhibition of glutamate binding at membranes of the central nervous system
Lovastatin	- inhibition of HMG-CoA reductase
	- apoptosis of the prostate cancer cell line LNCaP
	- prevention of membrane localization of Ras in the myeloid cell line U-937
	induction of cyclin-dependent kinase inhibitors in human mammary cells
	- inhibition of the transcription of Scr-genes in macrophages
	- inhibition of angiogenesis (tumour stimulated blood vessel growth via increase of anti-tumour activity of TNF-α)
	- inhibition of mesangial proliferation via increased P27Kip1 protein degradation
Lovastatin + farnesol	- accelerated degradation of HMG-CoA reductase
Lovastatin + geranylgeraniol	- counteraction of the inhibition of cell proliferation by lovastatin in C-6 glioma cells
	- potentiation of lovastatin inhibition of oncogenic H-ras processing
Lovastatin + limonene	- synergistic action on the inhibition of HMG-CoA reductase
Lovastatin + perillyl alcohol	- counteraction of the effect of lovastatin on HMG-CoA reductase
Mevalonic acid (MVA)	- activation of the orphan nuclear receptor LXR-α
	- blocking of the translation-dependent suppression of HMG-CoA reductase mRNA
Myrcene	- induction of liver mono-oxygenase
	- influence on sister-chromatid exchange induced by mutagens in V79 and HTC-cells
Nerolidol	- inhibition of neoplasia of the large bowel
	- antibiotic effect on malaria parasites
Perillic acid	- inhibition of farnesyltransferase and geranylgeranyl transferase
	- inhibition of Ras/MAP kinase-driven IL-2 production in human T lymphocytes
Perillyl alcohol	- inhibition of isoprenylation and other effects, the same as limonene, but in specific cases more effective
	- induction of G0/G1 arrest and apoptosis in Bcr/Abl-transformed myeloid cell lines
	- inhibition of the growth of murine B16 melanomas

* For primary targets compare with Tables 4.1 and 4.2.

Manumycin is an interesting new specific inhibitor of FTase, although it does not have a terpenoid structure. It selectively reduced p21 ras farnesylation and inhibited protein kinase 2 in human colon tumour cells (Colo 320-DM), followed by DNA cleavage and apoptosis (Di Paolo *et al.*, 2000). Manumycin does not affect cholesterol biosynthesis, and MVA did not protect the tumour cells from induced apoptosis. This is another promising example of an antibiotic with FTase inhibiting activity, a target of which may be tumour cells with a K-ras gene. In a review of FTase inhibitors Prendergast (2000) stresses that recent studies indicate that FTase inhibitors are not only involved in the inhibition of Ras prenylation. The anti-tumour effects of FTase inhibitors are also due to an effect on RhoB prenylation. Endosomal Rho protein functions in receptor activities.

Table 7.2. Mode of action and possible utilization of more complex terpenoids at the cell level *

Terpenoids	Mode of action/utilization
Andirobin group	
Methyl angolensate	- activity against *Plasmodium falciparum* (malaria parasite)
Dammarane group	
Dammara dienol	
Dammarene diol-II	- antiviral activity *in vitro* against Herpes simplex virus I and II
Dammarenolic acid	
Panaxadiol	- transcriptional activation of superoxide dismutase and catalase genes
Panaxatriol	- increase of the interleukin-1 production by influencing the gene expression of IL-1
Gedunin group	
Gedunin	- activity against *Plasmodium falciparum*
Glutinane group	
Celastrol	- inhibition of lipid peroxidation of rat liver mitochondria
Pristimerin	- anti-microbial, anti-inflammatory and anti-tumour activities
	- inhibition of the induction of inducible NO-synthase via inhibition of NF Kappa B activation
Glycyrrhetinic acid	- restoration of decreased leukocyte counts and cellular immunocompetence in low-dose γ-ray-irradiated mice
	- enhancement of glucocorticoid action
	- induction of liver microsomal cytochrome p450 in mice
	- anti-ulcer activity
	- inhibition of the growth of mouse B16 melanoma cells
Glycyrrhizin	- enhancement of Fas-mediated apoptosis in T cell lines

Terpenoids	Mode of action/utilization
Lupane group	
Betulinic acid	- inhibition of neuroectodermal tumours
	- apoptosis in glioma cells
	- inhibition of the human immunodeficiency virus (HIV) type 1 and 2
Oleanane group	
Oleanolic acid	- anti-ulcer activity
	- decrease of cytochrome p450 2E1 in mice
	- inhibition of HIV-1 replication in acutely infected H9 cells
	- inhibition of tumour growth in mice
	- induction of differentiation in mouse myeloid leukemia M1 cells
Taraxastane group	
Taraxasterol	- inhibition of induction of Epstein-Barr virus antigen
	- anti-microbial activity against *Staphylococcus aureus*
Taraxerane group	
Taraxerol	- anti-tumour action
	- inhibition of induction of Epstein-Barr virus antigen
Ursane group	
Ursolic acid	- action against chronic dermal inflammation
	- anti-ulcer activity
	- inhibition of chitin synthase II activity of *Saccharomyces cerevisiae*
	- inhibition of the cyclooxygenase-2 catalyzed prostaglandin biosynthesis

* For primary targets compare with Table 4.2.

It should be noted that rat kidneys, especially of males, are uniquely sensitive to hydrocarbons like D-limonene that is nephrotoxic to these animals. Cats can also be intoxified with D-limonene, linalool and crude citrus oil extracts that cause hypersalivation, muscle tremors, ataxia and hypothermia. Limonene does not exert a nephrotoxic effect in dogs, but can cause a.o. ulcerations of the oral mucosa, anemia, leukocytosis and prolongation of prothrombin time. One dog accidentally developed inthrathoracic and abdominal hemorrhage and died (Rosenbaum and Kerlin, 1995). This is mainly a drawback of the use of insecticidal dips containing D-limonene, but should be kept in mind with respect to the development of novel cancer therapies for domestics as well for human beings which might have increased sensitivity to limonene.

At the University of Wisconsin (USA) the group of Gould has initiated the production of perillyl alcohol by Endorex Corp. (OTC;ENDR), a diversified biotech company developing mainly immuno-pharmaceuticals.

The daily dose of formulated perillyl alcohol is 10-15 g. It may be applied against tumour types responding to limonene, of which 8g/m^2/day seems to be the maximum tolerated dose (Vigushin *et al.*, 1998), but as it causes regression at lower doses (Gould *et al.*, 1990), the risk of side effects is reduced. The near future will reveal if perillyl alcohol inhibits FTase or RhoB (Prendergast, 2000) and other targets may show up; we must leave this question open as this book now enters the production phase.

Ad (2). Table 4.1 gives an overview of terpenoids with established effects on the cell cycle leading to apoptosis. The majority of studies is still in the *in vitro* phase; we prefer to present at least *in vivo* Phase I and novel promising developments in this section. Shidoji *et al.* (1997) mention that geranylgeranoic acid (by activation of the cysteine protease cascade) has been proven to be effective in a double-blinded Phase II trial with post-operative hepatoma patients.

There exists some controversy on the apoptosis-inducing activity of ascorbic acid and vitamins K (isoprenoids). On one hand the increase of the intracellular ascorbic acid level can make cells resistant to oxidative stress-induced apoptosis. On the other hand, synergistic apoptosis-inducing effects have been found between ascorbic acid and vitamins K2 and particularly K3 (ratio ascorbic acid:vitamine K3 ≈ 100:1), and even between these compounds and other anti-tumour drugs. The multi-component aspects of therapeutic applications are discussed by Sakagami *et al.* (2000).

Primary leukemia cells respond to vitamin K2 by down-regulation of Bcl-2 and up-regulation of Bax protein expression with concomitant activation of caspase-3 (Nishimaki *et al.*, 1999). According to Blutt *et al.* (2000) 1,25-dihydroxy vitamin D3 (calcitriol) causes cell cycle arrest and apoptosis in LNCaP prostate cancer cells which is associated with down-regulation of Bcl-2 and Bcl-X(L), although Bax protein expression is not affected. There may be a prospectus for combination therapies in down-regulation of Bcl-2. Ahmed *et al.* (1999) report that although all-*trans*-retinoic acid (RA) alone failed to induce apoptosis in two human CD34-positive leukemia cell types, it enhanced cytarabine and fludarabine effects on CD34-positive cells. RA had an additive effect on Bcl-2 reduction induced by cytarabine, but not fludarabine, suggesting that it has a different mode of action when combined with fludarabine.

A prognostic study in acute promyelocytic leukemia indicated that a combination of chemotherapy and all-*trans*-RA is more effective than either treatment alone, in which persistent reverse transcription polymerase chain reaction is favourable for long-term survival prognosis (Hu *et al.*, 1999). Much remains to be done before long-term effects of retinoids applied in combination with cytostatics can be predicted for different types of tumours.

Chapter 7. Natural terpenoids to the benefit of human health   317

A cell line derived from cervical carcinoma containing 600 copies of the HP-V16 mRNA genome treated with all-*trans*- and 9-*cis*-retinoic acid showed an increase of up to 94% of G1 arrest and a reduction of the S phase of 80% (Narayanan *et al*., 1998). These data may approach 100% of G1 arrest in less aggressive tumour cells, but regression maybe expected unless combined with other anti-tumour agents.

Hormone-responsive mammary and prostate cancer cells seem to be more sensitive to vitamin E treatment than leukemia cells, when inhibition of DNA synthesis is taken as a criterion. We compared several studies to arrive at an estimated effective dose of 0.05-0.15 mM in the target tissues. This implies a daily intake of a few times 100 mg.

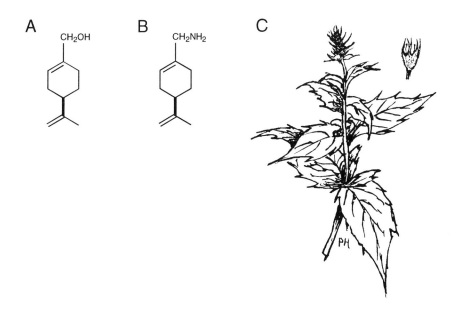

Fig. 7.4. The structures of **A**: perillyl alcohol; **B**: perillyl amine, two anti-tumour terpenes; **C**: lateral shoot of *Mentha spicata*, a source of perillyl alcohol.

Chemotherapy of pancreatic cancer is still very difficult. Perillyl alcohol, whether synthetic or from a natural source such as *Mentha spicata* (Fig. 7.4C) is one of the best options with respect to terpenoids. Burke *et al*. (1997) tested the relative *in vivo* anti-tumour activities of farnesol, geraniol and perillyl alcohol against pancreatic adenocarcinomas of Syrian golden hamsters. Farnesol and geraniol were much more active than perillyl alcohol, a result not expected from their effects *in vitro*. The same degree of inhibition of pancreatic tumour cells was obtained with 25 µM farnesol, 25 µM geranylgeraniol, 100 µM geraniol and 300 µM perillyl alcohol. Another improvement was the replacement of the OH group of perillyl alcohol by NH_2 (Fig. 7.4B). The inhibition of HMG-CoA reductase may be part of the success of the farnesol treatment, and as yet nothing can be said of the possible effects of combinations of e.g. geraniol and perillyl amine. Moreover, these are hamsters and not humans and the trials were of limited duration.

Selectively acting terpenoids are known from marine biology, such as dilopholide, a xenicane-type diterpene and other diterpenes isolated from the brown alga *Dilophus ligulatus*, that exert toxicity to human carcinoma and leukemia cells. Cembrane-type diterpenes from a soft coral, the sarcophytols inhibit tumour promotion in mouse skin. Sarcophytol A inhibits nitrosoamine-induced pancreatic carcinogens in hamsters and has antiprogressive effects on these tumours (Yokomatsu *et al.*, 1996). Similar results can be obtained with combinations of carotenoids, such as β-carotene, canthaxanthin and astaxanthin that exert promising effects on mammary tumours, for the time being in mice (Chew *et al.*, 1999).

Ad (3). In chapter 4 the transition of section 4.1.5 to 4.1.6 is marked by a discussion of retinoids as possible inducers of (re)differentiation of tumour cells. Retinoic acid has been demonstrated to induce human neuroblastoma SKNBE cells into a neuronal phenotype. It modulates expression of matrix metalloproteinase (Table 7.4). Differentiation into a neuronal phenotype was typified by expression of a growth-associated p-43 oncogene (Chambaut-Guerin *et al.*, 2000). Retinoids appear to be more effective when combined with other inducers of differentiation. Retinoic acid had a synergistic effect on differentiation of acute promyelocytic leukemia with bufalin, a steroid from the Chinese toad venom preparation Chan'su (Yamada *et al.*, 1998). A primary culture of cells was taken from patients. All-*trans*-retinoic acid (ATRA) significantly augmented the differentiating activity of phenylbutyrate on a myeloid leukemia cell line. Combination of ATRA (1 µM) and phenylbutyrate (0.5 mM) increased the expression of the myelomonocytic marker CD11b eightfold (Yu *et al.*, 1999). It should be noted that ATRA is antagonistic at low concentrations and synergistic at higher ones. According to these authors ATRA can be an improvement of treatment with phenylbutyrate alone, of which it has been difficult to realize therapeutic levels of 1-2 mM, but that in combination with ATRA is now accessible for prolonged treatments.

In combination with retinoids vitamin D metabolites have already been shown to be potentially useful terpenoids in the differentiation of human malignant gliomas (Magrassi *et al.*, 1995). James *et al.* (1999) state that differentiation therapy has been accepted as a less cytodestructive strategy in the treatment of hyperproliferative disorders. They mention the use of ATRA in the treatment of acute promyelocytic leukemia and underline the importance of combinations of vitamin D derivatives and retinoids such as ATRA and 9-*cis*-retinoic acid to treat established leukemia cell lines. Another important aspect is the potentially effect of vitamin D compounds on apoptosis induced by 9-*cis*-RA in HL-60 cells. The combination effectively induces the expression of Bcl-2 (Elstner *et al.*, 1996) and Bax. Insertion of members of the Bax family into the mitochondrial membrane induces release of cyt. c and apoptotic cell death (Pawlowski and Kraft, 2000).

In a review of prevention and therapy of cancer by dietary monoterpenes Crowell (1999) brings to the attention that treatment of chemically-induced mammary tumours with monoterpenes results in tumour redifferentiation, associated with increased expression of the mannose-6-phosphate/insulin-like growth factor II receptor and transforming growth factor β 1.

In conclusion, terpenoids, which are incorporated in differentiation therapies can both act as inhibitors of malignant development in stem cells and be potent eradicators of malignant cell types which are in the post-initiation phase.

Ad (4). It has been known for several years that oxysterols can induce apoptosis, concomitant with a rapid decrease in Bcl-2 protein (see section 4.1.8). Both 7-ketocholesterol and 25-hydroxycholesterol induced apoptosis in vascular smooth muscle cells, a process that is inhibited by exogenous cholesterol in the case of 25-hydroxycholesterol. Cholesterol does not have any apoptotic effect itself (Nishio and Watanabe, 1996), so at first sight down-regulation of HMG-CoA reductase and MVA depletion could be the cause of apoptosis. However, as 7-ketocholesterol-induced apoptosis was not affected, the authors concluded that 25-hydroxycholesterol-induced apoptosis occurs by a different mechanism than that of 7-ketocholesterol. It is possible that 25-hydroxycholesterol has a more specific effect on the lipoprotein pool; in addition, as has been discussed in section 4.1.8 HMG-CoA reductase in tumour cells can be resistant to the normal feedback regulation (Elson *et al.*, 1999).

Lovastatin induces apoptosis of benign prostatic hyperplasic cells *in vitro*, an effect that is counteracted by MVA and geranylgeraniol, but not with 7-ketocholesterol (Padayatty *et al.*, 1997). MVA is essential for cell proliferation because it is a precursor of farnesyl-PP and geranylgeranyl-PP and therefore it is not surprising that farnesol and geranylgeraniol can restore protein prenylation upon MVA depletion (Crick *et al.*, 1998). This indicates that compounds, which interfere with protein prenylation such as derivatives of geranylgeraniol, could improve the effect of lovastatin. Crick *et al.* (1998) demonstrated that a (phosphonoacetamido)oxy derivative of geranylgeraniol prevented progression of C6 glioma cells into the S phase, even when additional MVA or geranylgeraniol was supplied. These are valuable developments, because a drawback of lovastatin and comparable HMG-CoA inhibitors when used as monotherapy to treat cancer is their significant toxicity at high doses (Agarwal *et al.*, 1999).

Although the therapeutic advantage of HMG-CoA inhibitors in the treatment of cancer patients depends on the type of tumour, its metastases and on previous treatments, documentation on the usage of these compounds is rapidly growing. Lovastatin inhibits proliferation of myeloid leukemic cells via G1 arrest, associated with p27 induction and is regarded to be an effective inducer of apoptosis in HL-60 cells (Park *et al.*, 1999).

With respect to the toxicity at higher doses, combination of e.g. lovastatin with more regular chemotherapeutics could be of benefit. In four colon cancer cell lines lovastatin selectively inhibited geranylgeranylation of target proteins, resulting in decreased expression of Bcl-2, but increased expression of Bax. What is more, pre-treatment with lovastatin significantly increased apoptosis induced by 5-fluorouracil or cisplatin, suggesting that lovastatin could be combined with 5-FU or cisplatin as chemotherapy for colon cancers (Agarwal *et al.*, 1999). To our opinion it would be of particular interest to know if lovastatin in combination with the chemotherapeutics mentioned (and maybe with adriamycin and manumycin) could be developed into an effective combination treatment (partly intraperitoneal) of colon cancer hepatic metastases, a therapy that may be supported by the intake of Ω-3-fatty acids and perillyl alcohol (Broitman *et al.*, 1996). Care has to be taken that the selected anti-tumour compounds have a comparable mode of action, e.g. tyrosine kinase inhibitors and protein kinase inhibitors. Synergistic systems are to be preferred. Specific tumour types have their own targets to which the anti-tumour compounds must exert their inhibiting action. These targets can be oncogenes, receptors, transcription factors and growth factors.

Recent studies have revealed that the β-lactone ring of lovastatin (Fig. 7.5) inhibits the proteasome degradation of cyclin-dependent kinase inhibitors p21 and p27. That is, only the β-lactone form (pro-drug) has this effect. While the open-ring form of lovastatin and pravastatin (an analogue with 100% open ring) inhibits HMG-CoA reductase, the pro-drug selectively inhibits the proteasome (Rao *et al.*, 1999b). These properties are similar to those of the proteasome inhibitor lactacystin. The unique mechanism of the β-lactone form of lovastatin could be used to develop a novel generation of lovastatin-based anti-tumour compounds, especially useful in combination therapies. Wang and Macaulay (1999) demonstrated that lovastatin inhibits medulloblastoma proliferation not because it is a general HMG-CoA

Chapter 7. Natural terpenoids to the benefit of human health

inhibitor, but because it blocks p21 ras processing. In four medulloblastoma cell lines lovastatin, although causing G1 arrest, induced HMG-CoA reductase gene-up regulation. The antibiotic manumycin A that inhibits FTase induced apoptosis in a time- and dose-dependent way.

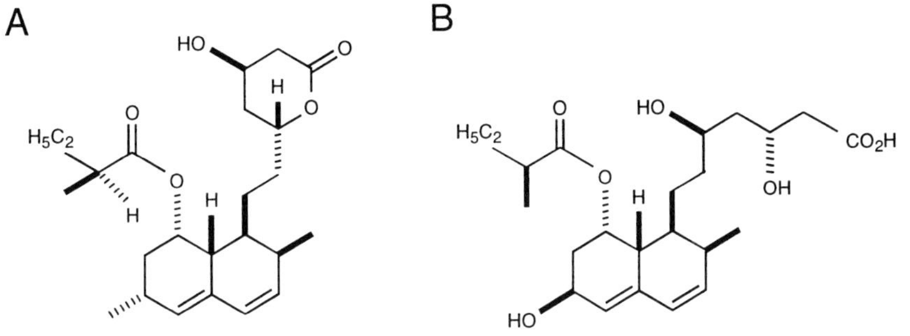

Fig. 7.5. The structures of **A**: lovastatin and **B**: pravastatin, two statins with anticholesterolemic action and candidates to combination therapies.

Monoterpenes and isoprenoid intermediates (Table 7.1) are interesting candidates to combination therapies in which one (or more) component(s) inhibit(s) HMG-CoA reductase. Farnesol effectively induces apoptosis in leukemic T- and B-lymphocytic cell lines, and spares primary hemopoietic cells (Rioja *et al.*, 2000). Mo and Elson (1999) found that β-ionone, a pure isoprenoid, suppresses the growth of B16 melanoma cells, mammary adenocarcinoma (MC F-7), human leukemic (HL-60) and human colon carcinoma (Caco-2) cells. Both β-ionone and a mixed isoprenoid, γ-tocotrienol, suppress HMG-CoA reductase activity and interfere with post-translational processing of lamin B, an activity essential to the assembly of daughter nuclei. According to Mo and Elson (1999) this interference renders neosynthesized DNA available to endonucleases, resulting in DNA damage and apoptosis. The authors suggest that various isoprenoid constituents of plant products with their additive and potentially synergistic actions on tumours could be the basis of favourable effects of fruit and vegetable diets on cancer development. This may even be of importance for members of the same class of terpenoids. Buhagiar *et al.* (1999) discussed the advantages of using a mixture of monoterpenes as present in essential oil of the conifer *Tetraclinis articulata* over a single compound. To this we add that a more selective usage of natural terpenoids (included in the diet or separately administered) *in combination* with MVA depletion and other selective inducers of apoptosis such as FTase inhibitors of herbal origin (Kwon *et al.*, 1999) merits increased interest in the near future.

Ad (5). The different strategies to treat cancers as presented in Box 7.2 cannot always be combined unless a careful selection is made of the compounds belonging to a certain category. Inhibition of angiogenesis leads to depletion of oxygen, which is often concomitant with insensitivity to feedback regulation of MVA production in tumour cells (see a.o. 4.1.8, Elson *et al.*, 1999 and previous discussions in section 7.1.8). This is of importance to the choice of HMG-CoA reductase inhibitors if combined with novel inhibitors of angiogenesis. As far as natural terpenoids are concerned, the *in vivo* effects of ursolic acid have to be awaited. Two ginsenoide saponins, 20(R)-ginsenoside-Rb2 and 20(S)-ginsenoside-Rg3, that are effective inhibitors of lung metastases produced by B16-BL6 melanoma and colon 26-Ms.1 carcinoma in mice significantly inhibited adhesion of fibronectin and laminin and inhibited invasion of melanoma cells into the reconstituted basement membrane (Mochizuki *et al.*, 1995). Although the ginsenosides did not repress the size of the original tumour, they exhibit anti-angiogenesis activity in lung metastases. These are promising developments suggesting that terpenoids could be incorporated in treatments, which include anti-angiogenesis factors. Table 7.3 gives potentially effective terpenoids to inhibit angiogenesis.

Ad (6). The G-proteins, discussed in this book at several places, come again in the picture within the context of receptors of growth hormone releasing hormones (GHRH receptors). The GHRH receptor is a G-protein-coupled receptor which brings us back to olfaction (see 4.1.2) and the mode of action of repellents and deterrents of herbivores (5.1.3) and to the impact of messengers on G-proteins in general, as discussed in chapters 3, 4, 5 and 6. Kineman (2000) discusses the possible modes of action of these antagonists in a commentary paper on the effect of a GHRH receptor antagonist on the proliferation of human small cell pulmonary carcinoma lines. Upon binding to its receptor in the hypothalamus, GHRH activates the α-subunit, in this way stimulating adenylyl cyclase, resulting in increasing intracellular cAMP that binds to and activates a different set of subunits of PKA, releasing catalytic subunits of which the nucleus is the target. Upon arrival at the nuclear membrane another binding protein, CREB, is activated and stimulates growth hormone gene transcription. The growth hormone is released because binding of GHRH to the nuclear receptor also causes influx of cellular Ca^{2+}. In the target organ (e.g. the liver), the growth hormone can stimulate proliferation genes, including type I IGF-I receptors present in tumour cells, or maybe they increase the effectivity of locally produced growth factors. See also Frohman and Kineman, 1999.

Chapter 7. Natural terpenoids to the benefit of human health

Table 7.3. Inhibition of angiogenesis by terpenoids and their combinations

Terpenoids	Effects on angiogenesis
All-*trans*-retinoic acid	- inhibition of angiogenesis in mice with transplanted KB cells
	- inhibition *in vivo* of angiogenesis in squamous cell carcinoma
Lovastatin	- increase of anti-tumour activity of tumour necrosis factor-α against tumour cells transformed with v-Ha-ras oncogene via inhibition of angiogenesis
Retinoic acid + α-interferon Retinoic acid + 1,25-dihydroxy vitamin D3	synergistic inhibition of angiogenesis of: ♦ human squamous cell carcinoma ♦ oncogenic Human papilloma virus type 16 OR 18 ♦ mammary carcinoma T47D ♦ vulval carcinoma A431
Retinoids Interleukin-12 1,25-dihydroxy vitamin D3	- synergistic action against angiogenesis in BALB/C mice with human tumour cell lines
Ursolic acid Oleanolic acid	- inhibition of angiogenesis in chick embryo chorioallantoic membrane and of proliferation of bovine aortic endothelial cells
Vitamin E (α-tocopherol)	- inhibition of angiogenesis and tumour growth factor-α in hamster carcinogenesis
β-carotene	- inhibition of angiogenesis in HPV-transformed cell lines (SKv-t, HeLa) in mice
1-α-hydroxy vitamin D3 1,25-dihydroxy vitamin D3	- inhibition of angiogenesis in colon carcinogenesis - activity against micrometastases in murine renal cell carcinoma - inhibition of angiogenesis in the transgenic retinoblastoma mouse model
13-*cis*-retinoic acid + β-interferon	- synergistic action against angiogenesis of Kaposi's sarcoma

According to a valuable scenario of Kineman (2000) GHRH antagonists could inhibit these messenger cascades at several sites, some of which have already been elucidated, the first step being of course pituitary GH production. GHRH antagonists appear to inhibit proliferation of a number of human cancer lines *in vitro*, associated with a decline in tumour IGF-II production (for details see Table 1 in Kineman, 2000). From a plethora of studies on human tumour cell lines, which respond to GHRH, substantial evidence has been provided that GHRH antagonists block the action of locally produced GHRHs. It is intriguing that a.o. the proliferation of H-69 small cell lung carcinoma is inhibited by vasoactive intestinal peptide (VIP) and forskolin (the second being an activator of adenylyl cyclase, resulting in increasing cAMP levels) as well as by GHRH antagonists. Such an "independent" GHRH receptor antagonist is JV-1-36 and we expect that novel GHRH antagonists soon will be discovered.

Steroid hormones act as growth hormones, which rises the question of the relation between GHRH antagonists and modulators of steroid receptors. The polyunsaturated γ-linolenic acid may obtain a valuable contribution in cancer therapy as an antagonist of steroid hormone receptors. Kenny et al. (2000) found that mammary cancer patients who took 2.8 g γ-linolenic acid/day achieved a significantly faster clinical response to tamoxifen than patients treated with tamoxifen alone. Although further research has to be done the authors propose γ-linolenic acid as an adjuvant to primary tamoxifen (in endocrine-sensitive mammary carcinoma). This is another example of a combination therapy.

The role of calcitriol (1,α-25-dihydroxycholecalciferol) in the inhibition of colon and prostate cancer is as yet epidemiological (Smith et al., 1999). It is of great interest to study the effect of calcitriol on GHRH receptors and to compare these effects with those on the expression of vitamin D receptors in different types of malignant cells.

Ad (7). In section 4.1.7 we discussed the mode of action of taxol and its derivatives on cancer cells and its potential to interfere with mitosis of many cell types. In fact, taxol has a structure remembering certain vitamin D3 derivatives, which makes it interesting to investigate the effects of other vitamin D3 compounds alone or in combination with taxol. Koshizuka et al. (1998) examined the ability of a novel 20-epi-vitamin D3 analogue (CB 1093), cisplatin (CDDP = cis, diamino dichlorplatinum) and taxol alone or in combination to inhibit mammary cancer cells (MCF-7) implanted in immunodeficient mice. In the animals infiltrating poorly differentiated adenocarcinomas developed. Although both CB 1093 and CDDP were less effective than taxol, the combination of CB 1093 + CDDP + taxol was more potent than taxol alone. No additive effect was seen by combining only two compounds. There were no cross-reactive toxicities. The authors conclude that combination of taxol (paclitaxel) with the other two vitamin D3 compounds holds promise in the treatment of mammary tumours.

One of the recently elucidated effects of taxol is serine phosphorylation of Bcl-2. In acute myeloblastic leukemia ATRA also causes Bcl-2 phosphorylation. However, the phosphorylated isoform from taxol-treated cells was slightly longer than the native isoform and both phosphorylated and native Bcl-2 were present. Bcl-2 phosphorylated after treatment with taxol or all-*trans*-retinoic acid retains its capacity to dimerise with Bax (Hu et al., 1998).

Retro-retinoids (metabolites of vitamin A) affect F-actin reorganization, which is an early event in apoptosis triggered by anhydro retinol (AR). The effects can be reversed with retinol (vitamin A). Actin depolymerization and apoptosis is also inhibited by Bcl-2, indicating that cytoskeleton reorganization is downstream of Bcl-2-related processes (Korichneva and Hammerling, 1999). AR is a natural antagonist of retinol, which makes AR or its analogues promising compounds for combination therapies based on taxol.

Combination of taxol (paclitaxel) with an immunomodulator (AS 101) which in itself has anti-tumour effects (by activating Raf and MAPK) has a dose-dependent, synergistic action on B16 melanoma in mice (Kalechman et al., 2000). The combination activated C-raf1, MAPK, ERK1 and ERK2. This activation was essential for the induction of p21waf and apoptosis. Combinations of taxol and its derivatives with other medication based on the strategies of Box 7.2, in particular (1), (4) and (6) could enlarge its therapeutic window while limiting its side effects e.g. on oral and intestinal mucosa.

Ad (8). Monoclonal antibodies are rapidly developing into a new generation of anti-tumour substances. Their mode of action is not completely elucidated and may include an inhibition of signalling pathways and effector-cell mediated cytotoxicity or selective activation of apoptosis in cancer cells. Clynes *et al.* (2000) demonstrated that engagement of Fcγ receptors on effector cells is a dominant component of the *in vivo* activity of antibodies against tumours. Cell-mediated cytotoxicity is dependent on the FcγRIII molecule on effector cells. The inhibitory FcγRIIB molecule seems to be a potent regulator of antibody-dependent cell-mediated cytotoxicity. Clynes *et al.* (2000) demonstrated that mouse monoclonal antibodies as well as the humanized therapeutic agents trastuzumab (Herceptin®) and rituximab (Rituxan®) bind to the activating (RIII) and inhibitory (RIIB) antibody receptors on myeloid cells. There was no tumour arrest in mice deficient in RII receptors, and those deficient in RIIB receptors showed much more antibody-dependent cell-mediated cytotoxicity. These results indicate that an effective anti-tumour antibody should bind preferably to activation Fc receptors and minimally to the inhibitory receptors such as FcγRIIB.

Rituxan® (IDEC Pharmaceuticals, San Diego, and Genentech Inc., San Francisco) has shown single-agent activity in Phase I and II clinical trials against B-cell non-Hodgkin's lymphoma (NHL). There appear to be both direct and immune-related effects, although the latter are not the cause of temporary B-cell depletion (Maloney, 1999). According to this author the excellent activity of rituximab is very promising for usage with or following chemotherapy. Davis et al. (1999) likewise conclude after treatment of 31 patients with 375 mg/m^2 rituximab, four times per week (intravenous infusions) that rituximab outpatient therapy is safe and shows significant clinical activity in patients with bulky relapsed or refractory B-cell NHL. It should be noted that these were low-grade follicular lymphomas, considered incurable. There was a response of 43% of the patients. According to Kosits and Callaghan (2000) rituximab is efficacious and safe for recurrent or refractory B-cell, low-grade NHL. The experience obtained with rituximab so far makes it a candidate for a first choice, maybe in combination with other agents which are not sufficiently effective if applied as a monotherapy, of have never been tested on lymphomas. Of these, the triterpenic acids isolated from lichen species with their effect on lymphocytes could be a useful group (Ogmundsdottir et al., 1998).

Another interesting terpenoid-related polyphenol is curcumin, that causes growth arrest and apoptosis of BKS-2 immature B-cell lymphoma by down-regulation of a number of cell cycle progression promoting genes: egr-1, c-myc, Bcl-XL, NF Kappa B and p53 (Han et al., 1999). This is not NHL, but (a) in combination with rituximab curcumin may affect expression of one of these important genes in NHL, and (b) a combination of rituximab with curcumin might be more effective in the treatment of B-cell lymphoma.

Summarizing, terpenoids can be employed in all major strategies to combat cancer (Box 7.2). In nature, as has been shown in chapter 5.1, chemical warfare is often based on multicomponent defense, in which the contribuants have different modes of action. That is why sections 6.2.2, 6.2.3 and chapter 7.1, although dealing with different disciplines, are closely related. The best possible usage of terpenoids in medicine may be in *multicomponent treatments*. At present, some terpenoids like ursolic acid, retinoids, vitamins A and E, D-limonene (and lovastatin, used in the first place to treat cholesterolemia) are applied in so-called "alternative medicine". This is a terrible denomination, not unlike alternative pest management, that once heralded modern integrated pest management (IPM) systems, based on the integration of different methods. The only alternative property of some of these terpenoids may be that they can act (apart from the targets mentioned in Tables 4.1 and 4.2 and their modes of action discussed in the present chapter) on metalloproteinases (Table 7.4), discussed in chapter 2. Metalloproteinases are early enzymes involved in the evolution of terpenoid production. To continue our discussion, a good medicine is *never* alternative, but it can be utilized as an additive treatment, or it may be integrated in phase-directed medication, in which the subsequent phases of a disease (cancer) are treated with the most adequate (combination of) medicines. As there is a role for terpenoids in IPM, so could they be recruited for multicomponent and phase-specific treatments, to which we propose the terminology Multi-Step-Treatment, or **MST**. Medication based on only one ingredient of an MST treatment will most probably give disappointing results and will only increase the number of those opposing unconventional cancer treatments (Ernst and Cassileth, 1999).

Table 7.4. Terpenoids with inhibitory action on metalloproteinases *

Terpenoids	Mode of action
All-*trans*-retinoic acid	- modulation of human melanoma cell invasion and proliferation by matrix metalloproteinase-1 inhibition and transforming growth factor-β inhibition
	- transcriptional repression of the human collagenase-1 (MMP-1) gene in MDA231 mammary cancer cells
Panaxadiol	- induction of the down-regulation of matrix
Panaxatriol	metalloproteinase-9 in a highly metastatic HT1080 human fibrosarcoma cell line
Ursolic acid	- induction of the down-regulation of the matrix metalloproteinase-9 gene in HT1080 human fibrosarcoma cells
Vitamin E (α-tocopherol)	- inhibition of protein kinase C together with matrix metalloproteinase-1 (collagenase) in human skin fibroblasts

* Compare with Tables 4.1, 4.2 and Fig. 7.9.

There may be a role for terpenoids in MST in the case of drug resistance. This is a well-recognized problem in cancer therapy. Cancer cells can become resistant to drugs for a number of reasons, an important one being the expression of P-glycoproteins that swipe the drugs from a cell. According to Blagosklonny (1999) this is not necessarily a disadvantage. Following a "two-drug" strategy, the first drug may be ineffective against the target cells, as long as it is a substrate for MRP or PgP pumps. In addition, it can be cytostatic but of course not completely cytotoxic. That is the task of the second drug that is not effective if used as monotherapy. Blagosklonny (1999) mentions combinations of adriamycin and etopoxide as first drugs with paclitaxel or epothilones as second one to treat leukemia. Moreland *et al.* (1999) showed that aphidicolin increases the sensitivity of mismatch repair deficient cell lines to two anti-tumour drugs, a process that is denominated "sensitizing". The antisense strategy has recently been developed against Polo-like kinase1 mRNA (Elez *et al.*, 2000).

The different effect of low and higher doses of agents affecting cell cycle progression have already been mentioned in this section. This is not only true for terpenoids. It does not come as a surprise, that Kong *et al.* (2000) report that natural phenolics and isothiocyanates at low doses *stimulate* MAPKs (mitogen-activated protein kinases), ERK2, JNK1, and p38 leading to expression of survival genes (c-fos, c-jun) and those expressing detoxifying enzymes. At higher concentrations these substances activate caspase cascades, resulting in apoptosis. This is of great importance not only for the dose-response relations in combination therapies, but also for treatments with additional dietary factors. Isoprenoid vitamins like vitamin A, D3, E and K2/K3 can mostly be provided in adequate dose, but this is much more difficult for dietary factors such as curcumin, eugenol, genistein, lycopene, D-limonene, Ω-3-fatty acids and ursolic acid, to stay with terpenoids. Some of these plant-derived terpenoids act synergistically, often not more than additive and in those cases one must make a choice between the risk of advising too low doses that may result in the *activation* of survival genes and possibly detoxification of the main cytostatic used, and leaving possible effective synergistic combinations unused.

Although our understanding of the mode of action of terpenoid antitumour compounds is rapidly increasing, there is also confusion - FTases perchance acting on Rho proteins and hardly on Ras (!) (Prendergast, 2000) - and as yet advising the optimal combination or MST therapy is like heading for treasure island in mainly uncharted waters. To our opinion this situation will change erelong. First, Ras and Rho are closely related, and the search for FTases acting on Ras continues with great energy (Frame and Balmain, 2000). Second, to our relief, we realize that we are not in an isolated position with regard to our vision on multicomponent treatments. Lackey *et al.* (2000) explored interactions between different classes of anti-tumour compounds, including anti-estrogens, retinoids, tyrosine kinase inhibitors and possible roles of monoterpenes, and in which way their synergistic interaction could be exploited in cancer treatment. Thanks to our improving knowledge of the effect of dietary components on carcinogenesis this also holds for dietary components (Manson *et al.*, 2000).

We hope that a.o. chapters 3.5, 4.1, 5.2 and 7.1 and the selection of references therein contribute to a multidisciplinary approach of cancer research and development of novel therapies. We also hope that the unfruitful discussions on the validity of dietary treatments, adjuvants, psychotherapeutics, etc. will discharge into a scientific approach as proposed by Shureiqi *et al.* (2000) based on a multistep clinical assessment of chemopreventive and chemoprotective efficacy.

7.2 Cosmetics

Practical aspects of anti-microbial effects of essential oils have been dealt with in section 7.1.1, with reference to section 4.2.1. Since ancient times essential oils have been included in balms, ointments, lotions, perfumes and frankincense. Their antibiotic, analgesic and stimulating effects contributed to their usage in cosmetics, while perfumes acquired their own position. Because there is little to see and much to smell they obtained a divine effluence and were very expensive. In Western Europe they were imported via the so-called frankincense route, running from Oman to Palestine with a passage to the mysterious city of Petra in South-Jordan. Many odorous compounds were distributed from there to the Roman and later to the Christian World, were they were reluctantly accepted. Once introduced, the delightful essential oils conquered the Greek, Roman and Christian civilisations and attained their status of luxury they still have. At excessive festivities wealthy Romans used to have perfumed pigeons flying through the rooms.

The present volume is on the function of natural terpenoids as messengers; in cosmetics terpenoids act as messengers in perfumes (sex attractants) and dermatological preparations (pharmacological effects).

7.2.1 Sex attractants

This book has a serious character and in fact does not deal with terpenoids utilized "just for fun", however tempting this may be. There exists substantial literature on this subject, covering the interests of both professionals and amateurs. The hidden functions as human semiochemicals, however, should not be neglected.

In "The Perfumed Garden", an Arabian treatise, probably written between 1394 and 1433 and brought into a comprehensive form by Shaykh Nefzawi and translated into English by Sir Richard Burton (ed. Walton, A.H., 1964) there are many recipes to increase the happiness of men and women, including cosmetics. We cannot resist in quoting one concerning terpenes: "If a woman wants to disappear bad exhalations from her vulva, she must pound red myrrh (probably *Commiphora myrrha*), then sift it and kneed the powder with myrtle-water (probably *Myrtus communis*) and rub her sexual parts with this wash. All disagreeable emanation will disappear". Resins of *Commiphora* species could have interfered with vaginal yeast infections, such as *Candida* spp. To procure additional astringent effects extracts of the walnut tree were used (juglone!).

The use of perfumes as sex attractants is as old as Cleopatra, who received the Roman statesman Marcus Antonius on a ship with perfumed sails, being massaged herself with a delicate, contrasting fragrance. Apart from resins these perfumes included essential oils of plants mentioned in Table 6.2, enriched with Nardus oil from India and cinnamon originating from China.

In the world of perfumery the terminology is somewhat different from ours. The ingredients of a perfume are denominated "notes", and these may consist of a number of components. A perfumer balances between art and science (many components are synthetic odours these days). Those who want to know more of the seductive powers of perfumes will find good reading in Newman's essay "Perfume, the essence of illusion" (1998).

In the British Museum in London one can have a look at the "beauty box" of Queen Thuthu († about 1400 BC), once placed in her sepulchral chamber and accommodating small containers for oils, ointments and balms. Plinius Secundus Major (23-79) describes precious perfumes developed by the Greek, a.o. by Megaleion, who composed a cosmetic oil based on wild roses (probably *Rosa gallica officinalis*), crocus and cinnamon. Apart from their effects on human behaviour essential oils were utilized to clean the skin before the process of soap-making was invented. These mixed functions of cleaning and perfuming made essential oils, according to the Greeks, a gift of Aeone, a nymph of Venus. The almost ritual usage of essential oils dissolved in neutral plant oils has often been depicted (Fig.7.6).

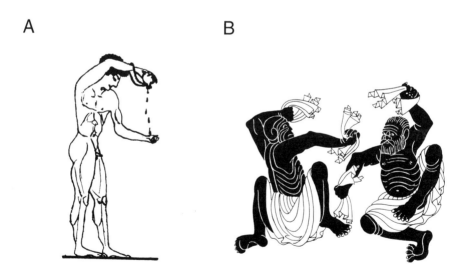

Fig. 7.6. Terpenoids for sports and spirit. **A**: A Greek athlete (500 BC) uses aromatic oil for skin care; **B**: The hallucinogenic effect of *Cannabis*, used by medieval Asiatic monks for ritual purposes maybe partly based on tetrahydro cannabinol. Other cannabinoids, terpenoids and flavonoids can reduce this effect.

Because the different contribuants of a perfume do not have the same degree of volatility there would be a rapid change in composition upon contact with the skin. Therefore, a fixative, neutral substance (formerly mostly musk) is added. Perfumes applied as sex attractants can be composed in such a way that longer lasting notes have a different effect than the shorter ones – leaving exotic fragrances like rose, sandal-wood, cedar, cade and balsams.

7.2.2 Dermatological preparations

There is no clear boundary between cosmetic and medicinal terpenoids. An example is the anti-microbial and anti-inflammatory effects of terpenes discussed above. Small burns of the skin (a hand hold above a candle) can be treated with chamomile extracts (after the cold-water method). Even better is an ointment prepared with azulenes extracted from *Matricaria chamomilla* and *Achillea millefolium* containing chamazulene and bisabolene dissolved in a plant oil. The azulenes accelerate repair of the epithelium (also for cats and dogs). As a rule, the teratogenic effects of terpenoid alcohols are less than that of the aldehydes. This holds for cinnamyl alcohol, citronellol, farnesol, geraniol, nerol, etc. In some cosmetics empyeumatic oils are used, obtained by dry distillation of wood. These oils may cause skin irritation and their content of benzopyrenes can reach critical levels. Bouhlal *et al.* (1988) report that this problem is considerably less when normal (hydro)distillation techniques are applied. As an example they mention the essential oil of *Juniperus oxycedrus* (cade), that has a pleasant odour and is safe when obtained via steam distillation or solvent extraction.

Several patents have been registered on fixatives and preservatives of essential oils for cosmetics. They have no messenger functions, so we refer to the appropriate bureaux in respective countries. They range from glycerol, paraffin oil, methylcellulose, fatty acid derivatives and lactoferrins to more solid materials such as ground chalk and talcum. In section 7.2.2 we briefly explore the human skin from feet to scalp:

Foot-cares commonly comprise tannin, tannic acid or gallotannic acid as astringents, to which one could add the bark of the walnut-tree (juglone). Formol ($\approx 2\%$) is the basic disinfectant, sometimes replaced by acetyl-salicylic acid ($\approx 1\%$). This composition may be enriched with lavender oil (1-2%), camphor (0.1-0.15%), menthol (0.2-0.3%) and oil of citrus fruits (0.3-0.5%).

There are some anecdotal reports on the beneficial effect of 1,8-cineole on psoriasis; maybe it is of greater importance that 1,8-cineole greatly increases skin penetration of polar compounds by interaction with lipids of the stratum corneum (Williams and Barry, 1991). In Western Europe a cream based on flowers of *Arnica montana* (Asteraceae) is reputed for centuries against bruises and wound infections. A tincture of this herb was already used by Paracelsus' father, who practiced first aid in the 15th century as Master Wilhelm von Hohenheim, close to the town Einsiedeln in Switzerland. *Arnica* species contain a.o. sesquiterpene lactones, of which helenalin (Fig. 7.7) effectively reduces oedemas. Creams based on a tincture of the roots of *Potentilla tormentilla* (Rosaceae) can be applied to treat epidermal clefts. Apart from tannins, the roots are rich in essential oils.

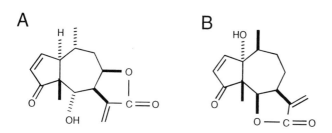

Fig. 7.7. The structures of **A**: helenalin and **B**: parthenin, two related sesquiterpene lactones present in *Arnica* spp.

Bisabolol, like azulenes, extracted from chamomile seems to be effective against acne. Isoborneol, a monoterpene present in many herbal oils showed activity against herpes simplex virus 1 (HSV-1). It causes immediate inactivation, whereas 0.06% completely inhibits viral replication (Armaka *et al.*, 1999). It probably inhibits the glycosylation of viral polypeptides. This could be an excellent ingredient to treat herpes simplex labialis. A Russian preparation for the same purpose is based on the oil of *Achillea millefolium*, mentioned above and this could be due to azulenes (Kumanova *et al.*, 1989).

An informative book on do-it-yourself cosmetics, based on many herbs containing essential oils is by Neuhold (1997). To our own experience nice after-shaves can be composed with essential oils of coriander as a fixative (Table 6.1), and a variation of lighter or heavier notes, ranging from citrus oil, *Alchemilla vulgaris* (alchemille), rosmarin or basil to *Juniperus* spp., vanilla, nutmeg or clove (traces!) and olibanum.

The preservative properties of those cosmetics which are low in antimicrobial oils can be improved by the essential oil of thyme (3%), that renders the product resistant to a number of bacteria, but less so to yeast like *Candida albicans* (Manou *et al.*, 1998). Dandruff is often caused by a fungal infection. There exist (non-terpenoid) anti-dandruff shampoos for prolonged treatment; with respect to lotions the classical approach is an alcoholic solution containing extracts of the roots of *Lappa major*, *L. minor* (Asteraceae), burdock and salicylic acid.

Rosmarin oil (Tables 6.1. and 6.2) is a component of hair-stimulating lotions. Lotions for prevention of hair loss often contain terpenoids from rosemary and birch with camphor and fenugreek (apart from stinging nettle). Of course there are lotions containing estrogens, vitamin A, even retinoids, to stay with terpenoids, but these are not free of risks. We finish this section with a recipe of the famous French surgeon baron Guillaume Dupuytren (1777-1835). Fill a 500-ml bottle halfway with wood shavings of choice (birch, pine, cade) and add ethanol 50% until full. Close the bottle and put it on a sunlit bench. Shake regularly. Another bottle of 100 ml is filled with two tablespoons of dried rosemary and ethanol 90%. A third bottle is used to make a tincture of a teaspoon ground nutmeg and 25 ml ethanol 96%. After two weeks the liquids are filtered and put together in a well-closed bottle. The extracts are to be thoroughly mixed. A few droplets are to be massaged into the scalp every day. In the 1960s the senior author of this book added some diethyl stilbo-estrol to this solution for male use, but great care should be taken with steroid hormones, especially in combination with UV radiation. In the original recipe Dupuytren used vodka in the first bottle, but the only reason for this probably lays in a certain date: 1812. This, we think, says enough.

7.3 Terpenoid analogues and derivatives

In the preceding part of chapters 6.2 and 7.1 a number of terpenoid analogues and derivatives have already been discussed. This is inevitable because these compounds contribute to our understanding of the mode of action of natural terpenoids. Some terpenoids that occur in nature are produced synthetically, as mentioned at the beginning of section 6.2.2. The same holds for analogues and more complicated derivatives and as this book deals with natural terpenoids we keep this section concise and focus on agriculture and medicine.

7.3.1 Applications in agriculture

An extended section of the proceedings of a Conference on Insect Chemical Ecology was dedicated to the application of pheromones and juvenoids with many contributions on synthetic compounds (Hrdý, ed., 1991). In section 2 of these proceedings phytoecdysteroids and their derivatives are discussed, section 3 deals with the prospects of practical application of pheromones, juvenoids and their analogues. Synthese routes for perillenal, ipsdienol, (E)-thujan-4-ol, including a two-step iterative method for chain elongation are described by Baeckström *et al.* (1991). Chapter 4 of Harborne's "Ecological Biochemistry" (1993) gives a survey of natural and synthetic plant estrogens and ecdysteroids.

Monofluoro analogues of methyl eugenol (fluorine to replace hydrogen at the 3-C site) have been developed and tested as an attractant to males of *Bactrocera dorsalis* (Diptera), the oriental fruit fly, by Liquido *et al.* (1998). (E)-1,2-dimethoxy-4-(3-fluoro-2-propenyl) benzene was as attractive to both laboratory-raised and wild fruit fly males. This compound showed reduced toxicity and intrachromosomal recombination in yeast compared to methyl eugenol. As appears from sections 5.1.3, 5.1.4 and 5.2.3 it is a long way from allomones to kairomones. When replacing the methyl group by 3-fluoropropyl does not induce neurological adaptations, a long agricultural life is to be expected.

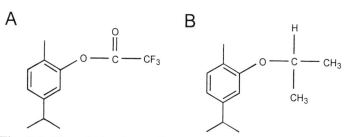

Fig. 7.8. Two derivatives of carvacrol with insecticidal action. **A**: carvacryl trifluoroacetate; **B**: carvacryl isopropyl ether.

Derivatives of monoterpenes are relatively easy to make, but one needs good bioassays to test them. Tsao *et al.* (1995) synthesized a number of acyl and ether derivatives of well-known monoterpenes. Catalyzators were pyridine in the case of acyl, starting with monoterpene alcohols and acids, and benzyltributylammonium bromide in the case of phenolic monoterpenes. A different method was used for ether derivatives of aliphatic alcoholic monoterpenes. Acyl derivatives had a significantly improved acute, fumigant, larvicidal and ovicidal effect against various insects, among which *Musca domestica*, the housefly and *Diabrotica virgifera virgifera* (western corn rootworm), but were less successful in other cases (Fig. 7.8A). The improvement of ether derivatives was limited. Carvacryl isopropyl ether (Fig. 7.8B) showed greatly improved fumigation activity against *Blattella germanica*, the German cockroach. The greater lipophylicity of esters and ethers in comparison to terpenes might contribute to more rapid penetration.

ethers in comparison to terpenes might contribute to more rapid penetration. Hydrolyzation in the insect body could reduce their structure to the natural compound, although a different effect on e.g. cyt. p450 monooxygenase, involved in the biosynthesis of insect hormones may be considered.

We suggest that bioassays of newly developed terpenoid derivatives include phytotoxic effects. As mentioned earlier, phenolic terpenoids tend to be potent fungicides and insecticides, but they are often more phytotoxic, and camphor, 1,8-cineole and pulegone are considered to belong to the most phytotoxic monoterpenes, but one should also be careful with L-carvone, eugenol and thymol (Lee et al., 1997).

7.3.2 Applications in medicine

What has been said for insect control also holds for medicines based on natural terpenoids. In the previous sections attention has been called for many analogues and derivatives. Here some interesting developments are shortly referred to, following the items of Box 7.2. This section on analogues is in close connection with section 7.1.8. To support our view, Fig. 7.9 gives the possible application of natural terpenoids and their derivatives to attack cancer cells. It can be compared with the *reversed* version of Fig. 6.5; the arrows now pointing inwards.

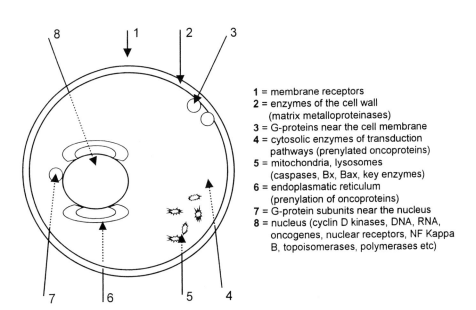

Fig. 7.9. Possible utilization of natural terpenoids and their analogues on targets positioned at different levels of a tumour cell.

Ad (1). End (1999) predicted a rapidly increasing number of FTase inhibitors. Many recent studies deal with analogues of natural FTase inhibitors (or should we say: RhoB inhibitors? (Prendergast, 2000)). Eummer et al. (1999) evaluate novel limonene phosphate and farnesyl-PP analogues, some of which have no effect at all. Donaldson et al. (1999) present the promising compound XR3054, structurally related to limonene, that inhibits Ras farnesylation and MAP kinase activation and inhibits proliferation of prostate and colon carcinoma cell lines, but is relatively inactive against a panel of mammary carcinoma cell lines. Zujewski et al. (2000) tested the FTase inhibitor R115777 on 27 patients with advanced cancer. The recommended dose for Phase II trials is 2 x 500 mg/day for 5 consecutive days, followed by 9 days of non-treatment. The preparation could be applied in metastatic colon cancer and may be combined with other chemotherapeutics. Other studies involve S-farnesyl thiosalicylic acid, mainly acting on Ras membrane association, and canventol, that inhibits protein prenylation and may act more specific than the sarcophytols, natural terpenoids from soft corals (see 7.1.8).

Ad (2). A synthetic thioether analogue of vitamin K (2-(2-mercaptoethanol)-3-methyl-1,4-naphthoquinone) or compound Cpd5 has potent inhibitory activity against several cell-cycle-relevant protein tyrosine phosphatases (PTPases). Cpd5 inhibits PTPases by sulfhydryl acylation and dephosphorylation of specific proteins, blocking kinase activity of these proteins. Human hepatoma cell lines reacted with growth inhibition (Nishikawa et al., 1999). The main target of Cpd5 seems to be Cdc25 phosphatases, that carry cysteine in the catalytic site (Tamura et al., 2000).

Aphidicolin is a fungal-derived tetracyclic diterpene antibiotic. Cinatl et al. (1999) evaluated the anti-tumour effects of a water-soluble ester aphidicolin glycinate on established human neuroblastoma xenografts in nude mice. Systemic intraperitoneally administration (60 mg/kg, 2 times/day, 10 consecutive days) suppressed tumour growth but did not cure the disease. Intratumoral injections (40 mg/kg, 2 times/day, 4 days) resulted in complete tumour regression. The animals remained free of tumour tissue for more than 90 days, and no tumour cells were microscopically detectable. This makes aphidicolin derivatives promising for further evaluation as a treatment for neuroblastoma.

Ad (3). A synthetic derivative of all-*trans*-retinoic acid, N-(4-hydroxyphenyl) retinamide (4-HPR) is a potent activator of transforming growth factor-β (TGF-β) in mammary carcinoma cells and induces arrest of DNA synthesis and apoptosis (Herbert et al., 1999). Human mammary cancer cells defective in TGF-β-signalling were refractive to 4-HPR-induced apoptosis.

A problem with retinoids is that some tumour cells fail to complete differentiation to a level responding to apoptotic signals. In such cases combinations of apoptosis-inducing drugs acting through different receptors and signalling pathways could be effective. Elstner et al. (1996) combined a potent 20-epi-vitamin D3 analogue (KH 1060) with 9-*cis*-retinoic acid to test their effect on HL-60 leukemic cells and found that it decreased Bcl-2 expression and induced apoptosis. Upon treatment the cells developed a myelomonocytic phenotype.

Tocoretinate, an ester bound retinoic acid, and tocopherol has been applied for skin ulcers but also for sclerotic skin cancers. Tocoretinate would seem to down-regulate the tensile tension of fibroblasts in collagen gel matrices by inducing tenascin-C expression. Tenascin-C is involved in cellular detachment and tissue remodelling. Tocoretinate could be topically applied for sclerotic skin diseases.

Ad (4). Apomine (SR-45023A) is a new representative of the family of cholesterol synthesis inhibitors. In comparison to other HMG-CoA reductase inhibitors apomine-induced growth inhibition of a number of tumour cell lines was not reverted by MVA, which it has in common with farnesol (Flach et al., 2000). Moreover, apomine and farnesol induced caspase-3 activity at concentrations similar to their IC50 values for inhibition of cell proliferation. Of a number of n-alkyl esters of gallic acid dodecyl gallate showed the most potent inhibition of cholesterol biosynthesis. It particularly inhibits rat squalene epoxidase (Abe et al., 2000). Dodecyl gallate probably binds specifically to the enzyme and scavenges reactive oxygen species required for the epoxidation reaction. A farnesol derivative also acted as a potent inhibitor. This is a novel development in which esters of gallic acid are developed into effective depletors of MVA.

Ad (5/6). A new estrogen agonist with an activity of about an order of magnitude greater than genistein has been isolated from heartwood of *Anaxagorea luzonensis* (Annonaceae). Kitaoka et al. (1998) prepared various flavonoids with isopentenyl side chains in the A-ring and evaluated these on their ability to bind with an estrogen receptor. The presence of an 8-isopentenyl group is important for receptor binding. Movement of the isopentenyl group from position 8 to 6 resulted in loss of the activity. This study opens a new horizon for prenylflavonoids. Methoxyestrogens are a new class of drugs that can inhibit tumour growth and angiogenesis. Purohit et al. (2000) synthesized a novel methoxyestrogen, 2-methoxyestrone-3-O-sulphamate (2-MeOEMATE) and compared its ability to inhibit proliferation of mammary tumour cells with that of 2-methoxyoestrone (2-MeOE1). The new compound was more effective against MCF-7 mammary carcinoma cells (estrogen-positive) than 2-MeOE1, but was also effective against receptor-negative MDA-MB-231 cells. These cells rounded up and became detached, indicating apoptosis. In an *in vivo* study (rats), 2-MeOE1 had little effect on tumour growth whereas the same dose (20 mg/kg/day, 11 days) of 2-MeOEMATE resulted in almost complete regression of 2/3 of the tumours.

Ad (7). The 2-methoxyestrogens mentioned in (5/6) may exert an antimitotic effect on cells by stabilizing microtubules in a similar way as taxol. One of the effects of taxol (paclitaxel) is inhibition of glucose uptake. Singh *et al.* (2000) examined the effect of exposing MCF-7 mammary carcinoma cells to 2-MeOEMATE or paclitaxel for 24 h and found that these compounds inhibited glucose uptake with 50 and 22% respectively, and of insulin-stimulated glucose uptake with 36 and 51%. This could be an additional effect of 2-methoxyestrogens in the treatment of mammary carcinomas.

We hope to have demonstrated in this chapter that those working at very different aspects of terpenoids can learn much from colleagues of other disciplines. Plant physiologists, entomologists, biochemists and physicians would greatly benefit upon the introduction of multidisciplinary projects concerned with natural terpenoids.

Chapter 8

Prospectus and suggestions for further research

8.1 Search for new bioactive terpenoids

There is a search for natural bioactive terpenoids by scientists of several disciplines. Biologists seek for bioactive essential oils and other plant-derived terpenoids, biochemists try to elucidate biosynthetic routes and to reveal the structure and function of the corresponding enzymes, gene technologists trace the responsive genes. Biologists and pharmacologists organize expeditions to rain forests and other regions of the world in search for new compounds. Not everyone has to go a long way: Indian scientific journals regularly report the isolation of bioactive terpenoids not seen before. Many novel terpenoids are discovered by marine biologists. In chapter 4.1 and section 7.1.8 terpenoids with anti-tumour functions have been reported to originate from marine biology. Once their modes of action have been elucidated they can be tested for different purposes. The marine pulmonates *Trimusculus reticulatus* and *T. conica* (Trimusculidae) have a unique lifestyle. The adults are essentially sessile and feed by filtering a mucous net produced by subepithelial glands on the mantle and sides of the foot. Being sessile they cannot escape from predators such as seastars, and their way of collecting and ingesting food (mostly microalgae) makes that they have to cope with intruders of their habitat. There is little space in the intertidal zone and much competition. Manker and Faulkner (1996) investigated the distribution and function of secondary metabolites from trimusculid limpets and found two diterpenes highly concentrated in the mantle, foot and mucus of *T. reticulatus* (Fig. 8.1). These compounds, and related diterpenes isolated from *T. conica* were tested for their repugnant effect on a seastar, but appeared to be less active than the complete mucous when tested alone.

However, one of the diterpenes (Fig. 8.1A) was highly toxic to veliger larvae of a sabellarid tube worm that forms large reefs in the intertidal zone in proximity to colonies of *T. reticulatus*. Upon contact with this compound larvae of the tube worm fail to metamorphose, including the inability to shed provisional setae or rotating the larval tentacles anteriorly. These marine diterpenes offer an interesting comparison to the clerodane diterpenes discussed before (klein Gebbinck, 1999).

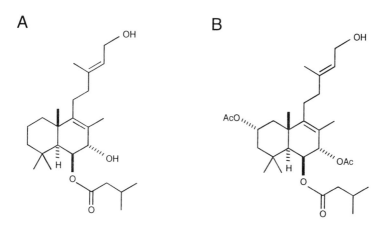

Fig. 8.1. Two diterpenes with possible defensive functions found in the marine pulmonate *Trimusculus reticulatus*. **A**: 6β-isovaleroxylabda-8,13-dien-7α,15-diol; **B**: 2α,7α-diacetoxy-6β-isovaleroxylabda-8,13-dien-15-ol.

The effect of terpenoids from *Artemisia annua* on malaria parasites has been mentioned earlier (sections 6.1.1, 7.1.1 and 7.1.4). Lee et al. (2000) have discovered novel artemisin derivatives with anti-tumour effects (Table 6.1). Other *Artemisia* species produce different terpenoids in addition to those of *A. annua*, although those already known may be present in lower concentrations. Zheng (1994) isolated new bisabolene derivatives from *A. sieberi* and silphiperfolane terpene acids from *A. chamaemelifolia*. This is another example of novel terpenoids isolated from plant species related to species known to produce valuable terpenoids.

8.1.1 Incorporation of terpenoids into new methodologies

BIOLOGY AND AGRICULTURE

A better understanding of the mode of action facilitates the selection of useful terpenoids. Synthesis of model compounds, followed by physiological (and in case of animals, behavioural) studies can reveal the active group and

Chapter 8. Prospectus and suggestions for further research

provide information on its selectivity. Simmonds and Blaney (1996) compared the mode of action of the complete azadirachtin molecule and a model compound containing the basic decalin fragment. Klein Gebbinck (1999) synthesized a number of model compounds based on the clerodane skeleton as present in azadirachtin and found that bioactive groups are present in both halves of the molecule. Some more simplified models were even more potent antifeedants than fragments more closely mimicking the corresponding substructure present in the original azadirachtin molecule. Many compounds were active against larvae of *Pieris brassicae*, the large cabbage white butterfly, but less than the natural molecule. Especially the presence of a C-3a hydroxyl group, although expected (klein Gebbinck, 1999), did not increase the activity of the models. The effect of the models on other insect species was even less.

Oda *et al.* (1995) used an interface bioreactor for microbial oxidation of citronellol (Fig. 8.2A). This was performed on the interface between a nutrient agar plate and decane. Many yeast species end up with (S)-citronellic acid as the oxidation product (Fig. 8.2B). Some bacteria e.g. *Geotrichum candidum* accumulated very high concentrations of citronellic acid in spite of its toxicity. According to Oda *et al.* (1995) the interface bioreactor is superior to a dispersion system (organic-aqueous two-liquid phase system).

The group of Tunen in the Netherlands is involved in biotechnological production of carvone in limonene-biosynthesizing plants, like caraway. In addition to stimulation of limonene hydroxylase (e.g. via isolation of the gene that encodes limonene synthase), inhibition of limonene hydroxylase could result in a high yield of limonene, a valuable monoterpene in cancer research. A new technique of systemic treatment of plants with neem preparations is discussed by Duthie-Holt *et al.* (1999). Pine trees can be protected against *Dendroctonus ponderosae*, the mountain pine beetle, by application of an emulsifiable formulation of neem seed extract (20 x 10^3 ppm) into a basal axe frill around the root collar of lodgepole pines. The authors expect little chance of resistance development both because of the multiple effect of azadirachtin and the fact that only few trees would have to be treated in a given year. In sub-Saharan Africa a big international project has been initiated for the use of neem as a source of natural pesticides and other useful products (Saxena, 1999).

Most probably the use of (terpenoid) biopesticides will increase the need for surfactants and other additives. Many terpenoids need surface-active agents for wetting, emulsifying and dispersing. A new generation of modern surfactants is based on vegetable oils condensed with glucose units. Modern surfactants have low toxicity and are biodegradable. An extensive report on

A

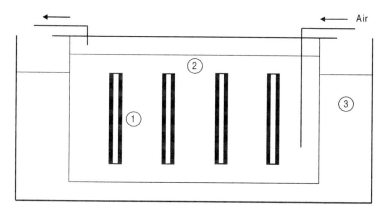

① = agar plates with immobilized micro-organisms
② = reaction solvent (decane with a reagent)
③ = thermostat

B

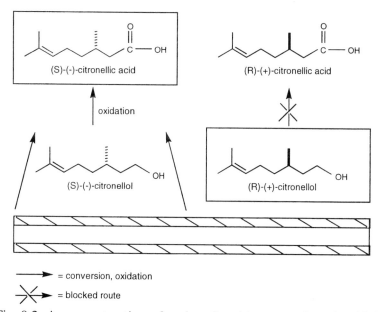

⟶ = conversion, oxidation
⤳̸ = blocked route

Fig. 8.2. **A**: reconstruction of an interface bioreactor for microbial oxidation of citronellol; **B**: enantio-selective oxidation of racemic citronellol into (S)-(-)-citronellic acid.

Chapter 8. Prospectus and suggestions for further research 343

surfactants and other additives in agricultural formulations is by Hewin International (1999).

MEDICINE
Our understanding of the role of natural terpenoids in cell cycle progressing is continuously increasing. They may become integrated in three compartments involved in induced cell death: the plasma membrane, the mitochondrion and the endoplasmatic reticulum (ER). The plasma membrane-associated caspase cascade is now known to be initiated by initiator caspase-9, the cyt. c/mitochondrial cascade by initiator-caspase 8, and as Mehmet (2000) depicts in his essay on caspases there is an ER-associated procaspase, denominated caspase-12. It would seem that caspase-12 is procaspase and caspase in one molecule, activated by Ca^{2+} as a pro-apoptotic messenger. It remains to be awaited if activated caspase-12 acts indirectly, e.g. on caspase-3, directly as an amyloid β-peptide, or both. The specific effects of terpenoids on different phases of the cell cycle as discussed in chapter 4.1 and section 7.1.8 open possibilities for treatment of two ailments at a time. An example would be the treatment of Alzheimer's disease with inhibition of gene expression of procaspase-12, in combination with e.g. FTase (or RhoB) inhibitors of anti-apoptotic proteins of cancer cells.

Novel terpenoid derivatives can be integrated in new generations of medical treatments. The farnesol derivative S-(E,E)-farnesyl thiosalicylic acid does not act as an FTase inhibitor, but dislodges H-ras from cell membranes and K-ras 4B, leading to degradation of Ras and inhibition of MAPK activity (Elad *et al.*, 1999). Farnesyl thiosalicylic acid can target K-ras 4B (12V) that is insensitive to FTase inhibitors, which speaks for its potential as a novel anticancer therapeutic.

Chemical synthesis of the complete taxol molecule is a tedious process with many difficult steps. Therefore, alternatives with a somewhat simpler structure, but with similar activity are most welcome. Service (1996) reports the synthesis of epothilone A and discodermolide, two terpenoids with taxol properties: binding on the protein target of microtubules. Strains of the fungus *Pestalotiopsis microspora* occur in *Taxodium distichum*, the bald cypress. They live as endophytes in the bark, phloem and xylem. Many produce taxol that was shown to be identical to the taxol produced by the trees (Li *et al.*, 1996). The fungi form a genetically unstable group, but their genetical flexibility may be used to adaptation to other host plants and to facilitate production of taxol or active derivatives. The fungus is also associated as an endophyte with *Taxus wallachiana* and isolates produce taxol in liquid culture. They also make large amounts of ergosterol. By selectively inhibiting sterol biosynthesis (MAD route) taxol production can be greatly increased (Li *et al.*, 1998). The authors explain this by assuming that the geranylgeranyl-PP pool is directed towards taxol synthesis.

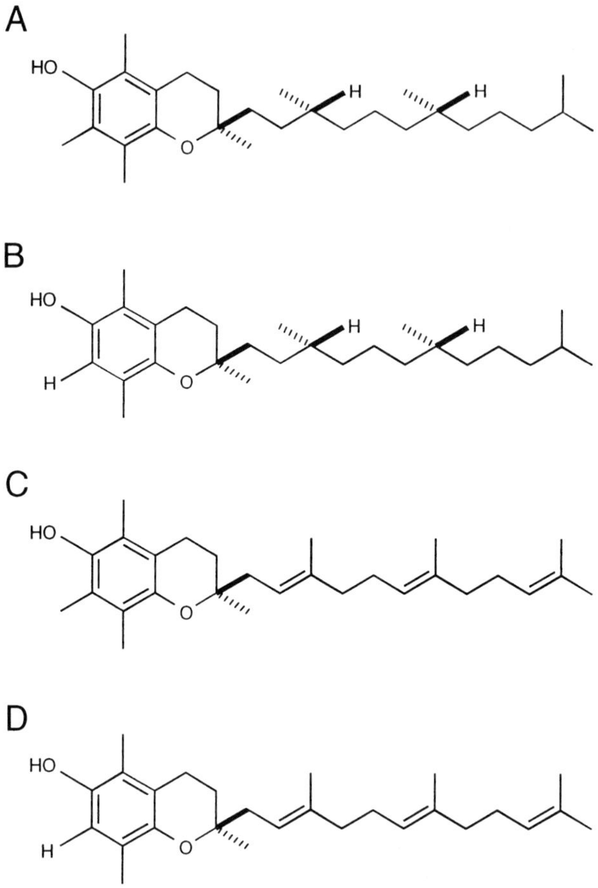

Fig. 8.3. Structures of tocopherols (vitamin E isomers) and closely related tocotrienols. **A**: RRR-α-tocopherol; **B**: RRR-β-tocopherol; **C**: α-tocotrienol; **D**: β-tocotrienol.

Simple derivatives of vitamin E, such as vitamin E acetate and naturally occurring tocotrienols (Fig. 8.3) have shown to be effective inducers of apoptosis in human mammary tumour cells, especially estrogen-responsive cells (Yu *et al.*, 1999b). Tocotrienols have an unsaturated farnesyl sidechain. Palm oil is rich in tocotrienols which makes it interesting as a dietary factor to support chemotherapy. Guthrie *et al.* (1997) found that tocotrienols are also effective against receptor-negative MDA-MB-435 mammary carcinoma cells, whereas α-tocopherol, another component of palm oil, was ineffective. Surprisingly, combinations of tocotrienols, α-tocopherol and tamoxifen were found to act synergisticly. The best combination against MCF-7 cells (estrogen-responding) was a 1:1 treatment with γ- or δ-tocotrienol and tamoxifen. According to Nesaretnam *et al.* (1998) tocotrienols do not act via an estrogen-receptor-mediated pathway and do not increase levels of growth-inhibiting insulin-like growth factor binding protein (IGFBP) in MCF-7 cells, suggesting a different mode of action than that of retinoids. Where therapies with antiestrogens and retinoic acid fail, combination of tamoxifen and tocotrienols (with additional dietary palm oil) could be considered.

Companies producing biochemicals offer a rapidly increasing range of kits to study apoptosis (such as Alexis Corporation, Switzerland): antigens and antibodies, caspase substrates and inhibitors and specific reagents, e.g. sphingosine 1-phosphate, an active inhibitor of apoptosis (used to test compounds overruling survival factors).

In addition to its effect on malaria parasites, artemisin and its derivatives could act on an intriguing target in *Plasmodium falciparum*, a parasitic malaria-causing protozoan with MAI metabolism. In this parasite the MAI route of isoprenoid biosynthesis is probably located in the apicoplast (Jooma *et al.*, 1999b). Apicoplasts are related to plant plastids and appear to have been acquired by endosymbiosis of a green alga (Fichera and Roos, 1997). These authors stress that antibiotics, which are supposed to target protein synthesis in the apicoplast cause death of the parasite.

Artemisin, first empirically applied by Paracelsus, and its derivatives discovered by Chinese scientists (Li and Wu, 1998) are now widely utilized to treat drug-resistant malaria caused by *P. falciparum* and have shown to be particularly active against the apicoplastic Translationally Controlled Tumour Protein (TCTP) (Bhisutthiban *et al.*, 1999). The fact that there may be a relationship between MAI metabolism and human tumour protein homologues in *P. falciparum* brings MAI routes with their DOXP pathway into the spotlight of cancer research.

8.1.2 Future of terpenoids in registration requirements

AGRICULTURE AND FORESTRY

All over the world there is an impetus to develop and market environmentally safe plant protection products. Not only is there an increasing chemophobia among users and crop consumers, accelerated re-registration in a.o. The United States and Europe will lead to withdrawal of a large number of pesticide products, a process that is already ongoing.

Terpenoid semiochemicals can be used to monitor and control many herbivorous insect species (sections 5.1.1 and 6.2.2). However, due to the current regulatory status of pheromones and other semiochemicals, natural terpenoids require registration in many countries. Weatherston and Minks (1995) in their survey of global aspects of semiochemical regulation propose a structure/activity approach to expedite and harmonize pheromone regulation.

The registration procedures for biocontrol agents are unclear, differ enormously between countries and are often too time-consuming to be interesting to agrochemical companies, even more because semiochemicals still occupy a low market share. In some countries such as Chile even pheromone traps had to be registered. In an informative appendix Weatherston and Minks (1995) provide a world-wide classification and legislation of pheromones. Some lights are glowing in the distance: in Canada semiochemicals are not considered to be conventional insecticides. In Denmark, Italy, Switzerland and the USA a reduced (but not identical) set of data is required for semiochemicals.

It has been known for a considerable period that monoterpenes can become biodegraded by several routes in which oxidation processes are important (Rogers, 1978). In aerated basins both mono- and diterpenes were degraded. Resins can be destroyed by yeast and fungi. For semiochemicals, released in the air, there is photo-oxidation and biodegradation upon return to the earth in wet and dry depositions. It is somewhat amusing to experience the concern of those who regard terpenoid semiochemicals as a form of air pollution. The annual global emissions of forest CBSCs (carbon-based secondary compounds) alone is estimated as 175×10^6 tons! Such emissions cause the blue haze, typical in forest areas on a hot summer day, which is initiated by photochemical reactions between terpene hydrocarbons and ozone and have inspired many artists to visualize the illusion of distance.

The introduction of terpenoid pheromones, allomones and kairomones in IPM programmes has been discussed before (sections 6.2.1, 6.2.2 and 6.2.3) with special attention to the "translation" of laboratory data into field expertise. Way and van Emden (2000) opine that "priority should be given to application of the right kinds of applied ecological and associated behavioural work in real situations in the field. At present, the balance is wrong, with too high priority given to fashionable technologies". As these authors correctly state: the IPM tool box has never been fuller.

The future for terpenoids in IPM lies in two opposing fields: (1) that of specific semiochemicals, particularly with coleopteran and hemipteran pheromones (lure-and-kill, and repellents effective in specific plant-insect relations), and (2) in crops with a large pest complex, in which neem products have already shown to be an effective, yet environmentally safe application of terpenoids. Terpenoids increase the window of multi-component host plant resistance. In addition, they may give a wink to classic, broad-spectrum insecticides of which the environmental stress can be greatly reduced in lure-and-kill systems (terpenes being the lure).

PHARMACOLOGY

Drugs derived from the herbal category are now an increasing field of interest, probably because their active constituents become identified and their modes of action elucidated. The period of about 1960-2000 was not the most favourable one for herbal pharmacognosy. Synthetic and upgraded drugs took over what centuries of empirical knowledge had brought together and resulted in a dimmed interest in herbal medicines, particularly in the USA and Europe. In the 1950s students of medicine and pharmacology at the University of Amsterdam could be observed drawing and studying plants for several hours per week, including a course on "Microscopy of pharmacological products". The following generation seems to have forgotten that many backyards and gardens harbour plants, which can eradicate complete families – let alone utilize them for whatever purpose. But times change, and in 1990 the USA made a start with the regulation of herbs which still leaves a lot to be wished for – see e.g. The Herbal Regulatory Dilemma in "Herbs of Choice", by Tyler (1994), who also provides a Proposal for reform. The German Commission E monographs (more than 300) cover a great deal of the herbal remedies sold in Germany, and in other parts of Europe. It is the opinion of the authors of the present volume that the discussion of some of these remedies does not always reflect the degree of effectivity they are considered to have by experienced practicians, but it is now generally approved that it is not in the advantage of a pharmacologist or physician of today to turn a blind eye towards herbal medicines, in which the previous chapters have shown terpenoids to take an important position.

A promising new development is the use of betulinic acid as a chemotherapeutic agent in the treatment of neuroectodermal tumours, covering neuroblastoma, medulloblastoma, glioblastoma and Ewing's sarcoma cells (Fulda *et al.*, 1999). This terpenoid, mentioned earlier in this volume (e.g. 7.1.8), is a.o. present in *Prunella vulgaris* (Lamiaceae) (Ryu *et al.*, 2000), and in *Glycyrrhiza squamuloza* roots, closely related to *G. glabra*. Betulinic acid is a novel herbal mitochondrion-targeted cytotoxic drug. Whereas some conventional chemotherapeutic agents elicit mitochondrial permeabilization indirectly, in this way counteracting oncoproteins of the Bcl-2 family and supporting those of the Bax family, betulinic acid (and other experimental drugs) act directly on the mitochondrial membrane

(Costantini *et al.*, 2000). Gene therapy using a plant gene is a different new strategy with promising results in intracerebral gliomas (Cortes *et al.*, 1998).

8.2 A multidisciplinary approach

This book is the result of a multidisciplinary study, and we have been advocating multidisciplinary research on terpenoids from the first page. We have tried to point out the snags and pitfalls of superspecialization, although we are aware that today it is an inevitable phenomenon.

There are few branches of medicine in which the increasing knowledge of phytochemistry and molecular biology of terpenoids has required such a profound influence as cancer research. Still, there are many gaps that might be bridged by combined projects of different disciplines. The following section will give a few examples.

8.2.1 Bacteria, fungi, plants, insects and humans

Fig. 8.4 shows the pentagram the reader has seen before. How can specialists engaged in different biological aspects of terpenoids assist each other? There is a wealth of literature in journals and books. There is the electronic web, providing millions of references. But the human brain cannot cope with the excess of information unless it becomes filtered. The question is: how to select adequately? We quote Goethe once more, now in his statement "Man sieht nur was man weiß" or: we only recognize what we know already. In the computer era, this statement is still valid. Upon introducing a number of keywords one gets so many titles that additional filters must be applied, but what is left does not guarantee that nothing has been overlooked, and did we use the proper keywords? Fig. 8.4 shows major topics for possible interdisciplinary studies on terpenoids, of which we mention a few:

The bacterial outer membrane is a drug barrier (Hancock, 1997). When Gram developed his differentiating stain in 1884 he could only have had an inkling of the differences in membranes between his two classes. The Archaea with their strong, tough and elastic cell walls (Beveridge, 1999) belong to the Gram-negative bacteria and are able to produce terpenoids via the MAD route (see 2.1.1). Cooperation between pharmacologists and bacteriologists could spark future research on compartmentation and membrane anchoring of terpenoids.

As stated by Gao and Kwaik (2000) our understanding of the mechanisms by which intracellular bacterial pathogens induce or inhibit apoptotic pathways has rapidly advanced in the past few years. As an example, *Shigella flexneri* that causes bacillary dysentery induces apoptosis in

Chapter 8. Prospectus and suggestions for further research

macrophages. This aspect of host-pathogen interaction is an intriguing basis for a multidisciplinary study, and Gao and Kwaik (2000) put some relevant questions for future research. The inititation of apoptosis likewise applies to inducible responses in plant defense, such as apoptosis involved in the hypersensitive response (McDowell and Dangl, 2000). In addition, there seems to be a common putative active site for the human glioma pathogenesis-related protein (GliPR) and the plant pathogenesis-related protein P14a, indicating a functional link between the human immune system and plant defense (Szyperski et al., 1998).

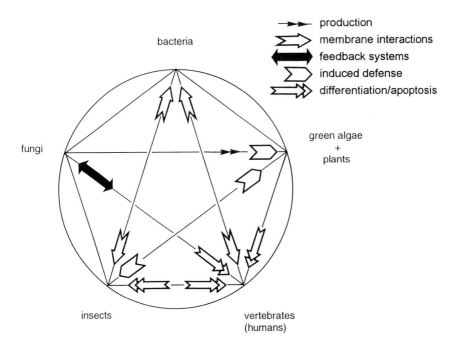

Fig. 8.4. Suggested interdisciplinary studies on terpenoids. Arrows point to the disciplin that may profit.

Bacteriologists cannot only provide information on biodegradation of terpenoids; they can also indicate steps in the production of valuable components with novel properties. An example is the biodegradation of lantadene A, a pentacyclic triterpenoid produced by lantana plants and acting as a hepatotoxin (Sharma et al., 1997). *Pseudomonas pickettii* can biodegrade this terpenoid into compounds with interesting bioactivity.

Many terpenoids with defensive functions against micro-organisms or herbivores are continuously present. Others have to be mobilized upon attack by invaders. With respect to defense induced by herbivorous insects there are several unanswered questions. Comparable reactions may occur at the onset of infection of plants by fungi. Some phytoalexins are sesquiterpenes, which makes the review by Hammerschmidt (1999) on 60 years of phytoalexin research intriguing reading for mycologists, plant pathologists and entomologists.

Key enzyme inhibitors of MVA biosynthesis such as lovastatin are derived from fungi. Fungi offer us much more. *Colletotrichum trifolii* is a filamentous pathogen with hyaline conidia occurring in *Medicago* spp. (alfalfa or lucerne) causing dark blackish-brown spots (anthracnose disease). Truesdell *et al.* (1999) cloned a gene encoding a Ras homologue (CT-ras) from this fungus. This fungal Ras exhibited extreme amino acid similarity to Ras proteins from lower and higher eukaryotes. What is more, a change in a single amino acid resulted in mutationally activated CT-ras induced tumour formation in mice by cellular transformation of mouse (NIH 3T3) fibroblasts. What happened in *C. trifolii* itself? Mutationally activated CT-ras induced defects in polarized growth and abnormal hyphal proliferation. Facts presented in section 2.1.2 speak for a closer relationship between fungi and animals than between fungi and plants. Ras GTPase links extra-cellular mitogens to intracellular mechanisms controlling cell proliferation, not only in fungi and vertebrates, but also in insects.

In the developing *Drosophila* wings cells lacking Ras accumulated in G1 and underwent apoptosis due to cell competition. Ras promoted G1/S transition. Terpenoid interference with Ras could affect wing development in insects, and this may explain why we found inhibition of wing formation in aphids upon exogenous application of terpenes like (E)-β-farnesene (Gut *et al.*, 1991). A homologue of the Ras GAP-binding protein G3BP occurring in *Drosophila* has been identified by Pazman *et al.* (2000). This binding protein, denominated Rasputin is required as an effector of Ras signalling during the development of the insect eye. It is symbolic that the authors used the name Rasputin for this protein that seems as mysterious as the Russian monk of that name, who had a great influence on the Russian Tsar family of Romanov.

Genes control other genes. Polymorphism as seen in aphids can be based on modification in regulatory genes that control the expression of structural genes. According to Carrol (2000) "it is now recognized that development involves a hierarchy of genetic control, including the precise timing and the position of *Hox* genes themselves, and the regulation of a cascade of downstream genes, together with the interaction with the products of other *Hox* genes, commonly producing broadly pleiotrophic results in many tissues and structures". With respect to the effect of terpenoids on protein prenylation, the question rises in which cases terpenoids affect gene expression of structural genes or of regulatory genes.

A comparison of the effects of terpenoids on gene expression in green algae and plants would greatly attribute to our knowledge of differentiation processes. Within this context it is obvious that tracing the phylogeny of proteins, which became involved in gene encoding and knowledge of the specific amino acid residues that were the targets of selection is of great importance to our understanding of the evolution of enzymes involved in terpenoid production. According to Chang and Donoghue (2000) it is now possible to recreate inferred ancestral proteins in the laboratory and study the function of these molecules. These authors provide likelihood models for detecting positively selected amino acid sites in a protein, and a model of evolutionary likelihood ratio tests. This is where four disciplines from four worlds could meet: physiologists, biochemists, pharmacologists and gene technologists in mycology, botany, entomology and zoology.

8.2.2 Terpenoids and model systems

In section 8.2.1 we have forwarded the seductive idea of integrating messenger functions of terpenoids in multidisciplinary research. Upon comparing the messenger functions of terpenoids in organisms belonging to different taxa one envisages a cornucopia of possibilities to incorporate these functions in model studies. As shown in chapter 2, the biosynthesis of terpenoids with their corresponding genes and regulatory mechanisms is a consolidated, ancient system, avoiding waste of nitrogen. What about terpenoids and the origin of eukaryotes? The undeniable chimerism of eukaryotic genomes (Katz, 1999) may stimulate integration of studies of the cytoskeleton, gene exchange and biosynthetic routes. Terpenoids with their MAI and MAD routes would offer excellent objects to investigate genes involved in transcription, DNA replication and translation that according to Katz (1999, and others quoted in that paper) unite Archaea with eukaryotes. A study of terpene production in Archaezoa (early diverged, primitive amitochondriate eukaryotes) could be most elucidating.

Following the pentagram of Fig. 8.4, we could subsequently suggest four more model systems, but that would do not justice to the burgeoning field of molecular biology. Instead, we advocate only one more: terpenoids and developmental processes in insects, more particularly, in aphids. Why insects? That insects could be used as biochemical models is nothing new. Law and Wells reviewed already in 1989 the diversity of problems which can be studied in insects with their short life cycles, resilience to surgical techniques, intestinal detoxification systems, dietary dependence on steroids, to mention a few differences with mammals, although there are many similarities with mammalian biochemistry. The fact that there is no need to cage and kill vertebrates is certainly an advantage. Terpenoids exert profound effects on a.o. Ras proteins. Recent understanding of the way Ras and downstream pathways can influence induction of apoptosis suggest that the regulatory pathways in mammals and insects might be more similar than previously thought (McNeill and Downward, 1999; Prober and Edgar, 2000).

The similarity between G-proteins in insect sensilla and membranes of vertebrate of vertebrate cells has been a reiterated theme in this volume. Colon cancer and the wing discs of flies both depend on APC, the adenomatous polyposis coli gene and there exists a human homologue of the *Drosophila* discs large (dlg) tumour suppressor (Peifer, 1996).

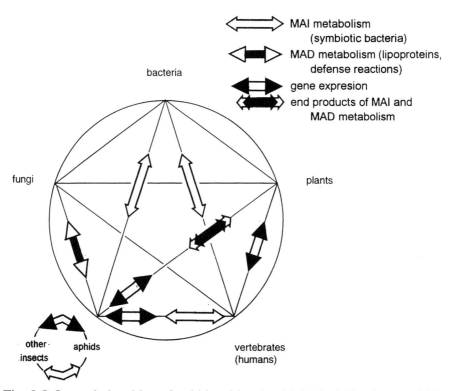

Fig. 8.5. Interrelationships of aphids with other biological kingdoms, which can be used in multidisciplinary studies of messenger functions of terpenoids.

Why aphids? Aphids have a tight relationship with the five biological kingdoms of our pentagram (Fig. 8.5). Their symbiotic relationship with micro-organisms (endosymbiotic *Buchnera* bacteria) have been dealt with several times in this volume. Their contribution to the presence of MAI metabolism in insects has been discussed in chapter 2.2. MAI metabolism with its DOXP pathway may produce IPP in vertebrates (via a.o. *E. coli*), and activate T cells (Jomaa *et al.*, 1999). In Fig. 8.5 this phenomenon is reflected in the triangle bacteria/aphids/humans.

Aphids can become infected with fungi, a.o. by *Entomophthora* and *Verticillium* spp. that penetrate the integument, develop in the hemocoel and invade all the host tissues (Latgé and Papierok, 1988). Many fungus species are highly specific to aphids which makes it interesting to study the biochemical basis of specificity, in particular with respect to the role of lipases, proteases and chitinases and lipoproteins produced by the fungi (MAD metabolism).

Nearly all known branches of MVA metabolism occur in aphids (see a.o. chapters 2.2, 3.2, 4.2, 5.1 and 6.2). Natural terpenoids influence feeding behaviour, the response to environmental factors, growth and development and polymorphism (Gut *et al.*, 1991; Harrewijn *et al.*, 1996). Because many aphid species reproduce parthenogenetically during the summer season, the individuals of a colony are genetically identical (and can be kept in that condition under laboratory conditions). Nevertheless they show wing dimorphism, making them ideal objects to study the effect of terpenoids on gene expression. A possible effect of farnesene isomers on polymorphism has been discussed in sections 4.1.2 and 5.1.1.

Aphids have a high level of terpenoids in their hemolymph with different ratios of isomers depending on their morphs and forms (van Oosten *et al.*, 1990; sections 2.2.2, 4.1.2 and 5.1.1). As mentioned above, the wing discs gene of *Drosophila* is a homologue of the human suppressor gene p53. Matsumine *et al.* (1996) have studied the activity of this suppressor protein to conclude that insect wing development and human tumour suppression have closely related signalling pathways. It could be of great value to study the effect of chemotherapeutics on these genes and the proteins they encode. This study could be extended to the process of wing formation in aphids, in view with our own experience that terpenoids are involved in polymorphism of these insects (van Oosten *et al.*, 1990). Von Dehn found already in 1963 that farnesol inhibits wing formation in *Aphis fabae*. This unexpected effect of an isoprenoid intermediate on polymorphism was published in a German journal. Whether or not an aphid takes the alate course of development, wing buds are always present in the embryos (Harrewijn, 1976). In comparison to *Drosophila*, aphids show few genetic defects or malformations and according to Russian research (anonymous) seem to have an excellent DNA repair mechanism.

Aphids can actively ingest nutrients (and terpenoids) from chemically defined diets or phloem sap and other plant juices with their well-developed feeding apparatus. Electrophysiological recordings of feeding activities of different aphid species suggests that the ancestors of modern aphid species were already equipped with the present feeding apparatus in the early Cretaceous, that is at the onset of the development of angiosperms with their diversified terpenoid production (Harrewijn *et al.*, 1998).

Aphids are easy to rear and can be fed with a few droplets of liquid diet. They survive many treatments including microinjections (even their nymphs), facilitating comparison of the effects of natural terpenoids with synthetic molecules (Harrewijn and Kayser, 1997). Ingestion of terpenoid allomones of plants, which have been incorporated in chemically defined diets, and administering terpenoids via microinjection provide excellent ways to discriminate between semiochemical effects and those on physiology of the aphids.

An intriguing example of the relationship between host plants and aphids is that of chamomile plants (*Matricaria chamomilla* and *M. matricarioides*, Asteraceae) and a number of aphid species, which live on these plants. Many aphid species are specialists; nevertheless quite a number of species can be found on the aerial parts of chamomile (e.g. *Aphis fabae, A. vandergooti, A. euonymi, Brachycaudus cardui, B. helichrysi, Macrosiphoniella tanacetaria, M. tapuskae, Myzus persicae*). A few thrive on the roots, like *Pemphigus fuscicornis* and *Protaphis dudichi* (on the root collar).

Chamomile plants produce a wealth of farnesene isomers and the same holds for the aphids. Although (E)-β-farnesene is an alarm pheromone of a.o. *M. persicae*, the aphids seem to be very happy on these plants, on which they can evoke an alarm reaction in conspecifics regardless the plants own farnesene production. The relationship between chamomile and aphids forms a basis for a number of multidisciplinary studies in which entomologists, plant pathologists, bacteriologists and those involved in gene expression could take part. Fig. 8.6 depicts five different levels of a chamomile plant with its MAI + MAD metabolism in relation to the secret life of aphids with their differently organized MAI and MAD metabolism. Each level of this microcosm rises questions for multidisciplinary research:

(1) Occasionally one finds an aphid on the flowers. Why so few? Plant growth substances, nutritional factors, (E)-β-farnesene, other terpenoids such as bisabolol, bisabololoxide, chamazulene? Inhibitors of alarm compounds in the flowers are as yet unknown, so (E)-β-farnesene may optimally protect generative parts (see Harrewijn *et al.*, 1995).

(2) Aphids with their MAI (symbionts) and MAD metabolism feed on phloem sap. They produce (E)-β-farnesene (alarm pheromone), (Z,E)-α-farnesene and (E,E)-α-farnesene. Aphids show two types of wing dimorphism. In autumn, short days induce the production of winged migrating sexuals, probably associated with a low level of JH3 (Hardy *et*

al, 1985). Is the high level of (E,E)-α-farnesene in sexuales a *cause* or a *result* of sexual dimorphism (Gut *et al.*, 1991)? Which feedback systems may be involved? External factors (e.g. crowding, temperature, nutritional factors) cause derepression of the wing discs gene, resulting in dispersal of the winged morph. The ratio of farnesene isomers in winged (alate) morphs differs from the unwinged (apterous) morphs. What causes these differences? Is there a role of the symbionts (MAI metabolism)?

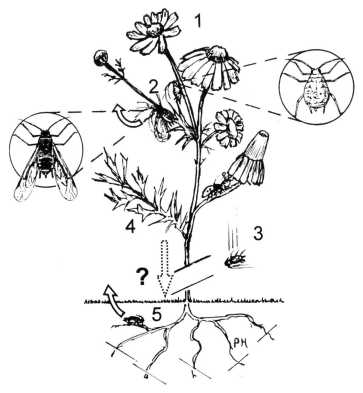

Fig. 8.6. The relation of MAI and MAD metabolism to the secret life of aphids on a chamomile plant. For explanation see text (aphids not drawn to scale).

(3) An aphid attacked by a predator releases (E)-β-farnesene via its siphunculi. A conspecific (sometimes an individual of a related species) receives the alarm message and drops down. Other aphid species use different terpenes as alarm pheromones. Why is the alarm pheromone so specific within a species?

(4) The vegetative parts contain (E)-β-farnesene, but also the alarm pheromone inhibitors caryophyllene and humulene. Did the aphids select these *Matricaria* spp. or did the plants select the aphids?
(5) The roots produce alarm pheromone without inhibitors; in some *Matricaria* spp. (E)-β-farnesene is present in the roots in high purity. However, this does not induce dispersal via the alarm reaction. If there is a low vapour pressure the aphids may not perceive these terpenes at all. What induces the root aphids to migrate to the primary host when winter comes?

Summarizing, the present model may serve to indicate that terpenoids not only act as semiochemicals, but have an as yet hardly investigated role in the transduction of signals that operate in the mutual relationship between host plants, aphids and their environment.

The onset of the 21st century is marked by an increasing integration of disciplines. The omnipresence of messenger functions of terpenoids in the regulation of growth and development of plants, insects and vertebrates is now well documented. Insects, like vertebrates, are dependent on G-proteins and oncoproteins; genes encoding such proteins, and tumour suppresser genes have a high degree of homology. Plants produce terpenoids, which a.o. repel insects and at the same time are toxic to human tumours.

Our present understanding of these phenomenons has already resulted in cooperation between plant and insect physiologists, biochemists, pharmacologists and oncologists (e.g. Bushunow *et al.*, 1999).

Novel techniques develop explosively. Antisense (Nita *et al.*, 2000; Simoes-Wust *et al.*, 2000; Terada *et al.*, 2000), telomerase inhibition (Dhaene *et al.*, 2000; Koga *et al.*, 2000), monoclonal antibodies, techniques to study anti-angiogenesis factors (Bertolini *et al.*, 2000), topoisomerase inhibitors, DNA-polymerase inhibitors etc., leading to unconventional therapies (Cortes *et al.*, 1998; Kaiser, 2000).

We expect an extended role of natural terpenoids in the development of a novel discipline: Comparative Evolutionary Biochemistry, branching into genetics, molecular cell biochemistry and metabolics - to the honour of science and to the benefit of humans and mammals.

Epilogue

It is the dawn of the third millennium. In the past millennium our attitude toward life has considerably changed. No longer are we in search of the elixir of life; we realize that in higher forms of life mortality has been introduced in favour of a high survival rate during the reproductive period. Early adulthood escapes genetic defects, which disclose themselves in a postreproductional period, especially if it is long, but it does not outselect them. As a result, we have to envisage that terpenoid hormones like estradiol and testosteron, essential for a vital sexual life, eventually become turned against us and increase the risk of malignancies of the reproductive organs. This makes terpenoids both friends and foe; it will be a challenge for coming scientists to obtain more insight into dynamics of ligand and receptor with respect to terpenoids. Increased knowledge of the action spectrum and pharmacokinetic profile of a terpenoid will greatly accelerate its application in biology (agriculture) and medicine. Even a simple terpene like perillyl alcohol exerts a number of side effects if incorporated in an anti-tumour treatment (Ripple *et al.*, 2000).

In this book we introduced the terms MAI and MAD metabolism to discriminate between a mevalonic acid (MVA) independent and MVA dependent route for the biosynthesis of isoprene units, and ultimately for terpenes and terpenoids. Although some MAI and MAD route endproducts have a role in the induction of apoptosis, MAI and MAD should be used within the context of terpenoid biosynthesis; our MAD is not to be confused with a different (and indeed more crazy) type of MAD used by the military: Mutual Assured Destruction, or total destruction resulting from Second Strike Capability with nuclear weapons.

At the beginning of the second millennium terpenoids were utilized to control insects in stored products, in medicine and in perfumes, although the unravelling of their basic chemical structure had to wait until the 19th century.

We realize today that nature was ahead of us in many ingeneous developments; terpenoids have been selected as secret weapons in the struggle for life and have acquired many functions in chemical communication of life forms. Unfortunately, biologists have not come up with an adequate definition of life, although there is a display of characteristics of life on this planet. Perhaps one of the most important is chemical communication; between organelles, within organisms and between organisms there is a never lasting stream of messengers on all thinkable levels. Natural terpenoids exert messenger functions from Archaea and Bacteria to mammals, and not in plants alone, although the decoration of "the plant language" by Penuelas *et al.* (1995) is fully justified.

Genes encode enzymatic proteins that make terpenoids, involved in chemical ecology. Terpenoids with their C-5 building blocks are produced in almost endless variation; the process of function acquisition is still ongoing. Many copies of a terpenoid biosynthesized via adjustable codes can be released into liquid media and into the air. Mankind has discovered that many copies can be made of a message existing of exchangeable codes, it only had to wait until Gutenberg (1398-1468) invented a way to multiply such a message in printing a text with exchangeable letter blocks.

Terpenoids are indispensable in embryogenesis, as demonstrated by developmental anomalies reemphasizing the functional redundancy of retinoid receptors (Smith *et al.*, 1998). Terpenoids have made an important contribution to the development of eukaryotes. Eukaryotes are chimeric organisms; they did not evolve via vertical transmission between generations but as an admixture of several independent lineages (Katz, 1999). According to this author the presence of a cytoskeleton is a defining feature of eukaryotes. Doolittle (1998) opines that the cytoskeleton enabled gene transfer from many donors into the nucleus of host cells, although to our present knowledge precursors of tubulin and actin are not encoded by any of the archaeal genomes. The relation between Archaea, Eubacteria and eukaryotes has been discussed in chapter 2. The recent studies on the mode of action of artemisin derivatives on a protein of the malaria parasite *Plasmodium falciparum*, which is a Translationally Controlled Tumour Protein (TCTP) is an example of a terpenoid with a target in a cell organelle (apicoplast) with MAI metabolism (Bhisutthiban *et al.*, 1999). The apicoplast is incorporated in a protozoan in which MAD routes are operating.

We expect that a better understanding of the evolution of MAI and MAD metabolism in prokaryotes and eukaryotes will contribute to a further constructon of the tree of life.

Epilogue

Some modern cancer therapies have the microtubuli of vertebrate cells as a target. Here we meet terpenoids again, now to combat tumour cells in their ancient Achilles' heel: the cytoskeleton. It is one strategy out of a number of promising new cancer treatments utilizing terpenoids.

No longer can we envision terpenoids to be involved in one-way messages within or between organisms. The five biological kingdoms depicted in the pentagram of this volume are in continuous chemical intercourse. We hope that recognition of this phenomenon sets a trail to be followed by scientists of different disciplines, united in comparative studies of the messenger functions of terpenoids, and that it may become a binding principle of those engaged in chemical ecology.

P. Harrewijn, A.M. van Oosten and P.G.M. Piron

References

Abbanat, D.R. and Ferry, J.G., 1990. Synthesis of acetyl coenzyme A by carbon monoxide dehydrogenase complex from acetate-grown *Methanosarcina thermophila*. J. Bacteriol. 172: 7145-7150.

Abe, I., Seki, T. and Noguchi, H., 2000. Potent and selective inhibition of squalene epoxidase by synthetic galloyl esters. Biochem. Biophys. Res. Comm. 270: 137-140.

Aboushadi, N., Engfelt, W.H., Paton, V.G. and Krisans, S.K., 1999. Role of peroxisomes in isoprenoid biosynthesis. J. Histochem. Cytochem. 47: 1127-1132.

Adany, I., Yazlovitskaya, E.M., Haug, J.S., Voziyan, P.A. and Melnykovych, G., 1994. Differences in sensitivity to farnesol toxicity between neoplastically and non-neoplastically derived cells in culture. Cancer Lett. 79: 175-179.

Adam, K.P. and Croteau, R., 1998. Monoterpene biosynthesis in the liverwort *Conocephalum conicum*: demonstration of sabinene synthase and bornyl diphosphate synthase. Phytochemistry 49: 475-480.

Adam, K.P., Thiel, R., Zapp, J. and Becker, H., 1998. Involvement of the mevalonic acid pathway and the glyceraldehyde-pyruvate pathway in terpenoid biosynthesis of the liverworts *Ricciocarpos natans* and *Conocephalum conicum*. Arch. Biochem. Biophys. 354: 181-187.

Adami, E., Marazzi-Uberti, E. and Turba, C., 1962. Anti-ulcer action of some natural and synthetic terpenic compounds. Med. Exptl. 7: 171-176.

Adams, J.M. and Cory, S., 1998. The Bcl-2 protein family: arbiters of cell survival. Science 281: 1322-1326.

Adekenov, S.M., Dzhazin, K.A., Bloszyk, E., Zapolskya-Dovnar, G.M., Kulyjasov, A.T. and Drozdz, B., 1996. Antifeedant activity of sesquiterpene lactones of Kazakhstan's plants. In: Abstracts ISCE 13th Annual Meeting (eds P. Drašar and I. Valterová): 77-78.

Adekenov, S.M., Zapolskya-Dovnar, G.M. and Belyaev, N.P., 1996b. Repellent activity of sesquiterpene lactones of *Tanacetum vulgare* L. In: Abstracts ISCE 13th Annual Meeting (eds P. Drašar and I. Valterová): 79.

Agarwal, B., Bhendwal, S., Halmos, B., Moss, S.F., Ramey, W.G. and Holt, P.R., 1999. Lovastatin augments apoptosis induced by chemotherapeutic agents in colon cancer cells. Clin. Cancer Res. 5: 2223-2229.

Aguilera, J.A., Garcia-Molina, V., Linares, A., Arce, V. and Garcia-Peregrin, E., 1987. Posthatching evolution of *in vivo* mevalonate metabolism in chick liver. Biochem. Int. 2: 375-383.

Ahn, K.S., Hahm, M.S., Park, E.J., Lee, H.K. and Kim, I.H., 1998. Corosolic acid isolated from fruit of *Crataegus pinnatifida* var. *psilosa* is a protein kinase C inhibitor as well as a cytotoxic agent. Planta Med. 64: 468-470.

Ahmed, N., Laverick, L., Sammons, J., Baumforth, K.R. and Hassan, H.T., 1999. Effect of all-*trans*-retinoic acid on chemotherapy-induced apoptosis and down-regulation of Bcl-2 in human myeloid leukaemia CD34 positive cells. Leuk. Res. 23: 741-749.

Akino, T., Turushima, T. and Yamaoka, R., 1995. 3-Formyl-7,11-dimethyl-(2E,6Z,10)-dodecatrienal - an antifungal compound in the mandibular gland of the ant, *Lasius fuliginosis* Latreille. (in Japanese). Nippon Nogeikagaku-Kaishi 69: 1581-1586.

Alcaraz, M.J. and Rios, J.L., 1991. Pharmacology of diterpenoids. In: Ecological Chemistry and Biochemistry of Plant Terpenoids (eds J. Harborne and F.A. Tomas-Barberan). Clarendon Press, Oxford: 230-263.

Alexander, H.J., 1987. Inhibition of *Pineus floccus* colonization by volatile compounds found in leaf tissue of red spruce. Virginia J. Sci. 38: 27-34.

Amano, T., Nishida, R., Kuwahara, Y. and Fukami, H., 1999. Pharmacophagous acquisition of clerodendrins by the turnip sawfly (*Athalia rosae ruficornis*) and their role in mating behaviour. Chemoecology 9: 145-150.

Anderson, M. and Bromley, A.K., 1987. Sensory system. In: Aphids, Their Biology, Natural Enemies and Control, Vol. A (eds A.K. Minks and P. Harrewijn), Elsevier Science Publ. Amsterdam: 153-161.

Andriamampandry, M., Freund, M., Wiesel, M.L., Rhinn, S., Ravanat, C., Cazenave, J.P., Leray, C. and Gachet, C., 1998. Diets enriched in (n-3) fatty acids affect rat coagulation factors dependent on vitamin K. C.R. Acad. Sci. III 321: 415-421.

Anthony, M.L., Zhao, M. and Brindle, K.M., 1999. Inhibition of phosphatidylcholine biosynthesis following induction of apoptosis in Hl-60 cells. J. Biol. Chem. 274: 19686-19692.

Ariazi, E.A., Satomi, Y., Ellis, M.J., Haag, J.D., Shi, W., Sattler, C.A. and Gould, M.N., 1999. Activation of the transforming growth factor β signaling pathway and induction of cytostasis and apoptosis in mammary carcinomas treated with the anticancer agent perillyl alcohol. Cancer Res. 59: 1917-1928.

Arigoni, D., Eisenreich, W., Latzel, C., Sagner, S., Radykewicz, T., Zenk, M.H. and Bacher, A., 1999. Dimethylallyl pyrophosphate is not the committed precursor of isopentenyl pyrophosphate during terpenoid biosynthesis from 1-deoxyxylulose in higher plants. Proc. Natl. Acad. Sci. USA 96: 1309-1314.

Armaka, N., Papanikolaou, E., Sivropoulou, A. and Arsenakis, M., 1999. Antiviral properties of isoborneol, a potent inhibitor of herpes simplex virus type 1. Antiviral Res. 43: 79-92.

References

Aruna, R. and Siuaramakrishnan, V.M., 1990. Plant products as protective agents against cancer. Indian J. Exp. Biol. 28: 1008-1011.

Askenazi, A. and Dixit, V., 1998. Death receptors: signaling and modulation. Science 281: 1305-1308.

Audsley, N., Weaver, J. and Edwards, J.P., 1999. Juvenile hormone synthesis by corpora allata of tomato moth, *Lacanobia oleracea* (Lepidoptera: Noctuidae), and the effects of allatostatins and allatotropin *in vitro*. Eur. J. Entomol. 96: 287-293.

Bach, T., 1995. Some new aspects of isoprenoid biosynthesis in plants - a review. Lipids 30: 191-202.

Back, K. and Chappell, J., 1996. Identifying functional domains within terpene cyclases using a domain-swapping strategy. Proc. Natl. Acad. Sci. USA 93: 6841-6845.

Baeckström, L., Li, L. and Kontek, B., 1991. Modification and synthesis of isoprenoids directed towards insect related compounds. Proc. Conf. Insect Chem. Ecol. Tabor (ed. I. Hrdý), Academia Praha/SPB Acad. Publ. The Hague: 129-135.

Baker, R., Parton, A.H. and Howse, 1982. Identification of an acyclic diterpene alcohol in the defense secretion of soldiers of *Reticulitermes lucifugus*. Experientia 38: 297-298.

Bala, A.E.A., Delorme, R., Kollman, A., Kerhoas, L., Einhorn, J., Ducrot, P.H. and Augé, D., 1999. Insecticidal activity of daphnane diterpenes from *Lasiosiphon kraussianus* (Meisn) (Thymelaeaceae) roots. Pestic. Sci. 55: 745-750.

Ballock, R.T., Heydemann, A., Wakefield, L.M., Flanders, K.C., Roberts, A.B. and Sporn, M.B., 1994. Inhibition of the chondrocyte phenotype by retinoic acid involves upregulation of metalloprotease genes independent of TGF-β. J. Cell Physiol. 159: 340-346.

Barbieri, M., 1998. The organic codes. The basic mechanism of macroevolution. Riv. Biol. 91: 481-513.

Barbosa, P., Krischik, V.A. and Jones, C.G. (eds), 1991. Microbial Mediation of Plant-Herbivore Interactions. John Wiley & Sons, New York, 530 pp.

Bard, M., Albrecht, M.R., Gupta, N., Guynn, C.J. and Stillwell, W., 1988. Geraniol interferes with membrane functions in strains of *Candida* and *Saccharomyces*. Lipids 23: 534-538.

Bardon, S., Picard, K. and Martel, P., 1998. Monoterpenes inhibit cell growth, cell cycle progression, and cyclin D1 gene expression in human breast cancer cell lines. Nutr. Cancer 32: 1-7.

Barnola, L.F., Hasegawa, M. and Cedeno, A., 1994. Monoterpene and sesquiterpene variation in *Pinus caribaea* needles and its relationship to *Atta laevigata* herbivory. Bioch. Syst. Ecol. 22: 437-445.

Bartelt, R.J., 1999. Sap beetles. In: Pheromones of Non-Lepidopteran Insects Associated with Agricultural Plants (eds J. Hardie and A.K. Minks), CABI Publ., Wallingford: 69-89.

Barthelman, M., Chen, W., Gensler, H.L., Huang, C., Dong, Z. and Bowden, G.T., 1998. Inhibitory effects of perillyl alcohol on UVB-induced murine skin cancer and AP-1 transactivation. Cancer Res. 58: 711-716.

Barton, H., 1959. Zum Wirkungsmechanismus des 1,4-Dimethyl-7-isopropylazulens auf die Tumorzelle. Acta Biol. Med. Germ. 6: 555-567.

Bassa, B.V., Roh, D.D., Vaziri, N.D., Kirschenbaum, M.A. and Kamanna, V.S., 1999. Effect of inhibition of cholesterol synthetic pathway on the activation of Ras and MAP kinase in mesangial cells. Biochim. Biophys. Acta 1449: 137-149.

Bassam, R.S., Pal, A., Croft, S.L., Taylor, R.J.K. and Field, M.C., 1999. The farnesyltransferase inhibitor manumycin A is a novel trypanocide with a complex mode of action including major effects on mitochondria. Mol. and Biochem. Parasit. 104: 67-80.

Baumann, P., Baumann, L., Lai, C.Y., Rouhbaksh, D., Moran, N.A. and Clark, M.A., 1995. Genetics, physiology and evolutionary relationships of the genus *Buchnera*: intracellular symbionts of aphids. Annu. Rev. Microbiol. 49: 55-94.

Beekman, A.C., Barentsen, A.R., Woerdenbag, H.J., van Uden, W., Pras, N., Konings, A.W., el-Feraly, F.S., Galal, A.M. and Wikstrom, H.V., 1997. Stereochemistry-dependent cytotoxicity of some artemisinin derivatives. J. Nat. Prod. 60: 325-330.

Beekman, A.C., Wierenga, P.K., Woerdenbag, H.J., van Uden, W., Pras, N., Konings, A.W., el-Feraly, F.S., Galal, A.M. and Wikstrom, H.V., 1998. Artemisinin-derived sesquiterpene lactones as potential antitumour compounds: cytotoxic action against bone marrow and tumour cells. Planta Med. 64: 615-619.

Bejarano, E.R. and Cerdá-Olmedo, E., 1992. Independence of the carotene and sterol pathways of *Phycomyces*. FEBS Lett. 306: 209-212.

Bell, G.D. and Doran, J., 1979. Gall stone dissolution in man using an essential oil preparation. British Med. Journal 1: 24.

Bennis, F., Favre, G., Haillard, F. le and Soula, G., 1993. Importance of mevalonate-derived products in the control of HMG-CoA reductase activity and growth of human lung adenocarcinoma cell line A549. Int. J. Cancer 55: 640-645.

Benkö, S., Macher, A., Szarvas, F. and Tiboldi, T., 1961. Effect of essential oils on the atherosclerosis of cholesterol-fed rabbits. Nature 190: 731-732.

Benyakir, D., Bazar, A. and Chen, A., 1995. Attraction of *Maladera matrida* (Coleoptera, Scarabaeidae) to eugenol and other lures. J. Econ. Ent. 88: 415-420.

Berenbaum, M.R., 1988. Allelochemicals in insect-microbe-plant interactions; agents provocateurs in the coevolutionary arms race. In: Novel Aspects of Insect-Plant Interactions (eds P. Barbosa and D.K. Letourneau), John Wiley & Sons, New York: 97-123.

Bergman, P. and Bergström, G., 1996. Chemical communication in bumblebee premating behaviour. In: Abstracts ISCE 13th Annual Meeting (eds P. Drašar and I. Valterová): 43-44.

Bergström, G., 1987. On the role of volatile chemical signals in the evolution and speciation of plants and insects: why do flowers smell and why do they smell differently? In: Proc. 6th Int. Symp. on Insect-Plant Relationships (eds V. Labeyrie, G. Fabres and D. Lachaise), Junk Publ. The Hague: 321-327.

Bergström, G. and Löfqvist, J., 1970. Chemical basis for odour communication in four species of *Lasius* ants. J. Insect Physiol. 16: 2353-2375.

Bernays, E.A., 1996. Selective attention and host-plant specialization. In: Proc. 9th Int. Symp. on Insect-Plant Relationships (eds E. Städler, M. Rowell-Rahier and R. Bauer), Entomol. exp. appl. 80: 125-131.

Bernays, E.A., 1999. When host choice is a problem for a generalist herbivore: experiments with the whitefly, *Bemisia tabaci*. Ecol. Entomol. 24: 260-267.

Bertolini, F., Fusetti, L., Mancuso, P., Gobbi, A., Corsini, C., Ferrucci, P.F., Martinelli, G. and Pruneri, G., 2000. Endostatin, an antiangiogenic drug, induces tumor stabilization after chemotherapy or anti-CD20 therapy in a NOD/SCID mouse model of human high-grade non-Hodgkin lymphoma. Blood 96: 282-287.

Beveridge, T.J., 1999. Structures of Gram-negative cell walls and their derived membrane vesicles. J. Bacteriol. 181: 4725-4733.

Bhati, A., 1950. Essential oil from the resin of *Commiphora mukul* Hook. Ex Storks. J. Indian Chem. Soc. 27: 436.

Bhisutthibhan, J., Philbert, M.A., Fujioka, H., Aikawa, M. and Meshnick, S.R., 1999. The *Plasmodium falciparum* translationally controlled tumor protein: subcellular localization and calcium binding. Eur. J. Cell Biol. 78: 665-670.

Biardi, L. and Krisans, S.K., 1996. Compartmentalization of cholesterol biosynthesis. Conversion of mevalonate to farnesyl diphosphate occurs in the peroxisomes. J. Biol. Chem. 271: 1784-1788.

Bifulco, M., Laezza, C. and Aloj, S.M., 1999. Inhibition of farnesylation blocks growth but not differentiation in FRTL-5 thyroid cells. Biochemie 81: 287-290.

Bino, R.J., Dafni, A. and Meeuse, A.D.J., 1984. Entomophily in the dioecious gymnosperm *Ephedra aphylla* Fork. (= *E. alte* C.A. Mey.), with some notes on *E. campylopedra* C.A. Mey. I. Proc. Kon. Akad. Wetensch. Amsterdam C 87: 1-13.

Birch, A.J., English, R.J., Massy-Westropp, R.A. and Smith, H., 1958. Studies in relation to biosynthesis. Part XV. Origin of terpenoid structures in mycelianamide and mycophenolic acid. J. Chem. Soc. London, Vol. 1958 I: 369-375.

Blagosklonny, M.V., 1999. Drug-resistance enables selective killing of resistant leukemia cells: exploiting of drug resistance instead of reversal. Leukemia 13: 2031-2035.

Block, G., Patterson, B. and Subar, A., 1992. Fruit, vegetables, and cancer prevention: a review of the epidemiological evidence. Nutr. Cancer 18: 1-29.

Blount, J.D., Houston, D.C. and Møller, A.P., 2000. Why egg yolk is yellow. Trends Ecol. Evol. 15: 47-49.

Blum, M.S., Crewe, R.M., Kerr, W.E., Keith, L.H., Garrison, A.W. and Walker, M.M., 1990. Citral in stingless bees: isolation and function in trail laying and robbing. J. Insect Physiol. 16: 1637-1648.

Blum, M.S., Everett, D.M., Jones, T.H. and Fales, H.M., 1991. Arthropod natural products as insect repellents. In: Naturally Occurring Pest Bioregulators (ed. P.A. Hedin) ACS Symp. Series, Am. Chem. Soc. Washington 449: 15-26.

Blutt, S.E., McDonnell, T.J., Polek, T.C. and Weigel, N.L., 2000. Calcitriol-induced apoptosis in LNCaP cells is blocked by overexpression of Bcl-2. Endocrinology 141: 10-17.

Bochar, D.A., Brown, J.R., Doolittle, W.F., Klenk, H.P., Lam, W., Schenk, M.E., Stauffacher, C.V. and Rodwell, V.W., 1997. 3-Hydroxy-3-methylglutaryl coenzyme A reductase of *Sulfolobus solfataricus*: DNA sequence, phylogeny, expression in *Escherichia coli* of the hmgA gene, and purification and kinetic characterization of the gene product. J. Bacteriol. 179: 3632-3638.

Bochar, D.A., Stauffacher, C.V. and Rodwell, V.W., 1999. Sequence comparisons reveal two classes of 3-hydroxy-3-methylglutaryl coenzyme A reductase. Mol. Genet. Metab. 66: 122-127.

Boelens, M.H., 1991. Spices and condiments. II. In: Volatile Compounds in Foods and Beverages (ed. H. Maarse). Marcel Dekker Inc., New York etc.: 449-479.

Bohlmann, J. and Croteau, R., 1999. Diversity and variability of terpenoid defences in conifers: molecular genetics, biochemistry and evolution of the terpene synthase gene family in grand fir. Novartis Found. Symp. 223: 132-145; Discussion 146-149.

Bohlmann, J., Phillips, M., Ramachandiran, V., Katoh, S. and Croteau, R., 1999. cDNA cloning, characterization, and functional expression of four new monoterpene synthase members of the Tpsd gene family from grand fir (*Abies grandis*). Arch. Biochem. Biophys. 368: 232-243.

Booden, M.A., Baker, T.L., Solski, P.A., Der, C.J., Punke, S.G. and Buss, J.E., 1999. A non-farnesylated Ha-Ras protein can be palmitoylated and trigger potent differentiation and transformation. J. Biol. Chem. 274: 1423-1431.

Borg-Karlson, A.-K., Persson, M., Sjödin, K., Norin, T. and Ekberg, I., 1996. Relative amounts and enantiomeric compositions of monoterpenes in different tissues of *Picea abies* and *Pinus sylvestris*. In: Abstracts ISCE 13th Annual Meeting (eds P. Drašar and I. Valterová): 47.

Bouhlal, K., Meynadier, J.M., Peyron, J.L., Peyron, L., Marion, J.P., Bonetti, G. and Meynadier, J., 1988. The cade (*Juniperus oxycedrus*) in dermatology. J. Parfumes Cosmet. Aromes 83: 73-76.

Bouwmeester, H.J., Davies, J.A.R. and Toxopeus, H., 1995. Enantiomeric composition of carvone, limonene, and carveols in seeds of dill and annual and biennal caraway varieties. J. Agric. Food Chem. 43: 3057-3064.

Bouwmeester, H.J., Gershenzon, J., Konings, M.C. and Croteau, R., 1998. Biosynthesis of the monoterpenes limonene and carvone in the fruit of caraway. I. Demonstration of enzyme activities and their changes with development. Plant Physiol. 117: 901-912.

Bouwmeester, H.J., Verstappen, F.W.A., Posthumus, M.A. and Dicke, M., 1999. Spider mite-induced (3S)-(E)-nerolidol synthase activity in cucumber and lima bean. The first dedicated step in acyclic C11-homoterpene biosynthesis. Plant Physiol. 121: 173-180.

Bowers, W.S., 1985. Phytochemical disruption of insect development and behavior. In: Bioregulators for Pest Control (ed. P.A. Hedin), Am. Chem. Soc.: 225-237.

Bowers, W.S., Nault, L.R., Webb, R.E. and Dutky, S.R., 1972. Aphid alarm pheromone: isolation, identification, synthesis. Science 177: 1121-1122.

References

Bowers, W.S., Nishino, C., Montgomery, M.E., Nault, L.R. and Nielson, M.W., 1977. Sesquiterpene progenitor, germacrene A: an alarm pheromone in aphids. Science 196: 680-681.
Bradfute, D.L. and Simoni, R.D., 1994. Non-sterol compounds that regulates cholesterogenesis. (Analogues of farnesylpyrophosphate reduce 3-hydroxy-3-methyl glutaryl-coenzyme A reductase levels). J. Biol. Chem. 269: 6645-6650.
Brembeck, F.H., Schoppmeyer, K., Leupold, U., Gornistu, C., Keim, V., Mossner, J., Riecken, E.O. and Rosewicz, S., 1998. A phase II pilot trial of 13-*cis*-retinoic acid and interferon-α in patients with advanced pancreatic carcinoma. Cancer 83: 2317-2323.
Brodsgaard, H.F., 1990. The effect of anisaldehyde as a scent attractant for *Frankliniella occidentalis* (Thysanoptera: Thripidae) and the response mechanism involved. IOBC/WPRS Bulletin XIII/5: 36-38.
Broitman, S.A., Wilkinson, J. 4th, Cerda, S. and Branch, S.K., 1996. Effects of monoterpenes and mevinolin on murine colon tumor CT-26 *in vitro* and its hepatic "metastases" *in vivo*. Adv. Exp. Med. Biol. 401: 111-130.
Brooke, M. de L., 1999. How old are animals? Tree 14: 211-212.
Brown, K.S., 1999. Deep green rewrites evolutionary history of plants. Science 285: 990-991. Discussed by Niklas *et al.* and replied by Brown in Science 285: 1673.
Brown, K., Havel, C.M. and Watson, J.A., 1983. Isoprene synthesis in isolated embryonic *Drosophila* cells. II. Regulation of 3-hydroxy-3-methylglutaryl coenzyme A reductase activity. J. Biol. Chem. 258: 8512-8518.
Brown, J.T., Hegarty, P.K. and Charlwood, B.V., 1987. The toxicity of monoterpenes to plant cells. Plant Sci. 48: 195-201.
Brown, M.S. and Goldstein, J.L., 1980. Multivalent feedback regulation of HMG-CoA reductase, a control mechanism coordinating isoprenoid synthesis and cell growth. J. Lipid Res. 21: 505-517.
Buchbauer, G., Jirovetz, L., Jäger, W., Dietrich, H., Plank, C. and Karamat, E., 1991. Aromatherapy: evidence for sedative effects of the essential oil of lavender after inhalation. Zeitschr. Naturf. Section C Biosci. 46: 1067-1072.
Buchbauer, G., Jirovetz, L., Jäger, W., Plank, C. and Dietrich, H., 1993. Fragrance compounds and essential oils with sedative effects upon inhalation. J. Pharm. Sci. 82: 660-665.
Buckle, J., 1993. Aromatherapy - a clinical trial. Acta Hortic. 332: 253-264.
Bucknall, R.A., Moores, H., Simms, R. and Hesp, B., 1973. Antiviral effects of aphidicolin, a new antibiotic produced by *Cephalosporium aphidicola*. Antimicr. Agents Chemother. 4: 294-298.
Buesa, C., Martinez-Gonzalez, J., Casals, N., Haro, D., Piulachs, M.D., Belles, X. and Hegardt, F.G., 1994. *Blattella germanica* has two HMG-CoA synthase genes. Both are regulated in the ovary during the gonadotrophic cycle. J. Biol. Chem. 269: 11707-11713.
Buhagiar, J.A., Podesta, M.T., Wilson, A.P., Micallef, M.J. and Ali, S., 1999. The induction of apoptosis in human melanoma, breast and ovarian cancer cell lines using an essential oil extract from the conifer *Tetraclinis articulata*. Anticancer Res. 19: 5435-5443.

Bui, E.T., Bradley, P.J. and Johnson, P.J., 1996. A common evolutionary origin for mitochondria and hydrogenosomes. Proc. Natl. Acad. Sci. USA 93: 9651-9656.

Burger, B.V., Nell, A.E., Spies, H.S., Roux, M. le and Bigalke, R.C., 1999. Mammalian exocrine secretions. XIII. Constituents of preorbital secretions of bontebok, *Damaliscus dorcas dorcas,* and blesbok, *D. d. phillipsi.* J. Chem. Ecol. 25: 2085-2097.

Burgett, M., 1980. The use of lemon balm (*Melissa officinalis*) for attracting honey bee swarms. Bee-world 61: 44-46.

Burke, Y.D., Stark, M., Roach, S.L., Sen, S.E. and Crowell, P.L., 1997. Inhibition of pancreatic cancer growth by the dietary isoprenoids farnesol and geraniol. Lipids 32: 151-156.

Bushunow, P., Reidenberg, M.M., Wasenko, J., Winfield, F., Lorenzo, B., Lemke, S., Himpler, B., Corona, R. and Coyle, T., 1999. Gossypol treatment of recurrent adult malignant gliomas. J. Neurooncol. 43: 79-86.

Buttery, R.G., Kamm, J.A. and Ling, L.C., 1984. Volatile components of red clover leaves, flowers, and seed pods: possible insect attractants. J. Agric. Food Chem. 32: 254-256.

Buttery, R.G., Xu, C. and Ling, L.C., 1985. Volatile components of wheat leaves (and stems): possible insect attractants. J. Agric. Food Chem. 33: 115-117.

Byers, J. A., 1995. Host-tree chemistry affecting colonization in bark beetles. In: Chemical Ecology of Insects 2 (eds R.T. Cardé and W.J. Bell): 154-213.

Caldwell, G.A., Naider, F. and Becker, J.M., 1995. Fungal lipopeptide mating pheromones: a model system for the study of protein prenylation. Microbiol. Rev. 59: 406-422.

Campbell, C.A.M., Dawson, G.W., Griffiths, D.C., Pettersson, J., Pickett, J.A., Wadhams, L.J. and Woodcock, C.M., 1990. Sex attractant pheromone of damson-hop aphid *Phorodon humuli* (Homoptera, Aphididae). J. Chem. Ecol. 16: 3455-3465.

Carl, W. and Emrich, L.S., 1991. Management of oral mucositis during local radiation and systemic chemotherapy: a study of 98 patients. J. Prosthet. Dent. 66: 361-369.

Carrol, R., 2000. Towards a new evolutionary synthesis. Tree 15: 27-32.

Cavalier-Smith, T., 2000. Membrane heredity and early chloroplast evolution. Trends Plant Sci. 5: 174-176.

Cerda, S.R., Wilkinson, J. 4th and Broitman, S.A., 1995. Regulation of cholesterol synthesis in four colonic adenocarcinoma cell lines. Lipids 30: 1083-1092.

Cerda, S.R., Wilkinson, J. 4th, Branch, S.K. and Broitman, S.A., 1999. Enhancement of sterol synthesis by the monoterpene perillyl alcohol is unaffected by competitive 3-hydroxy-3-methylglutaryl-CoA reductase inhibition. Lipids 34: 605-615.

Céspedes, C.L., Calderón, J.S., Gomez-Garibay, F., Segura, R., King-Diaz, B. and Lotina-Henssen, B., 1999. Phytogrowth properties of limonoids isolated from *Cedrela ciliolata.* J. Chem. Ecol. 25: 2665-2676.

Chambaut-Guerin, A.M., Herigault, S., Rouet-Benzineb, P., Rouher, C. and Lafuma, C., 2000. Induction of matrix metalloproteinase MMP-9 (92-kDa gelatinase) by retinoic acid in human neuroblastoma SKNBE cells: relevance to neuronal differentiation. J. Neurochem. 74: 508-517.

Chandler, R.F., Hooper, S.N., Hooper, D.L., Jamieson, W.D. and Lewis, E., 1982. Herbal remedies of the Maritime Indians: sterols and triterpenes of *Tanacetum vulgare* L. (tansy). Lipids 17: 102-106.

Chang, B.S.W. and Donoghue, M.J., 2000. Recreating ancestral proteins. Tree 15: 109-114.

Chang, J.Y. and Liu, L.Z., 1998. Neurotoxicity of cholesterol oxides on cultured cerebellar granule cells. Neurochem. Int. 32: 317-323.

Chang, J.Y., Phelan, K.D. and Chavis, J.A., 1998. Neurotoxicity of 25-OH-cholesterol on sympathetic neurons. Brain Res. Bull. 45: 615-622.

Chapman, R.F., 1999. It's all in the neurones. Entomol. exp. appl. 91 (Proc. SIP-10): 259-265.

Chapman, R.F., Bernays, E.A. and Simpson, S.J., 1981. Attraction and repulsion of the aphid, *Cavariella aegopodii*, by plant odors. J. Chem. Ecol. 7: 881-888.

Chew, B.P., Park, J.S., Wong, M.W. and Wong, T.S., 1999. A comparison of the anticancer activities of dietary β-carotene, canthaxanthin and astaxanthin in mice *in vivo*. Anticancer Res. 19: 1849-1853.

Chi, S., Kitanaka, C., Noguchi, K., Mochizuki, T., Nagashima, Y., Shirouzu, M., Fujita, H., Yoshida, M., Chen, W., Asai, A., Himeno, M., Yokoyama, S. and Kuchino, Y., 1999. Oncogenic Ras triggers cell suicide through the activation of a caspase-independent cell death program in human cancer cells. Oncogene 18: 2281-2290.

Choudhury, S.N., Ahmed, R.Z., Barthel, A. and Leclercq, P.A., 1998. Essential oil composition of *Cinnamomum bejolghota* (Buch-Ham.) sweet: a secondary muga (*Antheraea Assama* W/w) food plant from Assam, India. Sericologia 38: 473-478.

Chuah, C.H., Goh, S.H. and Tho, Y.P., 1986. Soldier defense secretions of the genus *Hospitalitermes* in Peninsular Malaysia. J. Chem. Ecol. 12: 701-712.

Cinatl, J. Jr, Cinatl, J., Kotchetkov, R., Driever, P.H., Bertels, S., Siems, K., Jas, G., Bindseil, K., Rabenau, H.F., Pouckova, P., Doerr, H.W. and Schwabe, D., 1999. Aphidicolin glycinate inhibits human neuroblastoma cell growth *in vivo*. Oncol. Rep. 6: 563-568.

Clare, K., Hardwick, S.J., Carpenter, K.L., Weeratunge, N. and Mitchinson, M.J., 1995. Toxicity of oxysterols to human monocyte-macrophages. Atherosclerosis 118: 67-75.

Clark, L.J., Hamilton, J.G.C., Chapman, J.V., Rhodes, M.J.C. and Hallahan, D.L., 1997. Analysis of monoterpenoids in glandular trichomes of the catmint *Nepeta raemosa*. The Plant Journal 11: 1387-1393.

Clarke, S., 1992. Protein isoprenylation and methylation at carboxyl-terminal cysteine residues. Annu. Rev. Biochem. 61: 355-386.

Clarke, S.R., Salom, S.M., Billings, R.F., Berisford, C.W., Upton, W.W., McClellan, Q.C. and Dalusky, M.J., 1999. A scentsible approach to controlling southern pine beetles: two new tactics using verbenone. J. Forestry 97: 26-31.

Clegg, R.J., Middleton, B., Bell, D.G. and White, D.A., 1982. The mechanism of cyclic monoterpene inhibition of hepatic 3-hydroxy-3-methyl glutaryl coenzyme A reductase *in vivo* in the rat. J. Biol. Chem. 257: 2294-2299.

Clynes, R.A., Towers, T.L., Presta, L.G. and Ravetch, J.V., 2000. Inhibitory Fc receptors modulate *in vivo* cytotoxicity against tumor targets. Nat. Med. 6: 443-446.

Cohen, L.H., Pieterman, E., van Leeuwen, R.E., Du, J., Negre-Aminou, P., Valentijn, A.R., Overhand, M., van der Marel, G.A. and van Boom, J.H., 1999. Inhibition of human smooth muscle cell proliferation in culture by farnesyl pyrophosphate analogues, inhibitors of in vitro protein: farnesyl transferase. Biochem. Pharmacol. 57: 365-373.

Collins, R.P., 1976. Terpenes and odoriferous materials from microorganisms. Proc. Symp. Natural Flavors and Odors, Connecticut, 1975. Lloydia 39: 20-24.

Collins, R.P. and Halim, A.F., 1970. Production of monoterpenes by the filamentous fungus *Ceratocystis variospora*. Lloydia 33: 481.

Concepcion, J.L., Gonzalez-Pacanowska, D. and Urbina, J.A., 1998. 3-Hydroxy-3-methyl-glutaryl-CoA reductase in *Trypanosoma (Schizotrypanum) cruzi*: subcellular localization and kinetic properties. Arch. Biochem. Biophys. 352: 114-120.

Conly, J.M., Stein, K., Worobetz, L. and Rutledge-Harding, S., 1994. The contribution of vitamin K2 (menaquinones) produced by the intestinal microflora to human nutritional requirements for vitamin K. Am. J. Gastroenterol. 89: 915-923.

Corbet, S.A., Danahar, G.W., King, V., Chalmers, C.L. and Tiley, C.F., 1995. Surfactant-enhanced essential oils as mosquito larvicides. Entomol. exp. appl. 75: 229-236.

Cornelius, M.L., Grace, J.K. and Yates, J.R., 1997. Toxicity of monoterpenoids and other natural products to the Formosan subterranean termite (Isoptera: Rhinotermitidae). J. Econ. Entomol. 90: 320-325.

Correll, C.C. and Edwards, P.A., 1994. Identification of farnesol as the non-sterol derivative of mevalonic acid required for the accelerated degradation of 3-hydroxy-3-methylglutaryl-coenzyme A reductase. J. Biol. Chem. 269: 17390-17393.

Cortes, M.L., de Felipe, P., Martin, V., Hughes, M.A. and Izquierdo, M., 1998. Successful use of a plant gene in the treatment of cancer *in vivo*. Gene Ther. 5: 1499-1507.

Cosse, A.A., Todd, J.L., Millar, J.G., Martinez, L.A. and Baker, T.C., 1995. Electroantennographic and coupled gas chromatographic responses of the mediterranean fruit fly, *Ceratitis capitata*, to mate-produced volatiles and mango odor. J. Chem. Ecol. 221: 1823-1836.

Costantini, P., Jacotot, E., Decaudin, D. and Kroemer, G., 2000. Mitochondrion as a novel target of anticancer chemotherapy. J. Natl. Cancer Inst. 92: 1042-1053.

Couillaud, F. and Rossignol, F., 1991. Characterization and regulation of HMG-CoA synthase in *Locusta migratoria* corpora allata. Arch. Insect Biochem. Physiol. 18: 273-283.

Cowan, M.M., 1999. Plant products as antimicrobial agents. Clin. Microbiol. Rev. 12: 564-582.

Cox, A.D., Garcia, A.M., Westwick, J.K., Kowalcyk, J.J., Lewis, M.D., Brenner, D.A. and Der, C.J., 1995. The CAAX peptidomimetic compound B581 specifically blocks farnesylated, but not geranylgeranylated or myristylated, oncogenic Ras signalling and transformation. J. Biol. Chem. 269: 19203-19206.

Crawley, M.J., 1989. Insect herbivores and plant population dynamics. Annu. Rev. Entomol. 34: 531-564.

Crick, D.C., Andres, D.A., Danesi, R., Macchia, M. and Waechter, C.J., 1998. Geranylgeraniol overcomes the block of cell proliferation by lovastatin in C6 glioma cells. J. Neurochem. 70: 2397-2405.

Crick, D.C., Andres, D.A. and Waechter, C.J., 1997. Novel salvage pathway utilizing farnesol and geranylgeraniol for protein isoprenylation. Biochem. Biophys. Res. Comm. 237: 483-487.

Crick, D.C., Waechter, C.J. and Andres, D.A., 1994. Utilization of geranylgeraniol for protein isoprenylation in C6 glial cells. Biochem. Biophys. Res. Comm. 205: 955-961.

Crock, J., Widung, M. and Croteau, R., 1997. Isolation and bacterial expression of a sesquiterpene synthase cDNA clone from peppermint (*Mentha* x *piperita*, L.) that produces the aphid alarm pheromone (E)-β-farnesene. Proc. Natl. Acad. Sci. USA 94: 1233-1238.

Croft, K.P.C., Jüttner, F. and Slusarenko, A.J., 1993. Volatile products of the lipoxygenase pathway evolved from *Phaseolus vulgaris* (L.) leaves inoculated with *Pseudomonas syringae* pv *phaseolicola*. Plant Physiol. 101: 13-24.

Crook, D.J. and Mordue (Luntz), A.J., 1999. Olfactory responses and sensilla morphology of the blackcurrant leaf midge *Dasineura tetensi*. Entomol. exp. appl. 91: 37-50.

Croteau, R., 1988. Catabolism of monoterpenes in essential oil plants. In: Flavors and Fragrances; A World Perspective (eds B.M. Lawrence, B.D. Mookherjee and B.J. Willis), Elsevier, Amsterdam, etc.: 65-84.

Croteau, R., Alonso, W.R., Koepp, A.E. and Johnson, M.A., 1994. Biosynthesis of monoterpenes: partial purification, characterization, and mechanism of action of 1,8-cineole synthase. Arch. Biochem. Biophys. 309: 184-192.

Crowell, P.L., 1999. Prevention and therapy of cancer by dietary monoterpenes. J. Nutr. 129: S775-778.

Crowell, P.L., Ren, Z., Lin, S., Vedejs, E. and Gould, M.N., 1994. Structure-activity relationships among monoterpene inhibitors of protein isoprenylation and cell proliferation. Biochem. Pharmacol. 8: 1405-1415.

Dairi, T., Motohira, Y., Kuzuyama, T. and Takahashi, S., 2000. Cloning of the gene encoding 3-hydroxy-3-methylglutaryl coenzyme A reductase from terpenoid antibiotic-producing *Streptomyces* strains. Mol. Gen. Genet. 262: 957-964.

Davis, T.A., White, C.A., Grillo-Lopez, A.J., Velasquez, W.S., Link, B., Maloney, D.G., Dillman, R.O., Williams, M.E., Mohrbacher, A., Weaver, R., Dowden, S. and Levy, R., 1999. Single-agent monoclonal antibody efficacy in bulky non-Hodgkin's lymphoma: results of a phase II trial of rituximab. J. Clin. Oncol. 17: 1851-1857.

Dawson, G.W., Griffiths, D.C., James, N.F., Mudd, A., Pickett, J.A., Wadhams, L.J. and Woodcock, C.M., 1987. Identification of an aphid sex pheromone. Nature 325: 614-616.

Dawson, G.W., Griffiths, D.C., Pickett, J.A., Smith, M.C. and Woodcock, C.M., 1984. Natural inhibition of the aphid alarm pheromone. Entomol. exp. appl. 36: 197-199.

Dawson, G.W., Griffiths, D.C., Merritt, L.A., Mudd, A., Pickett, J.A., Wadhams, L.J. and Woodcock, C.M., 1990. Aphid semiochemicals: a review and recent advances on the sex pheromone. J. Chem. Ecol. 16: 3019-3030.

De Kort, C.A. and Granger, N.A., 1981. Regulation of the juvenile hormone titer. Annu. Rev. Entomol. 26: 1-28.

Del Villar, K., Urano, J., Guo, L. and Tamanoi, F., 1999. A mutant form of human protein farnesyltransferase exhibits increased resistance to farnesyltransferase inhibitors. J. Biol. Chem. 274: 27010-27017.

Den Ouden, H., Bulsink, A. and Theunissen, J., 1996. Compounds repellent to *Delia radicum* (L.) (Dipt., Anthomyiidae). J. Appl. Ent. 120: 427-432.

Desjardins, A.E., Hohn, T.M. and McCormick, S.P., 1993. Trichothecene biosynthesis in *Fusarium* species: chemistry, genetics, and significance. Microbiol. Rev. 57: 595-604.

Devlin, D.R. and Borden, J.H., 1994. Efficacy of antiaggregants for the pine engraver, *Ips pini* (Coleoptera: Scolytidae). Can. J. Forest Res. 24: 2469-2476.

Devonshire, A. L., 1989. Resistance of aphids to insecticides. In: Aphids, their Biology, Natural Enemies and Control, Vol. C (eds A.K. Minks and P. Harrewijn), Elsevier Science Publ. Amsterdam: 123-139.

Dhaene, K., van Marck, E. and Parwaresch, R., 2000. Telomeres, telomerase and cancer: an up-date. Virchows Arch. 437: 1-16.

Dicke, M., van Beek, T.A., Posthumus, M.A., Ben Dom, N., Bokhoven, H. and de Groot, Æ., 1990. Isolation and identification of a volatile kairomone that affects acarine predator-prey interactions. Involvement of host plant in its production. J. Chem. Ecol. 16: 381-396.

Dicke, M., Gols, R., Ludeking, D. and Posthumus, M.A., 1999. Jasmonic acid and herbivory differentially induce carnivore attracting plant volatiles in Lima bean plants. J. Chem. Ecol. 25: 1907-1922.

Dicke, M., Sabelis, M.W., Takabayashi, J., Bruin, J. and Posthumus, M.A., 1990b. Plant strategies of manipulating predator-prey interactions through allelochemicals: prospects for application in pest control. J. Chem. Ecol. 16: 3091-3118.

Di Croce, L., Vicent, G.P., Pecci, A., Bruscalupi, G., Trentalance, A. and Beato, M., 1999. The promoter of the rat 3-hydroxy-3-methylglutaryl coenzyme A reductase gene contains a tissue-specific estrogen-responsive region. Mol. Endocrinol. 13: 1225-1236.

Di Paolo, A., Danesi, R., Nardini, D., Bocci, G., Innocenti, F., Fogli, S., Barachini, S., Marchetti, A., Bevilacqua, G. and Del Tacca, M., 2000. Manumycin inhibits ras signal transduction pathway and induces apoptosis in COLO320-DM human colon tumour cells. Br. J. Cancer 82: 905-912.

Disch, A., Schwender, J., Muller, C., Lichtenthaler, H.K. and Rohmer, M., 1998a. Distribution of the mevalonate and glyceraldehyde phosphate/pyruvate pathways for isoprenoid biosynthesis in unicellular algae and the cyanobacterium *Synechocystis* PCC6714. Biochem. J. 333: 381-388.

Disch, A., Hemmerlin, A., Bach, T.J. and Rohmer, M., 1998b. Mevalonate-derived isopentenyl diphosphate is the biosynthetic precursor of ubiquinone prenyl side chain in tobacco BY-2 cells. Biochem. J. 331: 615-621.

Dobson, H.E.M., 1991. Analysis of flower and pollen volatiles. In: Essential oils and Waxes (eds H.F. Linskens and J.F. Jackson), Springer-Verlag, Berlin: 231-251.

Dobson, H.E.M., Bergström, G. and Groth, J., 1990. Differences in fragrance chemistry between flower parts of *Rosa rugosa* Thumb. (Rosaceae). Israel J. Bot. 39: 143-156.

Dolara, P., Luceri, C., Ghelardini, C., Monserrat, S.A., Aiolli, S., Luceri, F. and Lodovici, M., 1996. Analgesic effects of myrrh. Nature 379: 29.

Dolle, P., Ruberte, E., Leroy, P., Morris-Kay, G. and Chambon, P., 1990. Retinoic acid receptors and cellular retinoid binding proteins. I. A systematic study of their differential pattern of transcription during mouse organogenesis. Development 110: 1133-1151.

Donaldson, M.J., Skoumas, V., Watson, M., Ashworth, P.A., Ryder, H., Moore, M. and Coombes, R.C., 1999. XR3054, structurally related to limonene, is a novel inhibitor of farnesyl protein transferase. Eur. J. Cancer 35: 1014-1019.

Doolittle, R.F., 1998. Microbial genomes opened up. Nature 392: 339-342.

Doolittle, W.F., 1996. Some aspects of the biology of cells and their possible evolutionary significance. In: Evolution of Microbial Life: 54th Symp. of the Society for General Microbiology (eds D.M. Roberts, P. Sharp, G. Alderson and M.A. Collins), Cambridge University Press: 1-21.

Doolittle, W.F., 1998. You are what you eat: a gene transfer ratchet could account for bacterial genes in eukaryotic nuclear chromosomes. Trends in Genetics 14: 307-311.

Douglas, A.E., 1992. Microbial brokers of insect-plant interactions. In: Proc. 8th Int. Symp. Insect-Plant Relationships (eds S.B.J. Menken, J.H. Visser and P. Harrewijn) Kluwer Acad. Publ. Dordrecht, etc.: 329-336.

Douglas, A.E. and Prosser, W.A., 1992. Synthesis of the essential amino acid tryptophan in the pea aphid (*Acyrthosiphon pisum*) symbiosis. J. Insect Physiol. 38: 565-568.

Dudareva, N., Cseke, L., Pichersky, E. and Blanc, V.M., 1996. Evolution of floral scent in *Clarkia*: novel patterns of S-linalool synthase gene expression in the *C. breweri* flower. Plant Cell 8: 1137-1148.

Dulac, C., 2000. The physiology of taste, vintage 2000. Cell 100: 607-610.

Duthie-Holt, M.A., Borden, J.H. and Rankin, L., 1999. Translocation and efficacy of a neem-based insecticide in Lodgepole pine using *Ips pini* (Coleoptera: Scolytidae) as an indicator species. J. Econ. Entomol. 92: 180-186.

Ebrahimi, F.A. and Chess, A., 1998. Olfactory G-proteins: simple and complex signal transduction. Curr. Biol. 8: R431-433.

Egan, S.E. and Weinberg, R.A., 1993. The pathway to signal achievement. Nature 365: 781-783.

Einhorn, L.H., Roth, B.J., Ansari, R., Dreicer, R., Gonin, R. and Loehrer, P.J., 1994. Phase II trial of vinblastine, ifosfamide, and gallium combination chemotherapy in metastatic urothelial carcinoma. J. Clin. Oncol. 12: 2271-2276.

Elad, G., Paz, A., Haklai, R., Marciano, D., Cox, A. and Kloog, Y., 1999. Targeting of K-Ras 4B by S-trans, trans-farnesyl thiosalicylic acid. Biochem. Biophys. Acta 1452: 228-242.

Elez, R., Piiper, A., Giannini, C.D., Brendel, M. and Zeuzem, S., 2000. Polo-like kinase 1, a new target for antisense tumor therapy. Biochem. Biophys. Res. Comm. 269: 352-356.

Ellis, M.B. and Ellis, J.P., 1985. Microfungi On Land Plants. An Identification Handbook. Croom Helm, London & Sydney, 818 pp.

Elson, C.E., Peffley, D.M., Hentosh, P. and Mo, H., 1999. Isoprenoid-mediated inhibition of mevalonate synthesis: potential application to cancer. Proc. Soc. Exp. Biol. Med. 221: 294-311.

Elstner, E., Linker-Israeli, M., Umiel, T., Le, J., Grillier, I., Said, J., Shintaku, I.P., Krajewski, S., Reed, J.C., Binderup, L. and Koeffler, H.P., 1996. Combination of a potent 20-epi-vitamin D3 analogue (KH 1060) with 9-*cis*-retinoic acid irreversibly inhibits clonal growth, decreases Bcl-2 expression, and induces apoptosis in Hl-60 leukemic cells. Cancer Res. 56: 3570-3576.

End, D.W., 1999. Farnesyl protein transferase inhibitors and other therapies targeting the Ras signal transduction pathway. Invest. New Drugs 17: 241-258.

Ernst, E. and Cassileth, B.R., 1999. How useful are unconventional cancer treatments? Eur. J. Cancer 35: 1608-1613.

Es-Saady, D., Simon, A., Ollier, M., Maurizis, J.C., Chulia, A.J. and Delage, C., 1996. Inhibitory effect of ursolic acid on B16 proliferation through cell cycle arrest. Cancer Lett. 106: 193-197.

Eummer, J.T., Gibbs, B.S., Zahn, T.J., Sebolt-Leopold, J.S. and Gibbs, R.A., 1999. Novel limonene phosphonate and farnesyl diphosphate analogues: design, synthesis, and evaluation as potential protein-farnesyl transferase inhibitors. Bioorg. Med. Chem. 7: 241-250.

Evans, K.A., 1992. The olfactory and behavioural response of seed weevils, *Ceutorhynchus assimilis*, to oilseed rape volatiles. Proc. 8th Int. Symp. on Insect-Plant Relationships (eds S.B.J. Menken, J.H. Visser and P. Harrewijn), Kluwer Acad. Publ. Dordrecht, etc.: 107-108.

Farnsworth, C.C., Gelb, M.H. and Glomset, J.A., 1990. Identification of geranylgeranyl-modified proteins in HeLa cells. Science 247: 320-322.

Feeny, P., 1991. Theories of plant chemical defense: a brief historical survey. Proc. 7th Int. Symp. on Insect-Plant Relationships (eds Á. Szentesi and T. Jermy), Akadémiai Kiadó, Budapest:163-175.

Feng, Q.L., Ladd, T.R., Tomkins, B.L., Sundaram, M., Sohi, S.S., Retnakaran, A., Davey, K.G. and Palli, S.R., 1999. Spruce budworm (*Choristoneura fumiferana*) juvenile hormone esterase: hormonal regulation, developmental expression and cDNA cloning. Mol. Endocrinol. 148: 95-108.

Ferrante, G., Richards, J.C. and Sprott, G.D., 1990. Structures of polar lipids from the thermophilic, deep-sea archaeobacterium *Methanococcus jannaschii*. Biochem. Cell Biol. 68: 274-283.

Ferry, J.G., 1999. Enzymology of one-carbon metabolism in methanogenic pathways. FEMS Microbiol. Rev. 23: 13-38.

Feyereisen, R., 1984. Regulation of juvenile hormone titer: Synthesis. In: Comprehensive Insect Physiology, Biochemistry and Pharmacology (eds G.A. Kerkut and L.I. Gilbert), Pergamon, Oxford: 391-429.

Fichera, M.E. and Roos, D.S., 1997. A plastid organelle as a drug target in apicomplexan parasites. Nature 390: 407-409.

Fischer, N.H., Williamson, G.B., Weidenhamer, J.D. and Richardson, D.R., 1994. In search of allelopathy in the Florida scrub: the role of terpenoids. J. Chem. Ecol. 20: 1355-1380.

Flach, J., Antoni, I., Villemin, P., Bentzen, C.L. and Niesor, E.J., 2000. The mevalonate/ isoprenoid pathway inhibitor apomine (SR-45023A) is antiproliferative and induces apoptosis similar to farnesol. Biochem. Biophys. Res. Comm. 270: 240-246.

Flechtmann, C.A.H., Dalusky, M.J. and Berisford, C.W., 1999. Bark and ambrosia beetle (Coleoptera: Scolytidae) responses to volatiles from aging loblolly pine billets. Environ. Entomol. 28: 638-648.

Flores, H.E., Vivanco, J.M. and Loyola-Vargas, V.M., 1999. 'Radicle' biochemistry: the biology of root-specific metabolism. Trends Plant Sci. 4: 220-226.

Floyd, M.A., Evans, D.A. and Howse, P.E., 1976. Electrophysiological and behavioral studies on naturally occurring repellents to *Reticulitermes lucifugus*. J. Insect Physiol. 22: 697-701.

Forman, B.M., Goode, E., Chen, J., Oro, A.E., Bradley, D.J., Perlmann, T., Noonan, D.J., Burka, L.T., McMorris, T. and Lamph, W.W., 1995. Identification of a nuclear receptor that is activated by farnesol metabolites. Cell 81: 687-693.

Forman, B.M., Ruan, B., Chen, J., Schroepfer, G.J. Jr and Evans, R.M., 1997. The orphan nuclear receptor LXRα is positively and negatively regulated by distinct products of mevalonate metabolism. Proc. Natl. Acad. Sci. USA 94: 10588-10593.

Frame, S. and Balmain, A., 2000. Integration of positive and negative growth signals during Ras pathway activation *in vivo*. Curr. Opin. Genet. Dev. 10: 106-113.

Franzios, G., Mirotsou, M., Hatziapostolou, E., Kral, J., Scouras, Z.G. and Mavragani-Tsipidou, P., 1997. Insecticidal and genotoxic activities on mint essential oils. J. Agric. Food Chem. 45: 2690-2694.

Frazier, J.L. and Hanson, F.E., 1986. Electrophysiological recording and analysis of insect chemosensory responses. In: Insect-Plant Interactions (eds J.R. Miller and T.A. Miller), Springer-Verlag, New York, etc.: 285-342.

Freije, J.M., Blay, P., Pendas, A.M., Cadinanos, J., Crespo, P. and Lopez-Otin, C., 1999. Identification and chromosomal location of two human genes encoding enzymes potentially involved in proteolytic maturation of farnesylated proteins. Genomics 58: 270-280.

Frey, J.E. and Bush, G.L., 1990. *Rhagoletis* sibling species and host races differ in host odor recognition. Entomol. exp. appl. 57: 123-131.

Frey, J.E., Cortada, R.V. and Helbling, H., 1994. The potential of flower odours for use in population monitoring of western flower thrips *Frankliniella occidentalis* Perg. (Thysanoptera: Thripidae). Biocontrol Sci. Techn. 4: 177-186.

Frohman, L.A. and Kineman, R.D., 1999. Hormonal control of growth. In: Handbook of Physiology (eds J.L. Kostyo and H.M. Goodman), Oxford University Press, New York: 189-221.

Fuchs, D.A. and Johnson, R.K., 1978. Cytologic evidence that taxol, an antineoplastic agent from *Taxus brevifolia* acts as a mitotic spindle poison. Cancer Treat. Rep. 62: 1219-1222.

Fulda, S., Jeremias, I., Steiner, H.H., Pietsch, T. and Debatin, K.M., 1999. Betulinic acid: a new cytotoxic agent against malignant brain-tumor cells. Int. J. Cancer 82: 435-441.

Fulton, G.J., Barber, L., Svendsen, E., Hagen, P.O. and Davies, M.G., 1997. Oral monoterpene therapy (perillyl alcohol) reduces vein graft intimal hyperplasia. J. Surg. Res. 69: 128-134.

Fulzele, D.P., Heble, M.R. and Rao, P.S., 1995. Production of terpenoid from *Artemisia annua* plantlet cultures in bioreactor. J. Biotechn. 40: 139-143.

Gangrade, S.K., Shrivastava, R.D., Sharma, O.P., Moghe, M.N. and Trivedi, K.C., 1990. Evaluation of some essential oils for antibacterial properties. Indian Perfumer 34: 204-208.

Gao, L.Y. and Kwaik, Y.A., 2000. The modulation of host cell apoptosis by intracellular bacterial pathogens. Trends in Microbiol. 8: 306-312.

Gardner, R.G. and Hampton, R.Y., 1999. A highly conserved signal controls degradation of 3-hydroxy-3-methylglutaryl-coenzyme A (HMG-CoA) reductase in eukaryotes. J. Biol. Chem. 274: 31671-31678.

Gardner, R., Cronin, S., Leader, B., Rine, J. and Hampton, R.Y., 1998. Sequence determinants for regulated degradation of yeast 3-hydroxy-3-methylglutaryl-CoA reductase, an integral endoplasmatic reticulum membrane protein. Mol. Biol. Cell 9: 2611-2626.

Geervliet, J.B.F., Vet, L.E.M. and Dicke, M., 1994. Volatiles from damaged plants as major cues in long-range host-searching by the specialist parasitoid *Cotesia rubecula*. Entomol. exp. appl. 73: 289-297.

Gelb, M. H., 1997. Protein prenylation, et cetera: signal transduction in two dimensions. Science 275: 1750-1751.

Gershenzon, J., 1994. Metabolic costs of terpenoid accumulation in higher plants. J. Chem. Ecol. 20: 1281-1328.

Gershenzon, J., Maffei, M. and Croteau, R., 1989. Biochemical and histochemical localization of monoterpene biosynthesis in the glandular trichomes of spearmint (*Mentha spicata*). Plant Physiol. 89: 1351-1357.

Gershenzon, J., McConkey, M.E. and Croteau, R.B., 2000. Regulation of monoterpene accumulation in leaves of peppermint. Plant Physiol. 122: 205-213.

Gershenzon, J., Murtagh, G.J. and Croteau, R., 1993. Absence of rapid terpene turnover in several diverse species of terpene-accumulating plants. Oecologia 96: 583-592.

Ghosh, S., Thorogood, P. and Ferretti, P., 1996. Regeneration of lower and upper jaws in urodeles is differentially affected by retinoic acid. Int. J. Dev. Biol. 40: 1161-1170.

Gibbs, B.S., Zahn, T.J., Mu, Y., Sebolt-Leopold, J.S. and Gibbs, R.A., 1999. Novel farnesol and geranylgeraniol analogues: a potential new class of anticancer agents directed against protein prenylation. J. Med. Chem. 42: 3800-3808.

Gilbert, L.I. and Schneiderman, H.A., 1958. The occurrence of substances with juvenile hormone activity in the adrenal cortex of vertebrates. Science 128: 844.

Gille, H. and Downward, J., 1999. Multiple Ras effector pathways contribute to G(1) cell cycle progression. J. Biol. Chem. 274: 22033-22040.

Giordano, W., Avalos, J., Fernández-Martin, R., Cerdá-Olmedo, E. and Domenech, C.E., 1999. Lovastatin inhibits the production of gibberellins but not sterol or carotenoid biosynthesis in *Gibberella fujikuroi*. Microbiol. 145: 2997-3002.

Giovannuci, E., Ascherio, A., Rimm, E.B., Stampfer, M.J., Colditz, G.A. and Willet, W.C., 1995. Intake of carotenoids and retinol in relation to risk of prostate cancer. J. Natl. Cancer Inst. 87: 1767-1776.

Giri, R.K., Parija, T. and Das, B.R., 1999. D-limonene chemoprevention of hepatocarcinogenesis in AKR mice: inhibition of c-jun and c-myc. Oncol. Rep. 6: 1123-1127.

Goldstein, J.L. and Brown, M.S., 1990. Regulation of the mevalonate pathway. Nature 343: 425-430.

Gols, G.J.Z., van Loon, J.J.A. and Messchendorp, L., 1996. Antifeedant and toxic effects of drimanes on Colorado potato beetle larvae. Entomol. exp. appl. 79: 69-76.

González-Coloma, A., Cabrera, M., Castanera, R. and Gutiérrez, C., 1995. Antifeedant and toxic effects of sesquiterpenes from *Senecio palmensi* to Colorado potato beetle. J. Chem. Ecol. 21: 1255-1270.

Gould, F., 1984. Role of behaviour in the evolution of insect adaptation to insecticides and resistant host plants. Bull. Ent. Soc. Am. 30: 34-41.

Gould, F., 1991. Arthropod behaviour and the efficacy of plant protectants. Annu. Rev. Entomol. 36: 305-330.

Gould, M.N., 1997. Cancer chemoprevention and therapy by monoterpenes. Environ. Health Perspect. 105 (S4): 977-979.

Gould, M.N., Maltzman, T.H., Tanner, M.A., Boston J.L., Hurt L.M. and Elson, C.E., 1987. Anticarcinogenic effects of terpenoids in orange peel oil. Proc. 78th Meeting Amer. Assoc. Cancer Res. 28: 153 (609).

Gould, M.N., Wacker, W.D. and Maltzman, T.H., 1990. Chemoprevention and chemotherapy of mammary tumors by monoterpenoids. In: Progress in clinical and biological research Vol 347. Mutagens and Carcinogens in the Diet (eds M.W. Pariza, H.U. Aeschbacher, J.S. Felton and S. Sato), Wiley-Liss Inc.: 255-268.

Graffi, A., Gummel, H., Jung, F., Krautwald, A. and Rapoport, S.M., 1959. Zum Wirkungsmechanismus des 1,4-dimethyl-7-isopropyl Azulens auf die Tumorzelle. Acta Biol. Med. Germ. 2: 555-567.

Greany, P.D., Styer, S.C., Davis, P.L., Shaw, P.E. and Chambers, D.L., 1983. Biochemical resistance of citrus to fruit flies. Demonstration and elucidation of resistance to the Caribbean fruit fly, *Anastrepha suspensa*. Entomol. exp. appl. 34: 40-50.

Green, D.R. and Reed, J.C., 1998. Mitochondria and apoptosis. Science 281: 1309-1312.

Greenberg, M.L. and Lopes, J.M., 1996. Genetic regulation of phospholipid biosynthesis in *Saccharomyces cerevisiae*. Microbiol. Rev. 60: 1-20.

Guerin, P.M. and Visser, J.H., 1980. Electroantennogram responses of the carrot fly, *Psila rosae*, to volatile plant compounds. Physiol. Entomol. 5: 111-119.

Gunderson, C.A., Samuelian, J.H., Evans, C.K. and Brattsten, L.B., 1985. Effects of the mint monoterpene pulegone on *Spodoptera eridania* (Lepidoptera: Noctuidae). Environ. Ent. 14: 859-863.

Gustafson, J.E., Liew, Y.C., Chew, S., Markham, J., Bell, H.C., Wyllie, S.G. and Warmington, J.R., 1998. Effects of tea tree oil on *Escherichia coli*. Lett. Appl. Microbiol. 26: 194-198.

Gut, J. and van Oosten, A.M., 1985. Functional significance of the alarm pheromone composition in various morphs of the green peach aphid, *Myzus persicae*. Entomol. exp. appl. 37: 199-204.

Gut, J., Harrewijn, P., van Oosten, A.M. and Piron, P.G.M., 1991. Terpenes as regulating factors in aphid morphology. Proc. Conf. Insect Chem. Ecol., Tábor (ed. I. Hrdý). Academia Praha/SPB Acad. Publ. The Hague: 347-360.

Guthrie, N., Gapor, A., Chambers, A.F. and Caroll, K.K., 1997. Inhibition of proliferation of estrogen receptor-negative MDA-MB-435 and -positive MCF-7 human breast cancer cells by palm oil tocotrienols and tamoxifen, alone and in combination. J. Nutr. 127(S): 544-548.

Gutiérrez, C., Fereres, A., Reina, M., Cabrera, R. and González-Coloma, A., 1997. Behavioral and sublethal effects of structurally related lower terpenes on *Myzus persicae*. J. Chem. Ecol. 23: 1641-1650.

Haag, J.D., Lindstrom, M.J. and Gould, M.N., 1992. Limonene-induced regression of mammary carcinomas. Cancer Res. 52: 4021-4026.

Haga, K., Suzuki, T., Kodama, S. and Kuwahara, Y., 1990. Secretion of Thrips. VI. Identification of ß-myrcene from *Thlibothrips isunoki* (Thysanoptera: Phlaeothripidae). Appl. Ent. Zool. 25: 138-139.

Hägele, B.F., Harmatha J., Pavlik, M. and Rowell-Rahier, M., 1996. Sesquiterpenes from the Senecioneae and their effect on food choice of the specialised leaf beetles *Oreina cacaliae, Oreina speciosissima* and the generalist snail *Arianta arbustorum*. Entomol. exp. appl. 80: 169-172.

Hägele, B.F. and Rowell-Rahier, M. 1996. Genetically based variation and influence of light on sesquiterpene, pyrrolizidine alkaloid, and nitrogen content of leaves of *Adenostyles alliariae* and *A. alpina* (Asteraceae). In: Abstracts ISCE 13th Annual Meeting (eds P. Drašar and I. Valterová): 55.

Hallahan, D.L. and West, J.M., 1995. Cytochrome P-450 in plant/insect interactions: geraniol 10-hydroxylase and the biosynthesis of iridoid monoterpenoids. Drug Metabol. Drug Interact. 12: 369-382.

Hammerschmidt, R., 1999. Phytoalexin: what have we learned after 60 years? Ann. Rev. Phytopath. 37: 285-306.

Hampton, R.Y. and Bhakta, H., 1997. Ubiquitin-mediated regulation of 3-hydroxy-3-methylglutaryl-CoA reductase. Proc. Natl. Acad. Sci. USA 94: 12944-12948.

Han, S.S., Chung, S.T., Robertson, D.A., Ranjan, D. and Bondada, S., 1999. Curcumin causes growth arrest and apoptosis of B cell lymphoma by downregulation of egr-1, c-myc, bcl-XL, NF-kappa B, and p53. Clin. Immunol. 93: 152-161.

Hancock, R.E.W., 1997. The bacterial outer membrane is a drug barrier. Trends in Microbiol. 37: 37-41.

Hanley, K., Jiang, Y., Crumrine, D., Bass, N.M., Appel, R., Elias, P.M., Williams, M.L. and Feingold, K.R., 1997. Activators of the nuclear hormone receptors PPAα and FXR accelerate the development of the fetal epidermal permeability barrier. J. Clin. Invest. 100: 705-712.

Harborne, J.B., 1993. Introduction to Ecological Biochemistry (4th edition). Acad. Press Ltd, London etc., 318 pp.

Hardcastle, I.R., Rowlands, M.G., Barber, A.M., Grimshaw, R.M., Mohan, M.K., Nutley, B.P. and Jarman, M., 1999. Inhibition of protein prenylation by metabolites of limonene. Biochem. Pharmacol. 57: 801-809.

Hardie, J., Baker, F.C., Jamieson, G.C., Lees, A.D. and Schooley, D.A., 1985. The identification of an aphid juvenile hormone and its titre in relation to photoperiod. Physiol. Entomol. 10: 297-302.

Hardie, J., Holyoak, M., Nicholas, J., Nottingham, S.F., Pickett, J.A., Wadhams, L.J. and Woodcock, C.M., 1990. Aphid sex pheromone components: age-dependent release by females and species-specific male response. Chemoecology 1: 63-68.

Hardie, J., Isaacs, R., Pickett, J.A., Wadhams, L.J. and Woodcock, C.M., 1994. Methyl salicylate and (-)-(1R,5S)-myrtenal are plant-derived repellents for black bean aphid, *Aphis fabae* Scop. (Homoptera: Aphididae). J. Chem. Ecol. 20: 2847-2855.

Hardie, J., Peace, L., Pickett, J.A., Smiley, D.W.M., Storer, J.R. and Wadhams, L.J., 1997. Sex pheromone stereochemistry and purity affect field catches of male aphids. J. Chem. Ecol. 23: 2547-2554.

Hardie, J., Visser, J.H. and Piron, P.G.M., 1994b. Perception of volatiles associated with sex and food by different adult forms of the black bean aphid, *Aphis fabae*. Physiol. Entomol. 19: 278-284.

Hardie, J., Visser, J.H. and Piron, P.G.M., 1995. Peripheral odour perception by adult aphid forms with the same genotype but different host-plant preferences. J. Insect Physiol. 41: 91-97.

Harker, M. and Bramley, P.M., 1999. Expression of prokaryotic 1-deoxy-D-xylulose-5-phosphatases in *Escherichia coli* increases carotenoid and ubiquinone biosynthesis. FEBS Lett. 448: 115-119.

Harrewijn, P., 1976. Role of monoamine metabolism in wing dimorphism of the aphid *Myzus persicae*. Comp. Biochem. Physiol. 55C: 147-153.

Harrewijn, P., 1993. Insect-plant relationships: attack and defense in time and space. Proc. Exper. & Appl. Entomol., N.E.V. Amsterdam (eds M.J. Sommeijer and J. van der Blom) 4: 3-20.

Harrewijn, P. and Kayser, H., 1997. Pymetrozine, a fast-acting and selective inhibitor of aphid feeding. *In situ* studies with electronic monitoring of feeding behaviour. Pest. Sci. 49: 130-140.

Harrewijn, P. and Noordink, J.P.W., 1971. Taste perception of *Myzus persicae* in relation to food uptake and developmental processes. Entomol. exp. appl. 14: 413-419.

Harrewijn, P., Piron, P.G.M., Gut, J. and van Oosten, A.M., 1991. The role of terpenoid end products in development of the aphid *Megoura viciae* (Buckton). Proc. Exper. & Appl. Entomol., N.E.V. Amsterdam (eds M.J. Sommeijer and J. van der Blom) 2: 74-79.

Harrewijn, P., van Oosten, A.M. and Piron, P.G.M., 1993. The impact of plant-derived terpenoids on behaviour and development of Aphididae. Proc. Exper. & Appl. Entomol., N.E.V. Amsterdam (eds M.J. Sommeijer and J. van der Blom) 4: 21-26.

Harrewijn, P., van Oosten, A.M. and Piron, P.G.M., 1994. Synergistic effect of natural sesqui- and monoterpenes on host acceptation and polymorphism of aphids. Proc. Exper. & Appl. Entomol., N.E.V. Amsterdam (eds H.J. Sommeijer and J. van der Blom) 5: 133-138.

Harrewijn, P., Minks, A.K. and Mollema, C., 1995. Evolution of plant volatile production in insect-plant relationships. Chemoecology 5/6, 2: 55-73.

Harrewijn, P., Piron, P.G.M. and van den Heuvel, J.F.J.M., 1996. The effects of natural terpenoids on behaviour and host plant acceptance of aphids. In: Abstracts ISCE 13th Annual Meeting (eds P. Drašar and I. Valterová): 56-57.

Harrewijn, P., Piron, P.G.M. and Ponsen, M.B., 1998. Evolution of vascular feeding in aphids: an electrophysiological study. Proc. Exper. & Appl. Entomol., N.E.V. Amsterdam (eds M.J. Sommeijer and P. J. Francke) 9: 29-34.

Harvey, S., Jemiolo, B. and Novotny, M., 1989. Pattern of volatile compounds in dominant and subordinate male mouse urine. J. Chem. Ecol.15: 2061-2072.

Havel, C.M., Rector II E.R. and Watson, J.A., 1986. Isopentoid synthesis in isolated embryonic *Drosophila* cells. Possible regulation of 3-hydroxy-3-methyl-glutaryl Coenzyme A reductase activity by shunted mevalonate carbon. J. Biol. Chem. 261 (22): 10150-10156.

Hay, R.K.M. and Waterman, P.G. (eds), 1993. Volatile Oil Crops: Their Biology, Biochemistry and Production. Longman Scientific & Technical, Harlow, UK, 185 pp.

Hedlund, K., Bengtsson, G. and Rundgren, S., 1995. Fungal odour discrimination in two sympatric species of fungivorous collembolans. Funct. Ecol. 9: 869-875.

Hegardt, F.G., 1999. Mitochondrial 3-hydroxy-3-methylglutaryl-CoA synthase: a control enzyme in ketogenesis. Biochem. J. 338: 569-582.

Heldin, C.H., 1995. Dimerization of cell surface receptors. Cell 80: 213-223.

Henrick, C.A., 1991. Juvenoids and anti-juvenile hormone agents: past and present. In: Proc. Conf. Insect Chem. Ecol., Tábor (ed. I. Hrdý), Academia Praha/SPB Acad. Publ. The Hague: 429-452.

Herbert, B.S., Sanders, B.G. and Kline, K., 1999. N-(4-hydroxyphenyl)retinamide activation of transforming growth factor-β and induction of apoptosis in human breast cancer cells. Nutr. Cancer 34: 121-132.

Herz, W., Pilotti, A.M., Soderholm, A.C., Shuhama, I.K. and Vichnewsky, W., 1977. New *ent*-clerodane-type diterpenoids from *Baccharis trimera*. J. Org. Chem. 42: 3913-3917.

Hewin International (market study), 1999. Surfactants and other additives in agricultural formulations. John Wiley & Sons, Inc. New York, 182 pp.

Hildebrand, D.F., Brown, G.C., Jackson, D.M. and Hamilton-Kemp, T.R., 1993. Effects of some leaf-emitted volatile compounds on aphid population increase. J. Chem. Ecol. 19: 1875-1887.

Hinson, D.D., Chambliss, K.L., Toth, M.J., Tanaka, R.D. and Gibson, K.M., 1997. Post-translational regulation of mevalonate kinase by intermediates of the cholesterol and nonsterol isoprene biosynthetic pathways. J. Lipid Res. 38: 2216-2223.

Hogg, G., 1961. Safe bind, safe find. The story of locks, bolts and bars. Phoenix House Ltd, London, 126 pp.

Hohl, R.J., 1996. Monoterpenes as regulators of malignant cell proliferation. Dietary Phytochemicals. In: Cancer research and treatment, Plenin Publ. Corp., New York: 137-146. Second ref: Adv. Exp. Med. Biol. 401: 137-146.

Hohl, R.J., Larson, R.A., Mannickarottu, V. and Yachnin, S., 1991. Inhibition of hydroxymethylglutaryl coenzyme A reductase activity induces a paradoxical increase in DNA synthesis in myeloid leukemia cells. Blood 77: 1064-1070.

Hohl, R.J. and Lewis, K.A., 1995. Differential effects of monoterpenes and lovastin on Ras processing. J. Biol. Chem. 270: 17508-17512.

Hohl, R.J., Lewis, K.A., Cermak, D.M. and Wiemer, D.F., 1998. Stereochemistry-dependent inhibition of Ras farnesylation by farnesyl phosphonic acids. Lipids 33: 39-46.

Holland, B. K., 1994. Prospecting for drugs in ancient texts. Nature 369: 702.

Honda, A., Salen, G., Nguyen, L.B., Xu, G., Tint, G.S., Batta, A.K. and Shefer, S., 1998. Regulation of early cholesterol biosynthesis in rat liver: effects of sterols, bile acids, lovastatin, and BM 15.766 on 3-hydroxy-3-methylglutaryl coenzyme A synthase and acetoacetyl coenzyme A thiolase activities. Hepatology 27: 154-159.

Hori, M., 1998. Repellency of rosemary oil against *Myzus persicae* in a laboratory and in a screenhouse. J. Chem. Ecol. 24: 1425-1432.

Hori, M., 1999. Antifeeding, settling inhibitory and toxic activities of labiate essential oils against the green peach aphid, *Myzus persicae* (Sulzer) (Homoptera: Aphididae). Appl. Entomol. Zool. 34: 113-118.

Hori, M. and Komatsu, H., 1997. Repellency of rosemary oil and its components against onion aphid, *Neotoxoptera formosana* (Takahashi) (Homoptera, Aphididae). Appl. Entomol. Zool. 32: 303-310.

Howse, P.E., Baker, R. and Evans, D.A., 1977. Multifunctional secretions in ants. Proc. 8th Int. Congr. Int. Union Study Soc. Insects: 44-45.

Hrdý, I. (ed.), 1991. Insect Chemical Ecology. Proc. Conf. Insect Chem. Ecol. (1990), Tábor. Academia Praha/SPB Acad. Publ. The Hague, 513 pp.

Hrutfiord, B.F. and Gara, R.I., 1989. The terpene complement of slow and fast growing Sitka spruce terminals as related to *Pissodes strobi* (PeckP (Col., Curculionidae)) host selection. J. Appl. Ent. 108: 21-23.

Hsu, H.Y., Yang, J.J. and Lin, C.C., 1997. Effects of oleanolic acid and ursolic acid on inhibiting tumor growth and enhancing the recovery of hematopoietic system postirradiation in mice. Cancer Lett. 111: 7-13.

Hu, J., Shen, Z.X., Sun, G.L., Chen, S.L., Wang, Z.Y. and Chen, Z., 1999. Long-term survival and prognostic study in acute promyelocytic leukemia treated with all-*trans*-retinoic acid, chemotherapy, and As2O3: an experience of 120 patients at a single institution. Int. J. Hematol. 70: 248-260.

Hu, Z.B., Minden, M.D. and McCulloch, E.A., 1998. Phosphorylation of Bcl-2 after exposure of human leukemic cells to retinoic acid. Blood 92: 1768-1775.

Huang, Y., Lam, S.L. and Ho, S.H., 2000. Bioactivities of essential oil from *Elettaria cardamomum* (L.) Maton. to *Sitophilus zeamais* Motschulsky and *Tribolium castaneum* (Herbst). J. Stored Prod. Res. 36: 107-117.

Hubell, S.P., Wiemer, D.F. and Adejare, A., 1983. An antifungal terpenoid defends a neotropical tree (*Hymenaea*) against attack by fungus-growing ants (*Atta*). Oecologia (Berl.) 60: 321-327.

Hudson, J.B. (ed.), 1990. Antiviral compounds from plants. Boca Raton/FL; CRC Press, 200 pp.

Huelin, F.E. and Coggiola, I.M., 1968. Superficial scald, a functional disorder of stored apples. IV. Effect of variety, maturity, oiled wraps and diphenylamine on the concentration of α-farnesene in the fruit. J. Sci. Food Agric. 19: 297-301.

Huelin, F.E. and Coggiola, I.M., 1970. Superficial scald, a functional disorder of stored apples. V. Oxidation of α-farnesene and its inhibition by diphenylamine. J. Sci. Food Agric. 21: 44-48.

Hylemon, P.B. and Harder, J., 1998. Biotransformation of monoterpenes, bile acids, and other isoprenoids in anaerobic ecosystems. FEMS Microbiol. Rev. 22: 475-488.

Imdorf, A., Bogdanov, S., Ochoa, R.I. and Calderone, N.W., 1999. Use of essential oils for the control of *Varroa jacobsoni* Oud. in honey bee colonies. Apidologie 30: 209-228.

Inglese, J., Glickman, J.F. and Lorenz, W., 1992b. Isoprenylation of a protein kinase. Requirement of farnesylation/alpha-carboxyl methylation for full enzymatic activity of rhodopsin kinase. J. Biol. Chem. 267: 1422.

Inglese, J., Koch, W.J. and Caron, M.G., 1992. Isoprenylation in regulation of signal transduction by G-protein-coupled receptor kinases. Nature 359: (6391) 147-150.

Ishikura, H., Mochizuki, T., Izumi, Y., Usui, T., Sawada, H. and Uchino, H., 1984. Differentiation of mouse leukemic M_1 cells induced by polyprenoids. Leuk. Res. 8: 843-852.

Isman, M., 1999. Pesticides based on plant essential oils. Pesticide Outlook 10: 68-72.

Itoh, H., Tanaka-Ueguchi, M., Kawaide, H., Chen, X., Kamiya, Y. and Matsuoka, M., 1999. The gene encoding tobacco gibberellin 3β-hydroxylase is expressed at the site of GA action during stem elongation and flower organ development. The Plant J. 20: 15-24.

Ivarsson, P. and Birgersson, G., 1995. Regulation and biosynthesis of pheromone components in the double spined bark beetle *Ips duplicatus* (Coleoptera, Scolytidae). J. Insect Physiol. 41: 843-849.

Ivarsson, P., Blomquist, G.J. and Seybold, S.J., 1996. *In vitro* biosynthesis of pheromone precursors from radiolabelled acetate in *Ips pini* (Say) and *Ips paraconfusus* Lanier

(Coleoptera: Scolytidae). In: Abstracts ISCE 13th Annual Meeting (eds. P. Drašar and I. Valterová): 116.

Iyengar, R., 1996. Gating by cyclic AMP: expanded role for an old signalling pathway. Science 271: 461-463.

Izumi, S., Takashima, O. and Hirata, T., 1999. Geraniol is a potent inducer of apoptosis-like cell death in the cultured shoot primordia of *Matricaria chamomilla*. Biochem. Biophys. Res. Comm. 259: 519-522.

Jackson, D.L., Jarosik, V. and Dixon, A.F.G., 1996. Resource partitioning and tolerance of monoterpenes in four species of spruce aphid. Physiol. Entomol. 21: 242-246.

Jacobs, D.R., Slavin, J. and Marquart, L., 1995. Whole grain intake and cancer: a review of the literature. Nutr. Cancer 24: 221-229.

Jager, W., Mayer, M., Platzer, P., Reznicek, G., Dietrich, H. and Buchbauer, G., 2000. Stereoselective metabolism of the monoterpene carvone by rat and human liver microsomes. J. Pharm. Pharmacol. 52: 191-197.

Jain, D.C. and Tripathi, A.K., 1993. Potential of natural products as insect antifeedants [review]. Phytoth. Res. 7: 327-334.

James, G.L., Goldstein, J.L., Brown, M.S., Rawson, T.E., Somers, T.C., McDowell, R.S., Crowley, C.W., Lucas, B.K., Levinson, A.D. and Marsters, J.C. Jr, 1993. Benzodiapine peptidomimetics: potent inhibitors of Ras farnesylation in animal cells. Science 260: 1937-1942.

James, S.Y., Williams, M.A., Newland, A.C. and Colston, K.W., 1999. Leukemia cell differentiation: cellular and molecular interactions of retinoids and vitamin D. Gen. Pharmacol. 32: 143-154.

Janowski, B.A., Willy, P.J., Devi, T.R., Falck, J.R. and Mangelsdorf, D.J., 1996. An oxysterol signalling pathway mediated by the nuclear receptor LXRα. Nature 383: 728-731.

Jansen, B.J.M., 1993. Total synthesis of insect antifeedant drimane sesquiterpenes. PhD thesis, Agricultural University Wageningen, 195 pp.

Jarvis, A.P., Johnson, S. and Moragan, E.D., 1998. Stability of the natural insecticide azadirachtin in aqueous and organic solvents. Pest. Sci. 53: 217-222.

Jayasuriya, H., Silverman, K.C., Zink, D.L., Jenkins, R.G., Sanchez, M., Pelaez, F., Vilella, D., Lingham, R.B. and Singh, S.B., 1998. Clavaric acid: a triterpenoid inhibitor of farnesyl-protein transferase from *Clavariadelphus truncatus*. J. Nat. Prod. 61: 1568-1570.

Jenkins, R.L., Loxdale, H.D., Brookes, C.P. and Dixon, A.F.G., 1999. The major carotenoid pigments of the grain aphid, *Sitobion avenae* (F.) (Hemiptera: Aphididae). Physiol. Entomol. 24: 171-178.

Jennings, W.G. and Tressl, R., 1974. Production of volatile compounds in the ripening Bartlett pear. Chem. Mikrobiol. Technol. Lebensm. 3: 52-55.

Jentsch, S., 1996. When proteins receive deadly messages at birth. Science 271: 955-957.

Jermy, T., 1990. Prospects of antifeedant approach to pest control - a critical review. J. Chem. Ecol. 16: 3151-3166.

Jermy, T., 1991. Evolutionary interpretations of insect-plant relationships - A closer look. Proc. 7th Int. Symp. on Insect-Plant Relationships (eds Á. Szentesi and T. Jermy), Akadémiai Kiadó, Budapest: 301-311.

Jermy, T., 1993. Evolution of insect-plant relationships - A devil's advocate approach. Entomol. exp. appl. 66: 3-12.

Jiang, R.J. and Koolman, J., 1999. Feedback inhibition of ecdysteroids: evidence for a short feedback loop repressing steroidogenesis. Arch. Insect Biochem. Physiol. 41: 54-59.

Jimenez, A., Mata, R., Perera-Miranda, R., Calderon, J., Isman, M.B., Nicol, R. and Arnason, J.T., 1997. Insecticidal limonoids from *Swietenia humilis* and *Cedrela salvadorensis*. J. Chem. Ecol. 23: 1225-1234.

Johnson, E.S., Kadam, N.P., Hylands, D.M. and Hylands, P.J., 1985. Efficacy of feverfew as prophylactic treatment of migraine. British Med. J. 291: 569-573.

Jomaa, H., Feurle, J., Luhs, K., Kunzmann, V., Tony, H.P., Herderich, M. and Wilhelm, M., 1999. Vγ9/Vδ2 T cell activation induced by bacterial low molecular mass compounds depends on the 1-deoxy-D-xylulose 5-phosphate pathway of isoprenoid biosynthesis. FEMS Immunol. Med. Microbiol. 25: 371-378.

Jomaa, H., Wiesner, J., Sanderbrand, S., Altincicek, B., Weidemeyer, C., Hintz, M., Turbachova, I., Eberl, M., Zeidler, J., Lichtenthaler, H.K., Soldati, D. and Beck, E., 1999b. Inhibitors of the nonmevalonate pathway of isoprenoid biosynthesis as antimalarial drugs. Science 285: 1573-1576.

Jones, A., 2000. Does the plant mitochondrion integrate cellular stress and regulate programmed cell death? Trends Plant Sci. 5: 225-230.

Jones, C.G. and Firn, R.D., 1991. On the evolution of plant secondary chemical diversity. Phil. Trans. Roy. Soc. London B 333: 273-280.

Jones, G., Venkatamaran, V., Manczak, M. and Schelling, D., 1993. Juvenile hormone action to suppress gene transcription and influence message stability. Dev. Genet. 14: 323-332.

Ju, Z. and Curry, E.A., 2000. Lovastatin inhibits α-farnesene synthesis without affecting ethylene production during fruit ripening in 'Golden Supreme' apples. J. Amer. Soc. Hort. 125: (in press).

Juergens, U.R., Stober, M. and Vetter, H., 1998a. The anti-inflammatory activity of L-menthol compared to mint oil in human monocytes *in vitro*: a novel perspective for its therapeutic use in inflammatory diseases. Eur. J. Med. Res. 3: 539-545.

Juergens, U.R., Stober, M., Schmidt-Schilling, L., Kleuver, T. and Vetter, H., 1998. Anti-inflammatory effects of eucalyptol (1,8-cineole) in bronchial asthma: inhibition of arachidonic acid metabolism in human blood monocytes *ex vivo*. Eur. J. Med. Res. 3: 407-412.

Juergens, U.R., Stober, M. and Vetter, H., 1998b. Inhibition of cytokine production and arachidonic acid metabolism by eucalyptol (1,8-cineole) in human blood monocytes *in vitro*. Eur. J. Med. Res. 3: 508-510.

Jurgens, U.J., Simonin, P. and Rohmer, M., 1992. Localization and distribution of hopanoids in membrane systems of the cyanobacterium *Synechocystis* PCC 6714. FEMS Microbiol. Lett. 71: 285-288.

Kaib, M., 1999. Termites. In: Pheromones of Non-Lepidopteran Insects Associated with Agricultural Plants (eds J. Hardie and A.K. Minks), CABI Publ. Wallingford: 329-353.

Kaiser, J., 2000. Controversial cancer therapy finds political support. Science 287: 2139-2141.

Kalechman, Y., Longo, D.L., Catane, R., Shani, A., Albeck, M. and Sredni, B., 2000. Synergistic anti-tumoral effect of paclitaxel (taxol) + AS101 in a murine model of B16 melanoma: association with Ras-dependent signal-transduction pathways. Int. J. Cancer 86: 281-288.

Kammerer, S., Arnold, N., Gutensohn, W., Mewes, H.W., Kunau, W.H., Hofler, G., Roscher, A.A. and Braun, A., 1997. Genomic organization and molecular characterization of a gene encoding HsPXF, a human peroxisomal farnesylated protein. Genomics 45: 200-210.

Karlberg, A.T., Magnusson, K. and Nilsson, U., 1992. Air oxidation of D-limonene (the citrus solvent) creates potent allergens. Contact Dermat. 26: 332-340.

Karlin, S., Brocchieri, L., Mrazek, J., Campbell, A.M. and Spormann, A.M., 1999. A chimeric prokaryotic ancestry of mitochondria and primitive eukaryotes. Proc. Natl. Acad. Sci. USA 96: 9190-9195.

Karlson, J., Borg-Karlson, A.K., Unelius, R., Shoshan, M.C., Wilking, N., Ringborg, U. and Linder, S., 1996. Inhibition of tumor cell growth by monoterpenes *in vitro*: evidence of a Ras-independent mechanism of action. Anticancer Drugs 7: 422-429.

Katz, L.A., 1999. The tangled web: gene genealogies and the origin of eukaryotes. Am. Naturalist 154: S137-145.

Kearney, M.L., Allan, E.J., Hooker, J.E. and Mordue, A.J., 1994. Antifeedant effects of *in vitro* extracts of the neem tree, *Azadirachta indica* against the desert locust (*Schistocerca gregaria* (Forskal)). Plant Cell Tissue & Organ Cult. 37: 67-71.

Keegans, S.J., Billen, J., Morgan, E.D. and Gökcen, O.A., 1993. Volatile glandular secretions of three species of new world army ants, *Eciton burchelli, Labidus coecus* and *Labidus predator*. J. Chem. Ecol. 19: 2705-2719.

Kenny, F.S., Pinder, S.E., Ellis, I.O., Gee, J.M., Nicholson, R.I., Bryce, R.P. and Robertson, J.F., 2000. Gamma-linolenic acid with tamoxifen as primary therapy in breast cancer. Int. J. Cancer 85: 643-648.

Kenrick, P. and Crane, P.R., 1997. The origin and early evolution of plants on land. Nature 389: 33-39.

Kevan, S.D., Nasr, M.E. and Kevan, P.G., 1999. Feeding menthol to honeybees (Hymenoptera: Apidae): Entry and persistence in haemolymph without causing mortality. Can. Entomol. 131: 279-281.

Khachik, F., Beecher, G.R. and Smith, J.C. Jr, 1995. Lutein, lycopene, and their oxidative metabolites in chemoprevention of cancer. J. Cell Biochem. 22S: 236-246.

Khan, M.A., 1988. Brain-controlled synthesis of juvenile hormone in adult insects. Entomol. exp. appl. 46: 3-17.

Kineman, R.D., 2000. Antitumorigenic actions of growth hormone-releasing hormone antagonists. PNAS 97: 532-534.

Kitaoka, M., Kadokawa, H., Sugano, M., Ichikawa, K., Taki, M., Takaishi, S., Iijima, Y., Tsutsumi, S., Boriboon, M. and Akiyama, T., 1998. Prenylflavonoids: a new class of non-steroidal phytoestrogen (Part 1). Isolation of 8-isopentenylnaringenin and an initial study on its structure-activity relationship. Planta Med. 64: 511-515.

klein Gebbinck, E.A., 1999. Synthesis of model compounds derived from natural clerodane insect antifeedants. PhD thesis, Agricultural University Wageningen, The Netherlands, 283 pp.

Klijnstra, J.W., Corts, K.A. and van Oosten, A.M., 1992. Toxic effects of farnesene on adult whiteflies. Med. Fac. Landbouww. Rijksuniv. Gent 57/2b: 485-491.

Koga, S., Hirohata, S., Kondo, Y., Komata, T., Takakura, M., Inoue, M., Kyo, S. and Kondo, S., 2000. A novel telomerase-specific gene therapy: gene transfer of caspase-8 utilizing the human telomerase catalytic subunit gene promoter. Hum. Gene Ther. 11: 1397-1406.

Kohl, N.E., Mosser, S.D., Jane-de-Solms, S., Giuliani, E.A., Pompliano, D.L., Graham, S.L., Smith, R.L., Scolnick, E.M., Oliff, A. and Gibbs, J.B., 1993. Selective inhibition of ras-dependent transformation by a farnesyltransferase inhibitor. Science 260: 1934-1937.

Kohl, N.E., Wilson, F.R., Mosser, S.D., Giuliani, E.A., Jane-de-Solms, S., Gomez, R.P., Lee, T.J., Smith, R.L., Graham, S.L., Hartman, G.D., Gibbs, J.B. and Oliff, A., 1994. Protein farnesyl transferase inhibitors block the growth of ras-dependent tumors in nude mice. Proc. Natl. Acad. Sci. 91: 9141-9145.

Kong, A.N., Yu, R., Chen, C., Mandlekar, S. and Primiano, T., 2000. Signal transduction events elicited by natural products: role of MAPK and caspase pathways in homeostatic response and induction of apoptosis. Arch. Pharm. Res. 23: 1-16.

Korichneva, I. and Hammerling, U., 1999. F-actin as a functional target for retro-retinoids: a potential role in anhydroretinol-triggered cell death. J. Cell Sci. 112: 2521-2528.

Kosits, C. and Callaghan, M., 2000. Rituximab: a new monoclonal antibody therapy for non-Hodgkin's lymphoma. Oncol. Nurs. Forum 27: 51-59.

Koshizuka, K., Koike, M., Kubota, T., Said, J., Binderup, L. and Koeffler, H.P., 1998. Novel vitamin D3 analog (CB1093) when combined with paclitaxel and cisplatin inhibit growth of MCF-7 human breast cancer cells *in vivo*. Int. J. Oncol. 13: 421-428.

Kraul, M.A. and Schmidt, F., 1957. Über die wachstumshemmende Wirkung bestimmter Extrakte aus Flores Chamomillae und eines synthetischen Azulenpräparates auf experimentelle Mäusetumoren. Arch. Pharmazie 62: 66-74.

Krieger, J., Strobel, J., Kroner, C., Boekhoff, I. and Breer, H., 1999. Biochemical studies of olfactory receptors and signal transduction. Abstracts 2nd Int. Symp. on Insect Pheromones, WICC Wageningen: 42-43.

Krieger, M., 1998. The "best" of cholesterols, the "worst" of cholesterols: a tale of two receptors. Proc. Natl. Acad. Sci. USA 95: 4077-4080.

Kubanek, J., Faulkner, D.J. and Andersen, R.J., 2000. Geographic variation and tissue distribution of endogenous terpenoids in the northeastern Pacific Ocean dorid nudibranch *Cadlina luteomarginata:* implications for the regulation of *de novo* biosynthesis. J. Chem. Ecol. 26: 377-389.

Kubo, I., 1992. Antimicrobial activity of green tea flavor components and their combination effects. Abstracts Papers Am. Chem. Soc. 203: 1-3.

Kuiper, G.G., Lemmen, J.G., Carlsson, B., Corton, J.C., Safe, S.H., van der Saag, P.T., van der Burg, B. and Gustafsson, J.A., 1998. Interaction of estrogenic chemicals and phytoestrogens with estrogen receptor beta. Endocrinology 139: 4252-4263.

Kumanova, R., Lambreva, M. and Dimitrova, Y., 1989. A new cosmetic stick for Herpes simplex labialis (in Russ.). Med. Biol. Inf. 6: 16-17.

Kupchan, S.M., 1970. Recent advances in the chemistry of terpenoid tumor inhibitors. Pure Appl. Chem. 21: 227-246.

Kurahashi, T., Lowe, G. and Gold, G.H., 1994. Suppression of odorant responses by odorants in olfactory receptor cells. Science 265: 110-120.

Kusamran, W.R., Tepsuwan, A. and Kupradinun, P., 1998. Antimutagenic and anticarcinogenic potentials of some Thai vegetables. Mutat. Res. 402: 247-258.

Kwon, B.M., Lee, S.H., Kim, M.J., Kim, H.K. and Kim, H.M., 1999. Isolation of farnesyltransferase inhibitors from herbal medicines. Ann. NY Acad. Sci. 886: 261-264.

Lackey, B.R., Gray, S.L. and Henricks, D.M., 2000. Synergistic approach to cancer therapy: exploiting interactions between anti-estrogens, retinoids, monoterpenes and tyrosine kinase inhibitors. Med. Hypotheses 54: 832-836.

Lafont, R., Bouthier, A. and Wilson, I.D., 1991. Phytoecdysteroids: structures, occurrence, biosynthesis and possible ecological significance. Proc. Conf. Insect Chem. Ecol., Tábor (ed. I. Hrdý), Academia Praha/SPB Acad. Publ. The Hague: 197-214.

Laing, D.G., Panhuber, H. and Slotnick, B.M., 1989. Odor masking in the rat. Physiol. Behav. 45: 689-694.

Lajide, L., Escoubas, P. and Mizutani, J., 1995. Termite antifeedant activity in tropical plants. 3. Termite antifeedant activity in *Detarium microcarpum*. Phytochemistry 40: 1101-1104.

Lam, L.K.T. and Zheng, B., 1991. Effects of essential oils of glutathione S-transferase activity in mice. J. Agric. Food Chem. 39: 660-662.

Landolt, P.J. and Averill, A.L., 1999. Fruit flies. In: Pheromones of Non-Lepidopteran Insects Associated with Agricultural Plants (eds J. Hardie and A.K. Minks), CABI Publ. Wallingford: 3-25.

Lange, B.M. and Croteau, R., 1999. Isopentenyl diphosphate biosynthesis via a mevalonate-independent pathway: isopentenyl monophosphate kinase catalyzes the terminal enzymatic step. Proc. Natl. Acad. Sci. USA 96: 13714-13719.

Lange, B.M., Wildung, M.R., McCaskill, D. and Croteau, R., 1998. A family of transketolases that directs isoprenoid biosynthesis via mevalonate-independent pathway. Proc. Natl. Acad. Sci. USA 95: 2100-2104.

Langenheim, J.H., 1994. Higher plant terpenoids: a phytocentric overview of their ecological roles. J. Chem. Ecol. 20: 1223-1280.

Laska, M., Galizia, C.G., Giurfa, M. and Menzel, R., 1999. Olfactory discrimination and odor structure-activity relationships in honeybees. Chem. Senses 24: 429-438.

Laska, M., Liesen, A. and Teubner, P., 1999b. Enantioselectivity of odor perception in squirrel monkeys and humans. Am. J. Physiol. 277: R1098-1103.

Laska, M. and Teubner, P., 1999. Olfactory discrimination ability of human subjects for ten pairs of enantiomers. Chem. Senses 24: 161-170.

Latgé, J.P. and Papierok, B., 1988. Aphid pathogens. In: Aphids, their Biology, Natural Enemies and Control, Vol. B (eds A.K. Minks and P. Harrewijn), Elsevier Amsterdam: 323-335.

Laurent, G., MacLeod, K., Stopfer, M. and Wehr, M., 1999. Dynamic representation of odours by oscillating neural assemblies. Entomol. exp. appl. 91: 7-18.

Law, J.H. and Wells, M.A., 1989. Insects as biochemical models. J. Biol. Chem. 264: 16335-16338.

Lawler, I., Stapley, J., Foley, J. and Eschler, B.M., 1999. Ecological example of conditioned flavor aversion in plant-herbivore interactions: effect of terpenes of *Eucalyptus* leaves on feeding by common ringtail and brushtail possums. J. Chem. Ecol. 25: 401-415.

Leal, W.S., 1999. Scarab beetles. In: Pheromones of Non-Lepidopteran Insects Associated with Agricultural Plants (eds J. Hardie and A.K. Minks), CABI Publ. Wallingford: 51-68.

Leather, S.R., 1996. Resistance to foliage-feeding insects in conifers: implications for pest management. Integr. Pest Manag. Rev. 1: 163-180.

Lee, C.H., Hong, H., Shin, J., Jung, M., Shin, I., Yoon, J. and Lee, W., 2000. NMR studies on novel antitumor drug candidates, deoxoartemisinin and carboxypropyldeoxoartemisinin. Biochem. Biophys. Res. Commun. 274: 359-369.

Lee, K.H., 1999. Novel antitumor agents from higher plants. Med. Res. Rev. 19: 569-596.

Lee, S., Tsao, R., Peterson, C. and Coats, J.R, 1997. Insecticidal activity of monoterpenoids to western corn rootworm (Coleoptera: Chrysomelidae), twospotted spider mite (Acari: Tetranychidae) and house fly (Diptera: Muscidae). J. Econ. Entomol. 90: 883-892.

Lehmann, J.M., Kliewer, S.A., Moore, L.B., Smith-Oliver, T.A., Oliver, B.B., Su, J.L., Sundseth, S.S., Winegar, D.A., Blanchard, D.E., Spencer, T.A. and Willson, T.M., 1997. Activation of the nuclear receptor LXR by oxysterols defines a new hormone response pathway. J. Biol. Chem. 272: 3137-3140.

Lemaire, M., Nagnan, P., Clement, J.L., Lange, C., Escoubas, P., 1986. Geranyl-linalool (diterpene alcohol): a natural toxin from wood and from termites of the genus *Reticulitermes*. Actes des Collogues Insectes Sociaux 3: 103-107.

Lemaire, M., Nagnan, P., Clement, J.L., Lange, C., Peru, L. and Basselier, J.J., 1990. Geranyllinalool (diterpene alcohol). An insecticidal component of pine wood and termites (Isoptera: Rhinotermitidae) in four European ecosystems. J. Chem. Ecol. 16: 2067-2079.

Lemon, W.C. and Getz, W., 1999. Neural coding of general odors in insects. Ann. Entomol. Soc. Am. 92: 861-872.

Lenfeld, J., Motl, O. and Trka, A., 1986. Anti-inflammatory activity of extracts from *Conyza canadensis*. Pharmazie 41: 268-269.

Leong, T.S., Hwang, E.I., Lee, H.B., Lee, E.S., Kim, Y.K., Min, B.S., Bae, K.H., Bok, S.H. and Kim, S.U., 1999. Chitin synthase II inhibitory activity of ursolic acid, isolated from *Crataegus pinnatifida*. Planta Med. 65: 261-263.

Letchamo, W., Gosselin, A. and Holzl, J., 1995. Growth and essential oil content of *Angelica archangelica* as influenced by light intensity and growing media. J. Ess. Oil Res. 7: 497-504.

Levinson, H. and Levinson, A., 1998. Control of stored food pests in the ancient Orient and classical antiquity. J. Appl. Ent. 122: 137-144.

Lezzi, M., Bergman, T., Mouillet, J.F. and Henrich, V.C., 1999. The ecdysone receptor puzzle. Arch. Insect Biochem. Physiol. 41: 99-106.

Li, J.Y., Strobel, G.A., Sidhu, R., Hess, W.M. and Ford, E.J., 1996. Endophytic taxol-producing fungi from bald cypress, *Taxodium distichum*. Microbiology 142: 2223-2227.

Li, J.Y., Sidhu, R., Bollon, A. and Strobel, G.A., 1998. Stimulation of taxol production in liquid cultures of *Pestalotiopsis*. Mycol. Res. 102: 461-464.

Li, Y. and Wu, Y.L., 1998. How Chinese scientists discovered qinghaosu (artemisin) and developed its derivatives. What are the future perspectives? Med. Trop. 58: S9-12.

Lichtenthaler, H.K., 1998. The plant's 1-deoxy-D-xylulose-5-phosphate pathway for biosynthesis of isoprenoids. Fett/Lipid 100: 128-138.

Lichtenthaler, H.K., 1999. The 1-deoxy-D-xylulose-5-phosphate pathway of isoprenoid biosynthesis in plants. Annu. Rev. Plant Physiol. Plant Mol. Biol. 50: 47-65.

Ligueros, M., Jeoung, D., Tang, B., Hochhauser, D., Reidenberg, M.M. and Sonenberg, M., 1997. Gossypol inhibition of mitosis, cyclin D1 and Rb protein in human mammary cancer cells and cyclin-D1 transfected human fibrosarcoma cells. Br. J. Cancer 76: 21-28.

Lilley, R. and Hardie, J., 1996. Cereal aphid responses to sex pheromones and host-plant odours in the laboratory. Physiol. Entomol. 21: 304-308.

Linette, G.P., Li, Y., Roth, K. and Korsmeyer, J., 1996. Cross talk between cell death and cell cycle progression: Bcl-2 regulates NFAT-mediated activation. Proc. Natl. Acad. Sci. USA 93: 9545-9552.

Lingham, R.B., Silverman, K.C., Jayasuriya, H., Kim, B.M., Amo, S.E., Wilson, F.R., Rew, D.J., Schaber, M.D., Bergstrom, J.D., Koblan, K.S., Graham, S.L., Kohl, N.E., Gibbs, J.B. and Singh, S.B., 1998. Clavaric acid and steroidal analogues as Ras- and FPP-directed inhibitors of human farnesyl-protein transferase. J. Med. Chem. 41: 4492-4501.

Linskens, H.F. and Jackson, J.F. (eds), 1991. Essential Oils and Waxes. In Series: Modern Methods of Plant Analysis, Springer-Verlag, Berlin, 337 pp.

Liquido, N.J., Krimian, A.P., DeMilo, A.B. and McQuate, G.T., 1998. Monofluoro analogues of methyl eugenol: new attractants for male of *Bactrocera dorsalis* (Hendel) (Dipt., Tephritidae). J. Appl. Ent. 122: 259-264.

Liu, J., 1995. Pharmacology of oleanolic acid and ursolic acid. J. Ethnopharmacol. 49: 57-68.

Liu, M. and Simon, M.I., 1996. Regulation by c-AMP-dependent protein kinase of a G-protein mediated phospholipase C. Nature 382: 83-88.

Loder, A., 1972. Selected aspects of the biosynthesis of penicillins and cephalosporins. Postepy Hig. Med. Dosw. 26: 493-500.

Loguercio, L.L., Scott, H.C., Trolinder, N.L. and Wilkins, T.A., 1999. Hmg-CoA reductase gene family in cotton (*Gossypium hirsutum* L.): unique structural features and differential expression of hmg2 potentially associated with synthesis of specific isoprenoids in developing embryos. Plant Cell Physiol. 40: 750-761.

Lois, L.M., Campos, N., Putra, S.R., Danielsen, K., Rohmer, M. and Boronat, A., 1998. Cloning and characterization of a gene from *Escherichia coli* encoding a transketolase-like enzyme that catalyzes the synthesis of D-1-deoxyxylulose 5-phosphate, a common precursor for isoprenoid, thiamin, and pyridoxol biosynthesis. Proc. Natl. Acad. Sci. USA 95: 2105-2110.

Lopez, D., Chambers, C.M. and Ness, G.C., 1997. 3-Hydroxy-3-methylglutaryl coenzyme A reductase inhibitors unmask cryptic regulatory mechanisms. Arch. Biochem. Biophys. 343: 118-122.

Loreto, F., Ciccioli, P., Cecinato, A., Brancaleoni, E., Frattoni, M., Fabozzi, C. and Tricoli, D., 1996. Evidence of the photosynthetic origin of monoterpenes emitted by *Quercus ilex* L. leaves by ^{13}C labeling. Plant. Physiol. 110: 1317-1322.

Lösel, P.M., Lindemann, M., Scherkenbeck, J., Campbell, C.A.M., Hardie, J., Pickett, J.A. and Wadhams, L.J., 1996. Effect of primary host kairomones on the attractiveness of the hop aphid sex pheromone to *Phorodon humuli* males and gynoparae. In: Proc. 9th Int. Symp. on Insect-Plant Relationships (eds E. Städler, M. Rowell-Rahier and R. Bauer), Entomol. exp. appl. 80: 79-82.

Lowery, D.T., Eastwell, K.C. and Smirle, M.J., 1997. Neem seed oil inhibits aphid transmission of potato virus Y to pepper. Ann. Appl. Biol. 130: 217-225.

Lowery, D.T. and Isman, M.B., 1994. Insect growth regulating effects of neem extract and azadirachtin on aphids. Entomol. exp. appl. 72: 77-84.

Lowery, D.T. and Isman, M.B., 1996. Inhibition of aphid (Homoptera: Aphididae) reproduction by neem seed oil and azadirachtin. J. Econ. Entomol. 89: 602-607.

Luik, A., Ochsner, P. and Jensen, T.S., 1999. Olfactory responses of seed wasps *Megastigmus pinus* Parfitt and *Megastigmus rafni* Hoffmeyer (Hym., Torymidae) to host-tree odours and some monoterpenes. J. Appl. Ent. 123: 561-567.

Luis, J.G., Gonzalez, A.G., Andrés, L.S. and Mederos, S., 1992. Diterpenes from *in vitro*-grown *Salvia canariensis*. Phytochemistry 31: 3272-3273.

Lujan, H.D., Mowatt, M.R., Chen, G.Z. and Nash, T.E., 1995. Isoprenylation of proteins in the protozoan *Giardia lamblia*. Mol. Biochem. Parasit. 72: 121-127.

Lum, P.Y., Edwards, S. and Wright, R., 1996. Molecular, functional and evolutionary characterization of the gene encoding HMG-CoA reductase in the fission yeast, *Schizosaccharomyces pombe*. Yeast 12: 1107-1124.

Lutz, R.J., McLain, T.M. and Sinensky, M., 1992. Feedback inhibition of polyisoprenyl pyrophosphate synthesis from mevalonate *in vitro*. Implications for protein prenylation. J. Biol. Chem. 267: 7983-7986.

Maarse, H. (ed.), 1991. Volatile Compounds in Foods and Beverages. In Series: Food Science and Technology. Marcel Dekker Inc., New York etc., ≈ 750 pp.

Maarse, H. and Kepner, R.E., 1970. Changes in composition of volatile terpenes in Douglas fir needles during maturation. J. Agric. Food Chem. 18: 1095-1101.

Machida, K., Tanaka, T., Yano, Y., Otani, S. and Taniguchi, M., 1999. Farnesol-induced growth inhibition in *Saccharomyces cerevisiae* by a cell cycle mechanism. Microbiology 145: 293-299.

Machlis, L. and Rawitscher-Kunkel, E., 1967. Mechanisms of gametic approach in plants. In: Fertilization (eds B. Metz and A. Monroy), Vol. 2: 117-161.

Magange, M.E., Gries, G. and Gries, R., 1996. Repellancy of various oils and pine oil constituents to house flies (Diptera: Muscidae). Environm. Entomol. 25: 1182-1187.

Magee, T. and Marshall, C., 1999. New insights into the interaction of Ras with the plasma membrane. Cell 98: 9-12.

Magrassi, L., Butti, G., Pezzotta, S., Infuso, L. and Milanesi, G., 1995. Effects of vitamin D and retinoic acid on glioblastoma cell lines. Acta Neurochir. (Wien) 133: 184-190.

Maldonado, E., Hampsey, M. and Reinberg, D., 1999. Repression: targeting the heart of the matter. Cell 99: 455-458.

Maldonado-Mendoza, I.E., Vincent, R.M. and Nessler, C.L., 1997. Molecular characterization of three differentially expressed members of the *Camptotheca acuminata* 3-hydroxy-3-methylglutaryl CoA reductase (HMGR) gene family. Plant Mol. Biol. 34: 781-790.

Maloney, D.G., 1999. Preclinical and phase I and II trials of rituximab. Semin. Oncol. 26: 74-78.

Manfredi, J.J., Parness, J. and Horwitz, S.B., 1982. Taxol binds to cellular microtubules. J. Cell Biol. 94: 688-696.

Manjunatha, M., Pickett, J.A., Wadhams, L.J. and Nazzi, F., 1998. Response of western flower thrips, *Frankliniella occidentalis* and its predator *Amblyseius cucumeris* to chrysanthemum volatiles in olfactometer and greenhouse trials. Insect Sci. Appl. 18: 139-144.

Manker, D.C. and Faulkner, D.J., 1996. Investigation of the role of diterpenes produced by marine pulmonates *Trimusculus reticulatus* and *T. conica*. J. Chem. Ecol. 22: 23-35.

Manou, I., Bouillard, L., Devleeschouwer, M.J. and Barel, A.O., 1998. Evaluation of preservative properties of *Thymus vulgaris* essential oil in topically applied formulations under a challenge test. J. Appl. Microbiol. 84: 368-376.

Manson, M.M., Gescher, A., Hudson, E.A., Plummer, S.M., Squires, M.S. and Prigent, S.A., 2000. Blocking and suppressing mechanisms of chemoprevention by dietary constituents. Toxicol. Lett. 112/113: 499-505.

Markovic, I., Norris, D.M. and Webster, F.X., 1996. Volatiles involved in the nonhost rejection of *Fraxinus pennsylvanica* by *Lymantria dispar* larvae. J. Agric. Food Chem. 44: 929-935.

Marshall, C.J., 1993. Protein prenylation: a mediator of protein-protein interactions. Science 259: 1865-1866.

Martin, W. and Müller, M., 1998. The hydrogen hypothesis for the first eukaryote. Nature 392: 37-41.

Martinou, J.C., 1999. Apoptosis. Key to the mitochondrial gate. Nature 399: 411-412.

Masuda, Y., Nakaya, M., Nakajo, S. and Nakaya, K., 1997. Geranylgeraniol potently induces caspase-3-like activity during apoptosis in human leukemia U937 cells. Biochem. Biophys. Res. Comm. 234: 641-645.

Matsumine, A., Ogai, A., Senda, T., Okumura, N., Saton, K., Baeg, G., Kawahara, T., Kobayashi, S., Okada, M., Toyoshima, K. and Akaiyama, T., 1996. Binding of APC to the human homolog of the *Drosophila* Discs Large Tumor Suppressor Protein. Science 272: 1020-1023.

Mau, C.J. and West, C.A., 1994. Cloning of casbene synthase cDNA: evidence for conserved structural features among terpenoid cyclases in plants. Proc. Natl. Acad. Sci. USA 91: 8497-8501.

Mazzoni, I.E., Said, F.A., Alaoyz, R., Miller, F.D. and Kaplan, D., 1999. Ras regulates sympathetic neuron survival by suppressing the p53-mediated cell death pathway. J. Neurosci. 19: 9716-9727.

McAdams, H.H. and Arkin, A., 1999. It's a noisy business! Genetic regulation at the nanomolar scale. Trends Genet. 15: 65-69.

McBrien, H.L. and Millar, J.G., 1999. Phytophagous bugs. In: Pheromones of Non-Lepidopteran Insects Associated with Agricultural Plants (eds J. Hardie and A.K. Minks), CABI Publ. Wallingford: 277-304.

McConkey, M.E., Gershenzon, J. and Croteau, R.B., 2000. Developmental regulation of monoterpene biosynthesis in the glandular trichomes of peppermint. Plant Physiol. 122: 215-223.

McDowell, J.M. and Dangl, J.L., 2000. Signal transduction in the plant immune response. TIBS 25: 79-82.

McGee, T.P., Cheng, H.H., Kumagai, H., Omura, S. and Simoni, R.D., 1996. Degradation of 3-hydroxy-3-methylglutaryl-CoA reductase in endoplasmatic reticulum membranes is accelerated as a result of increased susceptibility to proteolysis. J. Biol. Chem. 271: 25630-25638.

McNeill, H. and Downward, J., 1999. Apoptosis: Ras to rescue in the fly eye. Curr. Biol. 9: R176-179.

Mehmet, H., 2000. Caspases find a new place to hide. Nature 403: 29-30.

Meigs, T.E. and Simoni, R., 1997. Farnesol as a regulator of HMG-CoA reductase degradation: characterization and role of farnesyl pyrophosphatase. Arch. Biochem. Biophys. 345: 1-9.

Melnykovych, G., Haug, J.S. and Goldner, C.M., 1992. Growth inhibition of leukemia cell line CEM-1 by farnesol: effects of phosphatidylcholine and diacylglycerol. Biochem. Biophys. Res. Comm. 186: 543-548.

Menken, S.B.J., Visser, J.H. and Harrewijn, P. (eds), 1992. Proc. 8th Int. Symp. Insect-Plant Relationships, Kluwer Acad. Publ., Dordrecht, etc., 424 pp.

Messchendorp, L., 1998. Terpenoid antifeedants against insects: a behavioural and sensory study. PhD thesis, Agricultural University Wageningen, The Netherlands, 135 pp.

Messchendorp, L., Gols, G.J.Z. and van Loon, J.J.A., 1996. Habituation and sensitization to drimane antifeedants in *Pieris brassicae* larvae. In: Abstracts ISCE 13th Annual Meeting (eds P. Drašar and I. Valterová): 97.

Messchendorp, L., van Loon, J.J.A. and Gols, G.J.Z., 1996b. Behavioural and sensory responses to drimane antifeedants in *Pieris brassicae* larvae. Entomol. exp. appl. 79: 195-202.

Messchendorp, L., Smid, H.M. and van Loon, J.J.A., 1998. The role of an epipharyngeal sensillum in the perception of feeding deterrents by *Leptinotarsa decemlineata* larvae. J. Comp. Physiol. A 183: 255-264.

Meyer-Warnod, B., 1984. Natural essential oils, extraction processes and application to major oils. Perfum. Flavor. 9: 93-103.

Miller, B., Heuser, T. and Zimmer, W., 1999. A *Synechococcus leopoliensis* SAUG 1402-1 operon harboring the 1-deoxyxylulose 5-phosphate synthase gene and two additional open reading frames is functionally involved in the dimethylallyl diphosphate synthesis. FEBS Lett. 460: 485-490.

Miller, D.R., Borden, J.H. and Lindgren, B.S., 1995. Verbenone: dose-dependent interruption of pheromone-based attraction of three sympatric species of pine bark beetles (Coleoptera: Scolytidae). Environ. Entomol. 24: 692-696.

Miller, J.R. and Cowles, R.S., 1990. Stimulo-deterrent diversion: a concept and its possible application to onion maggot control. J. Chem. Ecol. 16: 3197-3212.

Miller, J.R. and Miller, T.A. (eds), 1986. Insect-Plant Interactions. Springer-Verlag, New York, etc.: 342 pp.

Miquel, K., Pradines, A. and Favre, G., 1996. Farnesol and geranylgeraniol induce actin cytoskeleton disorganization and apoptosis in A 549 lung adenocarcinoma cells. Biochem. Biophys. Res. Comm. 225: 869-876.

Miyazawa, M., Watanabe, H. and Kameoka, H., 1997. Inhibition of acetylcholinesterase activity by monoterpenoids with a *p*-menthane skeleton. J. Agric. Food Chem. 45: 677-679.

Mo, H. and Elson, C.E., 1999. Apoptosis and cell-cycle arrest in human and murine tumor cells are initiated by isoprenoids. J. Nutr. 129: 804-813.

Mochizuki, M., Yoo, Y.C., Matsuzawa, K., Sato, K., Saiki, I., Tono-oka, S., Samukawa, K. and Azuma, I., 1995. Inhibitory effect of tumor metastasis in mice by saponins, ginsenoside-Rb2, 20(R)- and 20(S)-ginsenoside-Rg3, of red ginseng. Biol. Pharm. Bull. 18: 1197-1202.

Monaco, P., Previtera, L. and Mangoni, L., 1982. Terpenes in *Pistacia* plants: a possible defence role for monoterpenes against gall-forming aphids. Phytochemistry 21: 2408-2410.

Montano, M.M., Ekena, K., Delage-Mourroux, R., Chang, W., Martini, P. and Katzenellenbogen, B.S., 1999. An estrogen receptor-selective coregulator that potentiates the effectiveness of antiestrogens and repress the activity of estrogens. Proc. Natl. Acad. Sci. USA 96: 6947-6952.

Montgomery, M.E. and Nault, L.R., 1977. Comparative response of aphids to the alarm pheromone, (E)-ß-farnesene. Entomol. exp. appl. 22: 236-242.

Moon, R.C., McCormick, D.L. and Mehta, R.G., 1983. Inhibition of carcinogenesis by retinoids. Cancer Res. 43: 2469-2474.

Morales, J., Fishburn, C.S., Wilson, P.T. and Bourne, H.R., 1998. Plasma membrane localization of Gαz requires two signals. Mol. Biol. Cell 9: 1-14.

Mordue, A.J. and Blackwell, A., 1993. Azadirachtin: an update (review). J. Insect Physiol. 39: 903-924.

Mordue, A.J., Nisbet, A.J., Nasiruddin, M. and Walker, E., 1996. Differential thresholds of azadirachtin for feeding deterrence and toxicity in locusts and an aphid. In: Proc. 9th Int. Symp. on Insect-Plant Relationships (eds E. Städler, M. Rowell-Rahier and R. Bauer), Entomol. exp. appl. 80: 69-72.

Moreland, N.J., Illand, M., Kim, Y.T., Paul, J. and Brown, R., 1999. Modulation of drug resistance mediated by loss of mismatch repair by the DNA polymerase inhibitor aphidicolin. Cancer Res. 59: 2102-2106.

Mori, K., Nagao, H. and Yoshihara, Y., 1999. The olfactory bulb: coding and processing of odor molecule information. Science 286: 711-715.

Morris-Kay, G.M. and Sokolova, N., 1996. Embryonic development and pattern formation. FASEB J. 10: 961-968.

Motzer, R.J., Schwartz, L., Law, T.M., Murphy, B.A., Hoffman, A.D., Albino, A.P., Vlamis, V. and Nanus, D.M., 1995. Interferon α-2a and 13-*cis*-retinoic acid in renal cell carcinoma: antitumor activity in a phase II trial and interactions *in vitro*. J. Clin. Oncol. 13: 1950-1957.

Mulkern, G.B. and Toczek, D.R., 1972. Effect of plant extracts on survival and development of *Melanoplus differentialis* and *M. sanguinipes* (Orthoptera: Acrididae). Ann. Entomol. Soc. Am. 65: 662-671.

Mumcuoglu, K.Y., Galun, R., Bach, U., Miller, J. and Magdassi, S., 1996. Repellency of essential oils and their components to the human body louse, *Pediculus humanus humanus*. Entomol. exp. appl. 78: 309-314.

Murray, K.E., 1969. α-Farnesene: isolation from the natural coating of apples. Aust. J. Chem. 22: 197.

Myung, C.S., Yasuda, H., Liu, W.W., Harden, T.K. and Garrison, J.C., 1999. Role of isoprenoid lipids on the heterotrimeric G-protein γ subunit in determining effector activation. J. Biol. Chem. 274: 16595-16603.

Nagase, T., Kawata, S., Nakajima, H., Tamura, S., Yamasaki, E., Fukui, K., Yamamoto, K., Miyagawa, J., Matsumura, I. and Matsuzawa, Y., 1999. Effect of farnesyltransferase overexpression on cell growth and transformation. Int. J. Cancer 80: 126-133.

Nagnan, P. and Clement, J.L., 1990. Terpenes from maritime pine *Pinus pinaster:* toxins for subterranean termites of the genus *Reticulitermes* (Isoptera: Rhinotermitidae). Biochem. Syst. Ecol. 18: 13-16.

Nakanishi, M., Goldstein, J.L. and Brown, M.S., 1988. Multivalent control of 3-hydroxy-3-methylglutaryl coenzyme A reductase. J. Biol. Chem. 263: 8929-8937.

Nambara, E. and McCourt, P., 1999. Protein farnesylation in plants: a greasy tale. Curr. Opin. Plant Biol. 2: 388-392.

Narayanan, B.A., Holladay, E.B., Nixon, D.W. and Mauro, C.T., 1998. The effect of all-*trans*- and 9-*cis*-retinoic acid on the steady state level of HPV16E6/E7 mRNA and cell cycle in cervical carcinoma cells. Life Sci. 63: 565-573.

Naya, Y. and Prestwich, G.D., 1982. Sesquiterpenes from termite soldiers. Structure of amiteol, a new 5ß,7ß,10ß-eudesmane from *Amitermes excellens*. Tetrahedon Lett. 23: 3047-3050.

Nehlin, G., Valterová, I. and Borg-Karlson, A.K., 1996. Monoterpenes released from Apiaceae and the egg-laying preferences of the carrot psyllid, *Trioza apicalis*. Entomol. exp. appl. 80: 83-86.

Nelson, A.C. and Kursar, T.A., 1999. Interactions among plant defense compounds: a method for analysis. Chemoecology 9: 81-92.

Nes, W.D. and Venkatramesh, M., 1994. Molecular asymmetry and sterol evolution. In: Isopentoids and other natural products: Evolution and Function (ed. W.D. Nes). Am. Chem. Soc. Washington: 31-43.

Nesaretnam, K., Stephen, R., Dils, R. and Darbre, P., 1998. Tocotrienols inhibit the growth of human breast cancer cells irrespective of estrogen receptor status. Lipids 33: 461-469.

Neuhold, M., 1997. Naturkosmetik und Parfum - selbstgemacht. Leopold Stocker Verlag, Graz-Stuttgart, 128 pp.

Neunlist, S., Bisseret, P. and Rohmer, M., 1988. The hopanoids of the purple non-sulfur bacteria *Rhodopseudomonas palustris* and *Rhodopseudomonas acidophila* and the absolute configuration of bacteriohopanetetrol. Eur. J. Biochem. 171: 245-252.

Newman, C., 1998. Perfume, the essence of illusion. National Geographic Oct.: 94-129.

Nikolaevskii, V.V., Kononova, N.S., Pertsovskii, A.I. and Shinkarchuk, I.F., 1990. Effect of volatile oils on the course of experimental atherosclerosis (Russ.). Patol. Fiziol. Eksp. Ter. 5: 52-53.

Ninomiya, K. and Kimura, T., 1988. Male odors that influence the preference of female mice: roles of urinary and perputial factors. Physiol. Behav. 44: 791-795.

Nisbet, A.J., Woodford, J.A.T., Strang, R.H.C. and Conolly, J.D., 1993. Systemic antifeedant effects of azadirachtin on the peach-potato aphid *Myzus persicae*. Entomol. exp. appl. 68: 87-98.

Nishida, R., Shelley, T.E., Timothy, S., Whittier, S. and Kaneshiro, Y., 2000. α-Copaene, a potential rendezvous cue for the mediterranean fruit fly *Ceratitis capitata*? J. Chem. Ecol. 26: 87-100.

Nishikawa, Y., Wang, Z., Kerns, J., Wilcox, C.S. and Carr, B.I., 1999. Inhibition of hepatoma cell growth *in vitro* by arylating and non-arylating K vitamin analogs. Significance of protein tyrosine phosphatase inhibition. J. Biol. Chem. 274: 34803-34810.

Nishimaki, J., Miyazawa, K., Yaguchi, M., Katagiri, T., Kawanishi, Y., Toyama, K., Ohyashiki, K., Hashimoto, S., Nakaya, K. and Takiguchi, T., 1999. Vitamin K2 induces apoptosis of a novel cell line established from a patient with myelodysplastic syndrome in blastic transformation. Leukemia 13: 1399-1405.

Nishino, C., Bowers, W.S., Montgomery, M.E., Nault, L.R. and Nielson, M.W., 1977. Alarm pheromone of the spotted alfalfa aphid, *Therioaphis maculata* Buckton (Homoptera: Aphididae). J. Chem. Ecol. 3: 349-357.

Nishio, E. and Watanabe, Y., 1996. Oxysterols induced apoptosis in cultured smooth muscle cells through CPP32 protease activation and Bcl-2 protein downregulation. Biochem. Biophys. Res. Comm. 226: 928-934.

Nishizuka, Y., 1986. Studies and perspectives of protein kinase C. Science 233: 305-309.

Nita, M.E., Ono-Nita, S.K., Tsuno, N., Tominaga, O., Takenoue, T., Sunami, E., Kitayama, J., Nakamura, Y. and Nagawa, H., 2000. Bcl-X(L) antisense sensitizes human colon cancer cell line to 5-fluorouracil. Jpn. J. Cancer Res. 91: 825-832.

Nordlander, G., 1991. Host finding in the pine weevil *Hylobius abietis*: effects of conifer volatiles and added limonene. Entomol. exp. appl. 59: 229-237.

Nottingham, S.F., Hardie, J., Dawson, G.W., Hick, A.J., Pickett, J.A., Wadhams, L.J. and Woodcock, C.M., 1991. Behavioral and electrophysiological responses of aphids to host and non-host plant volatiles. J. Chem. Ecol. 17: 1231-1242.

Ntiamoah, Y.A., Borden, J.H. and Pierce, H.D. Jr, 1996. Identity and bioactivity of oviposition deterrents in pine oil for the onion maggot, *Delia antiqua*. Entomol. exp. appl. 79: 219-226.

Obeng-Ofori, D. and Reichmuth, C., 1997. Bioactivity of eugenol, a major component of essential oil of *Ocimum suave* (Wild.) against four species of stored-product Coleoptera. Int. J. Pest Manag. 43: 89-94.

Ochieng, S.A. and Hansson, B.S., 1996. Single antennal receptor neuron responses in the solitarious desert locust, *Schistocerca gregaria* (Orthoptera: Acrididae) to aggregation pheromone complex. In: Abstracts ISCE 13th Annual Meeting (eds P. Drašar and I. Valterová): 19.

Oda, S., Inada, Y., Kato, A., Matsudomi, N. and Ohta, N., 1995. Production of (S)-citronellic acid and (R)-citronellol with an interface bioreactor. J. Ferment. Bioeng. 80: 559-564.

Ogmundsdottir, H.M., Zoega, G.M., Gissurarson, S.R. and Ingolfsdottir, K., 1998. Antiproliferative effects of lichen-derived inhibitors of 5-lipoxygenase on malignant cell-lines and mitogen-stimulated lymphocytes. J. Pharm. Pharmacol. 50: 107-115.

Ohizumi, H., Masuda, Y., Nakajo, S., Sakai, I., Ohsawa, S. and Nakaya, K., 1995. Geranylgeraniol is a potent inducer of apoptosis in tumor cells. J. Biochem. 117: 11-13.

Ohta, S., Nishio, K. Kubota, N., Ohmori, T., Funayama, Y. Ohira, T., Nakajima, H., Adach, M. and Saijo, N., 1994. Characterization of a taxol-resistant human small-cell lung cancer cell line. Japan. J. Cancer Res. 85: 290-297.

Okada, S., Yabuki, M., Kanno, T., Hamazaki, K., Yoshioka, T., Yasuda, T., Horton, A.A. and Utsumi, K., 1999. Geranylgeranylacetone induces apoptosis in HL-60 cells. Cell Struct. Funct. 24: 161-168.

Oldham, N.J., Morgan, E.D., Gobin, B., Schoeters, J. and Billen, J., 1994. Volatile secretions of old world army ant *Aenictus rotundatus* and chemotaxonomic implications of army ant dufour gland chemistry. J. Chem. Ecol. 20: 3297-3305.

Oloke, J.K., Kolawole, D.O. and Erhun, W.O., 1989. Antimicrobial effectiveness of six paradols. 1: a structure-activity relationship study. J. Ethnopharmacol. 25: 109-113.

Olson, M.F., Ashworth, A. and Hall, A., 1995. An essential role for Rho, Rac and Cdc42G-TPases in cell cycle progression through G1. Science 269: 1270-1273.

Omkumar, R.V., Gaikwad, A.S. and Ramasarma, T., 1992. Feedback-type inhibition of 3-hydroxy-3-methyl glutaryl coenzyme A reductase by ubiquinone. Biochem. Biophys. Res. Comm. 184: 1280-1287.

Oosterhaven, K., 1995. Different aspects of S-carvone, a natural sprout growth inhibitor. PhD thesis, Agricultural University Wageningen, 152 pp.

Oosterhaven, K., Poolman, B. and Smid, E.J., 1995. S-carvone as a natural potato sprout inhibiting, fungistatic and bacteriostatic compound. Ind. Crops Prod. 4: 23-31.

Oren, M., 1999. Regulation of the p53 tumor suppressor protein. J. Biol Chem. 274: 36031-36034.

Ortego, F., Rodriguez, B. and Castañera, P., 1995. Effects of *neo*-clerodane diterpenes from *Teucrium* on feeding behavior of Colorado potato beetle larvae. J. Chem. Ecol. 21: 1375-1386.

Otto, J.C., Kim, E., Young, S.G. and Casey, P.J., 1999. Cloning and characterization of a mammalian prenyl protein-specific protease. J. Biol. Chem. 274: 8379-8382.

Ourisson, G., 1990. The general role of terpenes and their global significance. Pure Appl. Chem. 62: 1401-1404.

Ourisson, G. and Nakatani, Y., 1994. The terpenoid theory of the origin of cellular life: the evolution of terpenoids to cholesterol. Chem. Biol. 1: 11-23.

Ourisson, G., Rohmer, M. and Poralla, K., 1987. Microbial lipids betrayed by their fossils. Microbiol. Sci. 4: 52-57.

Padayatty, S.J., Marcelli, M., Shao, T.C. and Cunningham, G.R., 1997. Lovastatin-induced apoptosis in prostate stromal cells. J. Clin. Endocrinol. Metab. 82: 1434-1439.

Panini, S.R., Schnitzer-Polokoff, R., Spencer, T.A. and Sinensky, M., 1989. Sterol-independent regulation of 3-hydroxy-3-methyl glutaryl-CoA reductase by mevalonate-derived product in MeV-1 cells. J. Biol. Chem. 264: 11044-11052.

Paracelsus (von Hohenheim, T.B.P.A.), 1527. De modo pharmacandi. Collected Capita Selecta, Univ. Basel.

Paré, P.W. and Tumlinson, J.H., 1997. De novo biosynthesis of volatiles induced by insect herbivory in cotton plants. Plant Physiol. 114: 1161-1167.

Paré, P.W. and Tumlinson, J.H., 1998. Cotton volatiles synthesized and released distal to the site of insect damage. Phytochemistry 47: 521-526.

Paré, P.W. and Tumlinson, J.H., 1999. Plant volatiles as a defense against insect herbivores. Plant Physiol. 121: 325-331.

Parish, E.J., 1994. Evolution of the oxysterol pathway. In: Isopentenoids and other Natural Products - Evolution and Function (ed. W.D. Nes), ACS symp. Series: 109-125.

Park, J.A., Lee, K.Y., Oh, Y.J. and Lee, S.K., 1997. Activation of caspase-3-protease via Bcl-2 insensitive pathway during the process of ginsenoside Rh2-induced apoptosis. Cancer Lett. 121: 73-81.

Park, W.H., Lee, Y.Y., Kim, E.S., Seol, J.G., Jung, C.W., Lee, C.C. and Kim, B.K., 1999. Lovastatin-induced inhibition of HL-60 cell proliferation via cell cycle arrest and apoptosis. Anticancer Res. 19 (4B): 3133-3140.

Parker, R.A., Pearce, B.C., Clark, R.W., Gordon, D.A. and Wright, J.J., 1993. Tocotrienols regulate cholesterol production in mammalian cells by post-trancriptional suppression of 3-hydroxy-3-methylglutaryl-coenzyme A reductase. J. Biol. Chem. 268: 11230-11238.

Parmryd, I., Andersson, B. and Dallner, G., 1999. Protein prenylation in spinach chloroplasts. Proc. Natl. Acad. Sci. USA 96: 10074-10079.

Passerini, J. and Hill, S.B., 1993. Field and laboratory trials using a locally produced neem insecticide against the Sahelian grasshopper, *Kraussaria angulifera* (Orthoptera, Acrididae) on millet in Mali. Bull. Ent. Res. 83: 121-126.

Paton, A., 1991. A synopsis of *Ocimum* (Labiatae) in Africa. Kew Bull. 47: 403-435.

Paulsen, E., Andersen, K.E., Carlsen, L. and Egsgaard, H., 1993. Carvone: an overlooked contact allergen cross-reacting with sesquiterpene lactone? Contact Dermat. 29: 138-143.

Pawlowski, J. and Kraft, A.S., 2000. Bax-induced apoptotic cell death. Proc. Natl. Acad. Sci. USA 97: 529-531.

Pazman, C., Mayes, C.A., Fanto, M., Haynes, S.R. and Mlodzik, M., 2000. Rasputin, the *Drosophila* homologue of the RasGAP SH3 binding protein, functions in Ras- and Rgr-mediated signaling. Development 127: 1715-1725.

Peeper, D.S, Upton, T.M., Ladha, M.H., Neuman, E., Zalvilde, J., Bernards, R., DeCaprio, J.A. and Ewen, M.E., 1997. Ras signalling linked to the cell-cycle machinery by the retinoblastoma proteins. Nature 386: 177-181.

Peffley, D.M. and Gayen, A.K., 1997. Inhibition of squalene synthase but not squalene cyclase prevents mevalonate-mediated suppression of 3-hydroxy-3-methylglutaryl coenzyme A reductase synthesis at a posttranscriptional level. Arch. Biochem. Biophys. 337: 251-260.

Peifer, M., 1996. Regulating cell proliferation: as easy as APC. Science 272: 974-975.

Pengsuparp, T., Cai, L., Constant, H., Fong, H.H., Lin, L.Z., Kinghorn, A.D., Pezzuto, J.M., Cordell, G.A., Ingolfsdottir, K. and Wagner, H., 1995. Mechanistic evaluation of new plant-derived compounds that inhibit HIV-1 reverse transcriptase. J. Nat. Prod. 58: 1024-1031.

Penuelas, J., Llusia, J. and Estiarte, M., 1995. Terpenoids: a plant language. Tree 10: 289.

Persoons, C.J., 1977. Structure elucidation of some insect pheromones: a contribution to the development of selective pest control agents. PhD thesis, Agricultural University Wageningen, The Netherlands: 83 pp.

Persson, M., Sjödin, K., Borg-Karlson, A.K. and Norin, T., 1996. Enantiomeric compositions of monoterpene hydrocarbons in different tissues of four individuals of *Pinus sylvestris*. Phytochemistry 41: 439-445.

Pethe, V. and Shekhar, P.V., 1999. Estrogen inducibility of c-Ha-ras transcription in breast cancer cells. Identification of functional estrogen-responsive transcriptional regulatory elements in exon 1/intron of the c-Ha-ras gene. J. Biol. Chem. 274: 30969-30978.

Petras, S.F., Lindsey, S. and Harwood, H.J. Jr, 1999. HMG-CoA reductase regulation: use of structurally diverse first half-reaction squalene synthetase inhibitors to characterize the site of mevalonate-derived nonsterol regulator production in cultured IM-9 cells. J. Lipid Res. 40: 24-38.

Piccinini, F., Chiarra, A. and Villani, F., 1972. Effect of terpenes on the cell permeability of tetracycline. Advan. Antimicrob. Antineoplastic Chemoter. Proc. 7th Int. Congr. Chemother. (Ed. M. Hejzlar). University Park Press, Baltimore: 855-857.

Pickett, J.A. and Griffiths, D.C., 1980. Composition of aphid alarm pheromones. J. Chem. Ecol. 6: 349-359.

Picman, A.K., Blackwell, B.A. and Gershenzon, J., 1993. Further terpenoids of cultivated sunflower, *Helianthus anuus* (Asteraceae). Biochem. Syst. Ecol. 21: 647.

Picman, A.K., Rodriguez, E. and Towers, G.H.N., 1979. Formation of adducts of parthenin and related lactones with cysteine and glutathione. Chem. Biol. Interactions 28: 83-89.

Plarre, R. and Vanderwel, D.C., 1999. Stored-product beetles. In: Pheromones of Non-Lepidopteran Insects Associated with Agricultural Plants (eds J. Hardie and A.K. Minks), CABI Publ. Wallingford: 149-198.

Polak, E.H., Fombon, A.M., Tilquin, C. and Punter, P.H., 1989. Sensory evidence for olfactory receptors with opposite chiral selectivity. Behav. Brain Res. 31: 199-206.

Poole, A., Jeffares, D. and Penny, D., 1999. Early evolution: prokaryotes, the new kids on the block. Bioessays 21: 880-889.

Powell, G., Hardie, J. and Pickett, J.A., 1996. Effects of the repellent polygodial on stylet penetration behaviour and non-persistent transmission of plant viruses by aphids. J. Appl. Ent. 120: 241-243.

Powell, J.S. and Raffa, K.F., 1999. Effects of selected *Larix laricina* terpenoids on *Lymantria dispar* (Lepidoptera: Lymantriidae) development and behavior. Environ. Entomol. 28: 148-154.

Prendergast, G.C., 2000. Farnesyltransferase inhibitors: antineoplastic mechanism and clinical prospects. Curr. Opin. Cell Biol. 12: 166-173.

Prestwich, G.D., 1984. Defensive mechanisms of termites. Annu. Rev. Entomol 29: 201-232.

Prestwich, G.D., Spanton, S.G., Goh, S.H. and Thio, Y.P., 1980. New tricyclic diterpene propionate esters from a termite defense secretion. Tetrahedron Lett. 22: 1563-1566.

Prober, D.A. and Edgar, B.A., 2000. Ras1 promotes cellular growth in the *Drosophila* wing. Cell 100: 435-446.

Pryce, R.J., 1971. The occurrence of bound, water-soluble squalene, 4,4-dimethyl sterols, 4α-methyl sterols and sterols in leaves of *Kalanchoë blossfeldiana*. Phytochemistry 10: 1303-1307.

Purohit, A., Hejaz, H.A., Walden, L., MacCarthy-Morrogh, L., Packham, G., Potter, B.V. and Reed, M.J., 2000. The effect of 2-methoxyoestrone-3-O-sulphamate on the growth of breast cancer cells and induced mammary tumours. Int. J. Cancer 85: 584-589.

Raff, M., 1998. Cell suicide for beginners. Nature 396: 119-122.

Raffa, K.F., Berryman, A.A., Simasko, J., Teal, W. and Wong, B.L., 1985. Effects of grand fir monoterpenes on the fir engraver, *Scolytus ventralis* (Coleoptera: Scolytidae), and its symbiotic fungus. Environ. Entomol. 14: 552-556.

Raguraman, S. and Saxena, R.C., 1994. Effects of neem seed derivatives on brown planthopper symbiotes. Phytoparasitica 22: 299-308.

Rao, P.J., Kumar, K.M., Singh, S. and Subrahmanyam, B., 1999. Effect of *Artemisia annua* oil on development and reproduction of *Dysdercus koenigii* F. (Hem., Pyrrhocoridae). J. Appl. Ent. 123: 315-318.

Rao, S., Porter, D.C., Chen, X., Herliczek, T., Lowe, M. and Keyomarsi, K., 1999b. Lovastatin-mediated G1 arrest is through inhibition of the proteasome, independent of hydroxymethyl glutaryl-CoA reductase. Proc. Natl. Acad. Sci. USA 96: 7797-7802.

Rebollo, A. and Martínez-A., C., 1999. Ras proteins: recent advances and new functions. Blood 94: 2971-2980.

Ren, Z., Elson, C.E. and Gould, M.N., 1997. Inhibition of type I and type II geranylgeranyl-protein transferases by the monoterpene perillyl alcohol in NIH3T3 cells. Biochem. Pharmacol. 54: 113-120.

Ren, Z. and Gould, M.N., 1994. Inhibition of ubiquinone and cholesterol synthesis by the monoterpene perillyl alcohol. Cancer Lett. 76: 185-190.

Renoux, J.M. and Rohmer, M., 1986. Enzymatic cyclization of all-*trans* pentaprenyl and hexaprenyl methyl ethers by a cell-free system from the protozoon *Tetrahymena pryriformis*. The biosynthesis of scalarane and polyclohexaprenyl derivatives. Eur. J. Biochem. 155: 125-132.

Rhodes, J. and Evans, B.K., 1987. Method of treating functional bowel disorders by the administration of peppermint oil to the intestine. U.S. Patent 4,687,667.

Rice, P.J. and Coats, J.R., 1994. Insecticidal properties of several monoterpenoids to the house fly (Diptera: Muscidae), red flour beetle (Coleoptera: Tenebrionidae), and southern corn rootworm (Coleoptera: Chrysomelidae). J. Econ. Entomol. 87: 1172-1179.

Richard, D.S. and Gilbert, L.I., 1991. Reversible juvenile hormone inhibition of ecdysteroid and juvenile hormone synthesis by the ring gland of *Drosophila melanogaster*. Experienta 47: 1063-1066.

Richard, H.M., 1991. Spices and condiments. I. In: Volatile Compounds in Foods and Beverages (ed. H. Maarse), Marcel Dekker Inc., New York etc.: 411-448.

Richards, J.B. and Hemming, F.W., 1972. Dolichols, ubiquinones, geranylgeraniol and farnesol as the major sites of mevalonate in *Phytophthora cactorum*. Biochem. J. 128: 1345-1352.

Richman, A., Gijzen, M., Starratt, A.N., Yang, Z. and Brandle, J.E., 1999. Diterpene synthesis in *Stevia rebaudiana*: recruitment and up-regulation of key enzymes from the gibberellin biosynthetic pathway. The Plant Journal 19: 411-421.

Rilling, H.C., Breunger, E., Epstein, W.W. and Crain, P.F., 1990. Prenylated proteins: the structure of the isoprenoid group. Science 247: 318-319.

Rioja, A., Pizzey, A.R., Marson, C.M. and Thomas, N.S., 2000. Preferential induction of apoptosis of leukaemic cells by farnesol. FEBS Lett. 467: 291-295.

Rios, J.L., Recio, M.C and Villar, A., 1987. Antimicrobial activity of selected plants employed in the Spanish Mediterranean area. J. Ethnopharmacol. 21: 139-152.

Ripple, G.H., Gould, M.N., Arzoomanian, R.Z., Alberti, D., Feierabend, C., Simon, K., Binger, K., Tutsch, K.D., Pomplun, M., Wahamaki, A., Marnocha, R., Wilding, G. and Bailey, H.H., 2000. Phase I clinical and pharmacokinetic study of perillyl alcohol administered four times daily. Clin. Cancer Res. 6: 390-396.

Ripple, G.H., Gould, M.N., Stewart, J.A., Tutsch, K.D., Arzoomanian, R.Z., Alberti, D., Feierabend, C., Pomplun, M., Wilding, G. and Bailey, H.H., 1998. Phase I clinical trial of perillyl alcohol administered daily. Clin. Cancer Res. 4: 1159-1164.

Ritter, F.J., Persoons, C.J. and Gut, J., 1981. Pheromones: exocrine chemical messengers in the service of reproduction and communication. Organorama 18 (4): 3-9.

Roces, F., 1994. Odour learning and decision-making during food collection in the leaf-cutting ant *Acromyrmex lundi*. Insectes Sociaux 41: 235-239.

Rodriguez, E., 1976. Sesquiterpene lactones: chemotaxonomy, biological activity and isolation. Rev. Latinoamer. Quim. 8: 56-62.

Roger, A.J., 1999. Reconstructing early events in eukaryotic evolution. The Am. Natur. 154: S147-158.

Rogers, I.H., 1978. Environmental effects of terpenoid chemicals: a review. J. Am. Oil Chemist's Soc. 55: 113-118.

Rohmer, M., 1998. Isoprenoid biosynthesis via the mevalonate-independent route, a novel target for antibacterial drugs? Prog. Drug Res. 50: 135-154.

Rohmer, M. and Bisseret, Ph., 1994. Hopanoid and other polyterpenoid biosynthesis in Eubacteria. In: Isopentoids and other Natural Products: Evolution and Function (ed. W.D. Nes), Am. Chem. Soc. Washington: 55-89.

Roland, J., Denford, K.E. and Jimenez, L., 1995. Borneol as an attractant for *Cyzenis albicans*, a tachinid parasitoid of the winter moth, *Operophtera brumata* L. (Lepidoptera: Geometridae). Can. J. Entomol. 127: 413-421.

Romagni, J.G., Allen, S.N. and Dayan, F.E., 2000. Allelopathic effects of volatile cineoles on two weedy plant species. J. Chem. Ecol. 26: 303-313.

Rosenbaum, M.R. and Kerlin, R.L., 1995. Erythema multiforme major and disseminated intravascular coagulation in a dog following application of a d-limonene-based insecticidal dip. J. Am. Vet. Med. Assoc. 207: 1315-1319.

Ross, J.D. and Sombrero, J., 1991. Environmental control of essential oil production in Mediterranean plants. In: Ecological Chemistry and Biochemistry of Plant Terpenoids (eds J.B. Harborne and F.A. Tomas-Berberan), GB-Oxford; Clarendon Press: 83-94.

Rosser, D.S.E., Ashby, M.N., Ellis, J.L. and Edwards, P.A., 1989. Coordinate regulation of 3-hydroxy-3-methylglutaryl Coenzyme A synthase, 3-hydroxy-3-methylglutaryl Coenzyme A reductase and prenyltransferase. Synthesis but not degradation in HepG2 cells. J. Biol. Chem. 264: 12653-12656.

Roush, W., 1996. Regulating G-protein signalling. Science 271: 1056-1058.

Rowinsky, E.K., Windle, J.J. and von Hoff, D.D., 1999. Ras protein farnesyltransferase: a strategic target for anticancer therapeutic development. J. Clin. Oncol. 17: 3631-3652.

Rubio, I., Wittig, U., Meyer, C., Heinze, R., Kadereit, D., Waldmann, H., Downward, J. and Wetzker, R., 1999. Farnesylation of Ras is important for the interaction with phosphoinositide 3-kinase γ. Eur. J. Biochem. 266: 70-82.

Rumpho, M.E., Summer, E.J. and Manhart, J.R., 2000. Solar-powered sea slugs. Mollusc/algal chloroplast symbiosis. Plant Physiol. 123: 29-38.

Ryan, M.F. and Byrne, O., 1988. Plant-insect coevolution and inhibition of acetylcholinesterase. J. Chem. Ecol. 14: 1965-1975.

Ryu, S.Y., Oak, M.H., Yoon, S.K., Cho, D.I., Yoo, G.S., Kim, T.S. and Kim, K.M., 2000. Anti-allergic and anti-inflammatory triterpenes from the herb of *Prunella vulgaris*. Planta Med. 66: 358-360.

Sacchettini, J.C. and Poulter, C.D., 1997. Creating isoprenoid diversity. Science 277: 1788-1789.

Sadof, C.S. and Grant, G.G., 1997. Monoterpene composition of *Pinus sylvestris* varieties resistant and susceptible to *Dioryctria zimmermanni*. J. Chem. Ecol. 23: 1917-1927.

Sahin, M.B., Perman, S.M., Jenkins, G. and Clark, S.S., 1999. Perillyl alcohol selectively induces G0/G1 arrest and apoptosis in Bcr/Abl-transformed myeloid cell lines. Leukemia 13: 1581-1591.

Sakagami, H., Satoh, K., Hakeda, Y. and Kumegawa, M., 2000. Apoptosis-inducing activity of vitamin C and vitamin K. Cell Mol. Biol. (Noisy-Le-Grand) 46: 129-143.

Sakai, I., Hashimoto, S., Yoda, M., Hida, T., Oshawa, S., Nakajo, S. and Nakaya, K., 1994. Novel role of vitamin K2: a potent inducer of differentiation of various human myeloid leukemia cell lines. Biochem. Biophys. Res. Comm. 205: 1305-1310.

Sakai, T., Tanaka, T., Osawa, S., Hashimoto, S. and Nakaya, K., 1993. Geranylgeranyl acetone used as an antiulcer agent is a potent inducer of differentiation of various human myeloid leukemia cell lines. Biochem. Biophys. Res. Comm. 191: 873-879.

Salom, S.M., Grosman, D.M., McClellan, Q.C. and Payne, T.L., 1995. Effect of an inhibitor-based suppression tactic on abundance and distribution of the southern pine beetle (Coleoptera: Scolytidae) and its natural enemies. J. Econ. Entomol. 88: 1703-1716.

Salvesen, G.S. and Dixit, V.M., 1997. Caspases: intracellular signalling by proteolysis. Cell 91: 443-446.

Sato, K., Mochizuki, M., Saiki, I., Yoo, Y.C., Samukawa, K. and Azuma, I., 1994. Inhibition of tumor angiogenesis and metastasis by a saponin of Panax ginseng, ginsenoside-Rb2. Biol. Pharm. Bull. 17: 635-639.

Satomi, Y., Miyamoto, S. and Gould, M.N., 1999. Induction of AP-1 activity by perillyl alcohol in breast cancer cells. Carcinogenesis 20: 1957-1961.

Savage, T.J. and Croteau, R., 1993. Biosynthesis of monoterpenes: regio- and stereochemistry of (+)-3-carene biosynthesis. Arch. Biochem. Biophys. 305: 581-587.

Savage, T.J., Hatch, M.W. and Croteau, R., 1994. Monoterpene synthases of *Pinus contorta* and related conifers. A new class of terpenoid cyclase. J. Biol. Chem. 269: 4012-4020.

Saxena, R.C., 1999. Building awareness and facilitating the use of neem as a source of natural pesticides and other useful products in Sub-Saharan Africa. Phytoparasitica 27: 177-180.

Saxena, R.C. and Khan, Z.R., 1986. Aberrations caused by neem oil odour in green leafhopper feeding on rice plants. Entomol. exp. appl. 42: 279-284.

Saxena, J. and Mathela, G.S., 1996. Antifungal activity of new compounds from *Nepeta leucophylla* and *Nepeta clarkei*. Applied Env. Microbiol. 62: 702-704.

Scadding, S.R., 1999. Citral, an inhibitor of retinoic acid synthesis, modifies pattern formation during limb regeneration in the axolotl *Ambystoma mexicanum*. Can. J. Zool. 77: 1835-1837.

Scadding, S.R. and Maden, M., 1994. Retinoic acid gradients during limb regeneration. Dev. Biol. 162: 608-617.

Schafer, W.R., Kim, R., Sterne, R., Thorner, J., Kim, S.H. and Rine, J., 1989. Genetics and pharmacological suppression of oncogenic mutations in Ras genes of yeast and humans. Science 245: 379-385.

Scheffrahn, R.H., Hsu, R.C., Su, N.Y., Huffman, J.B., Midland, S.L. and Sims, J.J., 1988. Allelochemical resistance of bald cypress *Taxodium distichum*, heartwood to the subterranean termite *Coptotermes formosanus*. J. Chem. Ecol. 14: 765-776.

Schilcher, H., 1987. Die Kamille: Handbuch für Ärzte, Apotheker und andere Naturwissenschaftler. Wissenschaftliche Verlagsgesellschaft GmbH Stuttgart, 152 pp.

Schlyter, F. and Birgersson, G.A., 1999. Forest beetles. In: Pheromones of Non-Lepidopteran Insects Associated with Agricultural Plants (eds J. Hardie and A.K. Minks), CABI Publ. Wallingford: 113-148.

Schmelz, E.A., Grebenok, R.J., Galbraith, D.W. and Bowers, W.S., 1999. Insect-induced synthesis of phytoecdysteroids in spinach, *Spinacia oleracea*. J. Chem. Ecol. 25: 1739-1757.

Schmutterer, H., 1987. Insect growth-disrupting and fecundity-reducing ingredients from the neem and chinaberry trees. In: CRC Handbook of Natural Pesticides, Vol. III: Insect Growth Regulators, part B (eds E.D. Morgan and M.B. Mandava), CRC, Boca Raton, FL: 119-170.

Schmutterer, H., 1990. Properties and potential of natural pesticides from the neem tree, *Azadirachta indica*. Annu. Rev. Entomol. 35: 271-297.

Schmutterer, H. (ed.), 1995. The Neem Tree: Source of Unique Natural Products for Integrated Pest Management, Medicine, Industry and Other Purposes. VCH, Weinheim, Germany, 696 pp.

Schneiderman, H.A., Gilbert, L.I. and Weinstein, M.J., 1960. Juvenile hormone activity in micro-organisms and plants. Nature 188: 1041-1042.

Schoonhoven, L.M., 1999. Insects and plants: two worlds come together. Entomol. exp. appl. 91: 1-6.

Schoonhoven, L.M. and Blom, F., 1988. Chemoreception and feeding behaviour in a caterpillar: towards a model of brain functioning in insects. Entomol. exp. appl. 49: 123-129.

Schoonhoven, L.M. and Luo, L.E., 1994. Multiple mode of action of the feeding deterrent, toosendanin, on the sense of taste in *Pieris brassicae* larvae. J. Comp. Physiol. A 175: 519-524.

Schroeder. F.C., Gonzàlez, A., Eisner, T. and Meinwald, J., 1999. Miriamin, a defensive diterpene from the eggs of a land slug (*Arion* sp.). PNAS 96: 13620-13625.

Schulz, S., Buhling, F. and Ansorge, S., 1994. Prenylated proteins and lymphocytes proliferation: inhibition by D-limonene and related monoterpenes. Eur. J. Immunol. 24: 301-307.

Schwemmler, W. and Gassner, G. (eds), 1989. Insect Endocytobiosis: Morphology, Physiology, Genetics, Evolution. Boca Raton/FL: CRC Press, 272 pp.

Schwender, J., Seemann, M., Lichtenthaler, H.K. and Rohmer, M., 1996. Biosynthesis of isoprenoids (carotenoids, sterols, prenyl side-chains of chlorophylls and plastoquinone) via a novel pyruvate/glyceraldehyde 3-phosphate non-mevalonate pathway in the green alga *Scenedesmus obliquus*. Biochem. J. 15: 73-80.

Schwender, J., Zeidler, J., Groner, R., Muller, C., Focke, M., Braun, S., Lichtenthaler, F.W. and Lichtenthaler, H.K., 1997. Incorporation of 1-deoxy-D-xylulose into isoprene and phytol by higher plants and algae. FEBS Lett. 414: 129-134.

Schwender, J., Muller, C., Zeidler, J. and Lichtenthaler, H.K., 1999. Cloning and heterologous expression of a cDNA encoding 1-deoxy-D-xylulose-5-phosphate reductoisomerase of *Arabidopsis thaliana*. FEBS Lett. 455: 140-144.

Seitz, F., 2000. Decline of the generalist. Nature 403: 483.

Service, R.F., 1996. Tumor-killer made: how does it work? Science 274: 2009.

Seybold, S.J., 1998. The origin of bark beetle pheromones: biochemical studies with ecological and evolutionary implications. Proc. 2nd Int. Symp. on Insect Pheromones: 75.

Seybold, S.J., Quilici, D.R., Tillman, J.A., Vanderwel, D., Wood, D.L., Blomquist, G.J., 1995. *De novo* biosynthesis of the aggregation pheromone components ipsenol and ipsdienol by the pine bark beetles *Ips paraconfusus* Lanier and *Ips pini* (Say) (Coleoptera: Scolytidae). Proc. Natl. Acad. Sci. USA 92: 8393-8397.

Sharma, O.P., Dawra, R.K., Datta, A.K. and Kanwar, S.S., 1997. Biodegradation of lantadene A, the pentacyclic triterpenoid hepatotoxin by *Pseudomonas pickettii*. Lett. Appl. Microbiol. 24: 229-232.

Sharma, R.N. and Saxena, K.N., 1974. Orientation and developmental inhibition in the housefly by certain terpenoids. J. Med. Entomol. 11: 617-621.

Shelley, M.D., Hartley, L., Fish, R.G., Groundwater, P., Morgan, J.J., Mort, D., Mason, M. and Evans, A., 1999. Stereo-specific cytotoxic effects of gossypol enantiomers and gossypolone in tumour cell lines. Cancer Lett. 135: 171-180.

Shi, W. and Gould, M.N., 1995. Induction of differentiation in neuro-2A cells by the monoterpene perillyl alcohol. Cancer Lett. 95: 1-6.

Shidoji, Y., Nakamura, N., Moriwaki, H. and Muto, Y., 1997. Rapid loss in the mitochondrial membrane potential during geranylgeranoic acid-induced apoptosis. Biochem. Biophys. Res. Comm. 230: 58-63.

Shureiqi, I., Reddy, P. and Brenner, D.E., 2000. Chemoprevention: general perspective. Crit. Rev. Oncol. Hematol. 33: 157-167.

Sikkema, J., de Bont, J.A.M. and Poolman, B., 1995. Mechanisms of membrane toxicity of hydrocarbons (review). Microbiol. Rev. 59: 201-222.

Simmonds, M.S.J. and Blaney, W.M., 1996. Azadirachtin: advances in understanding its activity as an antifeedant. In: Proc. 9th Int. Symp. on Insect-Plant Relationships (eds E. Städler, M. Rowell-Rahier and R. Bauer), Entomol. exp. appl. 80: 23-26.

Simmonds, M.S.J., Blaney, W.M., Esquivel, B. and Rodriguez-Hahn, L., 1996. Effect of clerodane-type diterpenoids isolated from *Salvia* spp. on the feeding behaviour of *Spodoptera littoralis*. Pestic. Sci 47: 17-23.

Simoes-Wust, A.P., Olie, R.A., Gautschi, O., Leech, S.H., Haner, R., Hall, J., Fabbro, D., Stahel, R.A. and Zangemeister-Wittke, U., 2000. Bcl-xl antisense treatment induces apoptosis in breast carcinoma cells. Int. J. Cancer 87: 582-590.

Simonin, P., Jurgens, U.J. and Rohmer, M., 1996. Bacterial triterpenoids of the hopane series from the prochlorophyte *Prochlorothrix hollandica* and their intracellular localization. Eur. J. Biochem. 241: 865-871.

Simonsen, T., Nordoy, A., Sjunneskog, C. and Lyngmo, V., 1988. The effect of cod liver oil in two populations with low and high intake of dietary fish. Acta Med. Scand. 223: 491-498.

Singh, R.P., Devakumar, C. and Dhingra, S., 1988. Activity of neem (*Azadirachta indica* A. Juss.) seed kernel extracts against the mustard aphid, *Lipaphis erysimi*. Phytoparasitica 16: 225-230.

Singh, S.V., Pandey, S., Guddewar, M.B. and Malik, Y.P., 1997. Response of neem extractives against red cotton bug, *Dysdercus koenigii* on cotton seed. Indian J. Ent. 59: 41-44.

Singh, A., Purohit, A., Hejaz, H.A., Potter, B.V. and Reed, M.J., 2000. Inhibition of deoxyglucose uptake in MCF-7 breast cancer cells by 2-methoxyestrone and 2-methoxyestrone-3-O-sulfamate. Mol. Cell Endocrinol. 160: 61-66.

Singletary, K., 2000. Diet, natural products and cancer chemoprevention. J. Nutr. 130/2S (Suppl.): 465-466.

Siperstein, M.D., 1984. Role of cholesterogenesis and isoprenoid synthesis in DNA replication and cell growth. J. Lipid Res. 25: 1462-1467.

Skulachev, V.P., 1996. The "cell suicide protein" is hidden in the intermembrane space of mitochondria. On its release, it causes apoptosis (in Russian). Biokhimiya 61: 2060-2063. Translation: Biochemistry (Moscow) 61: 1477-1479.

Slaga, T.J., Fischer, S.M., Weeks, C.E., Klein-Szanto, A.J.P. and Reiniers, J., 1982. Studies on the mechanisms involved in multistage carcinogenesis in mouse skin. J. Cell Biochem. 18: 99-119.

Slàma, K., 1987. Insect hormones and bioanalogues in plants. Proc. 6th Int. Symp. on Insect-Plant Relationships (eds V. Labeyrie, G. Fabres and D. Lachaise), Junk-Kluwer Acad.: 9-16.

Smirle, M., Lowery, T. and Zurowski, C.L., 1996. Influence of neem oil on detoxication enzyme activity in the obliquebanded leafroller, *Choristoneura rosaceana*. Pest. Biochem. Physiol. 56: 220-230.

Smith, C.A., Svoboda, K.P. and Noon, M.M., 1997. Controlling the growth and quality of hydroponically-grown basil. Acta Hortic. 450: 479-486.

Smith, C.M., Khan, Z.R. and Pathak, M.D. (eds), 1994. Techniques for evaluating insect resistance in crop plants. Lewis Publishers Inc., Boca Raton, Florida, pp. 320.

Smith, D.C., Johnson, C.S., Freeman, C.C., Muindi, J., Wilson, J.W. and Trump, D.L., 1999. A Phase I trial of calcitriol (1,25-dihydroxycholecalciferol) in patients with advanced malignancy. Clin. Cancer Res. 5: 1339-1345.

Smith, M.A., Parkinson, D.R., Cheson, B.D. and Friedman, M.A., 1992. Retinoids in cancer therapy. J. Clin. Oncol. 10: 839-864.

Smith, R.M., Brophy, J.J., Cavill, G.W.K. and Davies, N.W., 1979. Iridodials and nepetalactone in the defensive secretion of the coconut stick insects, *Graeffea crouani*. J. Chem. Ecol. 5: 727-735.

Smith, S.M., Dickman, E.D., Power, S.C. and Lancman, J., 1998. Retinoids and their receptors in vertebrate embryogenesis. J. Nutr. 128: S467-470.

Smitt, R.L., Gottlieb, E., Kucharski, L.M. and Maguire, M.E., 1998. Menaquinone (vitamin K2) biosynthesis: localization and characterization of the *men* A gene from *Escherichia coli*. J. Bacteriol. 180: 2782-2787.

Søballe, B. and Poole, R.K., 1999. Microbial ubiquinones: multiple roles in respiration, gene regulation and oxidative stress management. Microbiology 145: 1817-1830.

Sokoloski, J.A. and Sartorelli, A.C., 1998. Induction of the differentiation of HL-60 promyelocytic leukemia cells by nonsteroidal anti-inflammatory agents in combination with low levels of vitamin D3. Leuk. Res. 22: 153-161.

Sohn, K.H., Lee, H.Y., Chung, H.Y., Young, H.S., Yi, S.Y. and Kim, K.W., 1995. Anti-angiogenic activity of triterpene acids. Cancer Lett. 94: 213-218.

Southwood, T.R.E., 1996. Insect-plant relations: overview from the symposium. In: Proc. 9th Int. Symp. on Insect-Plant Relationships (eds E. Städler, M. Rowell-Rahier and R. Bauer), Entomol. exp. appl. 80: 320-324.

Spurgeon, S.L. and Porter, J.W. (eds), 1983. Biosynthesis of isoprenoid compounds. Vol. 2, John Wiley & Sons, New York, 122 pp.

Sriskandan, S. and Cohen, J., 1999. Gram-positive sepsis. Mechanisms and differences from Gram-negative sepsis. Infect. Dis. Clin. North Am. 13: 397-412.

Srivastava, N.K., Misra, A. and Sharma, S., 1997. Effect of Zn deficiency on net photosynthetic rate, ^{14}C partitioning, and oil accumulation in leaves of peppermint. Photosynthetica 33: 71-79.

Stammati, A., Bonsi, P., Zucco, F., Moezelaar, R., Alakomi, H.L. and von Wright, A., 1999. Toxicity of selected plant volatiles in microbial and mammalian short-term assays. Food Chem. Toxicol. 37: 813-823.

Stanley, R.G., 1958. Terpene formation in pine from mevalonic acid. Nature 182: 738-739.

Stanley, G., Algie, J.E. and Brophy, J.J., 1986. 1,3,3-Trimethyl-2,7-dioxabicyclo (2,2,1) heptane: a volatile oxidation product of α-farnesene in Granny Smith apples. Chem. Ind. 16: 556.

Starks, M., Back, K., Chappell, J. and Noel, J.P., 1997. Structural basis for cyclic terpene biosynthesis by tobacco 5-epi-aristolochene synthase. Science 277: 1815-1820.

Stayrook, K.R., McKinzie, J.H., Barbhaiya, L.H. and Crowell, P.L., 1998. Effects of the antitumor agent perillyl alcohol and signal transduction in pancreatic cells. Anticancer Res. 18: 823-828.

Steele, C.L., Crock, J., Bohlmann, J. and Croteau, R., 1998. Sesquiterpene synthases from grand fir (*Abies grandis*). Comparison of constitutive and wound-induced activities, and cDNA isolation, characterization, and bacterial expression of delta-selinene synthase and gamma-humulene synthase. J. Biol. Chem. 273: 2078-2089.

Steele, J.C., Warhurst, D.C., Kirby, G.C. and Simmonds, M.S., 1999. *In vitro* and *in vivo* evaluation of betulinic acid as an antimalarial. Phytother. Res. 13: 115-119.

Steiner, A.A. (ed.), 1999. 7th Bibliograpy on Soilless Culture. ISOSC B.S.C. VII, Mulder Wageningen, The Netherlands, 414 pp.

Steinbrecht, R.A., Laue, M., Maida, R. and Ziegelberger, G., 1996. Odorant-binding proteins and their role in the detection of plant odours. In: Proc. 9th Int. Symp. on Insect-Plant Relationships (eds E. Städler, M. Rowell-Rahier and R. Bauer), Entomol. exp. appl. 80: 15-18.

Stephan, B.R., 1987. Differences in the resistance of Douglas fir provenances to the woolly aphid *Gilletteella cooleyi*. Silvae Genet. 36: 76-79.

Stowe, M.K., Turlings, C.J., Loughrin, J.H., Lewis, W.J. and Tumlinson, J.H., 1995. The chemistry of eavesdropping, alarm, and deceit. In: Chemical Ecology (eds T. Eisner and J. Meinwald), Nat. Acad. Press Washington, D.C.: 51-65.

Struhl, K., 1999. Fundamentally different logic of gene regulation in eukaryotes and prokaryotes. Cell 98: 1-4.

Sturm, S., Gil, R.R., Chai, H.B., Ngassapa, O.D., Santisuk, T., Reutrakul, V., Howe, A., Moss, M., Besterman, J.M., Yang, S.L., Farthing, J.E., Tait, R.M., Lewis, J.A., O'Neill, M.J., Farnsworth, N.R., Cordell, G.A., Pezzuto, J.M. and Kinghorn, A.D., 1996. Lupane derivatives from *Lophopetalum wallichii* with farnesyl protein transferase inhibitory activity. J. Nat. Prod. 59: 658-663.

Sun, J., Blaskovich, M.A., Knowles, D., Qian, Y., Ohkanda, J., Bailey, R.D., Hamilton, A.D. and Sebti, S.M., 1999. Antitumor efficacy of a novel class of non-thiol-containing peptidomimetic inhibitors of farnesyltransferase and geranylgeranyltransferase. I. Combination therapy with the cytotoxic agents cisplatin, taxol, and gemcitabine. Cancer Res. 59: 4919-4926.

Svoboda, J.A., Feldlaufer, M.F. and Weirich, G.F., 1994. Evolutionary aspects of steroid utilization in insects. ACS Symposium Series 562: 126-139.

Szyperski, T., Fernandez, C., Mumenthaler, C. and Wuthrich, K., 1998. Structure comparison of human glioma pathogenesis-related protein GliPR and the plant pathogenesis-related protein P14a indicates a functional link between the human immune system and a plant defense system. Proc. Natl. Acad. Sci. USA 95: 2262-2266.

Tachibana, A., Tanaka, T. and Taniguchi, M.O.S., 1996. Evidence for farnesol-mediated isoprenoid synthesis regulation in a halophilic archaeon, *Haloferax volcanii*. FEBS Lett. 379: 43-46.

Tada, M., Tsubouchi, M. and Matsuo, K., 1990. Mechanism of photoregulated carotenogenesis in *Rhodotorula minuta*. VIII. Effect of mevinolin on photoinduced carotenogenesis. Plant Cell Physiol. 31: 319-323.

Takahashi, S., Kuzuyama, T. and Seto, H., 1999. Purification, characterization, and cloning of a eubacterial 3-hydroxy-3-methylglutaryl coenzyme A reductase, a key enzyme involved in biosynhesis of terpenoids. J. Bacteriol. 181: 1256-1263.

Takahashi, S., Kuzuyama, T., Watanabe, H. and Seto, H., 1998. A 1-deoxy-D-xylulose 5-phosphate reductoisomerase catalyzing the formation of 2-C-methyl-D-erythritol 4-phosphate in an alternative nonmevalonate pathway for terpenoid biosynthesis. Proc. Natl. Acad. Sci. USA 95: 9879-9884.

Takano, T. and Ohgi, M., 1994. Effect of anion variations in nutrient solution on the growth and essential oil content and composition in perilla plants. Proc. Sino Intern. Colloquium Soilless Culture (Agric. Meijo University Nagoya, Japan): 106-112.

Takano, T. and Yamamoto, A., 1996. Effect of anion variations in a nutrient solution on basil growth, essential oil content, and composition. Acta Hortic. 426: 389-396.

Takasaki, M., Konoshima, T., Kozuka, M. and Tokuda, H., 1995. Anti-tumor-promoting activities of euglobals from *Eucalyptus* plants. Biol. Pharm. Bull. 18: 435-438.

Takasaki, M., Konoshima, T., Tokuda, H., Masuda, K., Arai, Y., Shiolima, K. and Ageta, H., 1999. Anti-carcinogenic activity of *Taraxacum* plant. II. Biol. Pharm. Bull. 22: 606-610.

Tamanoi, F., 1993. Inhibitors of Ras farnesyltransferases. Trends Biochem. Sci. 18: 349-353.

Tamasawa, N., Hayakari, M., Murakami, H., Matsui, J. and Suda, T., 1997. Reduction of oxysterol levels up-regulates HMG-CoA reductase activity in rat liver. Atherosclerosis 131: 237-242.

Tamura, K., Southwick, E.C., Kerns, J., Rosi, K., Carr, B.I., Wilcox, C. and Lazo, J.S., 2000. Cdc25 inhibition and cell cycle arrest by a synthetic thioalkyl vitamin K analogue. Cancer Res. 60: 1317-1325.

Tanaka, M., Tamura, K. and Ide, H., 1996. Citral, an inhibitor of retinoic acid synthesis, modifies chick limb development. Dev. Biol. 175: 239-247.

Tanaka, T., Sugiura, H., Inaba, R., Nishikawa, A., Murakami, A., Koshimizu, K. and Ohigashi, H., 1999. Immunomodulatory action of citrus auraptene on macrophage functions and cytokine production of lymphocytes in female BALB/c mice. Carcinogenesis 20: 1471-1476.

Taraye, M., Thompson, J.D., Escarré, J. and Linhart, Y.B., 1995. Intra-specific variation in the inhibitory effects of *Thymus vulgaris* (Labiatae) monoterpenes on seed germination. Oecologia 101: 110-118.

Taylor, W.E. and Vickery, B., 1974. Insecticidal properties of limonene, a constituent of citrus oil. Ghana J. Agric. Sci. 7: 61-62.

Teal, P.E., Meredith, J.A. and Gomez-Simuta, Y., 1999. Isolation and identification of terpenoid sex pheromone components from extracts of hemolymph of males of the Caribbean fruit fly. Arch. Insect Biochem. Physiol. 42: 225-232.

Terada, R., Nakajima, M., Isshiki, M., Okagaki, R.J., Wessler, S.R. and Shimamoto, K., 2000. Antisense waxy genes with highly active promoters effectively suppress waxy gene expression in transgenic rice. Plant Cell Physiol. 41: 881-888.

Terry, L.I., 1997. Host selection, communication and reproductive behaviour. In: Thrips as Crop Pests (ed. T. Lewis), CABI, University Press, Cambridge: 65-118.

Teulon, D.A.J., 1988. Pest management of the New Zealand flower thrips, *Thrips obscuratus* (Crawford) (Thysanoptera: Thripidae) on stonefruit in Canterbury, New Zealand. PhD thesis, Lincoln College, University of Canterbury, New Zealand, 221 pp. (+ addendum).

Thai, L., Rush, J.S., Maul, J.E., Devarenne, T., Rodgers, D.L., Chappell, J. and Waechter, C.J., 1999. Farnesol is utilized for isoprenoid biosynthesis in plant cells via farnesyl pyrophosphate formed by successive monophosphorylation reactions. Proc. Natl. Acad. Sci. USA 96: 13080-13085.

Thomas, O.O., 1989. Phytochemistry of the leaf and flower oils of *Tanacetum cilicium*. Fitoterapia 60: 131-134.

Thornberry, N.A. and Lazebnik, Y., 1998. Caspases: enemies within. Science 281: 1312-1316.

Tillman, J.A., Seybold, S.T., Jurenka, R.A. and Blomquist, G.J., 1999. Insect pheromones - an overview of biosynthesis and endocrine regulation. Insect Biochem. Mol. Biol. 29: 481-514.

Tingey, W.M., 1986. Techniques for evaluating plant resistance to insects. In: Insect-Plant Interactions (eds J.R. Miller and T.A. Miller), Springer-Verlag, New York, etc.: 251-284.

Tittiger, C., Blomquist, G.J., Ivarson, P. Borgeson, C.E. and Seybold, S.J., 1999. Juvenile hormone regulation of HMG-R gene expression in the bark beetle *Ips paraconfusus* (Coleoptera: Scolytidae): implications for male aggregation pheromone biosynthesis. Cell Mol. Life Sci. 55: 121-127.

Tomaschko, K.H., 1994. Ecdysteroids from *Pycnogonum litorale* (Arthropoda, Pantopoda) act as chemical defense against *Carcinus maenas* (Crustacea, Decapoda). J. Chem. Ecol. 20: 1445-1455.

Tornabene, T.G., Langworthy, T.A., Holzer, G. and Oro, J., 1999. Squalenes, phytanes and other isoprenoids as major neutral lipids of methanogenic and thermoacidophilic "archaebacteria". J. Mol. Evol. 13: 73-83.

Toxopeus, H. and Bouwmeester, H.J., 1993. Improvement of caraway essential oil and carvone production in the Netherlands. Ind. Crops Prod. 1: 295-301.

Tripathi, S.N., Shastri, V.V.S. and Satyavati, G.V., 1968. Experimental and clinical studies on the effect of guggulu (*Commiphora mukul*) in hyperlipaemia and thrombosis (Achayara). J. Res. Indian Med. 2: 10.

Truesdell, G.M., Jones, C., Holt, T., Henderson, G. and Dickman, M.B., 1999. A Ras protein from a phytopathogenic fungus causes defects in hyphal growth polarity, and induces tumors in mice. Mol. Gen. Genet. 262: 46-54.

Tsao, R., Lee, S., Rice, P.J., Jensen, C. and Coats, J.R., 1995. Insect-control: monoterpenoids and their synthetic derivatives as leads for new insect-control agents (review). In: Synthesis and Chemistry of Agrochemicals. IV (chapter 28), ACS Symp. Series 584: 312-324.

Tsuchiya, Y., Matsumoto, A., Shudo, K. and Okamoto, T., 1980. Identification of germacrene D, one of the volatile metabolites produced by Actinomycetes. (In Japanese). Yakugaku Zasshi 100: 468-471.

Tucker, A.O. and Tucker, S.S., 1988. Catnip and the catnip response. Econ. Bot. 42: 214-231.

Turlings, T.C.J., Tumlinson, J.H. and Lewis, W.J., 1990. Exploitation of herbivore-induced plant odors by host seeking parasitic wasps. Science 250: 1251-1253.

Tyler, V.E., 1994. Herbs of Choice. The Therapeutic Use of Phytomedicinals. Haworth Press, New York etc., 209 pp.

Uchiyama, T., Moriyama, S., Nabeya, Y., Kawaguchi, J., Kanetake, M. and Yamazaki, N., 1991. Synergistic termite controlling agents containing *Phellodendron* or *Citrus* extracts and pyrethroid, organophosphorus and/or carbamate insecticides. (in Japanese). Kokai Tokkyo Koho 3: 103.

Udagawa, Y., 1995. Some responses of dill and thyme, grown in hydroponics, to the concentration of nutrient solution. Acta Hortic. 396: 203-210.

Ultee, A., Gorris, L.G.M. and Smid, E.J., 1998. Bactericidal activity of carvacrol towards the food-borne pathogen *Bacillus cereus*. J. Appl. Microbiol. 95: 211-218.

Vaidya, S., Bostedor, R., Kurtz, M.M., Bergstrom, J.D. and Bansal, V.S., 1998. Massive production of farnesol-derived dicarboxylic acids in mice treated with the squalene synthase inhibitor zaragozic acid A. Arch. Biochem. Biophys. 355: 84-92.

Vajs, V., Todorovic, N., Ristic, M., Tesevic, V., Todorovic, B., Janackovic, P., Marin, P. and Milosavljevic, S., 1999. Guaianolides from *Centaurea nicolai*: antifungal activity. Phytochemistry 52: 383-386.

Van Beek, T.A. (ed.), 2000. *Ginkgo biloba*. Vol. 12 of: Medical and Aromatic Plants - Industrial Profiles. Harwood Academic, 550 pp.

Van Beek, T.A., Scheeren, H.A., Rantio, T., Melger, W.Ch. and Lelyveld, G.P., 1991. Determination of ginkgolides and bisobalides in *Ginkgo biloba* leaves and phytopharmaceuticals. J. Chromatogr. 535: 375-387.

Van den Heuvel, J.F.J.M., Bruyère, A., Hogenhout, S.A., Ziegler-Graff, V., Brault, V., Verbeek, M., van der Wilk, F. and Richards, K., 1997. The N-terminal region of the luteovirus readthrough domain determines virus binding to *Buchnera* GroEL and is essential for virus persistence in the aphid. J. Virol. 71: 7258-7265.

Van den Heuvel, J.F.J.M., Hogenhout, S.A., Verbeek, M. and van der Wilk, F., 1998. *Azadirachta indica* metabolites interfere with the host-endosymbiont relationship and inhibit the transmission of potato leafroll virus by *Myzus persicae*. Entomol. exp. appl. 86: 253-260.

Van den Heuvel, J.F.J.M., Verbeek, M. and van der Wilk, F., 1994. Endosymbiotic bacteria associated with circulative transmission of potato leafroll virus by *Myzus persicae.* J. Gen. Virol. 75: 2559-2567.

Van Duuren, B.L., Blazej, T., Goldsmidt, B.M., Katz, C., Melchionne, S. and Sivak, A., 1971. Cocarcinogenesis studies on mouse skin and inhibition of tumor induction. J. Natl. Cancer Inst. 46: 1039-1044.

Van Loon, J.J.A. and Schoonhoven, L.M., 1999. Specialist deterrent chemoreceptors enable *Pieris* caterpillars to discriminate between chemically different deterrents. Entomol. exp. appl. 91: 29-35.

Van Oosten, A.M., Gut, J., Harrewijn, P. and Piron, P.G.M., 1990. Role of farnesene isomers and other terpenoids in development of different morphs and forms of the aphid *Myzus persicae.* (Proc. IIIrd Int. Symp. on aphids, Kecskemét, Hungary) Acta Phytopath. et Entomol. Hungarica 25: (1-4) 331-342.

Veerman, A., 1980. Functional involvement of carotenoids in photoperiodic induction of diapause in the spider mite, *Tetranychus urticae.* Physiol. Entomol. 5: 291-300.

Venkataraman, V., O'Mahony, P.J., Manzcak, M. and Jones, G., 1994. Regulation of juvenile hormone esterase gene transcription by juvenile hormone. Dev. Genet. 15: 391-400.

Verma, A.K., Slaga, T.J., Wetz, P.W., Mueller, G.C. and Boutwell, R.K., 1980. Inhibition of skin tumor promotion by retinoic acid and its metabolite 5,6-epoxyretinoic acid. Cancer Res. 40: 2367-2371.

Vicente, M., Chater, K.F. and De Lorenzo, V., 1999. Bacterial transcription factors involved in global regulation. Mol. Microbiol. 33: 8-17.

Vigushin, D.M., Poon, G.K., Boddy, A., English, J., Halbert, G.W., Pagonis, C., Jarman, M. and Coombes, R.C., 1998. Phase I and pharmacokinetic study of D-limonene in patients with advanced cancer. Cancer research campaign Phase I/II clinical trials committee. Cancer Chemother. Pharmacol. 42: 111-117.

Villani, F., Piccinini, F., Favalli, L., Cristalli, S. and Chiarra, A., 1974. Influence of terpenes on pharmacokinetics of tetracycline. Progr. Chemother. (antibacterial, antiviral, antineoplast.), Proc. 8th Int. Congr. Chemother. 1: 669-672.

Visser, J.H., 1986. Host odor perception in phytophagous insects. Annu. Rev. Entomol. 31: 121-144.

Visser, J.H. and de Jong, R., 1988. Olfactory coding in the perception of semiochemicals. J. Chem. Ecol. 14: 2005-2018.

Visser, J.H. and Piron, P.G.M., 1995. Olfactory antennal responses to plant volatiles in apterous virginoparae of the vetch aphid *Megoura viciae.* Entomol. exp. appl. 77: 37-46.

Visser, J.H. and Piron, P.G.M., 1998. An open Y-track olfactometer for recording of aphid behavioural responses to plant odours. Proc. Exper. & Appl. Entomol., N.E.V. Amsterdam (eds M.J. Sommeijer and P.J. Francke), 9: 41-46.

Visser, J.H., Piron, P.G.M. and Hardie, J., 1996. The aphids' peripheral perception of plant-volatiles. In: Proc. 9th Int. Symp. on Insect-Plant Relationships (eds E. Städler, M. Rowell-Rahier and R. Bauer), Entomol. exp. appl. 80: 35-38.

Vité, J.P. and Baader, E., 1990. Present and future use of semiochemicals in pest management of bark beetles. J. Chem. Ecol. 16: 3031-3041.

Vitols, S., Angelin, B. and Juliusson, G., 1997. Simvastatin impairs mitogen-induced proliferation of malignant B-lymphocytes from humans: in vitro and in vivo studies. Lipids 32: 255-262.

Vogel, S., 1993. Betrug bei Pflanzen: Die Täuschblumen. In: Acad. Scient. et Lit. Moguntina. Franz Steiner Verlag, Stuttgart: 1-47.

Von Dehn, M., 1963. Hemmung der Flügelbildung durch Farnesol bei der schwarzen Bohnenlaus, Doralis fabae Scop. Naturwissenschaften 50: 578-579.

Von Niemeyer, H., Lenarduzzi, M. and Watzek, G., 1995. Zur Wirkung von Verbenon auf den Buchdrucker, Ips typographus L. (Col., Scolytidae). Anz. Schädlingskde, Pflanzenschutz, Umweltschutz. 68: 182-186.

Von Niemeyer, H. and Watzek, G., 1996. Test von Monoterpenen als Zusatz zu Pheroprax® bzw. Chalcoprax® in Pheromonfallen zum Fang des Buchdruckers, Ips typographus L. bzw. des Kupferstechers, Pityogenes chalcographus L. (Col., Scolytidae). Anz. Schädlingskde, Pflanzenschutz, Umweltschutz 69: 109-110.

Voziyan, P.A., Haug, J.S. and Melnykovych, G., 1995. Mechanism of farnesol cytotoxicity: further evidence for the role of PKC-dependent signal transduction in farnesol-induced apoptotic cell death. Biochem. Biophys. Res. Comm. 212: 479-486.

Wall, M.E., 1998. Camptothecin and taxol: discovery to clinic. Med. Res. Rev. 18: 299-314.

Walton, A.H., 1964. The Perfumed Garden of the Shaykh Nefzawi. Hamilton & Co (Stafford), 271 pp.

Wan, J., Wilcock, A. and Coventry, M.J., 1998. The effect of essential oils of basil on the growth of Aeromonas hydrophila and Pseudomonas fluorescens. J. Appl. Microbiol. 84: 152-158.

Wang, H.G., Takayama, S., Rapp, U.R. and Reed, J.C., 1996. Bcl-2 interacting protein, BAG-1 binds to and activates the kinase Raf-1. Proc. Natl. Acad. Sci. USA 93: 7063-7068.

Wang, K. and Ohnuma, S., 1999. Chain-length determination mechanism of isoprenyl diphosphate synthases and implications for molecular evolution. Trends Biochem. Sci. 24: 445-451.

Wang, T., Danielson, P.D., Li, B.Y., Shah, P.C., Kim, S.D. and Donahoe, P.K., 1996b. The p21ras farnesyltransferase α subunit in TGF-β and activin signalling. Science 271: 1120-1122.

Wang, W. and Macaulay, R.J., 1999. Apoptosis of medulloblastoma cells in vitro follows inhibition of farnesylation using manumycin A. Int. J. Cancer 82: 430-434.

Wang, Y., Blandino, G., Oren, M. and Givol, D., 1998. Induced p53 expression in lung cancer cell line promotes cell senescence and differentially modifies the cytotoxicity of anti-cancer drugs. Oncogene 17: 1923-1930.

Wani, M.C., Taylor, H.L., Wall, M.E., Coggon, P. and McPhail, A.T., 1971. Plant antitumor agents. VI. The isolation and structure of taxol, a novel antileukemic and antitumor agent from Taxus brevifolia. J. Am. Chem. Soc. 93: 2325-2327.

References

Wanner, K.W., 1997. Foliar and systematic applications of neem seed extract for control of spruce budworm, *Choristoneura fumiferana* (Clem.) (Lepidoptera: Tortricidae), infesting black and white spruce seed orchards. Can. Entomol. 129: 645-655.

Ward, B.B., Courtney, K.J. and Langenheim, J.H., 1997. Inhibition of *Nitrosomonas europaea* by monoterpenes from coastal redwood (*Sequoia sempervirens*) in whole-cell studies. J. Chem. Ecol. 23: 2583-2598.

Watson, G.M.F. and Tabita, F.R., 1997. Microbial ribulose 1,5-biphosphate carboxylase/oxygenase: a molecule for phylogenetic and enzymological investigation. FEMS Microbiol. Lett. 2: 13-22.

Watson, J.A., Havel, C.M., Lobos, D.V., Baker, F.C. and Morrow, C.J., 1985. Isoprenoid synthesis in isolated embryonic *Drosophila* cells. J. Biol. Chem. 260: 14083-14091.

Wattenberg, L.W., 1992. Inhibition of carcinogenesis by minor dietary constituents. Cancer Res. 52: 2085-2091.

Way, M.J. and van Emden, H.F., 2000. Integrated pest management in practice: pathways towards successful application. Crop Protection 19: 81-103.

Weatherston, I. and Minks, A.K., 1995. Regulation of semiochemicals - global aspects. Int. Pest Manag. Rev. 1: 1-13.

Weaver, D.K., Phillips, T.W., Dunkel, F.V., Weaver, T., Grubb, R.T. and Nance, E.L., 1995. Dried leaves from Rocky Mountain plants decrease infestation by stored-product beetles. J. Chem. Ecol. 21: 127-142.

Weisman, R.A., Christen, R., Los, G., Jones, V., Kerber, C., Seagren, S., Glassmeyer, S., Orloff, L.A., Wong, W., Kirmani, S. and Howell, S., 1998. Phase I trial of retinoic acid and cisplatinum for advanced squamous cell cancer of the head and neck based on experimental evidence of drug synergism. Otolaryngol. Head Neck Surg. 118: 597-602.

Werner, R.A., 1995. Toxicity and repellency of 4-allylanisole and monoterpenes from white spruce and tamarack to the spruce beetle and eastern larch beetle (Coleoptera, Scolytidae). Environ. Entomol. 24: 372-379.

Wibe, A., Borg-Karlson, A.K. Persson, M., Norin, T. and Mustaparta, H., 1998. Enantiomeric composition of monoterpene hydrocarbons in some conifers and receptor neuron discrimination of α-pinene and limonene enantiomers in the pine weevil, *Hylobius abietis*. J. Chem. Ecol. 24: 273-287.

Wigglesworth, V.B., 1976. Insects and the Life of Man. Collected essays on pure science and applied biology. Chapman and Hall, London, 217 pp.

Williams, A.C. and Barry, B.W., 1991. Terpenes and the lipid-protein-partitioning theory of skin penetration enhancement. Pharm. Res. 8: 17-24.

Williams, L.A.D. and Mansingh, A., 1996. The insecticidal and acaricidal actions of compounds from *Azadirachta indica* (A. Juss.) and their use in tropical pest management. Integr. Pest Manag. Rev. 1: 133-145.

Williams, S.N., Anthony, M.L. and Brindle, K.M., 1998. Induction of apoptosis in two mammalian cell lines results in increased levels of fructose-1,6-biphosphate and CDP-choline as determined by 31P MRS. Magn. Reson. Med. 40: 411-420.

Wills, R.B.H., Scott, K.J. and McBailey, W., 1977. Reduction of superficial scald in apples with monoterpenes. Aust. J. Agric. Res. 28: 445-448.

Wills, R.B.H. and Scriven, F.M., 1979. Metabolism of geraniol by apples in relation to the development of storage breakdown. Phytochemistry 18: 785-786.

Wilson, W.T., Cox, R.L., Moffat, J.O. and Ellis, M., 1990. Improved survival of honey bees (*Apis mellifera* L.) colonies from long term suppression of tracheal mites (*Acarapis woodi* Rennie) with menthol. Bee Sci. 1: 48-54.

Wink, M., 1988. Plant breeding: importance of plant secondary metabolites for protection against pathogens and herbivores. Theor. Appl. Genet. 75: 225-233.

Woerdenbag, H.J., 1986. *Eupatorium cannabinum* L. A review emphasizing the sesquiterpene lactones and their biological activity. Pharmaceutisch Weekblad - Scientific Edition 8: 245-251.

Woerdenbag, H.J., Hendriks, H. and Bos, R., 1991. *Eupatorium cannabinum* L.: Der Wasserdost oder Wasserhanf. Neue Forschungen an einer alten Arzneipflanze. Zeitschr. Phytotherapie 12: 28-34.

Woerdenbag, H.J., Malingré, T.M., Lemstra, W. and Konigs, W.T., 1987. Cytostatic activity of eupatoriopicrin in fibrosarcoma-bearing mice. Phytother. Res. 1: 76-79.

Woitek, S., Unkles, S.E., Kinghorn, J.R. and Tudzynski, B., 1997. 3-Hydroxy-3-methylglutaryl-CoA reductase gene of *Gibberella fujikuroi*: isolation and characterization. Curr. Genet. 31: 38-47.

Wu, C., Gunatilaka, A.A., McCabe, F.L., Johnson, R.K., Spjut, R.W. and Kingston, D.G., 1997. Bioactive and other sesquiterpenes from *Chilocyphus rivularis*. J. Nat. Prod. 60: 1281-1286.

Yaguchi, M., Miyazawa, K., Katagiri, T., Nishimaki, J., Kizaki, M., Toyama, K. and Toyama, K., 1997. Vitamin K2 and its derivatives induce apoptosis in leukemia cells and enhance the effect of all-*trans* retinoic acid. Leukemia 11: 779-787.

Yamada, K., Hino, K., Tomoyasu, Y., Honma, Y. and Tsuruoka, N., 1998. Enhancement by bufalin of retinoic acid-induced differentiation of acute promyelocytic leukemia cells in primary culture. Leuk. Res. 22: 589-595.

Yazlovitskaya, E.M. and Melnykovych, G., 1995. Selective farnesol toxicity and translocation of protein kinase C in neoplastic HeLa-53K and non-neoplastic CF-3 cells. Cancer Lett. 88: 179-183.

Yokomatsu, H., Hiura, A., Tsutsumi, M. and Satake, K., 1996. Inhibitory effect of sarcophytol A on pancreatic carcinogenesis after initiation by N-nitrosobis(2-oxypropyl)amine in Syrian hamsters. Pancreas 13: 144-159.

Yokoyama, S., Nozawa, F., Mugita, N. and Ogawa, M., 1999. Suppression of rat liver tumorigenesis by 25-hydroxycholesterol and all-*trans* retinoic acid: differentiation therapy for hepatocellular carcinoma. Int. J. Oncol. 15: 565-569.

Yu, K.H., Weng, L.J., Fu, S., Piantadosi, S. and Gore, S.D., 1999. Augmentation of phenylbutyrate-induced differentiation of myeloid leukemia cells using all-*trans* retinoic acid. Leukemia 13: 1258-1265.

Yu, W., Simmons-Menchaca, M., Gapor, A., Sanders, B.G. and Kline, K., 1999b. Induction of apoptosis in human breast cancer cells by tocopherols and tocotrienols. Nutr. Cancer 33: 26-32.

Yuba, A., Yazaki, K., Tabata, M., Honda, G. and Croteau, R., 1996. cDNA cloning, characterization, and functional expression of 4S-(-)-limonene synthase from *Perilla frutescens*. Arch. Biochem. Biophys. 332: 280-287.

Zakharová, O.S., 1988. Mevalonic acid and isoprenoids as regulators of proliferation in mammalian cells. Usp. Sovrem. Bio. 106: 442-453.

Zhao, K. and Singh, J., 1998. Mechanisms of percutaneous absoption of tamoxifen by terpenes: eugenol, D-limonene and menthone. J. Controlled Release 55: 253-260.

Zheng, G.Q., 1994. Cytotoxic terpenoids and flavonoids from *Artemisia annua*. Planta Med. 60: 54-57.

Zhou, B., Hiruma, K., Jindra, M., Shinoda, T., Segraves, W.A., Malone, F. and Riddiford, L.M., 1998. Regulation of the transcription factor E75 by 20-hydroxyecdysone and juvenile hormone in the epidermis of the tobacco hornworm, *Manduca sexta*, during larval molting and metamorphosis. Dev. Biol. 193: 127-138.

Zibareva, L., 2000. Distribution and levels of phytoecdysteroids in plants of the genus *Silene* during development. Arch. Insect Biochem. Physiol. 43: 1-8.

Zielinska, Z.M. and Laskowska-Bozek, H., 1980. Effects of juvenile hormone and related growth regulators on mammalian cell proliferation: a comparative study. Toxicol. Appl. Pharmacol. 55: 49-54.

Zujewski, J., Horak, I.D., Bol, C.J., Woestenborghs, R., Bowden, C., End, D.W., Piotrovsky, V.K., Chiao, J., Belly, R.T., Todd, A., Kopp, W.C., Kohler, D.R., Cow, C., Noone, M., Hakim, F.T., Larkin, G., Gress, R.E., Nussenblatt, R.B., Kremer, A.B. and Cowan, K.H., 2000. Phase I and pharmacokinetic study of farnesyl protein transferase inhibitor R115777 in advanced cancer. J. Clin. Oncol. 18: 927.

Zuzaga, J.L., Sinnett-Smith, J., van Lint, J. and Rozengurt, E., 1996. Protein kinase D (PKD) activation in intact cells through a protein kinase C-dependent signal transduction pathway. The EMBO J. 15: 6220-6230.

F.C.O.

General reading

Agrawal, A.A., Tuzun, S. and Bent, E. (eds), 1999. Induced plant defenses against pathogens and herbivores. APS Press (Am. Phytopath. Soc.), 403 pp. (ISBN 0-89054-242-2).
Bell, W.J. and Cardé, R.T. (eds), 1984. Chemical Ecology of Insects. Chapman & Hall, London, etc., 524 pp. (ISBN 0-412-99611-1).
Blumenthal, M. (ed.), 1998. The complete German Comission E monographs (Therapeutic Guide to Herbal Medicinals). Haworth Press, 684 pp. (ISBN 0-96-55555-0-X).
Cardé, R.T. and Bell, W.J. (eds), 1995. Chemical Ecology of Insects 2. Chapman & Hall, New York, etc., 432 pp. (ISBN 0-412-03951-6).
Cardé, R.T. and Minks, A.K, (eds), 1997. Insect Pheromone Research – New Directions Chapman & Hall, New York, London, etc., 684 pp. (ISBN 0-412-99611-1).
Charlwood, B.V. and Banthorpe, D.V., 1991. Terpenoids. Series; Methods in Plant Biochemistry Vol. 7. Acad. Press, London, 565 pp. (ISBN 0-12-461017-X).
Cooper, S., 1991. Bacterial Growth and Division: Biochemistry and Regulation of prokaryotic and Eukaryotic Division Cycles. Acad. Press, San Diego, 501 pp. (ISBN: 0-1218-7905-4).
Copping, L.G. (ed.), 1996. Crop Protection Agents from Nature. The Royal Soc. of Chem., 500 pp. (ISBN 0-85404-414-0).
Harborne, J.B., 1993. Introduction to Ecological Biochemistry, 4th Ed. Acad. Press Ltd, London, etc., 318 pp. (ISBN 0-12-324685-7).
Harborne, J.B. and Baxter, H., 1993. Phytochemical Dictionary. A handbook of bioactive compounds from plants. Taylor & Francis, London, etc., 791 pp. (ISBN 0-85066-736-4).
Hay, R.K.M. and Waterman, P.G. (eds), 1993. Volatile Oil Crops: Their Biology, Biochemistry and Production. Longman Science and technology, Essex, U.K., 185 pp. (ISBN 0-582-00557-4).
Hardie, J. and Minks, A.K. (eds), 1999. Pheromones of Non-Lepidopteran Insects Associated with Agricultural Plants. CABI Publishing, CAB International, Wallingford U.K., 466 pp. (ISBN 0-85199-345-1).

Hummel, H.E. and Miller, T.A. (eds), 1984. Techniques in Pheromone Research. Springer-Verlag, New York, etc., 464 pp. (ISBN 0-387-90919-2).

Kaufman, S.H. (ed.), August, J., Anders, M.W. and Coyle, J., 1999. Apoptosis: Pharmacological Implications and Therapeutic Opportunities (A Volume in the Advances in Pharmacology Series). Academic Press, 640 pp. (ISBN 0-1240-2355-X).

Khripach, V.A., de Groot, Æ. and Zhabinskii, V.N., 1999. Brassinosteroids: A New Class of Plant Hormones. Acad. Press, San Diego, 464 pp. (ISBN 0-1240-6360-8).

Langenheim, J.H., Silverstein, R.M. and Simone, J.B. (eds), 1994. Proceedings of the Symposium on Chemical Ecology of Terpenoids (19th Annual Meeting of the Int. Soc. of Chem. Ecol. (ISCE)). J. Chem. Ecol. 20: 1221-1406.

Maarse, H. (ed.), 1991. Volatile compounds in foods and beverages. Dekker, New York, etc., 764 pp. (ISBN 0-8247-8390-5).

Marks, D.B., Marks, A.D. and Smith, C.M., 1996. Basic Medical Biochemistry: A Clinical Approach. Lippincott, Williams & Wilkins. (ISBN 0-68305-595-X).

McIntyre, A., 1996. Flower power: Flower remedies for healing body and soul through herbalism, homeopathy, aromatherapy, and flower essences. Henry Holt and Co., New York, 287 pp. (ISBN 0-80504-216-4).

McNaughton, V., 2000. Lavender – The Grower's Guide. Garden Art Press, 212 pp. (ISBN 1-870673-36-0).

Miller, J.R. and Miller, T.A. (eds), 1986. Insect-Plant Interactions. Springer-Verlag, New York, etc., 341 pp. (ISBN 0-387-96260-3).

Nes, W.D. (ed.), 1994. Isopentenoids and other natural products. Evolution and Function. ACS Symposium Series nr 562 (American Chemical Society, Washington DC), 255 pp. (ISBN 0-8412-2934-1).

Ridgway, R.L., Silverstein, R.M. and Inscoe, M.N. (eds), 1990. Behavior-modifying Chemicals for Insect Management. Marcel Dekker Inc., New York, 761 pp. (ISBN 0-8247-8156-2).

Robbers, J.E. and Tyler, V.E., 1999. Tyler's Herbs of Choice. The therapeutic use of phytochemicals. Harworth Press, 306 pp. (ISBN 0-7890-0159-4).

Ryman, D., 1993. Aromatherapy: The complete guide to plant and flower essences for health and beauty. Bantam Books, New York, 373 pp. (ISBN 0-55337-166-5).

Schoonhoven, L.M., Jermy, T. and van Loon, J.J.A., 1998. Insect-Plant Biology: From Physiology to Evolution. Chapman and Hall, 409 pp. (ISBN 0-412-80480-8).

Schulz, V., Hänsel, R. and Tyler, V.E., 1997. Rational Phytotherapy. A Physicians' Guide to Herbal Medicine. Haworth Press, 306 pp. (ISBN 3-540-62648-4).

Smith, C.M., Khan, Z.R. and Pathak, M.D. (eds), 1994. Techniques for Evaluating Insect Resistance in Crop Plants. CRC press, Inc., 320 pp. (ISBN 0-87371-856-9).

Towers, G.H.N. and Stafford, H.A. (eds), 1990. Biochemistry of the Mevalonic Acid Pathway to Terpenoids (Recent Advances in Phytochemistry, Vol. 24). Plenum, New York, 341 pp. (ISBN 0-306-43604-3).

General reading 419

Walton, N.J. and Brown, D.E. (eds), 1999. Chemicals from Plants: Perspectives on Plant Secondary Products. World Scientific Pub. Co (London, Imperial College Press; Singapore, World Scientific), 425 pp. (ISBN 9-8102-2773-6).

Wess, J., 1999. Structure-Function Analysis of G-protein-Coupled Receptors. John Wiley & Sons, 412 pp. (ISBN: 0-4712-5228-X).

Weiss, E.A., 1997. Essential oil crops. CABI Publishing, CAB International, Wallingford, U.K., 600 pp. (ISBN 0-85199-137-8).

Out of print, but still recommended (try used book-stores or the internet, e.g. amazon.com.):

Clayton, R.B., 1969. Steroids and terpenoids. Series: Methods in Enzymology, Vol. 15. Acad. Press, New York, 903 pp. (ISBN 0-12-181872-1).

Goodwin, T.W. (ed.), 1970. Aspects of Terpenoid Chemistry and Biochemistry. Proc. Symp. Phytochem. Soc. Liverpool, Acad. Press, London/New York, 441 pp. (ISBN 0-12-289840-0).

Sukh Dev, 1989. CRC Handbook of Terpenoids: Acyclic, Monocyclic, Bicyclic, Tricyclic, and Tetracyclic Terpenoids. Boca Raton, CRC Press. (ASIN: 0849336112).

Templeton, W., 1969. An introduction to the chemistry of the terpenoids and steroids. Butterworths, London, 277 pp. (ASIN: 0-408-52600-9).

Glossary

Contains less common words and expressions encountered in the chapters of this volume and in titles and text of the references.

Adaptive radiation: the occurrence of numerous speciation processes within a monophyletic taxon during a rather short span of time.
Agonist (of messenger X): a substance that causes response in a target cell through binding with the specific receptor of messenger molecule X.
Allelochemical: see allomone.
Allomone: a plant metabolite (often toxic) that affects another species than the producer.
Allelopathic (metabolite): a plant substance that inhibits the development of a competitive plant species in a biotope.
Allopatric speciation: speciation occurring in populations located in different areas (biotopes) or living in isolation.
Angiogenesis: neovascularization (a normal process e.g. in damaged tissue that can also be induced by tumour cells).
Antagonist (of messenger X): a substance with a structure similar to an agonist. Binding takes place to the specific receptor of messenger X or to the messenger molecule itself. As a result there is no response from the receptor.
Antibiosis: toxic principle of an organism.
Antifeedant: a plant metabolite (often, but not necessarily toxic), which interferes with host acceptation.
Antixenosis: preference of one host above another.
Antineoplastic: antitumour, see antiproliferative.
Antiproliferative: inhibition of cell division.
Apoptosis: programmed cell death; essential in the development of multicellular organisms.
Aposymbionts: insects, which do not harbour symbionts.
Archezoa: early-diverging, primitively amitochondriate eukaryotes.

Arrhenotoky: parthenogenetic development in which males only are produced (e.g. in honeybees).

Autoecious: whithout annual alternation between primary and secondary hosts (switch between secundary hosts remains possible).

Carcinoma: tumour developing in epithelial tissue.

Caspase: an enzyme (heterodimeric protein) involved in the induction of apoptosis.

Cecidogenesis: deformations (e.g. galls) induced in a host plant by insect feeding activities.

Chiral selectivity: selectivity in (bio)chemical reactions or binding based on the presence of chiral centra in a chemical substance (terpenoid).

Clade: a branch of a phylogenetic tree, including all "twigs" of that branch; it corresponds to a monophyletic taxon.

Cladogram: a diagram of a phylogenetic tree showing the sequence of ramifications, but without indication of an absolute time scale.

Competitive inhibition: inhibition caused by an antagonist (often an analogue of a natural messenger molecule), which binds reversibly to the receptor of that messenger molecule. To be effective, the competitive inhibitor mostly needs a relatively high concentration.

Cytoskeleton: proteineous fibres reinforcing eukaryotic cells and enabling them to change their morphology (examples are actin and tubulin in muscle cells).

Cytosol: the fluid portion of the cytoplasm outside the organelles.

Cytostasis: inhibition of cell division by a mechanism interfering with a certain stage of the cell cycle.

Down regulation: the inhibition of a receptor enzyme, growth factor or gene by a specific regulating factor, e.g. by inhibiting binding sites and/or mRNA levels.

Dysplasia: the occurrence of abnormal cells of which the cell cycle can become deregulated, originating in a group of excessively proliferating cells.

Ecdysteroids: steroid hormones in insects inducing ecdysis.

Enantiomer: see Introduction of this volume.

Endomeiosis: a peculiar meiotic process during maturation of parthenogenic eggs in aphids.

Endocytobiosis: symbiotic relationship between an intracellular bacterial symbiont and an eukaryotic cell, e.g. located between the egg membrane and the rear pole of an egg cell.

Endopterygotes: insects with complete metamorphosis and a pupal stage. Young are called larvae.

Endosymbionts: bacterial symbionts housed in specialized cells of an eukaryote, e.g. in the mycetome of aphids, unlike the symbiotic flora in the intestines.

Essential oils: (mostly a blend of) volatile terpenes that can be dissolved in organic solvents, but form an oily substance already in the plants in which they are produced.

Exopterygotes: insects with only partial metamorphosis and no pupal stage. Young are called nymphs (and sometimes incorrectly larvae).

Generalist (feeder): an animal with a wide range of host plants.

Glioma: tumour of nonneuronal brain cells.

Gynoparae: parthenogenetic female insects, which produce sexual females (on a secondary host).

Glossary 423

Heteroecious: with an annual alternation between primary and secondary hosts.

Heterotrimeric: see Introduction of this volume.

Horizontal resistance: resistance against all known biotypes.

Hyperplasia: abnormally high rate of cell division in a certain area of a tissue that may lead to dysplasia.

Idiopathic constipation: retention of faeces for longer than the normal period of passage through the intestinal tractus (in humans 24-48 h).

In situ (cancer) tumour: a tumour that is still localized, but may develop into an invasive tumour.

Invasive (cancer) tumour: a tumour that has become malignant and renegades new tumours through metastasis.

Isoprene structure: see Introduction of this volume.

Isoprenylation: see prenylation.

K-strategists: animals, which maximize their competitive ability at the expense of their rate of reproduction.

Metastases: new tumours derived from a primary tumour by spread of invasive cancer cells through blood or lymphatic vessels to distant sites of the body.

Monophagous: with feeding restricted tot a single host species.

Morph: a form (e.g. winged or wingless) of an insect, which is genetically identical to a different form.

Microtubules: see cytoskeleton.

Neoplasm, neoplastical: synonymous with tumour and tumour development.

Nonpreference: an insect does not recognize a plant as a host.

Parasitoid: a natural enemy that lays eggs in an insect or one of its endopterygote stages, eventually killing it; in contrast to a predator that kills the insect by eating (parts of) it.

Parthenogenesis: the production of life offspring from unfertilized eggs that hatch already in the maternal body, as occurs in aphids except when they reproduce sexually.

Phagostimulant: a substance that induces feeding behaviour upon gustatory perception.

Phagocytosis: the ability to engulf other organisms (or cells) and to incorporate organelles.

Posttranscriptional: see transcriptional.

Posttranslational: see translational.

Oligophagous: with feeding restricted to a few (often related) species of host plant.

Oviparae: sexual females, which mate with males and lay fertilized eggs.

Polyphagous: feeding on a wide range of host plant species.

Prenylation: a process in the cell in which an isoprenoid structure of 15 or 20 carbon atoms is attached to a side chain of a protein. The so-called prenylated proteins are active in signal transduction in the cell.

Primary host: plant on which the sexual phase (original) of an insect occurs.

Prokaryotes: organisms without a cell nucleus.

Procaspase: a protein that can be devided into an inactive part and an active caspase molecule; one can speak of the activation of a procaspase.

Proto-oncogenes: cell growth-promoting genes that can become carcinogenic upon mutation.

Sarcoma: tumour of connective tissue.

Semiochemical: a substance used in chemical communication between organisms.

Soldiers: specific forms (morphs in aphids) of insects with body parts adapted to defend conspicifics.

Specialist (feeder): an animal with one or a very limited number of host plants.

Regression (of tumours): reduction in the size of tumours (and their metastases) caused by a natural process (activation of the immune system) or by chemotherapeutics.

R-strategists: animals, which maximize their rate of increase at the expense of competitive ability.

Sympatric speciation: speciation occurring in populations located in the same area (biotope).

Synergistic (effect): a situation (e.g. in chemotherapy) in which the combined effects of different drugs are greater than the sum of the effects of each drug. Often used to point out an inexpected additive effect, suggesting more than reality permits.

Synthase: an enzyme involved in biosynthesis of a (terpenoid) metabolite from a precursor.

Synthetase: the original nomination of a synthase, now obsolete.

Secondary host: (in heteroecious aphids) plant on which parthenogenetic reproduction takes place.

Synomone: semiochemical released by plants attacked by herbivores recruiting (one of) its natural enemies, in fact a "cry for help" metabolite.

Telescoping (of generations): the occurrence of more than one generation of offspring in a parthenogenetically reproducing female insect, because new embryos develop already within the unborn nymph (e.g. in aphids).

Thelytoky: parthenogenetic development in which females only are produced.

Transcription(al): a process in the cell in which the genetic information of DNA is used to create a specific sequence of bases in a RNA chain.

Transduction(al): a transmission step of a messenger cascade in which different parts of a cell are involved. Term used by physiologists; e.g. signal transduction in a synapse.

Transgression: (in pathology): a process in which a tumour becomes more malignant. (In geology: the spread of a sedimentary deposition over older layers).

Translation(al): a process in the cell in which the genetic information of a mRNA molecule controls the amino acid composition in protein synthesis.

Trichomes: glandular epidermal hairs on plants.

Tumour suppressor genes: encode for apoptosis-inducing proteins, but may lose their suppressive character upon mutation.

Turnover (rate): the biochemical process of conversion of a (terpenoid) molecule into another one.

Vascular feeding: intake of nutrients from a plant's vascular system (mostly phloem).

Vertical resistance: resistance exclusively against certain biotypes.

Virginoparae: female insects giving birth to young by parthenogenesis.

Viviparae: aphids of which the eggs develop in the maternal body; giving birth to live young.

Index

A

Abies, 200
A. grandis, 35, 87, 209, 242
abscisic acid, 1
Acanthoscelides obtectus, 277
Acarapis woodi, 270
acetylcholinesterase, 89, 176, 177, 278, 279
Achillea millefolium, 149, 331, 332
Acorus calamus, 260
Acromyrmex lundi, 219
Acyrthosiphon pisum, 37, 186, 203
Adelges cooleyi, 202
Adelgidae, 202, 273
adenocarcinoma, 95, 143, 144, 146, 155, 159, 168, 312, 317, 321, 324
Adenostyles alpina, 235
Aedes aegyptii, 273
Aenictus rotundatus, 195
Aeromonas hydrophilla, 290
Aesculus hippocastanum, 258, 304
Ageratum houstonianum, 283
agonist, 229, 247, 279, 337
Agrimonia eupatoria, 260
Ajuga remota, 51
A. reptans, 52
alanine, 63, 97
alarm pheromone, 39, 46, 124, 125, 176, 189, 191, 192, 193, 194, 202, 225, 247, 274, 284, 354, 355, 356
Alchemilla vulgaris, 260, 332
allomone, 1, 5, 55, 56, 85, 125, 128, 150, 151, 191, 197, 198, 199, 204, 205, 206, 216, 221, 234, 235, 237, 239, 240, 242, 243, 244, 246, 250, 251, 275, 286, 289, 334, 346, 354
Alzheimer, 155, 343
Amblypeeta lutescens lutescens, 189
Amblyseius cucumeris, 225
Ambystoma mexicanum, 118
Amitermes excellens, 192
Anastrepha suspensa, 187, 189
Anaxagorea luzonensis, 337
Anethum graveolens, 205, 256, 259, 260, 287
A. sowa, 256
Angelica archangelica, 259
angiogenesis, 147, 154, 159, 169, 310, 313, 322, 323, 337, 356
antagonist, 167, 190, 207, 208, 247, 267, 280, 284, 303, 318, 322, 323, 324, 325
antenna, 109, 123, 124, 126, 128, 186, 187, 196, 205, 219, 220, 222, 246, 247, 248, 250
Anthonomus eugenii, 187
anti-angiogenetic, see angiogenesis
anti-attractant, 287
antibody, 325, 345, 310, 356
Antidorcas marsupialis, 49
antifeedant
    insect, 45, 54, 55, 128, 148, 197, 200, 202, 206, 212, 214, 215, 245, 249, 255, 272, 276, 277, 285, 286, 287, 290, 341
    vertebrate, 85
antifeeding, 212, 214

antioxidant, 113, 118, 303, 310
antisense strategy, 327, 356
aphid, 37, 39, 46, 47, 49, 52, 81, 124, 126, 129, 146, 168, 171, 176, 181, 184, 189, 192, 202, 210, 216, 217, 225, 236, 242, 244, 247, 249, 254, 273, 276, 280, 284, 289,293, 350, 351, 352, 355, 356
aphidicolin, 147, 160, 169, 170, 327, 336
Aphis euonymi, 354
À. fabae, 49, 81, 186, 196, 202, 203, 205, 217, 218, 219, 274, 353, 354
A. gossypii, 280
A. vandergooti, 354
apicoplast, 298, 345, 358
apomine, 337
apoptosis, 6, 357
   insect cell, 191, 280, 348, 350, 352
   micro-organism, 107, 129, 348
   plant cell, 130, 349
   vertebrate cell, 92, 94, 102, 104, 129, 131, 133, 134, 135, 139, 141, 142, 143, 144, 145, 146, 147, 148, 150, 154, 155, 156, 159, 165, 166, 167, 175, 280, 310, 311, 312, 313, 314, 315, 316, 319, 320, 321, 325, 326, 328, 336, 337, 338, 344, 345, 347, 349, 352, 357
apple, 40, 83, 84, 85, 266
Archaea, 15, 26, 27, 29, 65, 348, 351, 358
Archaebacteria, 12, 13, 14, 15, 26, 27, 69, 230
Archamoeba, 60
archeukaryote, 12
Archaezoa, 27, 60, 351
Arianta arbustorum, 234
Arion, 197
Arnica montana, 332
aromatherapy, 308
aromatic
   herb, see plant
   oil, see essential oil
   plant, 5, 36, 37, 172, 174, 208, 209, 259, 260, 262, 271, 277, 278
   resin, 255
Artemisia, 5, 174, 255, 302, 340
A. absinthium, 256, 260, 277, 304
A. annua, 174, 204, 256, 299, 340
A. argyi, 256, 304
A. chamaemelfolia, 340
A. sieberi, 340
A. tridentata, 209

A. vulgaris, 256, 260, 304
arthritis, 297, 300, 306
aspartate, 63, 64
Aspergillus, 20, 291
A. niger, 291
A. ochraceus, 291
asthma, 304, 305
Athalia rosae ruficornis, 242
Atta cephalotes, 195
A. laevigata, 196
axolotl, 116, 117, 118
ayurvedic, 301, 306
Azadirachta, 267
A. indica, 54, 55, 200, 214, 267, 281, 285, 303
azadirachtin, 55, 268, 341
   antifeedant, 54, 178, 200, 210, 212, 271, 273, 276, 277, 282, 339
   effect on symbionts, 177, 210, 212, 281
   insecticide, 177, 211, 212, 214, 268, 277, 279, 281, 285
   production of, 54, 210, 268, 269

B

Bacillus cereus, 290, 291
B. subtilis, 173
B. thuringiensis, 279
Bactrocera dorsalis, 334
baculovirus, 97, 98, 135
Balsamodendron mukul, 255
bark beetle, 24, 25, 38, 68, 74, 187, 188, 209, 215, 221, 273, 276
bee-hive, 270
Bemisia tabaci, 204
betulinic acid, 147, 315, 347
Bifidobacterium, 157
bile acid, 53, 93, 115
binding protein, 54, 102, 105, 117, 231, 232, 251, 322, 345, 350
   odorant, 125, 126, 127, 228, 229, 230, 231, 244, 247, 251
   pheromone, 126, 244
biomarker, 145
bioreactor, 254, 341, 342
Biprorulus bibax, 189
bisabolene, 189, 191, 203, 256, 285, 331, 340
bisabolol, 42, 44, 254, 257, 332, 354
Blattella germanica, 87, 192, 273, 334

Index 427

body louse, see *Pediculus humanus h.*
Bombus pratorum, 189
Bombyx mori, 52, 230
borneol, 256, 258, 272, 283, 298, 332
Botryodiplodia theobroma, 305
Brachycaudus cardui, 354
B. helichrysi, 354
Brassica oleracea, 215
Brevicoryne brassicae, 205, 217, 218
bronchitis, 261, 305
Brucea antidysenterica, 258
B. javanica 258
Buchnera, 46, 177, 204, 210, 242, 352
bufalin, 318

C

CA, see corpora allata
cadinene, 44, 255, 257, 307
Cadlina luteomarginata, 85
calciferol, 23, 324
calcitriol, 165, 316, 324
Calliphora vicina, 229
camphene, 32, 256
camphor, 5, 217, 271, 292, 335
    in pharmacology, 258, 296, 298, 299, 306, 307, 333
    production of, 38, 45, 206, 209, 254, 258
    repellent, 202, 204, 206, 273, 274
Camptotheca acuminata, 66
Candida 329
C. albicans, 172, 332
cannabinol, 330
Cannabis, 330
Capsicum annuum, 196
caraway, see *Carum carvi*
carcinoma, 91, 98, 138, 145, 146, 148, 151, 153, 155, 156, 165, 311, 317, 318, 321, 322, 323, 324, 336, 337, 338, 345
Carcinus maenas, 197
carene, 33, 173, 206, 226
carnosol, 45
carotene, 17, 50, 53, 112, 147, 168, 308, 310, 318, 323
carotenoid, 1, 4, 13, 19, 23, 49, 50, 52, 53, 63, 64, 77, 83, 288, 309, 3010, 318
Carum carvi, 33, 61, 146, 171, 206, 218, 236, 256, 265, 276, 287, 292, 295, 298, 341

carvacrol, 173, 174, 208, 209, 256, 257, 258, 273, 280, 291, 294, 334
carveol, 33, 61, 256, 276, 283, 294, 304
carvone, 4, 36, 146, 170, 171, 236, 266, 268, 274, 276, 283, 289, 291, 292, 293, 294, 333, 335
    in agriculture, 8, 146, 170, 235, 281, 292, 294
    effect on insects, 146, 171, 205, 207, 218, 219, 235, 273, 284, 291
    production of, 33, 61, 146, 206, 218, 236, 237, 255, 256, 257, 265, 287, 293, 295, 339, 341
caryophyllene, 201, 309
    antibiotic, 173, 256, 257
    effect on insects, 187, 194, 196, 197, 206, 217, 219, 273, 277, 356
    production of, 174, 187, 217, 254, 256, 257, 278, 287, 288, 309
Caryophyllus aromaticus, 5, 171, 253, 256, 259, 277, 288, 298, 332
caspase, 107, 130, 131, 132, 133, 135, 141, 142, 144, 145, 150, 153, 154, 155, 165, 312, 316, 328, 337, 343, 345
Catharanthus roseus, 89
Cavariella aegopodii, 146, 171, 218, 236, 293
celastrol, 314
Centaurea nicolai, 291
Cephalosporium aphidicola, 169
Ceratitis capitata, 187, 196
Ceratocystis, 17, 175
C. variospora, 255
C. virescens, 255
Cetraria islandica, 163
Ceutorrhynchus assimilis, 222
Chaetosiphon fragaefolii, 212
chamazulene, 5, 42, 44, 163, 254, 257, 266, 298, 331, 354
chamomile, see *Matricaria*
Chamomilla officinalis, 261, 308
chewing-gum, 294
Chiloscyphus rivularis, 148
chloroplast, see plastid
cholesterol, 4, 23, 26, 52, 53, 90, 91, 93, 94, 95, 115, 166, 183, 191, 301, 302, 308, 311, 314, 337
    biosynthesis, 26, 39, 50, 53, 60, 69, 75, 76, 78
    homeostasis, 11, 25, 26, 69, 72, 78, 80, 83, 87, 90, 92, 93, 95, 101, 167, 300

cholesterolemia, 5, 78, 297, 300, 301, 304, 321, 326
Choristoneura fumiferana, 74, 214
C. rosaceana, 177, 281
Christmas tree, see *Picea abies*
chromatin, 102, 114
Chrysanthemum, 219, 225, 277
chrysanthenol, 255
chrysanthenone, 276
cineole, 95, 178, 292, 335
   antibiotic, 173, 174, 256, 257, 258, 298
   effects on insects, 206, 209, 217, 271, 273, 274, 280, 295
   in pharmacology, 298, 299, 305, 308, 331
   production of, 32, 33, 34, 38, 45, 178, 204, 209, 254, 255, 256, 257, 258, 274
Cinnamomum bejolghota, 38
C. camphora, 206
C. obtusifolium, 38
cisplatin, 134, 150, 151, 320, 324
citral, 118, 196, 220, 221
Citrobacter freundii, 173
citronella (oil), 195, 209, 295
citronellal, 173, 205, 208, 209, 217, 218, 257, 273, 295
citronellol, 17, 89, 173, 175, 205, 217, 255, 283, 292, 331, 341, 342
Citrus, 146, 170, 199, 200, 202, 209, 236, 256, 315, 331, 332
C. natsudaidai, 170
Cladosporium cladosporoides, 291
Clarkia breweri, 33
Clavariadelphus truncatus, 107
Claviceps purpurea, 17
Clerodendron infortunatum, 45
C. trichotomum, 242
cloister garden, 254
Clostridium, 27
C. sporogenes, 173
clove, see *Caryophyllus aromaticus*
coevolution, 183, 229, 239, 242
Coleoptera, 8, 24, 39, 113, 123, 182, 187, 189, 199, 220, 221, 239, 277, 278, 279, 346
colon carcinoma, 95, 113, 144, 146, 165, 167, 168, 176, 311, 314, 320, 321, 322, 323, 324, 336, 352
Colorado potato beetle, see *Leptinotarsa*
Colletotrichum trifolii, 140, 350

combination therapy, 148, 150, 151, 155, 165, 166, 167, 168, 263, 316, 320, 321, 324, 325, 328
Commiphora, 54, 277, 297, 299, 329
C. molmol, 299
C. mukul, 255, 298, 301, 306
C. myrrha, 54, 329
compactin, 68, 76, 78, 105, 188, 300
configuration, 3, 8, 45, 50, 53, 119
conifer, 7, 18, 32, 35, 87, 172, 201, 215, 221, 222, 279, 286, 321
Conocephalum conicum, 33
constipation, 308
convergent evolution, 233
Conyza canadensis, 260, 306, 307
copaene, 187, 217, 288
Coptotermes formosanus, 201, 277
Coriandrum, 5
C. sativum, 256, 260, 277, 308, 332
Coridothymus capitatus, 208, 256, 258
corpora allata, 47, 48, 49, 74, 76, 283
corticosterone, 51
Cotesia rubecula, 225
Crataegus pinnatifida, 149, 165
cubebene, 38
Culex pipiens, 295
Cuminum cyminum, 256
cuparene, 307
Curcuma domestica, 156
C. longa, 156
curcumene, 307
curcumin, 147, 156, 326, 328
Cyanobacteria, 15, 16, 22, 29, 49
cyclase, 11, 16, 23, 30, 33, 35, 41, 42, 44, 63, 64, 73, 94, 98, 157, 171, 322, 324
cymene, 34, 174, 206, 208, 256, 257, 258, 273, 296, 298
cytochrome, 76, 89, 112, 314, 315, 335
   in apoptosis, 107, 131, 132, 134, 154, 319, 343
   in evolution, 29, 60, 112
cytokinin, 1, 30, 31, 312
cytoskeleton, 27, 147, 155, 310, 325, 351, 358, 359
Cyzenis albicans, 271

D

Damaliscus dorcas dorcas, 183
D. dorcas phillipsi, 183

Index

dammarane group, 314
Dasineura tetensi, 124
Daucus carota, 205, 287
degrading enzyme, 64, 73, 74
Delia antiqua, 206, 273
D. radicum, 206, 273
Dendroctonus frontalis, 276
D. ponderosae, 341
D. rufipennis, 209
D. simplex, 209
Detarium microcarpum, 201
detergent, 8, 80, 268
detoxification, 90, 156, 160, 240, 242, 282, 328, 351
Diabrotica virgifera virgifera, 283, 334
diapause, 48, 49, 52
diarrhoea, 308
Dictammus albus, 37
differentiation, 6, 59, 99, 102, 107, 117, 129, 130, 140, 151, 153, 154, 155, 156, 157, 161, 166, 251, 282, 290, 310, 312, 315, 318, 319, 337, 351
    redifferentiation of tumour cells, 106, 141, 154, 155, 156, 310, 318, 319
dihydrotestosterone, 51, 54
Dilophus ligulatus, 318
dimorphism, 204, 353, 354, 355
Dioryctria zimmermanni, 209
Diploclisia glaucensis, 282
Diploptera punctata, 74
discodermolide, 343
dispersion, 32, 268, 341
distillation, 38, 263, 264, 265, 266, 268, 331
DNA synthesis, 5, 60, 89, 96, 101, 105, 107, 141, 144, 148, 151, 163, 166, 168, 317, 336
dolichol, 1, 4, 26, 69, 78, 83, 300
DOXP
    pathway, 13, 15, 16, 18, 22, 23, 29, 33, 35, 49, 50, 61, 95, 226, 234, 254, 298, 305, 345, 353
    synthase, 16, 61, 64, 298
Drosophila, 24, 208, 350, 352, 353
D. melanogaster, 74, 85, 115, 207, 280, 284
Dysdercus koenigii, 204, 286

E

EAG, see electroantennogram
electroantennogram, 126, 185, 186, 187, 205, 217, 218, 219, 220
Elettaria cardamomum, 256, 259, 278
embryogenesis, 117, 358
enantiomer, 2, 4, 30, 32, 35, 125, 129, 143, 150, 186, 187, 189, 190, 205, 222, 245, 251, 256, 265, 276, 278, 291, 296
endoplasmatic reticulum, 60, 72, 78, 86, 90, 134, 138, 343
endosymbiont, 25, 38, 46, 81, 82, 177, 193, 203, 204, 210, 281
enfleurage, 263, 264
Enterobacter, 173
Entomophthora, 353
Ephedra, 238
epothilone, 327, 343
ergosterol, 23, 343
Erigeron canadensis, 306
Escherichia coli, 13, 29, 41, 61, 62, 111, 130, 156, 157, 173, 174, 175, 191, 353
essential oil, 5, 17, 44, 198, 253, 254, 264, 265, 268, 269, 292, 294, 295, 300, 329, 332, 339
    antibiotic, 172, 173, 174, 235, 242, 290, 291, 298, 331
    effect on insects, 187, 194, 207, 208, 242, 271, 274, 277, 278, 279, 280, 282, 283, 284, 294, 295
    in perfumery, 253, 265, 329, 330
    in pharmacology, 5, 174, 254, 256, 278, 304, 306, 308, 309, 321, 331, 332
    production of, 7, 33, 37, 38, 41, 44, 57, 174, 175, 178, 182, 194, 207, 253, 254, 255, 256, 259, 261, 267, 331
estradiol, 51, 54, 115, 357
estrogen, 72, 73, 80, 115, 140, 149, 216, 328, 333, 334, 337, 338, 344, 345
Eubacteria, 12, 14, 15, 27, 29, 60, 61, 230, 358
Eucalyptus, 178, 257
eucalyptus oil, 295, 305
eugenol, 295, 298, 335
    effect on insects, 201, 206, 219, 220, 221, 273, 277, 334
    in pharmacology, 145, 299, 328
    insecticide, 201, 277, 278, 280, 283
    production of, 171, 256, 257, 258, 280
eukaryote, 5, 13, 14, 15, 24, 26, 27, 29, 55, 57, 60, 61, 65, 66, 72, 88, 89, 96, 100,

101, 102, 111, 114, 131, 140, 157, 226, 228, 350, 351, 358
eupatoriopicrin, 266
Eupatorium, 266
E. cannabinum, 266
extraction, 145, 262, 263, 264, 265, 266, 280, 331

F

5-fluorouracil, 318
faranal, 190
farnesene, 176, 181, 188, 201, 288, 350
  biosynthesis, 41, 81, 82, 85, 226
  effect on insects, 37, 46, 48, 177, 187, 189, 193, 195, 202, 203, 204, 205, 219, 222, 225, 247, 254, 273, 274, 284, 285, 288, 350, 353, 354, 355, 356
  in feedback systems, 72, 81, 83, 85
  in pharmacology, 308, 312
  production of, 40, 41, 44, 46, 192, 194, 255, 256, 257, 265, 266, 288, 354, 356
  semiochemical, 48, 49, 176, 184, 187, 188, 189, 190, 191, 192, 194, 195, 196, 197, 202, 217, 219, 223, 225, 239, 247, 271, 274, 354
farnesoic acid, 47, 76, 86, 120
farnesol, 47, 69, 71, 81, 86, 115, 135, 175, 201
  biosynthesis, 23, 41, 44, 78, 86
  effect on insects, 189, 196, 202, 203, 207, 282, 353
  in feedback systems, 69, 72, 76, 78, 80, 87, 94, 147, 292, 300, 302, 312, 313, 337
  in pharmacology, 106, 135, 141, 142, 143, 144, 146, 147, 176, 302, 308, 311, 312, 317, 318, 321, 331, 343
  in protein prenylation, 100, 105, 110, 144, 155, 319
  internal messenger, 81, 88, 89, 282, 353
  production of, 44, 69
  semiochemical, 189
farnesyl acetate, 95, 189, 312
fenchone, 257, 292, 298
fibrosarcoma, 101, 148, 151, 327
Flavimonas oryzihabitans, 173
Flavobacterium suaveolens, 173

flavouring agent, 294
flea-shampoo, 295
Foeniculum capidaceum, 257
F. vulgare, 257, 260, 308
frankincense, 277, 329
Frankliniella occidentalis, 219, 225
frontalin, 209
FTase inhibitor, 101, 107, 138, 140, 311, 314, 321, 336, 343
fumigant, 201, 277, 284, 334
Fusarium, 291
F. tricinctum, 291
fytoalexin, 64

G

gallstone, 308
gedunin, 292, 314
gene
  duplication, 57, 66, 87
  expression, 6, 35, 45, 54, 59, 68, 87, 89, 100, 101, 102, 105, 114, 115, 116, 117, 118, 135, 141, 143, 150, 154, 156, 166, 191, 207, 228, 242, 294, 310, 312, 314, 343, 351, 353, 354
generalist, 6
  feeder, 215, 217, 235, 250
  receptor, 244, 245, 247, 251
genistein, 328, 337
Geotrichum candidum, 341
geranial, 118, 175, 196, 257, 258
geraniol, 30, 89, 171, 172, 196
  antibiotic, 173, 312
  biosynthesis, 23
  effect on insects, 196, 200, 201, 203, 205, 208, 209, 217, 219, 220, 247, 273, 295
  in feedback systems, 69, 83, 84, 85
  in pharmacology, 142, 144, 146, 147, 176, 259, 311, 317, 318, 331
  internal messenger, 135
  production of, 17, 38, 255, 256, 257
  semiochemical, 187, 219 (?), 221, 247
geranylgeraniol, 39, 100, 105
  biosynthesis, 23, 78
  in feedback systems, 69, 93, 147, 313, 319
  in pharmacology, 135, 143, 144, 146, 147, 312, 313, 318, 320

Index 431

in protein prenylation, 100, 110, 140, 144, 155, 319
internal messenger, 191, 312
semiochemical, 195
geranylgeranoic acid, 146, 316
geranyllinalool, 191, 199, 273, 312
germacranolide, 266
germacrene, 38, 39, 173, 184, 185, 193, 225, 271
German Commission E, 300, 304, 347
Geum urbanum, 257, 260
Giardia lamblia, 100
Gibberella fujikuroi, 68, 77
gibberellic acid, 43, 44, 45, 64, 254
gibberellin, 1, 45, 68, 77, 116
gin, 45
ginger, see *Zingiber*
Ginkgo biloba, 45, 258, 267, 302, 303, 304, 310
ginkgolide, 45, 258, 267, 303, 304
ginseng, 150, 299
ginsenoside, 144, 150, 159, 258, 322
glomeruli, 126, 127, 128, 232, 233, 246, 248
glucoside (bond), 35, 36, 149, 234, 268
glycoprotein, 4, 78, 327
glycyrrhetinic acid, 314
Glycyrrhiza glabra, 307, 347
G. squamuloza, 347
glycyrrhizic acid, 307
glycyrrhizin, 307, 314
Gnetum, 238
gossypol, 144, 147, 150, 151, 176, 177
G-protein, 98, 114
heterotrimeric, 98, 99, 130, 228
in cell cycle/differentiation, 102, 103, 105, 126, 130, 141, 143, 146, 156, 158, 170, 171, 356
in odour perception, 123, 125, 126, 127
in protein prenylation, 98, 104, 143, 156, 170, 228
in signal transduction, 98, 99, 105, 114, 119, 121, 122, 123, 125, 126, 228, 231, 247, 251, 322, 352
Graeffea crouani, 185
Gram-negative bacteria, 13, 172, 174, 175, 290, 348
Gram-positive bacteria, 13, 172, 174, 175, 290
green alga, 13, 16, 19, 20, 64, 183, 184, 345, 351

GroEL, 177, 210
growing site, 174, 259
Gynaikothrips uzeli, 194
gynoparae, 81, 186, 202, 205, 216

H

25-hydroxycholesterol, 76, 78, 92, 94, 167, 168, 301, 319
20-hydroxyecdysone, 50, 51, 52, 74, 115, 229, 281, 282
26-hydroxysterol, 301
hair loss, 333
Haloferax volcanii, 64, 69, 71
Haplothrips leucanthemi, 194
Heliothis virescens, 74, 283
hepatoma, 146, 150, 167, 312, 316, 336
herpes simplex virus, 169, 314, 332
himachalene, 306, 307
HMG-CoA, 12, 26, 65, 68, 74, 76, 78, 80, 94, 95, 301, 320
cyclase, 11
lyase, 65
reductase, 11, 13, 14, 15, 17, 24, 39, 60, 64, 65, 66, 68, 69, 72, 73, 75, 76, 77, 78, 80, 81, 85, 86, 87, 88, 89, 90, 91, 92, 94, 95, 100, 105, 106, 107, 113, 138, 141, 147, 158, 159, 163, 166, 167, 168, 187, 292, 300, 301, 312, 313, 318, 319, 321, 322, 337
synthase, 11, 65, 66, 69, 76, 80, 81, 86, 87, 91, 92
homoterpene, 223
Hospitalitermes, 191
housefly, see *Musca*
human breast cancer, see mammary tumour
humulene, 194, 356
Humulus lupulus, 194, 260, 309
hydrodistillation, see distillation
hydrogenosome, 60
hydroponics, 259, 261
Hylobius abietis, 197, 221, 222
Hymenaea courbaril, 195
Hymenoptera, ????
hypercholesterolemia, see cholesterolemia
hypericin, 259, 309
Hypericum perforatum, 37, 259, 260, 304, 309
hyperplasia, 131, 143, 319

hypersensitive response, 349
Hyptis capita, 257
Hyssopus officinalis, 36, 220

I

integrated pest management, 185, 227, 272, 279, 287, 290, 326, 346
interneuron, 126, 246, 247, 248, 250
Inula conyza, 277
IPM, see integrated pest management
Ips duplicatus, 68, 188
I. paraconfusus, 24, 68, 87, 187, 188
I. pini, 24, 187, 275
I. typographus, 274, 276, 279
ipsdienol, 24, 25, 38, 187, 188, 334
ipsenol, 24, 25, 38, 187, 188, 275
irradiation, 149, 156, 169, 307
irritable bowel syndrome, see irritable colon syndrome
irritable colon syndrome, 308
Isatis tinctoria, 298
isobolographic analysis, 284
isoborneol, 256, 332
isomer, 2, 3, 4, 16, 30, 32, 37, 41, 44, 46, 49, 50, 81, 119, 120, 129, 181, 185, 188, 189, 192, 193, 195, 196, 201, 202, 216, 219, 247, 254, 274, 276, 290, 294, 295, 296, 344, 353, 354, 355
isoprenylation, see prenylation
isopulegol, 283
isothujone, 256, 258, 287

J

jasmonic acid, 226
JH, see juvenile hormone
Juglans nigra, 194
juglone, 194, 273, 285, 287, 329, 331
Juniperus communis, 44, 222, 257
J. oxycedrus, 44, 331
juvenile hormone, 5, 46, 47, 48, 68, 74, 76, 85, 113, 114, 115, 117, 153, 168, 187, 190, 197, 204, 280, 281
    JH1, 48, 74, 114, 115
    JH2, 48, 114, 190
    JH3, 47, 48, 68, 74, 75, 76, 81, 82, 85, 87, 113, 114, 115, 120, 153, 187, 196
juvocimene, 282

K

7-ketocholesterol, 78, 92, 94, 319
kairomone, 5, 38, 125, 127, 128, 182, 183, 187, 191, 198, 200, 205, 206, 216, 217, 221, 228, 236, 237, 240, 242, 243, 246, 247, 250, 252, 270, 275, 278, 287, 288, 334, 346
Kalanchoë blossfeldiana, 52
Klebsiella oxytoca, 173
K. pneumoniae, 173
Kraussaria angulifera, 200
kummel, 61, 171, 218, 236, 256, 276, 287, 292, 298

L

Labidus coecus, 195
L. predator, 195
Lacanobia oleracea, 49
Lactobacillus, 157
L. acidophilus, 17
lamin, 91, 107, 159, 167, 321
Lamium album, 260
L. galeobdolon, 260
lanosterol, 26, 51, 69, 93, 94
lantadene, 349
Lappa major, 333
L. minor, 333
Larix laricina, 206, 209
Lasiosiphon kraussianus, 280
Lasius carniolicus, 195
L. fuliginosis, 195
Lavandula angustifolia, 260, 309
L. officinalis, 260, 309
L. spica, 260
Leptinotarsa decemlineata, 129, 199, 215, 250, 251, 272, 273, 285
leukemia, 106, 135, 141, 143, 144, 151, 153, 155, 156, 312, 316, 317, 318, 319, 327
    lymphoid, 106, 138, 144, 166
    myeloid, 106, 107, 138, 144, 155, 166, 312, 315, 318, 324
    promyelocytic, 153, 156, 316, 318, 319
lichen, 18, 56, 163, 326
limb regeneration, 116, 117, 118
limonene, 4, 5, 32, 36, 146, 170, 172, 265, 291, 292, 295, 315, 316
    antibiotic, 146, 173, 242, 256, 257, 258

Index 433

biosynthesis, 33, 41, 61, 90, 295, 341
effect on insects, 146, 197, 199, 200, 202, 206, 207, 209, 226, 273, 283, 287, 295
in cell cycle/differentiation, 91, 139, 143, 145, 146, 147, 156, 313
in feedback systems, 138, 147
in pharmacology, 143, 144, 145, 146, 147, 148, 156, 159, 167, 176, 229, 256, 257, 258, 308, 311, 312, 326, 328
in protein prenylation, 96, 105, 106, 107, 138, 139, 143, 144, 145, 170, 229, 311, 312, 313, 336
in signal transduction, 170, 222, 247
production of, 32, 33, 35, 61, 206, 209, 222, 256, 257, 258, 265, 287
semiochemical, 186, 193, 195, 279
linalool, 174, 315
antibiotic, 172, 173, 174, 175, 207, 256, 257, 258, 290
biosynthesis, 33, 57, 223
effect on insects, 187, 199, 205, 206, 207, 217, 219, 223, 230, 271, 273, 274, 288
in cell cycle/differentiation, 176
in pharmacology, 298, 313
production of, 17, 33, 38, 172, 174, 175, 217, 226, 255, 256, 257, 258, 278
semiochemical, 184, 187, 189, 196, 239
linolenic acid, 226, 324
Liothrips kuwanai, 194
Lipaphis erysimi, 212
lipoprotein, 25, 53, 76, 78, 92, 95, 166, 167, 168, 300, 301, 304, 319, 354
liverwort, 19, 33
Locusta migratoria, 200, 277
Lolium multiflorum, 292
Lonicera japonica, 298
Lophopetalum wallichii, 150
lotion, 260, 299, 329, 332, 333
lovastatin, 66, 77, 78, 85, 87, 88, 95, 96, 101, 106, 159, 166, 167, 168, 300, 301, 313, 319, 320, 323, 350
low-density lipoprotein, 53, 78, 92, 95, 166, 167, 168
lure, 183, 221, 278, 346, 347
lycopene, 50, 310, 328
Lymantria dispar, 206
lysosome, 110, 134, 165, 166

M

maceration, 263, 264
Macrosiphoniella tanacetaria, 354
M. tapuskae, 354
Macrosiphum euphorbiae, 280
Maladera matrida, 189, 221
malaria, 5, 169, 174, 255, 298, 299, 313, 314, 340, 345, 358
mammary carcinoma, 91, 145, 146, 156, 159, 311, 321, 323, 324, 336, 337, 338, 345
mammary cancer/tumour, 80, 115, 143, 146, 149, 151, 156, 160, 163, 168, 176, 311, 312, 313, 317, 318, 319, 324, 327, 337, 344
Manduca sexta, 115
manumycin, 107, 314, 320, 321
Maprounea africana, 170
Matricaria, 5, 44, 254, 255, 257, 265, 298, 307, 331, 332, 354, 355, 356
M. chamomilla, 135, 254, 257, 261, 312, 331, 354
M. matricarioides, 257, 354
matricine, 42, 44, 254, 255, 257
Mayetiola destructor, 124
Medicago, 350
medicinal herb, 36, 50, 149, 169, 254, 300
medulloblastoma, 320, 321, 347
Megastigmus pinus, 200
M. rafni, 200
Megoura viciae, 49, 168, 184, 185, 186, 193, 203, 205, 217, 218
Melaleuca alternifolia, 174, 257
melanoma, 149, 151, 153, 159, 312, 313, 314, 321, 3202, 325, 327
Melissa, 5,
M. officinalis, 220, 221, 257, 261, 270, 308, 309
Mentha, 61
M. piperita, 35, 41, 257, 259, 261, 265, 308
M. pulegium, 36, 207, 257, 284
M. spicata, 35, 207, 257, 261, 271, 284, 317
menthol, 35, 36, 292
acaricide, 270, 271
antibiotic, 298
biosynthesis, 265
effects on insects, 203, 280

in cell cycle/differentiation, 138, 147
in feedback systems, 95, 147
in pharmacology, 298, 299, 305, 307, 308, 331
production of, 257
menthone, 35, 36, 145, 207, 257, 265, 284
mesodermic cell, 78
metal ion, 29, 30, 33, 41, 44, 63, 224, 299
metalloproteinase, 97, 100, 117, 318, 326, 327
metastase, 148, 153, 159, 167, 320, 322, 323
methyl angolensate, 314
mevalonate kinase, 53, 60, 65, 68, 83
mevinolin, 78, 95, 96, 166
microsome, 50, 54, 78, 100, 294, 314
microtubule, 163, 338, 343, 359
migraine, 261, 300
Minthostachys, 293
mitochondria,
in animals, 27, 28, 29, 60, 69, 86, 91, 107, 110, 111, 130, 131, 133, 134, 146, 154, 314, 319, 343, 347, 351
in plants, 19, 23, 26, 27, 28, 29, 60, 69, 107, 130, 131, 134, 351
mitosis, 59, 104, 163, 324
Monarda fistulosa, 209
monastic garden, 254
Monomorium pharaonis, 190
moult, 51, 83, 200, 212, 229, 230, 281
moulting hormone, 5, 26, 51, 52, 281
Mucor piriformis, 173, 290
mucositis, 307
multicomponent treatment, 285, 306, 326, 327, 328
multi-drug resistance, 113, 298
multi-stage defense, 273, 275
Multi-Step-Treatment, 327, 328
Musca domestica, 207, 273, 281, 282, 283, 284, 332, 334
mycetocyte, 56, 203
Mycobacterium smegmatis, 173
myrcene, 30, 172
effect on insects, 196, 202, 203, 205, 206, 209, 273
in cell cycle/differentiation, 144
in pharmacology, 313
in protein prenylation, 144
production of, 32, 35, 174, 205, 206, 256, 258, 287

semiochemical, 25, 188, 194, 195, 239, 247
myrcenol, 188, 305
Myrmicaria eumenoides, 195
myrtenal, 203
Myrtus communis, 329
Myzus persicae, 81, 129, 186, 193, 196, 197, 202, 203, 205, 210, 211, 212, 213, 217, 218, 249, 274, 276, 280, 354

N

- -3-fatty-acid, 113, 320, 328
Nasonovia ribisnigri, 211, 212
Nasutitermes, 46, 191
necrosis, 129, 135, 146, 170, 305, 323
neem, 177, 200, 210, 212, 214, 262, 264, 269, 276, 277, 281, 285, 286, 295, 303, 341, 346
neem extract, 200, 214, 262, 276, 277, 285, 286
Neotoxoptera formosana, 274
Nepeta, 168, 185
N. cataria, 185, 257
N. mussinii, 186
N. racemosa, 89, 186, 257
nepetalactol, 168, 185, 186, 187, 216, 244
nepetalactone, 168, 185, 186, 187, 244, 257
Nephotettix virescens, 214
nephrotoxicity, 315
neral, 118, 173, 175, 196, 257, 258
nerol, 38, 89, 175, 176, 196, 205, 221, 247, 305, 331
nerolidol, 17, 184, 189, 223, 224, 225, 313
neuroblastoma, 156, 318, 336, 347
neurotoxicity, 301
Nicotiana tabacum, 100
nimbin, 269, 277
Nitrosomonas europaea, 172
nonatriene, 223, 224, 225
non-Hodgkin's lymphoma, 325, 326

O

Ochetellus glaber, 195
ochraceolide, 144, 150
ocimene, 184, 187, 189, 195, 217, 222, 223, 225, 226, 256, 257, 278, 287, 288
Ocimum, 278

Index

O. basilicum, 259, 261, 282
O. cannum, 278
O. sanctum, 282
O. suave, 257, 278
octopamine receptor, 280
Odontotermes obesus, 201
Oedogonium, 183
oil gland, 33, 37, 38, 41, 44, 134, 178, 263, 264, 309
ointment, 299, 304, 306, 307, 329, 330, 331
oleanolic acid, 147, 149, 156, 159, 169, 299, 315, 323
oleoresin, 25, 35, 209, 213, 277
olfactometer, 186, 222, 225, 274, 289
oncoprotein, 133, 347, 356
oncosis, 134
Operophtera brumata, 272
oregano, 208, 256, 257, 258, 291
Oreina cacaliae, 221
Origanum majorana, 257, 261
O. maru, 36
O. vulgare, 208, 257, 258
Ostrinia nubilalis, 285
oviposition, 37, 205, 206, 207, 209, 227, 243, 249, 277, 287
oxysterol, 86, 90, 91, 92, 93, 94, 300, 301, 307, 319

P

palmitoylation, 97, 98, 99, 102
palm oil, 344, 345
panaxadiol, 314, 327
panaxatriol, 314, 327
pancreatic cancer, see p. tumour
pancreatic tumour, 138, 142, 146, 153, 311, 312, 317, 318
parasite, 63, 169, 174, 255, 261, 270, 298, 299, 300, 313, 314, 340, 345, 358
parasitoid, 24, 225, 271, 285
Parkinson, 155, 220
parthenin, 266, 322
Parthenium hysterophorus, 266
parthenolide, 261, 300
Pediculus humanus humanus, 208, 295
Pemphigus fuscicornis, 354
P. utricularius, 213
Penicillium, 291
P. decumbens, 17

P. expansum, 173, 290
P. griseofulvum, 17
P. ochrochloron, 291
perfume, 5, 8, 171, 191, 253, 264, 291, 329, 330, 331, 357
Perilla frutescens, 90, 259
perillene, 194, 229, 273
perillic acid, 105, 107, 138, 143, 144, 145, 146, 313
perillyl alcohol, 5, 96, 317
  effects on insects, 96, 229
  in cell cycle/differentiation, 138, 143, 144, 147, 156, 313
  in feedback systems, 91, 95, 138, 143, 147, 167, 313
  in pharmacology, 138, 143, 145, 146, 147, 156, 160, 167, 229, 257, 284, 311, 316, 317, 318, 320, 357
  in protein prenylation, 96, 105, 106, 107, 138, 144, 145, 150, 229, 311, 316
  production of, 207, 257, 315, 317
Periplaneta americana, 184, 185, 225
periplanone, 184, 185
peroxisome, 53, 60, 66, 69, 83, 86, 94, 110
Pestalotiopsis microspora, 343
Petasites albus, 221
P. hybridus, 221, 235
phagocythosis, 56
phellandrene, 32, 33, 35, 37, 191, 206, 209, 256, 257, 273, 287
pheromone, 8, 39, 48, 49, 68, 69, 74, 87, 113, 123, 126, 128, 183, 187, 188, 190, 219, 228, 230, 236, 237, 244, 250, 333, 334, 346
  aggregation-, 25, 183, 187, 188, 189, 216, 275, 278
  alarm-, 39, 46, 124, 125, 176, 189, 191, 192, 193, 194, 202, 225, 247, 274, 284, 354, 355, 356
  sex-, 39, 46, 125, 168, 182, 183, 184, 185, 186, 187, 189, 190, 191, 194, 205, 216, 230, 237, 239, 240, 242, 244, 247, 251, 278
  trail-, 190, 217
  -trap, 279, 346
phloem, 25, 187, 205, 210, 212, 213, 214, 215, 343, 353, 354
Phomopsis helianthi, 291
Phorodon humuli, 186, 187, 191, 216

photosynthesis, 16, 19, 35, 45, 50, 57, 118, 226
Phycomyces blakesleeanus, 77
phytoalexin, 237, 350
phytoecdysteroid, 50, 51, 52, 116, 179, 216, 230, 281, 282, 333
Phytophthora cactorum, 17
Phytoseiulus persimilis, 223, 225
phytosteroid, 52, 302
phytosterol, 50, 51, 52, 54, 302
Picea abies, 32, 33, 35, 188, 222
P. glauca, 209
P. rubens, 202
P. sitchensis, 202
Pieris, 225, 244, 246, 273
P. brassicae, 128, 215, 219, 341
P. rapae, 224
Pimpinella anisum, 258
pine, see *Pinus*
pinene, 32, 172, 283, 292
   antibiotic, 173, 174, 175, 256, 257, 258
   biosynthesis, 53
   effect on insects, 187, 196, 197, 200, 204, 205, 209, 217, 222, 273, 279, 280, 287
   in cell cycle/differentiation, 138
   in feedback systems, 147
   production of, 18, 32, 33, 35, 174, 187, 203, 204, 206, 208, 255, 256, 257, 258, 278, 287
   semiochemical, 186, 193, 195, 222, 239
Pineus floccus, 202
Pinus, 25, 33, 35, 53, 196, 197, 199, 201, 206, 207, 215, 275, 305, 312, 333, 341
P. attenuata, 18
P. caribaea, 196
P. pinaster, 199, 201
P. sylvestris, 35, 222
P. taeda, 188
piperitone, 202
Pissodes strobi, 209
Pistacia atlantica, 214
Pityogenes chalcographus, 279
Plasmodium, 298
P. falciparum 169, 298, 299, 314, 345, 358
P. vinckei, 298
plastid, 13, 18, 19, 22, 23, 29, 50, 55, 56, 61, 64, 65, 85, 87, 95, 100, 345
Plutella xylostella, 280, 285
Podocarpus nakaii, 50

pollinator, 8, 38, 182, 216, 220, 237, 238, 239, 243, 251, 270
polygodial, 54, 215, 219, 273, 276
polygonal, 215
polymorphism, 48, 49, 350, 353
potato, see *Solanum*
Potentilla tormentilla, 332
pravastatin, 166, 320, 321
precursor, 4, 11, 12, 18, 19, 21, 22, 24, 25, 26, 27, 29, 30, 43, 52, 59, 60, 61, 69, 81, 89, 120, 138, 157, 188, 196, 223, 319, 358
predator, 24, 192, 196, 197, 199, 217, 223, 225, 227, 242, 271, 285, 339, 355
prenylation of proteins, 96, 97, 98, 99, 100, 101, 102, 103, 104, 105, 107, 109, 110, 117, 118, 120, 125, 135, 138, 139, 141, 143, 144, 145, 146, 155, 156, 166, 167, 170, 228, 229, 247, 310, 311, 312, 313, 314, 319, 334, 336, 351
pristimerin, 314
procaspase, 130, 131, 132, 343
Prochlorothrix hollandica, 49
prokaryote, 19, 26, 27, 49, 57, 60, 88, 101, 102, 111, 358
prostate cancer, 165, 310, 311, 313, 316, 317, 324, 336
prostate hyperplasia, 319
Protaphis dudichi, 354
Proteus aeruginosa, 305
P. vulgaris, 173, 305
Prunella vulgaris, 347
Prunus, 186, 216
P. domestica, 216
Pseudomonas, 173
P. aeruginosa, 173, 291, 305
P. mevalonii, 64
P. pickettii, 349
Psila rosae, 220
Pulegium vulgare, 36, 257
pulegone, 30, 279, 294, 335
   antibiotic, 173, 204, 212, 242, 243, 257, 281
   biosynthesis, 53
   effect on insects, 37, 203, 204, 206, 207, 273, 279, 280, 283, 284
   in cell cycle/differentiation, 138
   in feedback systems, 147
   in pharmacology, 5, 37
   production of, 37, 207, 257, 284
   semiochemical, 192

Index

pulmonary
    carcinoma, 322
    emphysema, 305
Pycnogonum litorale, 197

Q

quercetin, 259
Quercus, 35
Q. garvyana, 272
Q. ilex, 35

R

radiotherapy, 156
Rasputin, 350
receptor protein, 126, 129, 140, 168, 231, 232, 247
resin, 7, 25, 35, 54, 203, 209, 213, 234, 255, 277, 298, 299, 301, 306, 309, 329, 330, 346
Reticulitermes, 199, 201
R. lucifugus, 46, 48, 191, 192
R. speratus, 201
retina, 119, 120
retinal, 52, 119, 120
retinoblastoma, 145, 323
retinoic acid,
    in differentiation, 94, 117, 118, 155, 318
    in pharmacology, 151, 153, 155, 156, 168, 316, 317, 318, 319, 323, 325, 327, 336, 337, 345
retinoid,
    in differentiation, 115, 116, 117, 118, 134, 147, 151, 153, 318
    in pharmacology, 151, 153, 155, 316, 318, 319, 323, 325, 326, 328, 333, 337, 345
    receptor, 89, 91, 117, 118, 119, 144, 147, 153, 358
    signal transduction (additional), 53, 229
Rhabdosia effusa, 258
R. excisa, 258
R. longikaurin, 258
R. sculponeata, 258
R. umbrosa, 258
Rhagoletis, 250
rheumatism, 255, 297, 306, 307
rhinaria, 124, 185, 189, 203

rhinitis, 305
Rhizobium, 291
RhoB (inhibition), 314, 316, 336, 343
rhodinol, 176
Rhodobacter, 291
Rhodopseudomonas, 49
rhodopsin, 119, 120, 122
Rhodotorula minuta, 83
Rhopalosiphum padi, 186, 187
Ricciocarpos natans, 33
Rosa gallica officinalis, 330
rosemary oil, 208, 209, 274, 275, 295, 305
Rosmarinus officinalis, 258, 261, 304, 333
Ruta graveolens, 258, 261

S

sabinene, 33, 35, 173, 205, 222, 255, 256, 257, 258, 273, 287
Saccharomyces cerevisiae, 66, 80, 110, 149, 172, 175, 315
saffron, 50
salannin, 212, 269, 277
Salmonella, 173, 302, 312
S. typhimurium, 290, 291
Salvia, 45, 214, 258
S. canariensis, 45, 267
S. officinalis, 34
santalene, 32, 202, 306, 307
santonin, 204
sarcophytol, 318, 336
Satureja thymbra, 208, 258
scald, 83, 84, 85
Scenedesmus, 13, 20
Schistocerca americana, 200
S. gregaria, 200, 210, 277
Schizosaccharomyces pombe, 66
Scolytus ventralis, 242
sedative, 5, 260, 261, 297, 308, 309
Senecio palmensi, 285
sensillum, 123, 126, 128, 199, 215, 219, 228, 230, 233, 245, 246
sepal, 239, 254
seudenol, 209
shampoo, 295, 332
Shigella flexneri, 348
S. shiga, 173
silphinene, 285
Simaba multiflora, 259
simvastatin, 166, 300, 301

Sinningia, 219
sirenin, 183, 184
Sitobion avenae, 52
S. fragariae, 186
Sitophilus zeamais, 278
sitosterol, 308
Sitotroga cerealella, 277
skin tumour, 151
Slavum wertheimae, 213
Smad, 145
Solanum berthaultii, 192, 288
S. tuberosum, 8, 170, 192, 210, 213, 215, 236, 276, 280, 288, 289, 292, 293, 294
soldier, 46, 48, 191, 192
solvent
    acetone, 263, 267, 278, 302, 307
    acetic ether, 307
    chloroform, 150, 263, 266, 268, 307
    dichloromethane, 268
    diethyl ether, 265, 267
    ethanol, 145, 263, 264, 265, 266, 267, 271, 286, 299, 304, 305, 306, 307, 333
    fat (neutral), 263, 264
    hexane, 263, 265, 266, 299
    lard, 264
    methanol, 51, 253, 262, 263, 265, 266, 267, 268, 269, 280, 306
    methyl ether, 267
    paraffin (oil), 218, 220, 305, 331
    petroleum ether, 286, 306
    water, 44, 172, 262, 263, 265, 266, 267, 268, 269, 302, 304, 306, 309, 329, 336
specialist (feeder), 215, 225, 240, 244, 245, 246, 247, 248, 250, 251, 354
Spinacia oleracea, 52
Spodoptera, 214, 225, 273
S. eridania, 206
S. exigua, 223, 225
S. littoralis, 214, 273
S. litura, 280
sprout inhibitor, 293
squalene, 16, 53
    biosynthesis, 16, 24, 26, 50, 53, 65, 69, 83
    in feedback regulastion, 80, 83, 86, 90, 91, 94, 107, 300, 311, 337
    in pharmacology, 308
    semiochemical, 183, 184
Staphylococcus aureus, 173, 291, 305, 315

steam distillation, 263, 265, 331
steroid hormone, 23, 25, 49, 50, 51, 52, 54
    biosynthesis, 26, 50, 53, 54, 93
    in cell cycle/differentiation, 140, 165, 229
    in pharmacology, 149, 165, 300, 324, 333
    mode of action, 50, 54, 80, 113, 115, 116, 117, 179, 183, 190, 324
sterol, 23, 36, 49, 51, 52, 54, 72, 183, 242
    biosynthesis, 4, 16, 18, 19, 24, 26, 29, 33, 50, 52, 53, 54, 55, 68, 77, 78, 96, 155, 312
    in cell cycle/differentiation, 147, 149, 160, 315
    in feedback systems, 24, 25, 53, 72, 78, 83, 86, 87, 90, 91, 95, 302, 312, 343
Stevia rebaudiana, 45
sticky trap, 219
stomatitis, 298
stored product insect, 277, 278, 357
Streptococcus faecalis, 173
Streptomyces, 14, 17
S. griseolosporeus, 17
suboesophageal ganglion, 248, 249
Sulfolobus, 27
S. solfataricus, 64
surfactant, 268, 295, 308, 341, 343
symbionin, 210
symbionts (see also endosymbionts),
    in evolution, 230
    of insects, 46, 52, 56, 63, 65, 68, 96, 168, 173, 175, 204, 210, 212, 242, 243, 281, 286, 289, 290, 354, 355
    of vertebrates, 56, 156
Synechococcus leopoliensis, 16

T

tamoxifen, 115, 145, 324, 345
Tanacetum, 255, 300
T. cilicium, 255
T. parthenium, 261, 300
T. vulgare, 255, 300
Taraxum japonicum, 160
taraxasterol, 147, 149, 160, 315
taraxerol, 147, 149, 160, 315
Taxodium distichum, 201, 343
taxol, 147, 160, 163, 164, 165, 259, 307, 324, 325, 338, 343, 344

Index

Taxus baccata, 163, 259
T. brevifolii, 163
T. mairei, 164, 259
T. wallachiana, 343
teratogenic effect, 331
termite, 46, 48, 173, 181, 190, 191, 192, 199, 201, 202, 217, 273, 277, 312
terpinene, 32, 35, 174, 206, 208, 214, 256, 257, 258, 287
terpineol, 173, 174, 175, 205, 256, 257, 274, 292, 295, 304, 305
Tetraclinis articulata, 321
Tetranychus urticae, 223, 255, 283
Therioaphis maculata, 193, 225
thiamine, 13, 14, 15, 61, 62
third trophic level, 227, 285
Thlibothrips isunoki, 194
threshold, 131, 178, 244, 245, 274
thrips, see Thysanoptera
Thrips obscuratus, 219
thujone, 173, 258, 273, 283, 287
thujopsene, 17
thyme oil, 174, 274, 298, 307, 332
thymol, 208, 294, 335
    acaricide, 271
    antibiotic, 173, 208, 257, 258, 291, 298
    effect on insects, 207, 208, 273, 280, 283
    in pharmacology, 307
    production of, 174, 208, 257, 258, 280
Thymus capitatus, 258
T. vulgaris, 37, 174, 236, 258, 259, 261, 274, 291
T. zygis, 258
Thysanoptera, 24, 39, 182, 194, 203, 206, 219, 229, 273
tocoretinate, 337
tocotrienol, 147, 159, 168, 321, 344, 345
Tomicus minor, 187
toosendanin, 285
toothpaste, 294
transcription,
    of genes, 2, 35, 38, 66, 74, 80, 86, 87, 89, 91, 93, 95, 98, 101, 102, 103, 113, 114, 117, 140, 313, 320, 322, 351
    post-transcriptional regulation, 80, 86, 89, 90, 91, 92, 94, 95, 168, 312
    transcriptional regulation, 87, 88, 89, 92, 113, 115, 130, 140, 147, 159, 187, 314, 327

transduction (of signals), 6, 29, 88, 96, 98, 99, 100, 101, 105, 110, 120, 123, 125, 139, 143, 144, 157, 158, 217, 228, 229, 243, 247, 356
translation,
of mRNA, 95, 96, 134
translational expression of genes, 35, 87, 94, 313, 345, 351, 358
Trialeurodes vaporariorum, 204, 255, 284
Tribolium castaneum, 278
Trichoplusia ni, 74, 113, 280
Trichosporium symbioticum, 242
Trigona subterranea, 196
Trimusculus conica, 339
T. reticulatus, 339, 340
Trioza apicalis, 205, 287
Triperygium regelii, 259
Triticum vulgare, 217, 278, 288, 292
Trypanosoma brucei, 107
T. cruzi, 65

U

ubiquinone, 4, 19, 23, 55, 78, 91, 111, 156
ursolic acid, 147, 149, 156, 159, 165, 169, 257, 298, 315, 322, 323, 327, 328

V

valeranone, 308
Varroa jacobsoni, 271
VASA, 242, 286, 290
verbenone, 187, 258, 273, 275, 276, 279
Verticillium, 353
Vibrio cholerae, 171, 173, 261, 291
V. vulnificus, 291
virus transmission,
    by insects, 192, 210, 216, 277
    in micro-organisms, 130, 235
    in vertebrates, 135, 169, 170, 260, 314, 315, 323, 332
vitamin,
    A, 52, 53, 116, 117, 119, 151, 310, 325, 326, 328, 333
    B, 17, 61
    D, 23, 53, 89, 144, 156, 165, 316, 318, 319, 323, 324, 328, 337
    E, 112, 168, 301, 310, 317, 323, 326, 327, 328, 344

K, 111, 112, 144, 147, 155, 156, 302, 316, 328, 336

W

warburganal, 54, 215, 219, 273
wing development, 350, 353

X

xylem, 25, 35, 214, 343

Y

yeast, 16, 171, 172, 173, 174, 175, 242, 290, 329, 332, 346
   cell cycle progression, 80, 88, 134, 148, 157, 175
   genes, 66, 175, 334
   HMG-CoA reductase, 66, 72, 73, 77
   in terpene production, 341
   MAD metabolism, 53, 65, 68, 72, 83, 111, 183
   MAI metabolism, 15, 63
   prenylation of proteins, 97, 100, 110
   sterol production, 72

Z

Zabrotes subfasciatus, 209, 273
zaragozic acid, 80, 86
Zymomonas mobilis, 12
Zingiber officinale, 41, 305
zingiberene, 41